AF616587

28th ITCC

PITTSBURGH-1981

Proceedings

28th International Technical Communication Conference

May 20-23, 1981 Pittsburgh, PA

SOCIETY FOR TECHNICAL COMMUNICATION

28th ITCC

PITTSBURGH-1981

Proceedings

28th International Technical Communication Conference

May 20-23, 1981 Pittsburgh, PA

Distributed by UNIVELT, INC., P.O. Box 28130, San Diego, CA 92128

SOCIETY FOR TECHNICAL COMMUNICATION

NOTICE

The papers published in these Proceedings were reproduced from originals furnished by the authors. The opinions and security of the information are the responsibility of the authors and not of the Society for Technical Communication.

ISBN 0-914548-34-4
STC-107-81

Printed in the United States of America

PREFACE

The *Proceedings of the 28th International Technical Communication Conference* might be viewed as the twenty-eighth installment in a series of "how-to" technical communication reference books. All of these twenty-eight books were written by professional communicators in response to needs expressed by other professionals. The *Proceedings* can offer information in areas where ITCC program committees have perceived existing literature to be inadequate.

The papers in this year's *Proceedings* were selected from abstracts that promised to contribute to new knowledge, to open new perspectives, to provide a more profound understanding of the profession, and to suggest a useful technique to assist in everyday publication operations. They support our theme, "Communication in the 80's: Getting to the Point."

This volume of the *Proceedings* reflects the judgment of the eight stem managers, whose experience and involvement urged them to select the papers. Paul A. Vehr of Honeywell and Cherrelyn A. Casey of IBM managed the Automated Composition Stem. Bertie E. Fearing of East Carolina University and Thomas L. Warren of Oklahoma State University coordinated the Education and Research Stem. Robert S. Burns and Joseph E. Smith, both of Pullman Swindell, took charge of the Graphics and Production Stem. Lois K. Thuss of the Johns Hopkins University Applied Physics Laboratory and Lacy R. Martin of Pullman Swindell organized the Writing and Editing Stem.

The 28th ITCC Program Committee appreciates the talents and energy of the Proceedings Chairman, Lark Gildermaster of On-Line Systems, Inc. Thanks to her organization, readers easily can find where to go for what they want to know.

Della A. Whittaker
Program Chairman

Joseph A. Rice
Deputy Program Chairman

Introduction and Acknowledgments

To facilitate referencing, the papers in this volume have been organized according to stem; within each stem, the papers have been sorted alphabetically by the first author's last name. To make the volume more useful, subject and author indexes have also been included.

The Proceedings has been compiled from camera-ready pages furnished by individual speakers prior to the conference. Their efforts in providing papers of professional quality are appreciated.

Special thanks for their help is extended to the members of the Proceedings Committee: Vivian Benton, Pat Henry, Laney Margolis, Larry Owen, and Lynne Stebbins Reilly. And, for their loyal assistance even though they are not STC members, I gratefully acknowledge the work of Elaine Huckestein, Cindy Bezeredi, Linda Smith, and Elizabeth Russell.

Lark Gildermaster
Chairman, Proceedings Committee

CITY OF PITTSBURGH

By virtue of the authority vested in me as Mayor of the City of Pittsburgh, I do hereby proclaim May 18 to 23, 1981 TECHNICAL COMMUNICATION WEEK in the City of Pittsburgh to focus public attention on the vital role technical communications plays in maintaining the free flow of clear, concise information upon which scientific and technological progress depends, to commend members of the Society for Technical Communication for their efforts in providing the important linkage between those who generate new knowledge and those who will benefit from it; to extend a cordial welcome to the technical communicators attending the 28th International Technical Communication Conference being held in our City this week and to wish this internationally-renowned organization continued growth and success in the years ahead.

IN WITNESS whereof I have hereunto set my hand and caused the Seal of the City of Pittsburgh to be affixed.

January 26, 1981

MAYOR

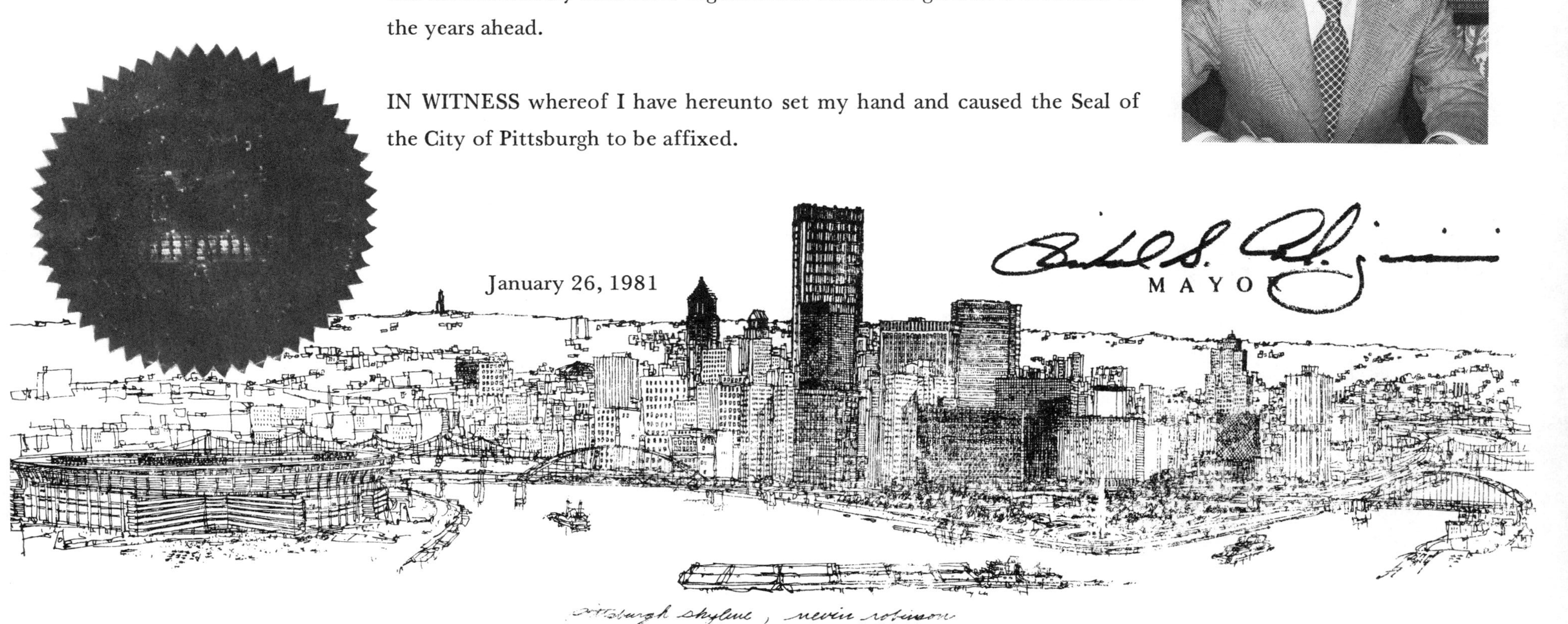

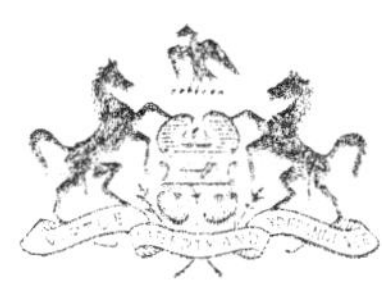

COMMONWEALTH OF PENNSYLVANIA
OFFICE OF THE GOVERNOR
HARRISBURG

GREETINGS:

As Governor of the Commonwealth of Pennsylvania, I am pleased to send warm regards to all who are attending the 28th International Technical Communication Conference, sponsored by the Society for Technical Communication. Welcome to Pittsburgh and the Keystone State!

Pennsylvania is proud to host this important conference which brings together nearly 1,000 technical communicators from throughout the nation and the world. Because technical communication plays a vital role in maintaining the free flow of clear, concise information upon which scientific and technological progress depends, this conference, with the theme "Communication in the 80's: Getting to the Point", serves a particularly useful function in providing a forum for the exchange of information, techniques and issues of concern to professionals in all technical communication media.

Fields as varied as business, government, industry, medicine and law are served by the writers, editors, educators, graphics and publications specialists and other professionals who provide the link between those who generate knowledge and those who benefit from it. The Commonwealth and the nation depend upon the knowledge of science and technology to advance the well being of all Americans.

On behalf of all the citizens of the Keystone State, I applaud the efforts of the Society for Technical Communication to advance the theory and practice of technical communication in all media. Best wishes for an enjoyable and enriching conference.

Dick Thornburgh
Governor

TABLE OF CONTENTS

A Automated Composition

E Education and Research

G Graphics and Production

W Writing and Editing

A
Automated Composition

Stem Managers

Paul A. Vehr
Technical Communicator
Process Management Systems Division
Honeywell, Inc.
Phoenix, Arizona 85023

Past Chairman, Phoenix Chapter

Cherrelyn Casey
Development Planner
Office Products Division
IBM
Boulder, Colorado 80302

Senior Member, STC
Vice President, Denver Metropolitan Chapter
of Women in Communications, Inc.

AUTOMATED COMPOSITION

The Automated Composition Stem is taking a look at the effects of Word Processing and Office Automation on Technical Communication. The world of the Technical Communicator has changed radically over the past several years.

I remember my first years in the field. We used pencils and ball-point pens by the gross and legal pads were a way of life. Then we got some electric typewriters. What an improvement! But thank goodness for correcting fluid and easy-erase paper. The people in graphics thought these were a real asset, too; they no longer had to be handwriting experts.

But everything still had to be rekeyboarded.

Then a couple of years ago we participated in a miracle — direct entry by the writers. While we were justifying the purchase of our system, I "guesstimated" a ten percent increase in writer productivity. The actual increase is just about four times the estimate.

About a year ago we added a phototypesetter to the system. We are still learning, but our publications have a more professional appearance. And typesetting means more and better looking data on fewer pages.

What does the future hold? Will we some day put on a "thinking cap" and let our thoughts do the input to the system? Only the future will tell. But several companies are working on voice-activated typewriters.

What do our panelists have to say about the now? The "Making Whoopie" gang is back to tell about "Making Whoopie Profitable." Other panelists will tell us about electronic mail in Alaska. About how to start your own publications company. About the uses of OCR. About typography. About the office of the future.

So come to the Automated Composition Stem panels to get a feel for where we are now and where we are going.

Playing Tag with Automated Text Processing

William B. Adams
IBM Corp. GPD

Development of Text Entry Markup for Automated Text Processing

Introduction

To make any document readable it is necessary to mark it up so that the reader can quickly and easily recognize the different elements of the document. These elements can then be organized in the mind of the reader to make intelligible sense. This markup is the accepted way to communicate written language to the reader so that we do not even think of it as markup. A brief review of this markup makes this point evident.

We indicate sentences by starting the first word with a capital letter and ending the last word with a period. We indicate paragraphs by a break in the running together of sentences plus either an extra line space between paragraphs or the indention of the first sentence. We indicate special paragraphs by inserting a name in front of them to identify their special meaning, such as "Note," "Caution," and "Danger." We indicate listed entries by prefixing them with numbers, letters, or bullets plus the indention of the left margin of the text. We indicate headings by special typography or additional line spacing above and below the text of the head, or both.

Typewriters were designed to make it easy to insert this style of markup: tabs were added to make it easy to set up standard width indentions; typography could be varied by the use of all uppercase letters, initial caps, or all lowercase; backspacing with overstrike capability allowed further variations in the typography to indicate headings and emphasize words or paragraphs.

Early Text Processing

During the early stages of automated text entry the same style of markup was used that was common on the typewriter, that is, use of spaces, capitalization, special words, indention indicated by tabs or spaces, etc. This was sufficient as long as the printers were only able to produce output that appeared very close to that of the input device.

This situation did not last very long. Developers began to add more functions to the printers. Programmers also added new capabilities to the programs. Pretty soon it was necessary to develop a series of commands that could be entered into the text stream to activate the new printer functions.

Simple Command Markup

To take advantage of the new capabilities, a new type of markup was added. This markup had to be keyed into the document in addition to the text. It was no longer necessary to insert several line spaces in a row: a single command could be entered to instruct the computer to leave several line spaces, or to go to a new column or page. In addition, other markup commands could be inserted to define text for a running head or foot that would be repeated on every page automatically.

Complicated Command Markup

The developers continued to add functions to the capabilities of the printer, and came up with a type of printer called a photocomposer. This machine could print extremely high-quality output equal in appearance to that of hand-set type. And, of course, the programmers for automated text processing were quick to meet the challenge.

New programs were written, which added the markup commands necessary to control these new printers. It was now possible to control spacing between lines, words, and even characters to as little as 1/800th of an inch. Hundreds of different typefaces were available. Typefaces could be intermixed on the same page or even within a line. Alignment, indention, tabbing, and other common functions could be used in almost limitless combinations.

As a result, the markup command sequence became formidable. A document that had a wide variety of headings and special paragraph elements might end up with more markup than text. This slowed down text entry greatly because these commands had to be entered interspersed within or between the lines of text. Also, different markup had to be used depending on what type of printer the output would be produced. In addition, the format specifications had to be understood by the key-entry person so that the proper markup was entered to match those specs. Keying errors were not only possible in text but became more frequent in markup, causing unacceptable printed output or, even worse, causing a program error that prevented getting any output at all.

Many efforts were made to simplify the procedure: some manufacturers added special keys, which would automatically insert several commands when pressed at the proper time, others added shorthand commands, which would call in several sequences automatically. Although the procedure was somewhat simplified as a result of these additions, the documents for printing on the newly developed printers still required specially trained operators for markup entry, special terminal capability to enter the text in a data base, and were printable only on the printer that could interpret that particular markup.

Clearly, this method was too slow for a production environment, and too cumbersome for documents that required constant update or modification. There had to be a better way.

Writer Entry

The writers' normal responsibility is to write down on paper the words that will convey to the reader the message they are trying to get across. As part of this writing they must indicate the format of the document as they write. This is done by entering all of the accepted markup that we have previously discussed. This is true whether they are writing a novel, a text book, or a maintenance manual. They should not be concerned with formatting the document to match company style or some other standard but alas, today it is inescapable.

Ideal Markup

Writers should only have to be aware of the words they write, the text elements that contain those words, and the logical placement of those elements within the appropriate document structure. They should be freed from the concern of formatting to meet specifications. The indication of the elements and the structure containing them should be reduced to the barest minimum to avoid distraction of the writers' train of thought.

Words are normally written in phrases or in sentences. It should not be necessary to type sentences strung together to indicate paragraphs. Instead, it would be more convenient if each sentence were an individual entry that could be entered beginning at the left margin. This would make it much easier to add or delete sentences during the course of initial writing.

Elements, such as paragraphs, are simply groupings of individual sentences which become strung together. All the writer would have to do is to define where one element ends and the next begins. This could be done very simply for all text elements.

Document structures encompass groupings of elements. These should be indicated within the stream of the text. For example, there should be no need during the writing process to start a new page to indicate a preface or individual chapters.

The mechanics of writing should also be considered. It is tiring work to push a pencil across a page while writing. It is also work to continuously press the keys of the typewriter. The ideal amount of key entry should consist only of the words in proper sequence and an indication of the text element and structure in which they reside.

Element and Structure Tagging

Text entry into a computer should be as simple as the ideal. Today's computer programs are capable of controlling complete formatting of a document. It should only be necessary to enter the words in their proper sequence and indicate by a computer-identifiable tag the element and structure in which they are contained. This type of text entry could be entered on almost any input device type connected to a computer and be formatted for any output device that is connected to a processor, regardless of its type or complexity.

Theory of GML Tag Markup

Basic tagging of a document should be used exclusively to identify the document structure and individual text elements as entered. The input should be completely free of formatting commands such as indent, add line space(s), or go to a new column or page. The omission of formatting commands prevents having to continually change or move the commands whenever words or paragraphs are added. It also allows printing on different printers, which may print a different number of words on a line or lines on a page.

The need for a type of program that could provide these desired capabilities was foreseen by IBM and a text-processing program named Document Composition Facility (DCF) was prepared. IBM Information Publishing in Tucson uses this program to format documents that support IBM products.

The Document Composition Facility Program

The DCF program is designed to be used with a General Markup Language (GML) tagged source document. The program allows a programmer to write a series of Application Processing Function (APF) mini-programs for the different elements. These APFs control the formatting of the document during printing, and are written to match the specifications for the particular style our company has established. They are contained in a library document and are called upon by the DCF program when a document is being processed. If several styles of documents are necessary, more than one APF may be written and stored in this library.

In addition to the APFs a "profile" document is also built for each style of document. The profile is what ensures that the proper APF is selected for the particular style of document needed. Both the APFs and the profiles are prepared once and then used over and over. The writers or text-entry operators never have to see or access them.

Document Organization Tags

In this discussion I will make an arbitrary distinction between structure and element tags, based on whether the tag is immediately followed by text or not. Structure tags have no text immediately associated with them. They only define structure, such as front matter, body, preface, index, or type of list. Element tags, while they imply a structure, are followed by, and thus identify, the text that follows them. These tags include those for heads, paragraphs, phrases, and other text items such as the document title and identifying number.

To use the DCF program, the structure and elements of a document are identified with what are called GML tags. These GML tags consist of a delimiter character, a name of up to eight characters, and an end character. They describe the structure of the document and identify each element in it. The document that contains the text and the GML tags is stored in the computer and is called the "source document."

Every document has an overall structure; for our discussion, let us assume we are describing a common text or resource document. The structure of this document could be broken down into four major parts: the front matter, the body, the appendix section, and the back matter. The front matter could be further broken down into a title page, a preface, and a table of contents. The body could be broken down into chapters, which, in turn, would be composed of smaller elements such as headings, paragraphs, list items, figures, etc. The appendix section could consist of several sections or chapters. The back matter could consist of a glossary, an index, and a back cover.

The GML tags that describe this organization would appear as shown in Figure 1.

```
:gdoc. (Start a general document)
   :frontm. (Start front matter section)
      :titlep. (Prepare title page)
         :title. (Identify title of document)
         :author. (Identify author name)
         :docnum. (Identify document number or ID)
      :etitlep. (Title page complete)
      :preface. (Start preface page)
         :p. (One of many paragraphs, etc.)
      :toc. (Start table of contents)
   :body. (Start body of document)
      :h1. (Start a new chapter with chapter heading)
         :h2. (One of many subheadings)
         :p. (One of many paragraphs, etc.)
   :appendix. (Start appendix section)
      :h1. (Start individual appendix)
         :h2. (One of many subheadings)
         :p. (One of many paragraphs, etc.)
   :backm. (Start back matter section)
      :glossary. (Start glossary)
      :index. (Start index)
      :backpg. (Start back page)
:egdoc. (End general document)
```

Figure 1. Document Organization Tags

GML Tag Categories

Our GML has a total of about 125 tags. This sounds like a lot of tags to have to memorize, and it would be if that was required, but the tags are broken down into several categories, and only the first category of approximately 30 tags needs to be memorized.

In the following lists, tag names that are followed by (e) require an ending tag to identify the end of the text element. An ending tag differs from an initial tag only by including the letter "e" between the initial delimiter and the first character of the tag name, for example, :ol. :eol.

Basic Tags

1. **Headings**

:h1.	Level 1 (Starts chapter)
:h2.	Level 2 (Heading within chapter)
:h3.	Level 3 ditto
:h4.	Level 4 ditto
:h5.	Level 5 ditto

2. **Paragraphs**

:p.	Simple
:n.	Note
:c. (e)	Caution
:d. (e)	Danger
:w. (e)	Warning
:li.	List item
:fn. (e)	Footnote

3. **List Types**

:ol. (e)	Start ordered list (Numbered)
:ul. (e)	Start unordered list (Bullets)
:sl. (e)	Start simple list (Indents only)
:nl. (e)	Start note list
:dl. (e)	Definition list
:dt.	Term (Used only with :dl. tag)
:dd.	Description (ditto)

4. **Phrases** (Used anywhere in the line)

:hp1. (e)	Highlighted phrase level 1
:hp2. (e)	Highlighted phrase level 2
:hp3. (e)	Highlighted phrase level 3
:q. (e)	Start quotation
:cit. (e)	Citation of another document name

5. **Cross-Reference**

:hdref.	Refer to a heading
:figref.	Refer to a figure
:liref.	Refer to a list item paragraph
:fnref.	Refer to a footnote

General Data Tags

:title.	Title of document
:security.	Security level of document
:docnum.	Order number of document
:date.	Date document was released or printed
:author.	Name of author(s)
:address. (e)	Start an address
:aline.	Identifies each line of the address (May also include name of company)
:copyr.	Year(s) of copyright

Figure Tags

:fig (e)	Start figure element (Includes space for paste-up, text items, or graphics)
:artwork.	Allow space for paste-up of segment of total figure
:lines.	Start series of 'as is' lines
:figcap.	Caption for figure (required if to be referred to)
:figdesc.	Additional description that will follow caption
:efig.	End of figure element

Miscellaneous Tags

:hp0. (e)	Highlighted phrase level 0 (same font as normal text)
:psc. (e)	Allows selective processing of specific process code
:xmp. (e)	Short 'as entered' examples

Text Entry in Source Document

Starting a Document

There are usually about 25 one-time-use tags required at the start of a document; other one-time-use tags may be needed between chapters. These tags may require some routine text, such as the name and address of your company, as well as some text that varies with each document, such as the title. Rather than your having to memorize and key in these tags and the routine text for every document, a small "form" document contains this information in outline form.

When you start a new document, this form document is copied into what will become the control document. With one copy command, all of these tags and the routine text are entered into the new document. Where unique text is required, the location is indicated by double question marks on the line. The tag name at the beginning of the line prompts you for the type of information required. Only this variable information must be keyed in.

In addition to the one-time-use tags and related text, the control document contains a series of embedded commands that merge individual chapters into a total document at formatting time. Interspersed among the embedded commands are the tags that designate the start of the body, the appendix, and back matter sections.

Entering the Text Documents

The text documents are broken up into small integral units, usually by chapters. Smaller units are easier to handle, and several can be updated simultaneously when time is short.

To start a source document, another form document is copied, which contains common entry lines used at the beginning of every document. These hold administrative and accounting information for the document, plus the initial head-1 tag that will be used to start a chapter. The comment lines at the top should be updated with the required information: document name, author, and date.

Begin entering the document with the text for the head-1 title, which will be the name of the chapter. The tag identifying each text element is always entered first, before any text. The text that goes with the tag is normally entered on the same line immediately following the period that ends the tag. It is only necessary to key in the tags of the basic set identified in "Basic Tags" on page 3 and follow the basic entry rules which follow.

Most of the remaining tags are used in groups and can be copied intact directly into the document from a form document.

Basic Entry Rules

1. Start all lines at the left margin. Do not use tabs or initial space characters.
2. Begin each sentence on a new line. Start a new line after the end of a sentence even if only one word is left on the line.
3. Enter tags in lowercase.
4. Enter either tags or text on every line. Do not leave any blank lines.
5. Use the appropriate tag for highlighting. Do not underscore or backspace and overstrike.
6. Use the correct tag for each element. Do not use an incorrect tag just because it gives a structure you desire.
7. Keep lines short (72 characters) and documents small for easy updating. Documents will be merged together at formatting time by the DCF program.
8. Do not start any text line with a period.

This type of text entry is simple. Typists can enter it close to the speed they can usually type on a typewriter. See "Basic Entry Rules" for an example of a keyed input document.

Running Head and Foot Information

You do not have to define the running head or foot of the page. The profile document points to the APF miniprogram which has already been programmed with the proper style of running head and foot. The text required is automatically picked up from data that is already keyed into the document in other places. One example of this is the chapter titles that appear at the bottom of every right-hand page. They are picked up from the head-1 title of the chapter and are changed each time a new chapter is started. Other information is picked up from the title page.

Advantages of GML Tagging

GML tagging provides a number of advantages to writers, editors, and text entry operators, which ultimately serve the end user better. Some of these advantages are described below.

Advantages to Writers

- Writers can concentrate on the text they are writing and forget format requirements.
- If writers generate the rough draft by hand, it is as easy to write in the tags as to use normal edit or proof marks.
- If writers normally type the document, it is easy to learn, recall, and type the GML tags with the other text characters.
- Writers have positive format control with a tag, because there is only one format interpretation for it.

- The document will be formatted consistently throughout. Every paragraph of the same type will have an identical format.

- Modifications, additions, and deletions are quickly and easily done because it is not necessary to renumber paragraphs or to correct page references.

- Individual pages can be seen formatted at the terminal for a quick check of formatting capability.

- It is easy to locate or scan head elements for order and parallelism.

- It is easy to compare or proof text items for consistency.

- It is easy to include special phrases that may change just before final printing by using symbols instead of actual text.

- It is possible to get fast turnaround (2 hours or less) for proof prints on a high-speed printer.

- Because of the fast turnaround, approval copies at the most recent level can be circulated.

- Even the first copies of a document can include a table of contents and figure list.

- Accurate cross referencing is available with automatic correction of page numbers when the document is updated.

- Index tags are available that allow writers to generate index entries at the same time they generate text for the document. The first draft of the document can have a simple though accurate index.

Advantages to Editors

- Editors will not be distracted by markup commands that obscure the text to be edited.

- There are no formatting commands to edit, which previously were taking up to 50% of the editors' time to verify or modify.

- Editors can now concentrate on editing for organization, language, and grammar.

- A fast spelling verifier that can accept mnemonics and other unique words.

- The identification of words not accepted as part of any limited vocabulary requirement.

Advantages to Text Entry Operators

- Tags composed of normal keyboard characters that allow fast power-typing of handwritten documents.

- Less correction of markup keying errors on writer-entered, GML tagged source documents.

- No lost time figuring out complicated command sequences used by the writer.

- No lost time figuring out commands to match format specification requirements.

- No lost time figuring out which markup commands to insert to make format runs on a specific printer.

- It is necessary to memorize only 30 tag names for normal input, which are short and are easily remembered.

- An additional 80 tags are available; these can be copied at will into the document with a simple copy command and modified where necessary. The tags act as prompts for entry of the variable document-related information that they identify.

- Documents are kept to short lengths (200 to 700 lines) for ease of handling, security, and backup capability.

- Assembly of the individual documents into a complete document is handled by a control document that contains general document information, plus the embedded commands to merge the separate units.

- There is no requirement to rekey headings or other text for use in the table of contents, which is generated automatically.

- A list of figures is generated automatically.

- Cross-referencing of text items requires no search and update.

- Updating can be done quickly: text can be moved, inserted into, and deleted from lists without having to modify any previous or following paragraphs.

- Spelling verification is very handy for catching typos and other spelling errors.

- Scan and search capabilities make it easy to verify that start and end tag pairs exist correctly.

- There is no need to type tabs, extra spaces, and backspace overstrike combinations.

- A common method of entry allows for any operator to assist or take over a job to help another operator without confusion or change in document style.

- It is easy to train new people how to enter a tagged source document.

Additional Advantages

- Experience required is minimal.

- Training is minimal. A secretary, writer, or engineer can be taught the basic system in a couple of hours.

- The documents are completely interchangeable.

  Individual sections can be entered by a variety of people, yet there will be no problem with consistent formats. If someone is sick, another person can take over and finish the document with no confusion or break in consistency. The documents can be transmitted to another branch and processed with no change in the source document.

  Sections of the document can be included in other documents of a different style without changes to the source document. When a section is relocated or copied to another document, the tagged index entries are moved along with each element.

- Automatic table of contents, figure list, and cross-referencing are available with no extra keying or last-minute changing of page numbers. Spelling checking identifies typos and other misspelled words.

Benefits for the End Users

GML tagging is especially beneficial to the end user because:

- The entire document is in a consistent format.

- Special paragraphs such as caution, danger, and note are easily recognizable.

- The table of contents is complete and contains accurate page locations.

- The list of figures is complete and accurate.

- Cross references to heads and figures, which may appear either before or following the location referenced, are complete and accurate.

- The legible type fonts do not have the problem of uneven impressions that could cause misreading of some characters.

Simple as Playing Tag

The new method of marking up text is almost as simple as a child's game of tag. You are it and the text elements are scattered all through your document. It is not necessary to capture each one and imprison them in elaborate commands so they will match specifications and a particular print device. All you must do to have good formatted output is to tag them properly.

MANAGING TEXT PROCESSING QUALITY

Jeannie Birndorf
IBM Information Systems Division
and
Jay Gillette
Colorado School of Mines

Jeannie Birndorf
IBM-ISD Boulder
Senior Associate Writer/Planner
P. O. Box 1900 (582/022)
Boulder, Colorado 80302
303-447-3280

Jay Gillette
Colorado School of Mines
Technical Communications Staff
Department of Humanities
Golden, Colorado 80401
303-279-0300

Advances in text processing techniques have complicated the responsibility for quality control in technical communications. The responsibility for quality control lies not only with the originator of the document, but also with everyone who processes that document. If everyone who works on a document works as an editor, mistakes can be caught and corrected at every level. Based on our experience, we conclude that instruction and review of professional editing techniques can improve final document quality and provide a basis for text quality control standards.

PROBLEM: MANAGING QUALITY CONTROL (JEANNIE BIRNDORF)

My work experience includes eleven years as a secretary, editorial assistant and editor, technical writer, and manager of word processors (correspondence secretaries). These assignments required proficiency on everything from a composer to work on a large CRT display screen, and the various levels of equipment in between.

From my experience as a writer, I went into the management of a word processing center, and brought many seasoned philosophies on how best to approach text processing. The frequent problem that I found in word processing was a lack of quality control that was multiplied by the:

- Sophisticated equipment
- Literal approach of some employees
- Emphasis on quantity versus quality
- Lack of a common learning base

Sophisticated equipment gives people a false sense of security about the quality of the work they produce. As a writer, I found that the only proficient way for me to work was to create and enter at the keyboard. However, while my fingers did a fairly good job of keeping up with my thought processes, my typing accuracy suffered because of the increased speed of the machine and the knowledge that I now had the ability to go back and "edit," that is, correct, my work.

I often found myself propagating errors because I had made a quick pass through a document on the display screen or the terminal paper, corrected errors as I found them, and then had not thoroughly proofread a hard copy of the document. I would also take the time to proofread a document and then not double check the corrections I had made, only to be embarrassed by an error found in a review cycle, or worse, in the printed document.

Another false sense of security arises if an automated spelling program is used. The automated spelling programs that I have used are only as good as the "dictionary" they use as a basis of comparison. Comparison programs cannot normally detect errors such as repeated words (that is, when you end one line of typing with the same word with which you begin the next line), nor can they detect a properly spelled but incongruously used word (that is, must for much).

The literal approach of some employees is a three-pronged problem. Some individuals are afraid to do something that is not included in their list of job responsibilities because of an alarming attitude that is creeping into the work force. This attitude can be summed up by the oft heard phrase, "That's not my job." Other individuals appear to have a willingness to do the job, but they lack the ability to anaylze what needs to be done.

The literal approach also seems to foster a lack of involvement in the work. As a secretary and an editorial assistant, I took pride in not being satisfied with a document unless I understood it, or at least, ensured that it was technically accurate by having it reviewed by an authority. However, I have seen many individuals type something verbatim, and never question the meaning of their work. These people either want no responsibility for the content of a document or have an unrealistic opinion of the ability of the originator. Whatever the rationale for this literal approach, the outcome is that one less pair of eyes is taking an active role in the quality of a document.

The age-old question of quantity versus quality has always been a concern of mine, and as a manager, I constantly fought the battle. While trying to be concerned with reasonable or normal turn-around times for work, I had a difficult time convincing some employees that fast isn't best if the document had typographical errors in it. These individuals didn't recognize that the typograhpical errors had to be identified by the originator of the document and then resubmitted for corrections. This takes far more time than proper quality control on the first pass.

Resolving these problems within word processing was further complicated by the varied backgrounds and educational experiences of the secretaries. For example, I had employees who had everything from a B.A. in Business Education to a high school education that included a typing class. The lack of a common learning base, combined with the sophisticated equipment, the literal approach of some employees, and the emphasis on quantity versus quality, created an environment that I felt could not be individually addressed.

I firmly believed that a course in proofreading and editing would realistically solve this problem. Because of my closeness to the situation, however, I did not want to be the instructor. I, therefore, sought an instructor for such a course through my local chapter of STC. They put me in contact with Jay Gillette who developed a course for all the secretaries in word processing at IBM Boulder.

SOLUTION: PROFESSIONAL EDITING TECHNIQUES (JAY GILLETTE)

Document quality control depends on two actions: errors must first be caught and then corrected. Through literally centuries of experience, the publishing industry has evolved techniques that effectively catch and efficiently correct errors. These techniques—editor's proofreading language and proofreader's correction symbols—can be taught in an intensive, one-day course and provide text processing employees with a common learning base for quality control. I believe it's unreasonable to demand quality without providing the training to achieve it.

The management at IBM Boulder sponsored this training early in 1980 for all its word processing employees. A May 1980 follow-up survey showed that 83% felt their proofreading improved as a result of the course skills, and 78% felt the course helped them improve the final quality of the documents they produced. The results were substantial enough that IBM's Information Development Department scheduled the course for its technical writers, editors, and editorial assistants.

Professional editing is a skill and an art. It requires a personality adept at handling details with attention and precision, and takes years to develop fully. Yet, some editing levels don't require the specialist. The dictionary's definition is a good beginning: to edit is to make original material suitable for publication or presentation.

This means that everyone who processes textual material serves as an editor, consciously or not. Unfortunately, there is little editorial training available for text processors. Even among professional editors, most training is artisanal style—the apprentice learns the trade directly at the hands of a more advanced editor.

Some kinds of editing can be done with basic training; the more advanced require a specialist's background. The Chicago *Manual of Style* classifies editing in general as either mechanical or substantive. In *The Levels of Edit*, the publications staff at Cal Tech's Jet Propulsion Laboratory suggests nine specific types of editing (in ascending order of complexity and expense): coordination, policy, integrity, screening, copy clarification, format, mechanical style, language, and substantive. I believe that with proper training, the less complex or mechanical editing levels can and should be carried out by everyone who processes a text or document.

One reason we proofread so poorly is that we read so well—we've learned from years of practice to read units at a time. We read words or whole phrases with one movement of the eye, and we read right over punctuation marks, noting their cues subconsciously, effortlessly. We are carefully taught **not** to enunciate words as we read, especially not to move our lips. This kind of reading is for speed first, comprehension second, proofreading, not at all.

The best proofreading could be called "prooftalking," using editor's proofreading language as its base. This language is a verbal shorthand (using mostly one-syllable sounds to signify punctuation marks), with the emphasis on enunciating everything on the page—every word, all punctuation, and even spaces in the text if necessary. This reading is thorough, designed to check the mechanics of a text rather than the substance. After you've cleared your final draft, it is the surest way to maintain final document quality.

Here's an example: "That's a interesting propos-ition!" sounds like this: "Doub quotes cap *that* pos *s a in-ter-est-ing propos* hyph *i t i o n* bang close quotes." When read out loud, two errors show up: the article should be "an," and "proposition" has a bad break. Because every element in the sentence receives attention, the reader catches even these small errors.

Traditionally, two-person proofreading teams use the language—the copyholder reads the original aloud, with the actual proofreader reading silently along in the new version. Some publications groups believe two-person proofreading is too expensive, and leave it to a single editor to sight proofread back and forth between the original and the new version.

I think that two effectively-trained editors can proofread a document almost as fast as a single person, with much greater reliability. In any case, certain classes of critical documents (like contracts or research reports with formulas) merit the highest quality proofreading we can give them, even if it takes a little longer or costs a little more.

Training in the proofreader's language also benefits single-person proofreading. Depending on the nature of the document, the editor can slow his own reading pace as much as he likes by using the language silently or out loud. No matter how distracted he may be, the language forces the mind's attention to focus on the elements of the text. It becomes much more difficult to skim the words, and you rarely find yourself having to reread lines you know you've been over but can't remember seeing.

Like any language, exposure isn't enough to become fluent. Yet because it is simple and logical, the proofreader's language can be picked up quickly. With the daily practice that text processing requires, fluency and speed come naturally.

Having caught the inevitable errors in a text, the problem remains to make sure they are effectively corrected. Here again we benefit by the publishing industry's long experience. The symbols that are called "proofreader's marks" in most dictionaries represent a basic yet efficient way to mark errors for correction.

Surprisingly enough, many text processors have never been instructed in this correction system. The result is that they may use homegrown or patchwork methods that they've personally developed (circling spelling mistakes, for example). While these methods may get the job done, their reliability and efficiency may be questioned.

The system's signs are simple and logical. Someone, long ago, made up a mark that suggests "paragraph" but it is guaranteed not to look like a capital P—it is actually a sort of backwards double-P. Again, the convention of circling a period makes this tiny element stand out. Merely circling a comma makes that into a period. You change a period quickly with a caret above and a tail to show it's now a comma (instead of blotting it out with a time-consuming circular dot that starts looking like a bullet).

Proofreader's correction symbols are largely standarized. With a few exceptions (dashes, for example), the marks are universal. If everyone understood the system, corrected material would be clear to everyone involved, from originator to processor, from processors to editors, from department to department. And if everyone was involved, there would be no blind reproduction—every stage would represent a quality control point.

Practice with these simple techniques, as with any skill, builds confidence. Eventually enough practice may bring out the intangible attitude we call the editor's eye, the second-nature precision that finds the mistake we'd all overlooked, the one that's so obvious now. That's managing text processing quality.

ACHIEVING NEW LEVELS OF PRODUCTIVITY THROUGH AUTOMATION

Robert J. Boeri
Data General Corporation

Robert J. Boeri
Data General Corporation
Mgr., Technical Communications
62 Alexander Drive
Research Triangle Park, NC 27709
(919) 549-8421
B.A., Business Administration,
Lowell University, 1975
M.B.A., Babson College, 1978
Award of Merit, 1977 STC
Publications Competition
Past Director, Carolina Chapter
of STC

Over the past several years, Data General employees—mainly from technical publications groups—have built text processing and typesetting programs, and other programs helpful to technical communicators. This paper shows how the Data General Technical Communications department at Research Triangle Park, N.C., adapted these to become more productive. This paper presumes you are familiar with text processing systems like RUNOFF or FORMAT.

AUTOMATING OUR PUBLICATIONS FUNCTION

In the fall of 1977, I was one of a small group relocated from Data General Corporation's headquarters in Westboro, MA to a new R&D facility in Research Triangle Park, NC. It was my job to build a technical publications department to support this new group. As some of you may know, starting from scratch has its frightening moments. It also is an exciting challenge. It gives you a rare opportunity to make fundamental decisions with long-lasting impact.

Since Data General is a computer company, one obvious decision was how to apply computer technology to the publications department. Technical publications personnel at D.G. had been developing ways to help boost productivity for some time, through automated text processing and typesetting systems. I wanted our department to apply this development, increase its effectiveness, and shape it to our own organizational needs.

Three central goals emerged:

- Save keystrokes
- Keep the system simple and flexible
- Automate drudgery.

Saving keystrokes meant avoiding duplication: not entering any information into the system redundantly, and using what was there in as many ways as possible.

Keeping the system simple and flexible meant just that; we did not want users to have to keep a manual by their side just to use it. It was especially important to make the medium transparent to publications personnel (and other users). I did not want writers to be concerned about typesetting concepts; indeed, by hiding the details of typesetting codes we can maintain uniform format styles. Flexibility also lets us defer certain production decisions as late in the cycle as possible. Examples of decisions I wanted to be able to postpone included: number of columns, header and trailer format, and the ultimate output device (letter-quality printer or typesetter).

Lastly, I wanted to automate drudgery. As we looked at the functions which none of us ever liked, we found more and more candidates for automation: tables, tables of contents, indexes, and many others. We wanted to focus our energies on technical communication, and let computers take care of the time-consuming details.

I am now going to tell you about our formatter, and show how some of its key concepts helped boost productivity. Following that, I'll discuss specific economies within different publications functions. Last of all, I will show one way to measure this productivity increase, and demonstrate some of our savings.

DESIGNING A PRODUCTIVE TEXT FORMATTER

We based our text processing and typesetting system on a formatter called RUNOFF. RUNOFF is the name commonly given to text processing systems which share several characteristics. Although I will spare you most of the detail, I must describe some of it.

Every writer uses the system, and types text into individual files using his or her computer terminal keyboard. To achieve certain formatting effects, writers also type commands into the files. Each command begins with a single percent sign, followed by the command's name, and left and right parentheses. Some RUNOFF commands require additional information. This information (called "arguments") goes between the parentheses.

One of the simplest RUNOFF commands is the one to begin a new paragraph:

%PAR()

This command has no arguments.

A RUNOFF command that requires one or more arguments is the %TITLE command. When I prepared this presentation, I used the command:

%TITLE("ACHIEVING NEW LEVELS OF PRODUCTIVITY THROUGH AUTOMATION")

Now you might think that having to use a command to begin this paper is a lot of extra effort. Why not simply type the paper's title and be done with it? True, if I simply type the title, it will be capitalized as I capitalized it and spelled as I spelled it. But I want more than that. I want the title to appear several lines from the top of the page, and I want it centered.

Moreover, if I decide to typeset the text, I will have to specify a type and pointsize. And I don't want to be bothered with these details. I'd much rather have the computer remember and supply them for me. Furthermore, if this were a chapter in a manual, that title would appear in my table of contents. Typing tables of contents is itself a special chore, requiring page numbers, special formats, etc. And remember: I don't want to have to type anything twice.

Lastly, I want all publications produced by the department to be uniform. If there are a dozen writers, I don't want each deciding such things as page layout, type style, etc. Letting each writer simply use the %TITLE command guarantees this uniformity. If we change our minds about the appearance or layout of a table of contents, or of a chapter's title page, all we must do is change the way the computer handles each %TITLE command. No chapter title pages, of which there may be hundreds, need be changed in any way.

From this simple example you see how we achieved our goals. We saved keystrokes; the chapter title is typed only once (and nobody had to supply extra information specifying type). We kept it simple; the %TITLE command suggests its own use, and it is easy to remember. We kept it flexible; we can have the computer achieve different formatting results for different kinds of output media (typesetter or letter-quality printer). We also can change our minds about that format without altering the original text. And we automated drudgery. Nobody ever has to prepare the table of contents; it is done automatically.

PACKAGING THE FILES

When a writer is ready to have the computer format a text file, he types a command to the computer. (This command is not part of the file.) RUNOFF then examines the text file (called a source file), letter by letter, from beginning to end; most text formatters work this way too. As RUNOFF reads the source file, it produces a formatted copy of the file (called a listing file). This beginning-to-end process is an important detail.

If your file is an entire large manual, formatting can take a long time. And there is a related problem. How would you like to be the production manager on the verge of a major release of documentation, with tight schedules (what other kind?). You have just received a torrent of large files from many writers, each with his or her own conventions for naming and keeping track of those files? Our solution to both of these problems is what I call "file packaging."

Before any writer begins a manual, among other things he or she does is to create a special "package" to contain the text files. This package is really a collection of small packages, connected together hierarchically. (We call this collection a "tree.") Please consult Figure 1.

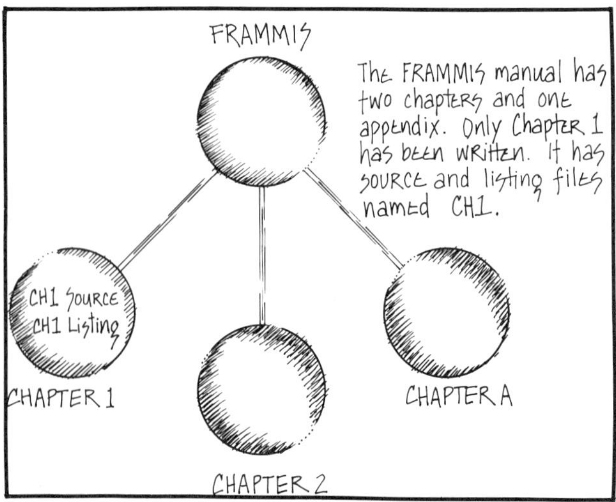

Figure 1. Tree of Manual Files

This tree has just two levels: several branches or packages at the lower level, and one at the top to tie them all together. For the most part, each of the lower-level packages contains one chapter or appendix. Each chapter or appendix consists of a set of files, each with standard names. The name of the package containing Chapter 1, for example, is simply CHAPTER1.

As you might suspect, we have the computer build these trees for us. All it needs in order to build the tree is the name of the manual, and the number of chapters or appendixes. It does the rest.

One obvious advantage to this structure is that it provides a clear separation of information. All of Chapter 1, for example, goes into one package. Another advantage is that we can format each chapter separately. This takes much less time than formatting the entire document each time there is a small change to only one chapter. Finally, when a manual is finished and ready for production, production personnel will know where to find each of the files. All these advantages boost productivity.

SPECIFIC KINDS OF SAVINGS

With this brief introduction to our automated text processing system, we can now examine some of its benefits. These benefits extend to each of the publications functions, and there are some benefits outside the publications organization, too.

Before I get to the benefits though, I will be candid. Our system has its shortcomings. Here are two: hyphenation and spell checking. RUNOFF does hyphenation, but to do the job as well as we would like requires more programming effort than we can afford. So, we don't allow hyphenation at the ends of lines, in typeset copy. Likewise, our spell checker detects bona fide spelling errors. Unfortunately, its vocabulary isn't big enough to distinguish errors from new terms. And to make it smart enough would require more resources than we think justified. So spell checking is best done by writers or editors.

The lesson we have learned so far is to apply automation where we can achieve the biggest productivity gains, at the lowest cost. As we become more clever, or resources become less expensive, we can then explore other ways to apply automation efficiently.

Writer Economies

The principal benefit of RUNOFF to writers is that they are freed from some mundane writing tasks. This lets them concentrate on the quality of the communication itself. Gone is the need to produce tables of contents, lists of tables, or lists of figures. The writer must design the index, and select index entries; the need to construct the index (each time the text is changed!) is gone.

Of course, a fundamental benefit is also the rapid turnaround of what used to be called "draft copy." The 1+ week turnaround for draft typing services is no more. Drafts are produced on the draft printer at around 200 lines per minute. Moreover, those drafts can be "typed" during the night shift, when others do not need the printer for more immediate jobs.

There are a host of other benefits. Apart from the more common text processing features (like flexible table creation and maintenance), our system has many advanced features. These include deferring referenced manual titles or part numbers, and producing certain kinds of central information for several documents.

Often a writer may want to refer to another manual, but has to remember every such reference. Perhaps the manual's title or eventual part number is not yet known. (Or, a very common situation: the product's name is not yet known.) Our system applies the programming notion of variables to these titles and part numbers. The system will automatically replace each variable with the real thing when it is known. For example, a sentence in the source file might read as follows:

See %INCLUDE("MANUAL1") for more information.

"INCLUDE" here means "replace MANUAL1 with whatever the real manual's title is." This would become:

See The FRAMMIS Reference Manual for more information.

This saves both the editor and the writer time and aggravation. The computer searches the source files and changes all references to "MANUAL1" to "FRAMMIS Reference Manual."

Similarly, a software writer often produces specific kinds of information which will be used in two or more places. One such type of information is program syntax expressions (like the description of the %PAR command earlier). Instead of repeating such a description each place it is needed, the writer does it only once, and "INCLUDEs" it each place it is needed. This lets us derive a language pocket guide, for example, from its parent language reference manual, with very little extra effort. Not only do we save effort here, but if the description must change, we change it only once and that takes care of every instance it is used.

Editor Economies

There are as many economies applied to editing jobs as to writing jobs. I described some of these earlier. Another is the ability to search for terms automatically, to check for proper spelling or special usage.

Another major editing benefit is the elimination of exhaustive proofreading: the text is not rekeyed since originally written. We still proofread (don't want to trust these computers too far!) but we can generally concentrate on subtle errors. One such error is the use of the article "an" where (perhaps due to a last-minute term replacement) the article "a" is appropriate. Another subtle error is a "see" reference which might be incorrect due to a change in figure size. That is, we might have to change a "see above" to "see below." Likewise, editors can do term searching automatically, to check for proper spelling or usage of special words.

Having each manual's copy online also lets us build a utility to do style analysis. Writers and editors use this utility to get a modified Flesch, reading grade-level score. Additionally, the utility prints advice about ways to correct sentences which are higher than some arbitrary grade level (e.g., grade 11). This helps editors because each writer can get this reading and advice in privacy, on his or her own CRT. Presumably, editing sessions can then deal with more sophisticated style issues.

Illustration and Typesetting Economies

I have mentioned that copy text is transported nearly "as is," and that most typesetter codes need not be keyed at all. Also, the system lets us guarantee uniformity of typeset styles, and gives us the flexibility to make global pasteup design decisions easily. Since the system can do either one or multi-column formatting, most pasteup labor is avoided too.

Another labor-saving technique we use is to typeset syntax statements in a monospace typestyle, and instruct RUNOFF not to bother formatting this information. Syntax information often depends on spacing and column alignment. Since we use monospace, this information is arranged the same on the draft (or letter-quality) printer as it will be on typeset copy.

Side Benefits

The system confers a great many benefits which may not be as easy to classify. One that we are developing is the ability to excerpt key information from a manual to use as a computer HELP facility. This lets us abstract key information from a manual, and display it on the user's terminal. Instead of having to keep the FRAMMIS manual by his or her side, the user can ask for key information from the computer. This information is guaranteed to be the same as in the manual, since it came from the same source!

Another benefit is that we can use some system facilities to help estimate the size of manuals which will be typeset. We can find out how large each completed manual's file is by issuing a simple keyboard command. Then, we apply a range of factors to this size to determine approximately how many pages the final typeset manual will be (in either one- or two-column format). This helps us schedule use of the typesetter, and gives us an estimate of typesetting, pasteup, and stat camera costs.

Likewise, we use this size information with periodic reports which writers submit (using RUNOFF and other utilities) to generate a broad range of schedules. The information in these reports ranges from very detailed chapter-level achievements to summary reports about high-level milestones.

Lastly, the whole R&D organization uses this system. Development personnel usually use RUNOFF to write their own memos, reports, etc. This cuts down enormously on the need for secretarial services. Each secretary services approximately 50 R&D personnel.

ESTIMATING THE SAVINGS

As I began to estimate our savings, I realized how difficult this really was. To do it, I first compared the proportions of technical communications staff to development personnel at our R.T.P. facility. For example, one question was "how many writers do we need to describe the work of 10 programmers?" I then compared these with similar ratios for our entire company. I know that staffing ratios in the rest of the company are lean. If our ratios were even smaller, I reasoned, then we were truly efficient! Our level of automation would have yielded the advantage.

Before getting to the details, let me emphasize how difficult it is to generalize about publications productivity. The following factors come to mind; I'm sure you can think of many others.

1) What kinds of technical manuals are being produced?

2) Are the documents simply reference manuals, or are they tutorial?

3) Do the writers also write support documents, for in-house use only?

4) Do the writers produce or review simple computer-generated messages?

5) Do the writers produce HELP or other information coming from the computer to the user?

6) What is the quality of the writers' source materials?

If only standard fare is being produced, the technical communicators' jobs are easier. I would consider a FORTRAN reference manual to be "standard fare," since there are many good models to follow. Non-standard fare might be a state-of-the-art relational data base reference. Also, writing, editing, and illustrating tutorial documents is more difficult than producing pure reference manuals. Thus more resources might be required; leaner ratios would not be appropriate.

By the third point I mean whether or not the writers help produce functional, design, or maintenance information; this describes to the programmers or engineers how or why they built the products as they did.

I have mentioned other communications that seem to come from the computer itself. Clearly *somebody* has to produce these. If that somebody is a writer, then more writers' time is needed.

Last and perhaps most important of all, the writer's job is made easier with good source material. My experience is that the best source material is usually none too good; sometimes the source material is as tenuous as "oral tradition" or old computer listings. Clearly, the poorer the source material the more technical communications resources are required.

Having made these disclaimers, I'll now get on to a discussion of the savings. Historical data shows the following proportions of technical communicators to R&D personnel:

- 1 writer : 5 programmers
- 1 editor : 4 writers
- 1 illustrator : 3 writers and editors

Illustrators, in this discussion, produce a full gamut of illustrations and they also do layout and pasteup. Our writers produce the full gamut of manuals, write internal documentation, and produce HELP information. Our editors offer full editing services, and also publish our in-house newsletter.

Our ratios, by contrast, were as follows:

- 1 writer : 8 2/3 developers (programmers and engineers)
- 1 hardware writer : 5 engineers
- 1 software writer : 9 programmers
- 1 illustrator : 6 writers and editors
- 1 editor : 5 writers

Since our personnel data was included in the "historical data," and our ratios are lean, we served to reduce the historical ratios. Removing this effect would widen the gap between the two sets of ratios. As you can see, the only comparable ratio is the hardware writer ratio (with the caveats I mentioned earlier).

In summary, the benefits of an automated and integrated text processing system are many. They save labor, reduce costs, and enrich everybody's jobs. Although you can question the extent of the savings from automation, there is no questioning the fact of these savings. And as text processing systems grow more powerful and less expensive, they will help you concentrate on quality as you reduce costs.

The Technical Writer and the Computer Terminal

John Butler

IBM Corp. GPD

This paper will describe the three writing environments in which I worked during the seventies. Primarily the emphasis will be on the third and present environment. The third environment is one in which the writer uses a video display terminal connected to a computer with a text processing system. The advantages of the present environment over the other two will be described in detail.

The paper will describe the evolution of the writing environment up to the present. The first environment described will be one in which the writer wrote and submitted (handwritten or typed) a manuscript to some one else for manuscript draft preparation. The second environment described will be one in which the writer used a typewriter type terminal connected to a computer with a text processing system to create manuscript drafts. The advantages of the second environment over the first will be described.

The third environment described will be one in which the writer uses the video display terminal connected to a computer with a text processing system. The text processing system described is one that is much easier to use than the one described in the second environment. The additional advantages of the present environment, over the first and second environments, will be described.

This paper is intended for those people in the first or second type of environments who want to migrate to the latest methods.

The Technical Writer and the Computer Terminal

A lot has been said about the use of the computer terminal especially the video display terminal as a writer's tool. Many people are opposed to the use of the terminal, partly because of their fear of change and partly because of their misunderstanding or unawareness of the advantages it offers. Some technical writers I have known feel that it is a waste of their time to enter their manuscripts into a computer terminal. However, they spend a great deal of time writing out drafts for somebody else to enter or type. No writer has more critical deadlines than a newspaper reporter and I refer to a popular television series as an example of the reporter's tools. On each desk is a video display terminal connected to an automated text-processing system. The technical writing profession has undergone some dramatic changes during the past ten years and I will describe those I encountered during that time in IBM.

IBM Writing Environments During the 1970's

As a technical writer during the 1970's I worked in three distinct environments in IBM. The first was an environment where I wrote a draft and submitted it to someone to type and later enter into a text-processing system. The second was an environment where I entered the manuscript into the text-processing system using an IBM 2741 terminal. This terminal was of the typewriter type and was connected to a text-processing system running on an IBM System/360 Model 40 computer. The third was an environment where I entered the infomation into the text-processing system using an IBM 3277 video display terminal. This terminal was connected to the IBM VM/370 system running on an IBM System/370 Model 168 computer. In the last two environments I was responsible for the entry of the entire manuscript and some of the basic formatting commands, such as paragraphs, bulleted lists and headings.

The First Environment

Submitting my draft to have a manuscript created by somebody else provided numerous disadvantages:

- Limited availability of data entry or typing support
- Human errors such as typos, spelling errors, skipped lines
- Inability of the operator to read the input
- Lost input

In addition to the delay at initial entry, more time was required for the correction and update cycles. The correction cycle was necessary to fix the initial input errors, and the update cycle was necessary to incorporate late occurring technical changes and recurring input errors. These correction and update cycles would often go on for some time which added to the completion time and the project costs.

The Second Environment

In the second environment I found that many of the disadvantages of having somebody else type or enter my manuscript were eliminated. Yes, there was some resistance to the change and there were some initial problems. The resistance was caused by fear and a lack of understanding of the text processing-system; the initial problems, by unfamiliarity with the system. The initial problems were overcome as the fear and misunderstanding went away with my learning and understanding the system. The initial resistance was overcome as I became familiar with the system and discovered its many advantages. One big improvement was that I became self-sufficient in creating a manuscript and had direct control over the amount of time it took and the number of mistakes made on initial entry.

In changing the process of manuscript creation to include the writer in the initial preparation, many advantages were added:

- Reduced costs of writing, editing, entry, update, and production
- Reduced number of entry mistakes
- Improved entry turnaround time
- Improved turnaround time for the correction and update cycles
- More timely and up-to-date manuscripts for review

These advantages resulted in a reduction of costs and completion time, both of which are of paramount concern to any information development project.

The Third Environment

The third environment provided additional advantages, because of the video display terminal and the text-processing system it was connected to. They proved to be easy to use: I could see a full page on the screen; I could update a manuscript quickly; I could move the cursor around and make changes much faster than on the 2741 terminal; I could move information around in the document and move through the document displaying pages on the screen; and I could format a manuscript and review that format at the screen before requesting a printed copy. Also, I had more time to concentrate on the contents of my manuscript and, at the same time, improve my speed, accuracy, and flexibility in its creation.

An Easy to Use Tool

If the writer has a video display terminal which is connected to a text-processing system the result is a valuable and easy to use tool. Initially there is resistance to its use, but as writers discover the advantages, they like the system and productivity is greatly improved.

How is productivity improved? Writers, even those who don't know how to type, can quickly create a manuscript in the computer as a data set. This data set can then be corrected and updated on the screen, and information can be moved around or rearranged very easily and quickly. The data set can be formatted and the format can be easily reviewed at the screen before the data set is printed. And at the end of the correction, update, or format cycle, the entire data set can then be reprinted.

Creating a data set with basic format commands using a video display terminal and a text-processing system takes about the same amount of time writers previously took to write out or type the manuscript. However, eliminated or reduced is the amount of time taken to correct errors made when someone else types or enters the writer's work. In addition, the writers have better turnaround time to correct, update, or reformat the manuscript. To summarize, the adavantages of this tool are:

- Easy to learn
- Easy to use
- Reduced costs

- Speed
- Flexibility
- Accuracy
- Elimination of unnecessary human intervention
- Format and review of manuscript at the screen

Using Basic Format Commands

When writers are provided with a computer terminal connected to a text-processing system that is easy to learn and easy to use they can input many, if not all, of the basic format commands, such as paragraphs, headings, lists, and so on. A system of this kind is available with IBM's Document Composition Facility (DCF). With the ability to enter basic format commands writers can, without involving anybody else, provide formatted copies of the manuscript for review. In my environment in IBM I can quickly (overnight) provide a complete updated and reprinted copy of the manuscript.

When writers create a data set that contains basic format commands, the completion time and cost of producing the camera-ready page are reduced. The production cycle will then only consist of entering the more difficult format commands and creating and pasting up the artwork and photographs on the camera-ready copy. In IBM today, one of the most costly activities in producing manuals is the production cycle, therefore, anything that can be done to reduce this cost is something of which we in IBM want to take full advantage.

Availability to Other Information Developers

An easy to use text-processing system accessed by a video display terminal can also be used by other information developers, such as programmers, engineers, managers, or scientists. These individuals can create their source documentation using the computer terminal and the text-processing system and further reduce the amount of information the writer must enter into the system. With the text-handling and editing capabilities of a text-processing system and a computer terminal the writer can easily take advantage of the source documentation already available.

Summary

The video display terminal and the text-processing system open up the computer to writers and other information developers, who can use them as a tool with numerous possibilities to increase their productivity, creativity, flexibility and accuracy. At the same time, writer and production costs are greatly reduced. The fact that programmers, engineers, managers or scientists create their source documentation in the text-processing system helps the writer. The writer can take advantage of this data and use much of it without having to reenter it into the system. Likewise, when the writer's manuscript is is entered into the text-processing system with the basic format commands, the production department requires less time to create camera-ready copy.

During the next decade there will be a dramatic increase in the number of people who will use the computer terminal and the text-processing system to create and maintain information. Some of the reasons for that increase are:

- Availability of small computers with text-processing capability at the local electronics store
- Increased use of computers in medium and small businesses
- Rising costs of manually handling (typing and retyping) drafts
- Need for accurate information on a timely basis
- Need for correcting and reproducing large volumes of information in a small amount of time

As a result of the changing environment in the information development business, the use of video display terminals and automated text-processing are rapidly becoming a must for survival. Of the numerous text-processing systems available in today's marketplace, IBM's Document Composition Facility (DCF) program product offers the capability to meet the demands of information development. Mr. William B. Adams, of IBM, is presenting a paper at this conference about this system and the Generalized Markup Language (GML) approach, the title of which is "Playing Tag with Automated Text-Processing."

WP AS A TOOL IN IMPROVING THE COST-EFFECTIVENESS OF TECHNICAL PUBLICATIONS

Anthony H. Firman
Firman Technical Publications, Inc.

Anthony H. Firman
Firman Technical Publications, Inc.
President
95 Church Street
Pembroke, MA 02359
617-826-5148

At the 27th ITCC in Minneapolis, in a set of three papers entitled "Making Whoopie!", Firman Publications described the professional use of word-processing (WP), and its general impact on technical publishing.

Since then we have been asked many questions, all of which basically come down to just three:
- *How much money do these WP-based creation and production techniques really save?*
- *How can we bring about these savings in our own company?*
- *How can we use these techniques to increase the effectiveness of our publications?*

In this second set of three papers, Firman Publications is presenting the answers to these questions. In answering these questions, we have tried to be as specific and practical as possible, and we have included examples wherever possible.

THE COST OF TRADITIONAL PUBLICATIONS

The cost of creating and producing a publication is based on many operations and many staffmembers. The various operations and the corresponding staffmembers are identified in Figure 1. (I am choosing to ignore the costs of reproduction and distribution, since they are largely out of our domain.)

For each of the staffmembers involved in these operations, there is a unit cost. Figure 2 shows some fairly typical costs for the staffmembers involved in publications, including both salary and overhead. "Overhead" is based on a flat rate of 100% of each staffmember's salary (which covers such factors as floor-space, utilities, furniture, and supplies); in some cases it also covers special job-related equipment. The "$/hour" is based on 235 8-hour days

	OPERATION	STAFFMEMBER
1	planning the document	writer
2	research	writer
3	writing	writer
4	editing	editor
5	reviewing	engineer
6	revision	writer
7	graphic design	illustrator
8	specifying type	illustrator
9	typing	typist
10	typesetting	typist
11	proofreading	proofreader
12	illustrating	illustrator
13	photography	photographer
14	reprographics	camera operator
15	layout and paste-up	junior illustrator
16	division management	division manager
17	group supervision	group supervisor

Figure 1. Operations in creating and producing publications.

STAFFMEMBER	SALARY	OVERHEAD	$/HOUR
writer	$25,000	$25,000	$26.50
illustrator	$20,000	$20,000	$21.25
editor	$25,000	$25,000	$26.50
engineer	$30,000	$30,000	$32.00
typist	$12,000	$12,000	$12.75
proofreader	$ 8,000	$ 8,000	$ 8.50
camera operator	$ 8,000	$10,000	$ 9.50
photographer	$20,000	$30,000	$26.50
jr illustrator	$10,000	$10,000	$10.75
div. manager	$35,000	$35,000	$37.25
grp. supervisor	$20,000	$20,000	$21.25

Figure 2. The total annual cost and cost/hour of the various staffmembers involved in creating and producing technical publications using traditional techniques.

per year. Note that the cost of engineering (or marketing) review is included in these figures, although admittedly it is often hard to identify.

The cost of a particular publication can be computed from figures such as these and from the hours that each staffmember puts in. However, this may not be easy. Your company may not have figures like these. You may not be able to identify the actual effort put into a particular publication by each staffmember. And you may not have access to the engineering-department figures for review time and cost.

Let us consider an example publication, and see how these costs work out to an average dollars/page figure. Our example is based on a maintenance manual for a piece of industrial equipment; there are about 100 pages; the copy is simply typed; there are a number of line illustrations and photographs; and the whole thing is a simple black-and-white design, bound in a ring-binder. Figure 3 shows the hours that each staffmember might put in to create and produce this manual, then multiplies these hours by the appropriate cost/hour figure to find the total cost/operation. By adding these figures up we get the total cost of the job ($23,929), and by dividing this by the number of pages we get the final cost/page ($239/page).

OPERATION	STAFFMEMBER	WORK HOURS	COST/ HOUR	TOTAL COST
planning	writer	20	$26.50	$ 530
research	writer	40	$26.50	$ 1,060
writing	writer	500	$26.50	$13,250
editing	editor	20	$26.50	$ 530
reviewing	engineer	20	$32.00	$ 640
revision	writer	80	$26.50	$ 2,120
graphic design	illustrator	20	$21.25	$ 425
typing	typist	50	$12.75	$ 638
proofreading	proofreader	20	$ 8.50	$ 170
illustrating	illustrator	100	$21.25	$ 2,125
photography	photographer	10	$26.50	$ 265
reprographics	camera operator	10	$ 9.50	$ 95
layout/paste-up	jr. illustrator	50	$10.75	$ 538
management	division mgr.	30	$37.25	$ 1,118
supervision	group super.	20	$21.25	$ 425
TOTAL:				$23,929
COST/PAGE:				$239/page

Figure 3. The hours that each staffmember might put in to create and produce a manual using traditional methods, multiplied by the appropriate cost/hour figure to find the total cost/operation.

This page cost - about $240/page - may surprise you. Perhaps you think it is very high, and you are looking for flaws in the data. You may find some differences between the figures I have used in this illustration and the actual figures that your company uses. However, let me caution you: many companies do not gather this type of data; or they gather the data, but they omit certain key factors such as "overhead - which we have to pay anyway".

In fact, a nationwide survey I made during the last half of 1980 shows that the cost of creating and producing technical manuals is generally in the range $200-300/page. This figure begins to make sense when you consider just one very basic factor: most writers require 4-8 hours to create each page of text, and at a cost of $25.50 per hour per writer, this cost alone is $106-212/page.

THE COST OF WP-BASED PUBLICATIONS

The professional use of WP in creating and producing technical publications - as described in "Making Whoopie!" - cuts costs significantly because it cuts the amount of work that must be done manually. The increased cost of equipment is still far below the savings in labor.

The various operations involved in creating and producing a publication that were listed in Figure 1 are listed again in Figure 4, together with a brief description of the impact WP has on each of them. Some of these operations are eliminated as separate job functions, or are reduced to trivial tasks that are incorporated into other operations: these include editing, typing, typesetting, and proofreading. Other operations are significantly reduced by the elimination of repetetive, time-consuming tasks: these include writing, which forms a major part of the creation process; the entire revision process; and layout/paste-up plus some aspects of illustration, which form much of the production process.

This impact on staffing levels appears immediately in the bottom line. We find that our methods of creating and producing publications result in costs that are substantially lower that the national average - usually in the range $100-200/page for technical manuals, and occasionally even lower*. As an example, Figure 5 compares the costs for the example 100-page manual of Figure 3 with the corresponding costs for this manual created and produced using WP-based techniques.

As you compare the itemized costs on Figure 5 you may notice that certain costs using WP-based techniques are higher than the corresponding costs using traditional techniques - even though I have claimed that the operation in question is simpler using WP-based techniques. There are two reasons for this:

1) When you use WP-based techniques you will employ fewer staff, but they will be of a higher caliber, and they will have more responsibility. Consequently, you will have to pay them more.
2) Overhead costs are increased by the cost of buying and operating (or renting) the WP system.

The important thing to realize is that although certain costs rise slightly, others fall considerably: ultimately, it is the famous bottom line that counts.

---

*Unlike many publications departments, Firman Publications is totally self-contained, so we cannot hide any costs. The one exception is the cost of our clients' review time; however, this is usually a relatively small part of the total cost.

OPERATION	IMPACT OF WP-BASED TECHNIQUES
1 planning	Simplified: outlines can be created and manipulated more easily.
2 research	No impact.
3 writing	The writer has more flexibility in manipulating content and format.
4 editing	Largely eliminated as a separate function.
5 reviewing	Simplified: fair printouts in near-final format are easier for non-visual reviewers to understand.
6 revision	Dramatically reduced by the memory database, CRT editing, and rapid fair-copy printout.
7 graphic design	Designer can experiment with different copy formats.
8 specifying type	No impact.
10 typesetting	Simplified if volume justifies connecting WP to phototypesetter.
11 proofreading	Eliminated: each reviewer effectively proofreads, and there is no retyping to check.
12 illustrating	Simplified: many diagrams are formatted on WP, and require only simple inking.
13 photography	No impact.
14 reprographics	No impact.
15 layout and paste-up	Greatly reduced: manuals are printed out in formatted pages by WP.
16 division management	Reduced, because individual writers are given more responsibility.
17 group supervision	Reduced or eliminated, because fewer, higher-grade staff require less supervision.

Figure 4. The impact of WP-based creation and production techniques on specific operations.

OPERATION	COSTS BASED ON TRADITIONAL TECHNIQUES	WP-BASED TECHNIQUES
1 planning	$ 530	$ 800
2 research	$ 1,060	$ 1,600
3 writing	$13,250	$ 6,000
4 editing	$ 530	$ 0
5 reviewing	$ 640	$ 640
6 revision	$ 2,120	$ 800
7 graphic design	$ 425	$ 600
9 typing	$ 638	$ 0
11 proofreading	$ 170	$ 0
12 illustrating	$ 2,125	$ 2,000
13 photography	$ 265	$ 265
14 reprographics	$ 95	$ 95
15 layout/paste-up	$ 538	$ 300
16 dept. management	$ 1,118	$ 400
17 group management	$ 425	$ 0
total:	$23,929	$13,500
cost/page	$239/page	$135/page

Figure 5. A comparison of the detailed costs of creating a 100-page manual using traditional techniques and using WP-based techniques. This example is based on the model used in Figure 3.

THE EFFECTIVENESS OF PUBLICATIONS

The effectiveness of a publication can be judged by the extent to which it makes other communication media redundant. Four factors can be identified: timeliness, usability, accuracy, and thoroughness.

Timeliness If a promotional publication is not timely - for example, if a product is ready for the marketplace before brochures are available - the vendor loses sales, simply because the marketplace is not informed about the product. If a support publication is not timely - for example, if a product reaches the field long before the manuals - then the user has no choice but to telephone the vendor whenever he/she has a question; repeated telephone calls asking the same questions of expensive sales and engineering people cost both the vendor and the user a lot of money: the vendor's costs rise, and the user loses patience - and buys from a different vendor next time.

Usability If a promotional publication is not easy to use, users rapidly lose interest, and turn to more informative vendors. If a support publication is not easy to use, users tire of searching fruitlessly, and again they fall back on the telephone as an easier solution to their problems.

Accuracy If a publication is not accurate, users discard it, and again fall back on calling the vendor. They also lose faith in the vendor.

Thoroughness If a publication is not thorough, users tire of searching fruitlessly for information they need. Once again, they fall back on telephoning the vendor; and once again, they lose faith in the vendor.

So in every case, an ineffective publication reduces the vendor's profits in two ways: by reducing sales and by increasing costs. It is unfortunate that both the lost sales and the increased costs - if their existance is recognized at all - are rarely quantifiable. Lost sales can be blamed on other factors; increased costs are buried in the overall cost of sales, engineering, and field service. However, if we can provide the vendor with effective publications, then sales improve, costs decline, and profits improve.

HOW WP CAN IMPROVE EFFECTIVENESS

The effectiveness of publications can be measured in terms of timeliness, usability, accuracy, and thoroughness - and WP-based techniques have an impact on all four.

Timeliness Perhaps the most obvious impact is on timeliness: WP significantly cuts the elapsed time between the conception of a publication and the distribution of printed copies. This is possible because much of the traditional time-consuming clerical work is eliminated. (A WP system can print out a perfect page in less then one minute: can your typist do that?)

Furthermore, WP can be used to build a database: all publications can be retained in disk-files, to be used as a source for updating old publications, and for creating new, related publications. As this this database grows, the ability to draw on it continually reduces the delivery time of new and updated publications.

(Incidentally, don't forget that every item that reduces delivery time by reducing labor also reduces the cost of labor.)

Usability The high-speed editing capability of WP functions as a "tool to think with" for the professional communicator, enabling him/her to develop communication techniques that make the reader's job easier. The communicator is not the only one who learns new techniques, however: the WP system itself can also learn new techniques in the form of special software that simplifies the communicator's task in the future*. These techniques also continually reduce the work required to create and produce a publication, so they continually reduce the delivery time.

WP also provides more attractive, more flexible printouts that conventional typing, which can make inexpensive manuals easier to use.

Accuracy Every draft copy produced by WP is a new, clean printout, in essentially the format of the final publication: it is not the traditional cut-and-pasted, marked-up, fourth-generation copy. This helps reviewers and editors see exactly what is going on, which in turn helps to ensure that "minor details" don't get overlooked. A handwritten "8" that looks like a "5" just won't get through any more.

Thoroughness Some of the new techniques that we have developed are more rigorous than traditional narrative text. Text mapping (which we use quite often for certain types of technical manual) is a case in point: its use of standardized functional headings tends to ensure that key information is not overlooked when the publication is being created. It is the use of WP that gives us the freedom to develop more thorough communication techniques.

CONCLUSIONS

WP-based techniques can significantly improve the cost-effectiveness of technical publications. These techniques simultaneously lower the costs and increase the effectiveness: they reduce costs by reducing the amount of work required to create and produce publications; and they increase effectiveness by improving the timeliness, usability, accuracy, and thoroughness of publications.

Furthermore, the impact of WP on the cost-effectiveness of publications is increased by its continued use: it results in the formation of a database of documents that continually reduces costs and delivery times still further; and it results in the development of operating techniques and special software that streamline operations.

In the long run, therefore, the cost of creating and producing the average page of text using WP-based techniques may be cut by as much as 50% over traditional methods.

---

*This is true if the WP system can be programmed by the user, which most (although unfortunately not all) systems can.

WORD PROCESSING/PHOTOCOMPOSITION . . . A MARRIAGE OF CONVENIENCE

by George E. Hoerter
Martin Marietta Denver Aerospace

George E. Hoerter
Chief, Editorial Services
Martin Marietta Denver Aerospace
P.O. Box 179 (S-2651)
Denver, Colorado 80201
(303) 977-4168
Member: Rocky Mountain Chapter-STC
Member: IABC/Colorado

Take one powerful, shared-logic word processor, interface it with a versatile phototypesetter, and what do you get? When Martin Marietta Denver Aerospace "married" two such systems, the result was a publications tool that took much of the pain out of producing large, high-quality, time-critical technical publications and visual aids and reduced production costs too.

This paper contrasts old production methods with the new methods developed at Denver Aerospace; discusses a method of controlling page count throughout the writing, editing, and production process to avoid last-minute rewriting to meet page limitations; and examines the equipment [a shared-logic word processor, an optical character recognition (OCR) unit, and phototypesetter] used in publishing technical documents and visual aids. Also included is a look into the future at more sophisticated equipment that will make the jobs of all technical communicators easier.

INTRODUCTION

The primary mission of Denver Aerospace's Editorial Services Section is to produce attractive, winning new-business proposals on a short turnaround schedule. These documents, by customer direction, are severely page limited and, therefore, require the use of production techniques that permit placing as much information on a page as possible, yet keep pages attractive and readable; phototypesetting was the logical solution to the problem.

Through years of practice in producing technical documentation and by acquiring modern, efficient equipment, Denver Aerospace has developed methods of operation for technical editors, production personnel, and engineering writers that ensure maintaining page limitations from the day writing begins. In the process, we have reduced overtime, the number of steps in the production process, schedule time, author review time, and, therefore, production costs.

THE OLD WAY . . .

Before the evolution of modern technology, producing a large, multivolume new-business proposal at Martin Marietta Denver Aerospace was a long and involved process. Each time the draft was revised, typists would rekey, proofread, and correct the copy. When the proposal team was satisfied with the draft, the volumes were turned over to Editorial Services to prepare the reproducible for printing.

After each volume was edited, another typing, proofreading, and correction cycle was necessary. At this point, the draft was retyped for final production on the IBM Magnetic Tape/"Selectric" Typewriter (MT/ST) for playout on IBM's Magnetic Tape/"Selectric" Composer (MT/SC) system. Corrections resulting from this retyping were added by going through a transfer operation on the MT/ST. This process involved transferring copy from the original tape to a new one by playing out the copy to the point of correction, skipping out copy to be corrected, inserting corrections, and continuing the transfer to the end of the tape. If any corrections were missed, the entire transfer process had to be repeated.

Typesetting the copy on the MT/SC was also a slow, cumbersome process. The operator had to sit at the unit during playout, make typeface changes by changing the type font, and manually make hyphenation decisions.

Top management comments had to be incorporated during the production cycle—causing either another transfer and playout cycle or a major cut-and-paste operation. As the scheduled date for shipping the proposal to the print shop got closer, the overtime hours required were extended, sometimes leading to an around-the-clock operation. With longer hours and extensive corrections, quality decreased and costs escalated.

As competition became keener for large procurements and because potential customers imposed more restrictions on bidders (e.g., page limitations and type size restrictions), the need for a change in our operating procedures became apparent. Emerging technology appeared to provide a solution to the problem.

A BETTER WAY . . . NOW

Because typesetting proposals and visual aids represents only a portion of our overall workload, in our design of a new production system we also had to address the problem of efficiently producing "typewriter-quality" reports. Economic factors played a large role in our selection process—we could not afford two separate units, but required one overall system that would work for all our documentation requirements.

Designing a Total System

In addition to input and output, we also recognized the need to build controls into the system to enable writers, editors, and production personnel to know where they stood from a page-count standpoint during all stages of writing and final production of typeset documents. We further recognized the need to eliminate the keyboarding redundancies inherent in our existing production system.

Word Processing Input

After examining our specific input requirements and surveying the marketplace, Denver Aerospace determined that Wang word processors best met our needs. We have installed approximately 35 systems with more than 140 keyboards. Word processing units, including Wang 20s, 25s, 30s, and OIS 140 systems are located throughout our facility in both centralized word processing centers and dispersed in project areas.

Wang units are used to capture keystrokes for most of our typeset documentation and as both input and output units for documents that formerly were prepared on typewriters (Fig. 1).

Scanned Input

Installing word processing equipment solved the rekeying problems for our major documentation producers, but did nothing to solve the problems in smaller project areas. To alleviate redundancies in these areas, we looked to technology again in the form of optical character recognition (OCR) units.

Figure 1
Word Processing Operators Input and Update Copy

The Dest Data/Word OCR units installed at Denver Aerospace can scan Prestige Elite, Letter Gothic 12, Courier 10, and OCR-B fonts—basically the only typefaces used in our facility. Now, producers of an occasional proposal or report or those who prepare visual aids for oral presentations can take advantage of word processing for editing, updating, and typesetting their work by preparing the draft in an OCR-readable typeface. The draft is then scanned onto the Wang word processor, which is used for editing and updating and for input to the typesetter (Fig. 2).

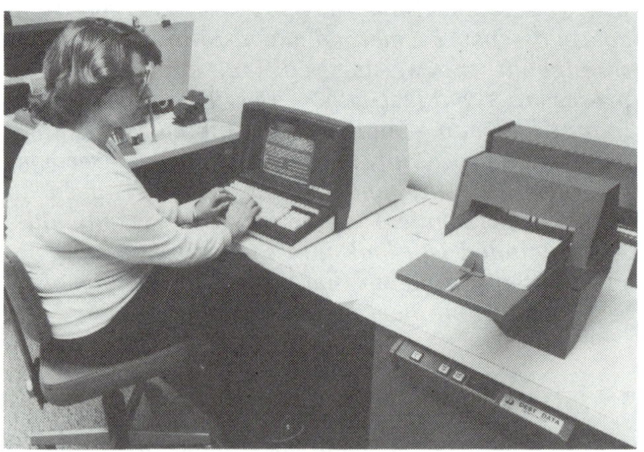

Figure 2
Typewritten Copy Is Scanned and Electronically Moved to the Word Processing System for Updating and Typesetting

Typesetting

For final output, we searched for a photocomposition system that could be interfaced with our word processing system and could also be used as a standalone system for preparing small typeset jobs. Denver Aerospace selected the Compugraphic Display Composition System (DCS) for use in the Editorial Services area.

In our configuration, the DCS comprises a VideoSetter Universal typesetter; two Unified Composers used to input and update files; two PreViews, which display copy to be typeset in the actual size, weight, and style type that will be used in the final product, to assist with page makeup; a hard-copy printer that provides review copies scanned from the PreView screen; a disk reader for offline typesetting; and a resin-coated (RC) processor (Fig. 3).

Wang Laboratories provided the interface device connecting our Wang 30 shared-logic system to the Compugraphic DCS. With this configuration, it is possible to input all typesetting commands from the seven keyboards attached to the Editorial Services Wang 30, which frees the two Unified Composers for making final corrections, checking hyphenation and line endings, and for making up pages with the PreViews.

System Controls

Key reasons for the success of the system designed at Denver Aerospace are the controls that were developed and implemented for its use.

Page Count

Editorial Services personnel developed a method for determining the amount of word processor or typewriter input that would equal one full typeset page. In developing this control method, we converted draft to a typeset page prepared in our standard format to determine the ratio of draft lines to finished typeset lines. When, because of customer direction, deviations to standard format are necessary (e.g., line spacing or type size), the ratio can be adjusted accordingly. Adjustments are also made depending on the method of presentation being used in a given document (i.e., artwork on facing pages or art embedded with text).

Before work on any proposal begins, the editor assigned to the job briefs the proposal team and provides instructions to persons who will be preparing draft. By following the instructions provided, the author of any section of the proposal will know how much copy he can write to keep within the page limitations imposed by the proposal manager. After his draft is prepared and typed, he can determine how much to cut or add to maintain his allocated page count. Also, an editor who is given too much copy for a particular section can work with the author to cut it before production begins—thereby eliminating the last-minute surprise of being over the page count and the painful process of reworking to recover.

Figure 3 Operator Uses the PreView to Check a File before Typesetting

Reviews

Another control is establishing where text reviews take place in the production cycle. With the cooperation of proposal teams, we have been able to schedule the last major author review after the technical editor has edited the draft and it has been updated by the word processing operator. This is the last point in the preparation process where major changes can be inserted without causing major production problems. After the document is typeset and pasted up, a project representative checks it for text flow and art placement. Any critical change will, of course, be made at this stage, but revisions are fewer if the major author review is scheduled before the document is typeset.

Results

Significant improvements in the proposal preparation process and in the final product have resulted from the "marriage" of word processing and photocomposition systems at Martin Marietta Denver Aerospace. Specifically, we have:

- **Reduced Redundancies**—Eliminated rekeyboarding entire drafts after each revision.
- **Reduced Proofreading Time for Both Authors and Editors**—Only changes need be proofed.
- **Reduced Production Time**—Eliminating production redundancies reduces both the time required to prepare the reproducible and the amount of last-minute corrections and major rework. Therefore, the authors can take more time to polish their draft, which results in a better end product.
- **Reduced Production Costs**—Less overtime is required because the draft does not have to be rekeyed for final production and because page count is controlled throughout the process. Cut-and-paste operations at the last minute are eliminated.

WHAT'S ON THE HORIZON?

Technology advances in publishing in recent years are only hints of what is to come. Automated publishing systems have already been introduced and are being used by large publishing entities such as major corporations, news magazines, and newspapers.

As technology is refined and more systems are placed in operation, prices will reach the point where smaller firms can take advantage of automated publishing systems. These systems will feature graphics terminals for preparing and modifying artwork and will eliminate the need for art paste-up—line art and halftones will be digitized and typeset in position along with text. Camera-ready output will give way to press plates or, through electronic printing, direct plain paper output.

These are systems that exist today—they are not "futures," but represent the latest state of the art in publishing (Fig. 4).

Many more tools will be available to make the job of tomorrow's technical communicator easier. However, tomorrow's technical communicator will face the same challenges we do today—how to best take advantage of technological advances to produce a better product at less cost in less time.

Figure 4
Future Systems Will Include Automated Illustration Terminals, Interactive Page Makeup Stations, and Will Output Fully Madeup Pages with Line and Tone Art in Place

ELECTRONIC MAIL:
NEW COMMUNICATIONS FRONTIER ON THE "LAST FRONTIER"

Charles H. Iliff and Constance C. Katasse

Anchorage Community College

Charles H. Iliff
Constance C. Katasse
Anchorage Community
College - Teachers
2533 Providence Avenue
Anchorage, Alaska 99504
(907) 263-1109/263-1160

Alaska, the largest State in the Union, has always been a land of barriers as well as an abundance of resources, ranging from furs, to gold, to oil. The population is still scattered in isolated pockets, determined primarily by the location of natural resources. Because of resulting communication problems, the University of Alaska recently implemented an Electronic Mail System. The advantage of electronic mail in Alaska is obvious: it provides dependable, "instantaneous" communication over enormous geographic distances. The major disadvantage is massive initial cost. Nevertheless, the University is committed to electronic mail and thusfar has been pleased with the benefits.

COMMUNICATIONS IN THE "GREAT LAND"

With a total area of approximately 587,000 square miles, the State of Alaska is about 2 1/2 times the size of Texas and about 1/5 the size of the rest of the United States. Alaska was granted statehood in 1959 and achieved international notoriety in the mid-1970's during the construction of the Trans-Alaska Oil Pipeline.

Located in the extreme northwestern part of North America, Alaska has more than 3 million lakes, 39 mountain ranges, 17 of the highest mountains in the United States, and over 28,000 miles of glaciers and ice fields. The longest distance east to west is about 24,000 miles; north to south, it is 1,420 miles.

On October 18, 1867, control of the Alaskan Territory was transferred by the Russians to the U. S. Government. For 60 years after, mail steamers provided most year-round communication, supplemented by dog teams for winter mail in the Arctic and Subarctic regions. The U. S. Army Signal Corps established the Alaska Communications System in 1900, toward the end of the Klondike Gold Rush, and in 1902, the telegraph from Eagle to Valdez was completed. It carried both government and commercial messages via an intricate combination of cable and telegraph links.

In 1923, the Alaska Railroad and the Richardson Highway were completed, linking coastal villages and towns to those in the Interior. In 1924, the first Alaskan airmail service was scheduled. Gradually, radio-telephone stations were established in remote areas by the Territorial Government and the Alaska Native Service. Light planes became more popular in the 1930's, and scheduled commercial airlines appeared by 1940.

During World War II, a large network of airfields was constructed, as well as the long-awaited Alaska Highway. It linked Alaskan roads with the Canadian highway system at Dawson Creek, British Columbia. By the late 1950's, "White Alice," an integrated microwave communications system, was providing communication services for all government agencies within the State.

Throughout the 1960's and 1970's, air travel and telephone services expanded dramatically. For the first time, many rural residents could communicate regularly with people who lived in the four major cities. Yet Alaska remains a land of immense distances and scattered "population pockets." Rivers flood periodically; and erosion by wind, water, and cold plagues telegraph and telephone lines. Jet aircraft can't fly in ice fog, and most experienced small-plane pilots are too wise to try. So for seven to nine months out of every calendar year, most Alaskans realize that "you can't get there from here." In addition, communication is always hampered by Alaska's land area. The State is so large that it includes four different time zones. When it's 9 A.M. in Sitka, it's 5 A.M. in Atka!

AN ELECTRONIC SOLUTION

Largely because of such communication barriers, in 1975 the University of Alaska Statewide System inaugurated electronic mail as part of its Honeywell Computer Network. "Mail" on the University of Alaska Computer Network (UACN) is a subsystem of the Time-Sharing System. Time-Sharing allows multiple users (up to a practical limit of 120) to simultaneously access the computer and share its resources from any campus of the University.

The UACN presently serves students, faculty, staff, and administrators throughout the State. The Network itself consists of minicomputers located on the Anchorage, Fairbanks, and Juneau campuses. Each minicomputer is connected to the large dual-processor Honeywell computer in Fairbanks. Kenai and Kodiak have access to Time-Sharing services through the Anchorage minicomputer.

The Network also has a variety of Time-Sharing terminals for public use, a few with graphics capability. Many of the terminals are available twenty-four hours a day, and access to others generally depends on building hours. The computer itself is available twenty-four hours a day, with occasional downtime for preventive maintenance.

Two unique aspects of the UACN are the extreme distance it spans and the fact that all users are treated equally. All users have access to the Time-Sharing System through minicomputer-driven "nodes," thus assuring physically equal facilities. Remote node users are charged nothing extra for communication, and all users are assigned equal priority on the System. Of the 5100 users, 3700 are students.

USING THE "MAIL"

The University of Alaska "Mail" subsystem is the first real computing experience for most users. To initiate a Time-Sharing session, the user must have an established account called a "Userid." The place to apply is the office of the closest computer node. Any student who applies for a Userid is initially given $100 worth of computer time, renewable each semester. If this proves insufficient, more can be obtained for a worthy project, as long as a faculty member supports the request.

After the Userid application has been submitted, the node supervisor usually offers a brief introduction to the Time-Sharing System. In a one-hour session, the new user can be taught how to sign on the terminal, execute rudimentary "Mail" commands, and thus be actively communicating. There is an on-line documentation system called "Explain," as well as Honeywell reference manuals for self help.

"Mail" permits any authorized user to send messages or files containing messages to any other user's "Mailbox." Each "Mailbox" is created automatically, and users do not need to know the technical details of computer storage. Every time a user signs on, s/he is notified of any unread messages. The user can select to read a message by printing it out at the terminal, and then decide whether to answer the message, retain it indefinitely, or delete it. In October 1980 alone, over 85,000 messages were sent.

"Mail" is used extensively for formal administrative communications among various parts of the University System, for making class homework assignments, and even for an informal student dating service. It allows immediate reply if the sender of the original message is still connected to the computer. Most important, "Mail" creates a uniquely conversational environment among users who would otherwise be separated by distance and time.

AN ACTUAL EXAMPLE

On June 17, Eric, a petroleum technology major at Kenai Peninsula Community College, decides to contact his sister, Sue, a graduate student at the Geophysical Institute in Fairbanks. He wants to arrange to view the midnight sun at Eagle Summit, 160 miles northwest of Fairbanks, during the Summer Solstice Celebration on June 21.

Eric could keep calling and calling Sue's telephone number, hoping to find her at home. But he knows that she spends most of her time in the campus laboratories. A successful telephone call would cost Eric $1.65 for the first three minutes plus $.55 for each additional minute. He could write a letter (one-way communication), but the usual delivery time is four days each way. The cost--a minimal $.15 for him and $.15 for Sue. So Eric quickly decides to use "Mail." At best, Sue is on the computer at the same time, and communication is instantaneous. At worst, the message is available the next time Sue signs on the computer. The cost--absolutely $0.00. The time lapse--variable, from zero to four hours.

Eric asks a friend to explain the "Mail" subsystem. Lori, who is a student in the college's computer program, provides him with a brief demonstration. In summary, this demonstration describes:

1. Obtaining a "Userid."

2. Learning how to "sign on" or activate the terminal.

3. Learning four commands used in "Mail."

   S - to send a message
   P - to print a message
   Q - to quit using mail, and if all else fails...
   H - to get help for a confused user

4. Learning how to "sign off" or deactivate the terminal.

In less than twenty minutes, Lori has demonstrated the use of the terminal, and Eric has sent his message on its way to Sue in Fairbanks. Later that afternoon, he signs on again and asks the computer to print any messages. In a matter of seconds, Eric learns that Sue has agreed to make his reservations for the bus ride to Eagle Summit, and she is looking forward to seeing him again.

PROBLEMS AND POSSIBILITIES

There are a few problems with the use of "Mail." Users can become compulsive. Furthermore, given the limited number of terminals, computer students with homework assignments and "Mail" users must compete for terminal time. Already there has been one "minor altercation" about whose use was more important. There have also been two unpleasant incidents in which one user stole another user's password and from the pirated account sent an obscene "Mail" message to over 1000 users chosen at random. Contrary to popular belief, no computer system is "security proof," and users must be made aware that electronic mail is no safer than paper mail.

Despite such problems, the future of electronic mail is bright. Minicomputer and microcomputer vendors are now actively marketing electronic mail systems, and the U. S. Postal Service is investigating the feasibility of such a system in selected areas of the country. Presently, a home computer user with a Radio Shack TRS-80 can establish an account with Western Union Electronic Mail, Inc. and send mailgrams to other users in the United States.

In February 1978, the British Post Office of Telecommunications implemented home computer use, including electronic mail and publishing, under the "Prestel" trademark. Electronic mail has been used successfully in embassy communications and large private companies. It is compatible with most common communication lines, such as WATS, satellites, PBX, and DDD. In addition, electronic mail has been integrated with other automated office technologies. Some obvious extensions of "Mail" use are electronic scheduling of appointments, telephone message-taking, dictation transcription, and memorandum distribution.

The range of communication possibilities by "Mail" is almost limitless, for both educational institutions and businesses. "Mail" presents a modern pen pal capability, with the added advantage of conversational mode. In Alaska, one group of correspondents, known as "Nodies Anonymous," has even facilitated the engagement of two of its members. Users can either hold intimate conversations or "tell the world" through this interactive communications network. Of course, when less personable communication is desired, "Mail" provides the perfect medium for relaying factual information.

The University of Alaska Statewide System is committed to using electronic mail to help solve its communication problem. There is, of course, a price. The computer hardware represents approximately $3 million, plus ongoing costs of a staff of 40 and $12,000 per month for two microwave lines from Fairbanks to Anchorage. Although "Mail" alone could not have justified these expenditures, it has become a very cost-effective addition. Any other organization which can justify the initial costs of a computer network should seriously consider the benefits of electronic mail.

THE TECHNICAL WRITER'S ROLE IN ADVANCED ELECTRONIC PUBLISHING ENVIRONMENTS

Nancy L. Karner
Xerox Corporation

Nancy L. Karner
Xerox Corporation
Technical Writer
1350 Jefferson Road
Rochester, New York 14623
(716) 427-1643
Member of Rochester STC Chapter
Charter Member of *Sigma Tau Chi*, (Writing Honorary)
Adm. VP of Xerox Communicators, (Toastmasters International Club)

As text processing equipment becomes more available within engineering and research areas, the technical writer has the opportunity to practice familiar roles and explore new areas of document development. The writer can prepare standards and should encourage their use to assure greater readability of technical documents. Training engineers to use the text editors wisely will result in better written documentation, improved formatting practices, and more efficient use of equipment and personnel.

How long does it take for a person to become addicted to a computer with text processing capability?

Not long if this person has easy access to a terminal and if the system is fast and efficient. An engineer who works with computer software is especially quick to get hooked.

Now imagine an entire group of software development engineers with access to a powerful and easy-to-use computer system. This system, primarily brought into the group for software development, also has a powerful text editor. Recognizing the capabilities of this editor, the engineers are delighted to use the text processing program to write, edit, print, and file their many documents.

Why might this present special challenges to a technical writer? Shouldn't the writer be delighted about the enthusiasm shown by non-writers toward producing technical documentation within this environment? Delighted, yes; challenged, unquestionably!

Unique Opportunities

In this internal technical environment, the engineers are often experiencing the ***process*** of writing for the first time. Although comfortable with the iterative nature of building computer code, they automatically groan over the process of documenting that code and explaining how it was developed. With a video terminal, the engineers find keying the material into the system especially fast; they may correct mistakes and re-work material without losing what was done before.

> Where it is possible for an author to work directly on a video terminal, the (electronic publishing) industry has found that (the author) feels he can do a better job, and that the copy which he writes should never have to be re-typed or re-proofread for new typographic errors before it is put into final form to be distributed to the "customer."*

Familiar with the computer system terminals with which they have designed and developed code, the engineers begin to build their documents, part by part. Experiencing the ease of inserting, deleting, and formatting text, the engineers become instant experts in producing staggering volumes of technical documentation. And here is the challenge for the technical writer.

AN OLD ROLE TO A NEW MEDIUM -- Setting Standards

Typing standards used to apply only to office secretaries. An engineer with a pencil, or one lucky enough to swipe a decrepit typewriter, didn't worry about formatting. His concern was to produce text which could be read by secretaries. But now the familiar and friendly terminal promises to turn out the engineer's vital information quickly with no erasures! Forgetting the editing function the secretary also supplied, the engineer is tempted to turn out his work to the world with no check at all. The technical writer must help set content and formatting standards to avoid what could be chaos.

* Winter, Don C. and Stewart, Donald E., **Future Trends in Electronic Publishing**, *Conference of the National Federation of Abstracting and Indexing Services*, Washington, D.C., March 6, 1979.

Electronic Form Files

The content and form standards may be pre-set into the electronic form files available from the text editing system. For example, the engineer may ask for an internal memo form. A copy of this form will then be visible and ready for text insertion (see *Figure 1*). When the engineer is satisfied with the memo, he can store it with its own name, leaving the original form intact.

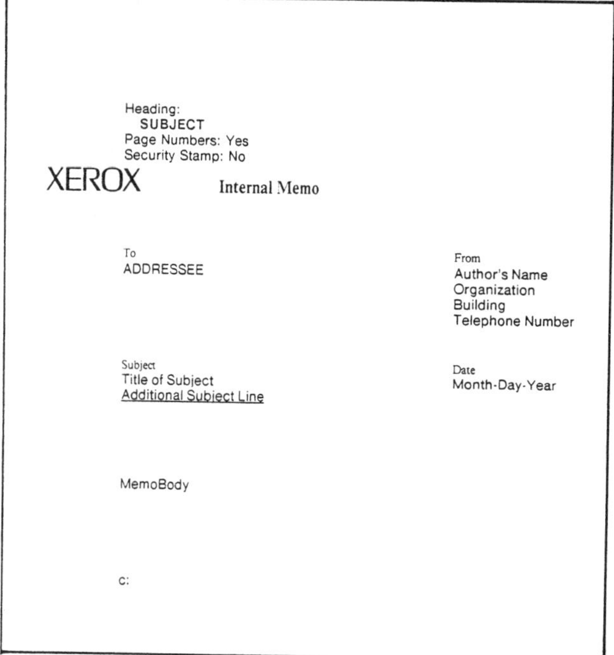

Heading:
SUBJECT
Page Numbers: Yes
Security Stamp: No

XEROX Internal Memo

To
ADDRESSEE

From
Author's Name
Organization
Building
Telephone Number

Subject
Title of Subject
Additional Subject Line

Date
Month-Day-Year

MemoBody

c:

Figure 1. Terminal Display of Internal Memo Form

Electronic form files can help to assure that necessary information is not forgotten. For example, the file can specifically ask for dates, accounting numbers, project names, resource figures, etc. These forms are pre-set, sometimes even locked up (programmed to prevent deletion of any part of the form), so the author must only select a place on the form and insert the information.

The form shows the engineer exactly what the text placement will be; and fonts, line leading, paragraph spacing, and tab settings are all pre-set. If the form is not locked, the author may change whatever he or she wishes to change but in the meantime is fully aware that the standard is not being followed. Although some engineers get a creative kick out of using Old English or Script fonts for headings and the resulting shock effect, most wish to convey the necessary information as quickly and effectively as possible.

Special Use Forms Save Time

Other examples of forms include progress reports, informal notes, labels, seminar notices, performance appraisals, newsletters, and technical report front pages. The form can be made up once, stored separately as a form file, and may then be used over and over again.

A group of forms which are especially effective in their saving of both time and effort are ones which allow the text processing for overhead transparencies, the overlay of this text into a 5" x 7" border, and the automatic printing of these pages. The finished slide, if kept to key words, will be easily read by most people in a conference room for 15-20 people (see *Figure 2*). Note the graphic symbol woven into the border of the transparency which identifies this slide as originating from the Electronics Division. The subject title and code (date and author's initials) at the bottom give additional information for whoever needs it.

Note that the large type size, the spacing within the border and between the lines, and the placement of the slide title and bottom line identification codes are all pre-set and automatic for the author. Again, this does not mean that the author cannot change the form, but an effective technical writer will also explain why this form gives optimum readability for technical presentations.

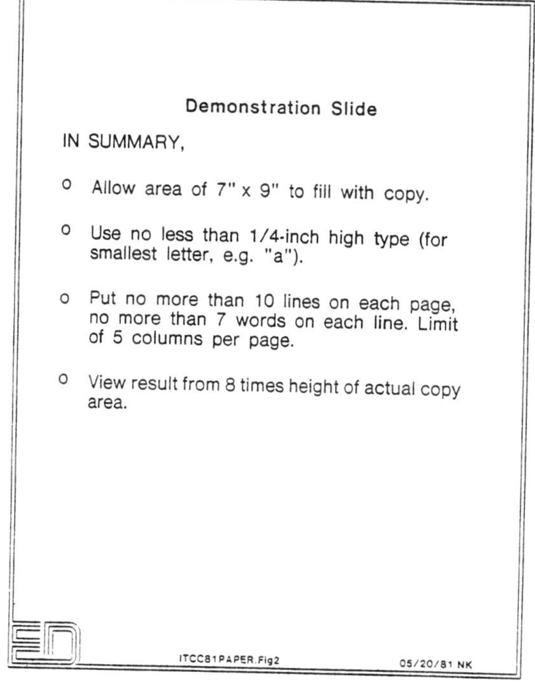

Figure 2. Overhead Transparency Copy from Printer.

ANOTHER FAMILIAR ROLE -- Teacher

Just as technical writers have done in the past, teaching the concepts of good style is still important. Knowing who you are writing to and how to organize the material to suit that audience best is as pertinent and critical as ever.

But what happens when engineers begin to write? Often the technical person will document the material for himself, not for the people who need it. The technical writer must always question who is getting the information, and then help tailor the document in the most effective way for that audience. Restructuring technical material with text processing is not as threatening or as drastic as it would be with conventional typewriter or older magnetic tape/card equipment.

Although it is not always possible to work with material in front of the engineer, nothing is more effective than showing document changes on the terminal screen and demonstrating the greater effectiveness of these changes. A persistent technical writer may actually persuade a die-hard engineer to stop printing the entire technical report in ***bold-face, italic*** type!

Style Manuals and Guidelines

Engineers and other technical people need to have examples to follow if they are to produce effective documentation. The group's technical writer can provide examples not only with electronic files but also with printed style manuals. In these manuals, the writer may emphasize writing style, vocabulary, grammar, guidelines in text formatting, use of graphics, and steps toward local hardcopy publishing.

Specific guidelines are needed for the naming of electronic files:

ProgressReportMay1981.NKDirectory.System7
or
ProgReport_NKMay81.7

Which is best? That depends on the system, the conventions already agreed upon, even on personal preference at times. Where the file is stored, either on a local or a remote system, will also affect the file's name. Control of files is especially important. Who will have read capability; who else will be able to both read and write on a file? How secure is the system, and how should individuals protect their files? Technical writers in these cases work with the system experts and the department managers to find and implement solutions.

The engineering groups have not usually had the full benefit of support departments. A graphics department can supply excellent artwork for the product or service before the paperwork is ready for the writers in customer services. But, even more than that, abstract ideas which need model representation for clear expression quite often get no graphic help at all. The originator of the idea, a person who protests to having no artistic ability, needs help in giving these ideas the best chance at being understood. The technical writer is often the best -- if not the only -- source of help, and she or he may tap support services in other parts of the organization.

Exploring New Areas

A writer in an engineering group is in a position to explore different types of documentation and to advise how these documents should be written. For example, how are requirements documents different from those describing design and specifications? Are there changes in format which would make them easier to understand? In the case of a software group, how can program code be translated into English? If a program design language is used, how may it be changed to better suit the writers in customer service areas?

To answer these questions, the technical writer is indeed a communicator, using contacts inside and outside the organization to find satisfying answers. The solutions she finds may affect the department greatly, and implementing these recommended solutions will require determination and tact.

More Hope Than Ever

Engineering departments are not noted for supplying large budgets to improve their technical documentation. In fact, few engineering departments even have a technical writer. This may be slowly changing. I have never felt that engineers fully deserved the reputation for being exceptionally poor writers. Like lawyers who are obsessed with being word precise, engineers and researchers also long to see their ideas represented accurately. The electronic publishing industry -- supplied with good video terminals, storage facilities, and printing devices -- may provide the way for technical writers and engineers to work together as they never have before.

Technical people can and do write better if they have the means to do it as an iterative process which they can see and control. And technical writers can help engineers do their jobs more efficiently and effectively by providing expertise with the electronic publishing system itself and with professional publications standards.

DOCUMENTING COMPUTER SYSTEMS USING TERMINALS FOR INPUT

Emanuel Katzin
Consultant

Emanuel Katzin
Documentation Consultant

5714 Birdwood
Houston, TX 77096
(713) 771-4287

Computer-oriented systems now reach into many facets of corporate operations. Where a few years ago the financial reporting functions were the only computerized operations, today production, scientific inventory control, risk analysis, mathematical modeling, material requirements planning, etc. are also computer based. Because of this an entire new group of people are involved with the computer, and must have documentation to know, understand and use the systems.

DEFINITION OF INDUSTRIAL AND COMMERCIAL DOCUMENTATION

Documentation is the written communication method for transferring and recording ideas, plans, details, methods, instructions, etc. It is widely used in business, industry and government for all types of activities — engineering, marketing, research, administration, production, etc.

Traditionally documentation is performed by classical technical writing. Today, however, with audiences becoming more diverse and subject matter growing more complex, additional media are used. Some of these are expanded use of graphics, enhanced classroom training, information recorded on sound tapes, and even sophisticated film and video tape/disc.

Each medium has unique characteristics and advantages. The choice of which to use is made by closely examining the subject matter, the audience, available resources such as people and equipment, and, of course, money.

The traditional written method, often called the User Manual, will be covered here. Emphasis is placed on computer-based systems, although many of the methods and ideas presented pertain equally to non-computerized systems.

Data processing is especially dependent on good documentation because of the complex and unforgiving nature of the computer. Data processing systems are composed of the required functions of the users (payroll, billing, inventory, etc.), the disciplines of system organization imposed on data, operations and people, and the need to perpetuate all of these combined efforts in English as well as in the computer language.

We will deal with the perpetuation in English.

IMPORTANCE OF DOCUMENTATION IN A SYSTEMS ENVIRONMENT

Recognizing the unique complexities of computer-based systems, we can easily understand the need for good and accurate documentation. Four major reasons stand out:

- Productivity
- Reference
- Continuity
- Consistency

Productivity

Business and government leaders decry the loss of productivity in the American industrial system. The majority of the American workforce today is white collar — clerical and middle management, and their job assignments daily become more complex and technical. Because of these complexities and the ever-increasing amount of knowledge, jobs have become more specialized. It is, therefore, virtually impossible:

- For one person to know an entire system or the complex of steps, or
- To assure that the entire system (all steps) have been performed — accurately and well, etc.

In addition the use of sophisticated electronic equipment — computer terminals facsimile machines, memory typewriters, word processors — has become widespread in the performance of many of these jobs.

In order to maintain, and certainly to increase, white collar productivity then, good and accurate documentation is a must.

Reference

Documentation serves as the *primary reference tool* used in a system-oriented organization. As such it records the concept of the system, the plan agreed to by the architects of the system, the details needed to operate the system, and the explanation needed by the user of the system.

In one respect it might be considered the *blueprint* of an administrative or business system, similar to the blueprint of a construction project. Typically, however, the documentation is used after the "construction" has taken place — primarily for reference purposes.

Continuity

Documentation, in a systems environment, can also be considered the prime source of life insurance for a system. Let's analyze a typical computer-based system in a business.

It usually takes several months — maybe a year or more — to complete the system analysis, design, programming, testing, implementation, parallel and production operations. This work is usually accomplished by a relatively *few* people in the organization. These few typically get promoted, move to another company or move on to other systems or duties in the same company. The technical maintenance of the system is usually relegated to relatively new programmer/analysts who themselves are looking for better things.

Virtually no one remains from the creators of the system. And yet the system probably has a life expectancy of 5-8 years. It could easily have cost $500,000 to $1 million.

Without documentation this system is basically dependent on *word of mouth* for perpetuation. The expense of documentation compared to the investment in the system is small, and should be considered, by any standard, as an integral part of the system.

Consistency

Because of good documentation, consistency in data, operations, instructions and results is maintained. Consistency will yield accuracy, which in turn means less time devoted to error correction. And anything which saves time certainly means less expenses and more profit. So, though it is difficult to calculate a dollar amount of savings, consistency has a real dollar value.

ADAC (add, change activity) DLAC (delete activity)
ADDP (add, change options) DLOP (delete options)
BGIN (go to beginning of program)

① ENTER COMMAND:	② DLAC

③
AFTER ENTRY OF ACTIVITY NUMBER, CHOOSE FIELD TO BE DELETED. CHOICES ARE:
ACT (entire activity) TARS (target start)
TARF (target finish) PRED (individual predecessors)

④ ENTER ACTIVITY NUMBER:	⑤ 100
⑥ ENTER FIELD:	⑦ ACT

B
1 2 3
7 8 9
You are Here

3-34

Applying good documentation techniques to meet these needs is an ever-widening effort. Providing documentation for the User groups is receiving more and more attention because of the application of computers to more areas of business (not just accounting functions) and because of distributed processing.

DOCUMENTING INTERACTIVE PROCESSING

One of the most prevalent and fastest growing means of entering User data is through CRT terminals connected to the computer. Unique problems arise in the presentation of this type of documentation, particularly when the System interacts with the User in a "conversation" mode. Some of the problems are:

- Having several options presented at one time, the User must have all the pertinent documentation for each of the options available.
- The documentation must help the User to follow the sequence of the presentation on the CRT screen.
- Using a computer terminal instead of pencil and paper can be more complex for the User.
- The User needs instructions for "finding his way" through the network of options and screens offered by the computerized system.

No single answer can be given to solve all these problems for all situations, of course. However, suggestions that have general applicability can provide ideas for organization and presentation.

To begin with, the basic elements of user documentation that apply to paper forms and batch-type computer systems apply equally here. For example, knowledge of the User, readability, simplicity of presentation, etc. retain their importance regardless of the type of processing. The major difference is the *style* of presentation.

One style that has proven quite successful, especially for "conversation" mode processing is shown below:

3.2.2 DELETING AN ACTIVITY (DLAC)

The [DLAC] command is used as follows:

Input Instructions

Field No.	Entered By	Instructions
①	Automat.	This is a request by the system to enter the input code selected from the menu.
②	User	Enter DLAC to delete either an entire activity or specific fields of data about an activity.
③	Automat.	This message and menu are then displayed on the terminal.
④	Automat.	This is a request by the system to enter the activity number you wish to be deleted from the master file.
⑤	User	Enter the activity number.
⑥	Automat.	This is a request by the system to enter the input code, chosen from the menu.
⑦	User	If you wish to delete the entire activity, enter *ACT* after entering the activity number. However, if you wish to delete only target start date or target finish date or predecessor activity information, enter the code shown in Field ③ after entering the activity number. Entering (CR) returns you to the menu at [2]. You cannot delete individual lag values. You cannot delete the entire predecessor and re-enter correct information.

3-35

The page on the left shows the screen exactly as it appears on the terminal. However, two enhancements are made to help the User:

1. Lines separating each of the fields are drawn to *delineate* the fields
2. A circled number is shown in each field to identify the field and denote the *sequence* of appearance.

In this way the User can see the screen exactly as it is displayed and at the same time have a definitive reference to each field.

On the facing page to the right the detailed instructions for each field are presented in the sequence that they appear. The circled number in the left column identifies the field from the screen. The center column tells who enters the data in the field — the User or the System. Detailed instructions, definitions or explanations are given in the large INSTRUCTIONS column.

Using this style of documentation helps the User follow the action of the system step-by-step — the "conversation" mode.

The little "YOU ARE HERE" diagram under the screen format is a help to the User in remembering where he is in the heirarchy of the system. Based on a previous diagram explaining the entire system the User can use this mini-diagram to get to the next desired screen or option.

Providing complete and easy-to-understand instructions for using the terminal is an obvious necessity, especially to new Users. This is generally done in the introductory part of the manual. An example is shown here:

STEP	ENTERED BY	SIGN-ON PROCEDURE
1	User	Turn on the terminal by pulling the POWER knob. Adjust the intensity (brightness) of the characters on the screen by turning the POWER knob. Key DATA Press (ENTER)
2	System	The following appears on the screen: ENTER OPERATOR I.D.
3	User	Key TK Operator Code Press (ENTER) Operator Code is your initials, maximum of 3 characters. For example, if your name were Dona A. Young, you would enter TKDAY. These characters will not appear on the screen.
4	System	The following appears on the screen: * * * ENTER COMMAND * * * ENTER SIGNOFF VERIFY

BATCH PROCESSING ON CRT TERMINALS

Systems that use CRT terminals to replace keypunches, essentially keeping "batch-type" processing, need a different style of User documentation. In this case the User is still entering data from some *source document* into a screen that probably resembles the previously used punched card or other data entry medium. It is therefore important that the User be able to easily see which data on the source document corresponds to and goes into which field on the CRT screen. In addition a complete set of instructions must be supplied for correctly entering the information. An example of documentation satisfying these needs is shown here:

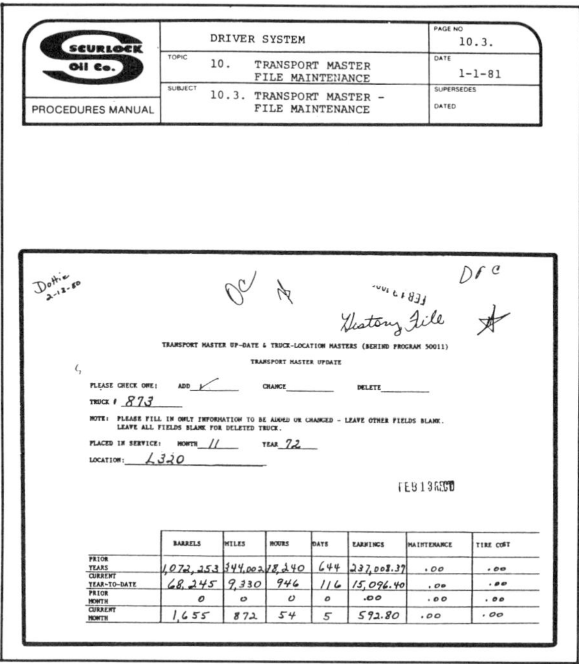

SCURLOCK Oil Co. — PROCEDURES MANUAL

DRIVER SYSTEM	PAGE NO. 10.3.
TOPIC 10. TRANSPORT MASTER FILE MAINTENANCE	DATE 1-1-81
SUBJECT 10.3. TRANSPORT MASTER - FILE MAINTENANCE	SUPERSEDES DATED

TRANSPORT MASTER UP-DATE & TRUCK-LOCATION MASTERS (BEHIND PROGRAM 50011)

TRANSPORT MASTER UPDATE

PLEASE CHECK ONE: ADD ✓ CHANGE ____ DELETE ____

TRUCK # 873

NOTE: PLEASE FILL IN ONLY INFORMATION TO BE ADDED OR CHANGED - LEAVE OTHER FIELDS BLANK. LEAVE ALL FIELDS BLANK FOR DELETED TRUCK.

PLACED IN SERVICE: MONTH 11 YEAR 72

LOCATION: L320

	BARRELS	MILES	HOURS	DAYS	EARNINGS	MAINTENANCE	TIRE COST
PRIOR YEARS	1,072,253	344,002	18,240	644	237,008.37	.00	.00
CURRENT YEAR-TO-DATE	68,245	9,330	946	116	15,096.40	.00	.00
PRIOR MONTH	0	0	0	0	.00	.00	.00
CURRENT MONTH	1,655	872	54	5	592.80	.00	.00

SCURLOCK Oil Co. — PROCEDURES MANUAL

DRIVER SYSTEM	PAGE NO. 10.3-1
TOPIC 10. TRANSPORT MASTER FILE MAINTENANCE	DATE 1-1-81
SUBJECT 10.3. TRANSPORT MASTER FILE MAINTENANCE	SUPERSEDES DATED

TRANSPORT MASTER-- FILE MAINTENANCE

TRUCK	PLACED IN SERVICE YR MO	LOCATION	A OR C
XXX	XX XX	XXXX	X

FIELD NAME	REQ	MAX CHAR	ALPH NUM	LDG O'S	COMMENTS
					THIS ENTRY IS USED TO ADD OR CHANGE THE PLACED-IN-SERVICE AND/OR THE LOCATION ON THE TRANSPORT MASTER UPDATE FORM.
TRUCK	YES	3	NUM	YES	THIS IS THE TRUCK NUMBER FOUND ON THE TRANSPORT MASTER UPDATE FORM. IF LESS THAN 3 DIGITS ARE SHOWN ENTER ALL THE DIGITS SHOWN AND PRESS SKIP. THIS WILL AUTOMATICALLY RIGHT JUSTIFY THE ENTRY, INSERT LEADING ZEROS AND MOVE THE CURSOR TO THE NEXT ENTRY.
PLACED IN SERVICE	NO	4	NUM	YES	THIS IS THE DATE THAT THE TRUCK WAS PLACED IN SERVICE. IT IS FOUND ON THE TRANSPORT MASTER UPDATE FORM. ENTER THE DATE IN THE FORMAT MMYY WHERE MM IS MONTH AND YY IS YEAR. THERE MUST BE 4 DIGITS. FOR EXAMPLE, IF THE DATE IS MARCH 1980, ENTER 0380.

An example of the source document is shown in the top page, resembling its position during the actual data entry operation. In the lower (facing) page is a resemblance of the input screen used, and below it are the detailed instructions. This particular style uses a *tabular* format although other formats might serve as well, depending on the content and audience.

Sometimes, to further help the User, a circled number is placed by each field on the source document and its corresponding field in the screen facsimile. Then an additional column called FIELD NUMBER is added to the Instructions format, and the corresponding numbers are printed beside each field name. This proves especially helpful when the screen facsimile has little resemblance to the source document or where confusing field names might be a problem for the User.

Although certainly not the final word, these styles and techniques of User documentation are proving helpful to many Users of computerized systems equipped with terminals. The combination of technology and ingenuity will inevitably spawn more and different methods.

MULTIPLE LOCATION GLOSSARY COORDINATION

Les J. Kizer
International Business Machines

Les J. Kizer

IBM
P. O. Box 390
Poughkeepsie, NY 12602
(914) 485-7548

Using VM/370-CMS and SCRIPT, a process was proposed and implemented for coordinating glossary entries in all manuals in which any one definition appears. This includes the master glossary and the definitions that are available to users upon request through interactive terminals. This facility was accomplished even though we have multiple programming locations, in several geographic locations.

PROBLEM STATEMENT

How does one keep terminology consistent for a large application programming project, especially when the application is periodically enhanced.

SITUATION

The application program is developed by numerous programmers, located in several geographic locations, with a very low ratio of writers and editors to programmers.

RECOMMENDATIONS

In the organization--a large design automation system--there are a number of task force type groups that work with our customers to determine how to improve the product. One of these groups is called the Information Working Group. As a result of the work done by this group, several thoughts about terminology definitions and glossaries have been recommended:

- All instruction manuals should have glossaries to define technical terminology introduced in that manual or that the user may be unfamiliar with.

- Many reference manuals should have similar glossaries.

- All technical terminology related to the product should be defined in a single master glossary.

- Technical terminology defined in the glossary of one manual should be identical to or a subset of the definition of the same terminology used in another manual.

CONCEPT

Since VM/370-CMS and SCRIPT are used, thinking followed lines that would permit implementation of the recommendations using the facilities of these programs. With the VM/370-CMS file naming conventions, the concept came very quickly to have each definition in a single VM/370-CMS file.

By using one file per definition:

- Any writer may insert any existing definition at any point in the text of a manual.

- The same definition will be inserted into the master glossary.

- The same definition could be selectively inserted into a glossary in any manual.

- The same definition could be generated into a partitioned data set to create an online glossary for users of interactive terminals.

See Figure 1.

VM/370-CMS limits the file name for any file to eight characters. With this limitation, one can be concerned with the

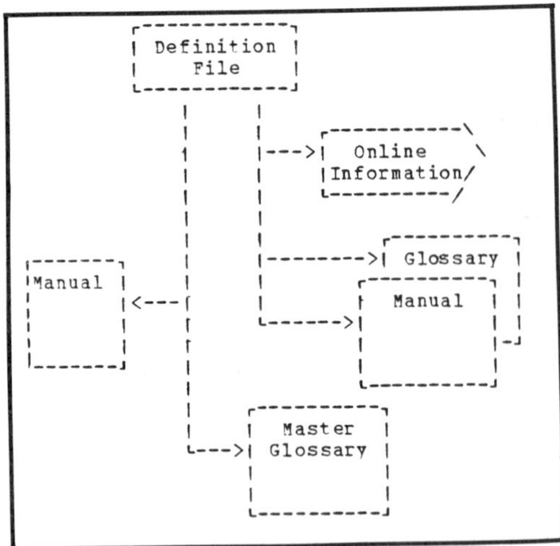

Figure 1. Multiple Use

the file naming convention. In addition to simply having a discrete name for each file, analysis discovered other considerations for the file naming convention to be used. These are:

- By sorting the file names, all of the technical terminology would be in alphameric sequence. In other words, no editorial sequencing of file names.
- When a programming location updated or developed a new definition that location was to become the designated technical owner for that definition.
- Locations, at least initially, needed to be able to assign their own file names for new definitions, with minimum contention. Contention would be resolved later by the central organization.

It was also felt that the file name must contain some visible characters to let those who work with these files know approximately where the file is in the alphameric sequence among all other definitions. The file naming convention that resulted from numerous discussions is shown in Figure 2.

<u>Characters 1 And 2</u>

All files with a file mode of SCRIPT containing "MG" as the first two characters are files that contain definitions that will be placed in the master glossary. These two characters have no other value than to identify these files as containing definitions for terminology.

<u>Characters 3 And 4</u>

These are the first two characters of the terminology being defined; for example,

```
Given the following glossary entry:

     Field Catalog:  a rule named A00001
     that contains . . . .

Then the file name might be:

     M G  F I  3 5 7  B
     |    |      |    |
     |    |      |    └ Location identifier.
     |    |      |
     |    |      | Assigned value to maintain
     |    |      └ sort sequence.
     |    |
     |    | First two characters of the
     |    └ glossary term being defined.
     |
     | Characters "MG" are reserved to
     └ identify glossary entries.
```

Figure 2. File Naming Convention

if the term being defined is "Data Set," these two characters would be "DA." When working from a terminal, these two characters help reduce the possible number of files that might contain a definition being looked for.

<u>Characters 5 Through 7</u>

Within all of the definition files that begin with the same four characters, such as "MGDA," these positions are used to place these files into a correct alphameric sort sequence. During the initial file name assignments, these numbers were spread from 000 to 999 to allow for future file names having the same first four characters.

<u>Character 8</u>

To indicate which programming location is the current technical owner of a definition, and not to disturb the sort sequence, the last character of the file name indicates the programming location that is the technical owner of the definition; such as "P" for Poughkeepsie.

It is possible to have two file names that are identical except for the last character; however, when this rare situation occurs, the central organization corrects it and informs the programming location.

<u>FILE FORMAT</u>

All definition files have the same basic format. Each definition is placed into its own file. Definitions are entered using the format shown in Figure 3.

```
.kp on
.sk
.of 5
Hits Ratio:
the ratio of the number of successful
references to main storage
to the total number of references.
.kp off
```

Figure 3. Format Of A Definition File

MASTER FILE

There is a master file that contains most of the formatting codes, front matter, and back matter for the master glossary. The master file inserts a single file that contains a list of SCRIPT imbed commands (.im) for all of the definition files.

This single file is produced by executing the VM/370-CMS command LISTFILE MG* SCRIPT * (EXEC. This command produces a listing of all of the definition files. This file is then sorted into alphameric sequence. The editorial assistant then changes the tokens (character string " &1 &2"; note the leading blank) to the SCRIPT imbed command (.im) and issues the FILE command. A script file is then created and the CMS EXEC brought into the newly created file. See Figure 4.

The SCRIPT/VS command is then used to produce the master glossary.

PROCESS

Whenever work is done by more than one location, and especially when located in cities remote from one another, one does need process rules and procedures.

The simplest rule is that all files will have the same format commands. This rule, of course, had to be negotiated and agreed upon by all programming locations and the central organization that coordinates the production of the master glossary.

The procedure is a little more complex. See Figure 5. In any given programming release, it is desirable to have a technical definition be identical throughout the library of publications supporting the application.

This means that new and revised definitions that are developed by each location must be transmitted to the central organization where they are collated, checked and then the entire replacement definition data base is transmitted to all locations so that a definition modified by one location is available to all others.

Master File:

CMS EXEC From Sorted LISTFILE Command

```
&1 &2 MGAB500E SCRIPT    G1
&1 &2 MGAC250E SCRIPT    G1
&1 &2 MGAC500C SCRIPT    G1
&1 &2 MGAC750E SCRIPT    G1
&1 &2 MGAE500E SCRIPT    G1
&1 &2 MGAL200C SCRIPT    G1
&1 &2 MGAL400C SCRIPT    G1
&1 &2 MGAL600C SCRIPT    G1
&1 &2 MGAL700P SCRIPT    G1
&1 &2 MGAL800C SCRIPT    G1
```

Imbed File:

Resulting SCRIPT File

```
.IM MGAB500E SCRIPT    G1
.IM MGAC250E SCRIPT    G1
.IM MGAC500C SCRIPT    G1
.IM MGAC750E SCRIPT    G1
.IM MGAE500E SCRIPT    G1
.IM MGAL200C SCRIPT    G1
.IM MGAL400C SCRIPT    G1
.IM MGAL600C SCRIPT    G1
.IM MGAL700P SCRIPT    G1
.IM MGAL800C SCRIPT    G1
```

Figure 4. Developing The Imbed File

The files that are received from the central organization are the ones that are inserted into the local manuals.

THE CHALLENGES

Multiple Definitions

The situation existed where a single term was defined in several places, and each definition varied.

The initial solution was to declare the definitions in the master glossary as the correct ones. From that point, each programming location could modify the definitions in the master glossary.

Multiple Manual Use With Variable Definition

Some terms have multiple definitions, some of which would not be appropriate for each of the manuals in which they might appear.

The solution was to code SCRIPT if commands (.if) that would bypass text not wanted for the manual into which the definition was being imbedded.

Pointers In Definitions To Associated Manuals

Some of the definitions in the master glossary carried the statement: "For further information, see Manual . . . ."

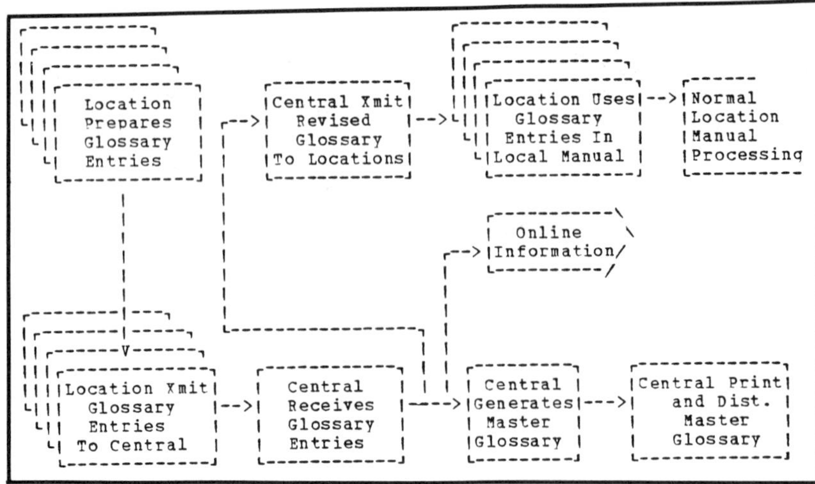

Figure 5. Glossary Preparation: Overview

Some of the definitions in manuals carried the statement "For further information, see page . . . ." These cross-references are excellent for the users. For imbedding such a definition file into several manuals, the file was programmed to contain SCRIPT if commands (.if) to bypass unwanted references.

END PRODUCT

Figure 6 shows an example of an end product.

ABOUT THE AUTHOR

Les J. Kizer is a Staff writer with IBM in Poughkeepsie, a senior member of STC, and a past STC chapter chairman, joining STC in 1961. During his career in teaching and publications, he has written training manuals, been an engineering writer, written and edited customer proposals, supervised programming project development documentation, written and planned IBM operating system documentation, and currently plans for, edits, and writes for documentation and education for the IBM Engineering Design System (EDS) in Poughkeepsie, New York. He has served on the Corporate Terminology Committee and on the Corporate Indexing Guidelines Committee.

HOLDING Screen Status: for a display terminal used as a virtual console under VM/370, an indicator located in the lower right of the screen that displays that the current contents of the screen remains on the screen until the user requests that the screen be erased. This status occurs either by pressing the Enter key, or is triggered by a message or warning displayed on the screen.

Hold Page Queue: in OS/VS and DOS/VS, a queue to which pages in real storage are initially assigned through operations such as page-in or page reclamation. See also active page queue, available page queue.

Hold Queue: (1) In OS/VS, a waiting list for jobs which initiation is to be delayed until the operator releases them from the queue. (2) In TCAM, a FEFO-ordered queue that is a part of the priority-level QCB for each destination QCB. If a terminal is intercepted (held), its messages are placed in the hold queue while messages for other terminals on this destination QCB are sent. See also intercepted terminal, intercepted station.

Home Address: an address written on a direct access volume, denoting a track's address relative to the beginning of the volume.

Home Loop: an operation involving only those input and output units associated with the local terminal.

Figure 6. End Product

GENERATING A MASTER INDEX

Les J. Kizer
International Business Machines

Les J. Kizer

IBM
P. O. Box 390
Poughkeepsie, NY 12602
(914) 485-7548

Using VM/370-CMS, SCRIPT, and an in-house index format program, a process was developed for producing a master index by merging indexes from all manuals containing indexes. This facility was accomplished even though there are multiple programming locations, in several geographic locations.

RECOMMENDATION

In the organization--an engineering design automation system--there are a number of task force groups that work with customers to determine how the product might be improved, including documentation. One of these groups is the Information Working Group. This group made several recommendations for indexing:

- All introductory, how-to, and reference manuals should have indexes, even reference manuals organized alphabetically by command names.
- The indexes from all manuals supporting the application program should be merged into a single, sorted master index.
- Indexing standard and guidelines should be followed.

PROBLEM STATEMENT

Obviously, indexes in individual manuals are an excellent way for the occasional user of a manual to locate specific information contained in that manual. But when the documentation supporting a project grows into a library of publications containing numerous manuals, how does the user of that library quickly locate information in a part of the library with which they are not familiar. For example, which manual does one read just to begin.

SITUATION

Many publications contained indexes, but these were maintained and formatted using a VM/370-CMS editor. In the process of automating, we obtained the index format program used for the VM/370 and CMS publications. This index format program considerably reduced the time needed to format indexes, thus all of the indexes were converted to the format needed by the format program. This predated SCRIPT/VS.

CONCEPT

The concept shown in Figure 1 is quite similar to the one used for VM/370 and CMS manuals. Very simply, the entries in the master index refer the user to the manual for that subject. The user then refers to the index in the manual to obtain a reference to the correct page number. The most obvious question asked about this approach was: "Why not have the page numbers in the master index and, therefore, save the customer one step in the process?"

Of course, we could; however, good logic prevailed. Any information retrieval mechanism that contains errors will tend not to be trusted and, consequently, will fall into disuse. In this case, we planned to publish the master index only when we had a full release of the application program. However:

- Functions are developed at other than release time that cause changes to various manuals in the library, which alters the pagination of existing manuals.

- More thorough information is developed on existing functions after the release has been shipped, which also alters the pagination of an existing manual.

- Manuals are rewritten and improved between releases of the application program, which also alters the pagination of an existing manual.

Therefore, any update to the library between releases tends to alter the pagination of existing manuals; thus page references in the master index could be incorrect, and wrong entries could cause our customers to distrust the master index.

Even though an existing manual is totally revised, the index entries in the revised manual have a high probability of being so similar to those in the previous edition as to be considered valid. In other words, by referring the customer to the indexes in the individual manuals, the exposure to errors is minimal and the integrity of the master index is upheld.

Data Reduction. Index entries sometimes have up to five levels, although the index format program formats only three levels. To merge this amount of detail into a master index tends to create long lists of items all referencing a single manual. This was considered superfluous. Consequently, we use only the first two levels of each index entry, something that was quite easy to implement using VM/370-CMS EXECs. This simple reduction reduced the number of pages in the master index by 26 percent, with no reduction in the validity and usefulness of the master index for information retrieval. The reduction in size is actually a plus for our customers.

BACKGROUND INFORMATION

Indexes were generated before the advent of SCRIPT/VS, which has an index facility. Our method of producing an index for a manual has been to develop a file containing one index entry per line, or record. The file is then processed using an index format program that would format multiple level index entries to agree with the standard on indexing, the IBM guidelines, and the refinements agreed to for the documentation of the engineering design system.

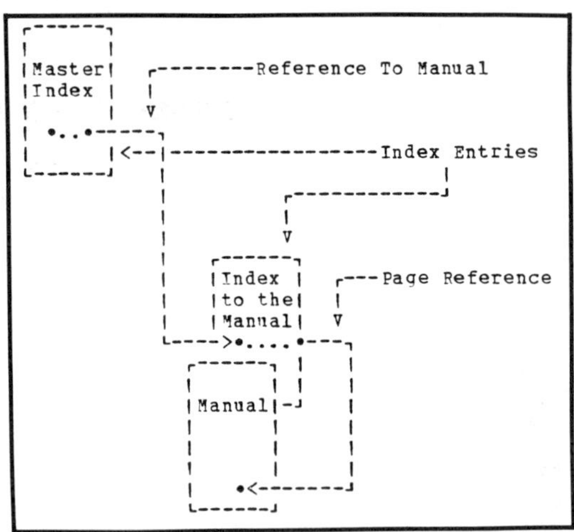

Figure 1. User Methodology

In developing the file for indexes, each level of the index entry was separated by a comma followed by a space. The page number was placed exactly two spaces beyond the index entry itself. As shown in Figure 2, set symbols defined in the document provide the actual page numbers.

COMPUTER PROCESS

Given the file shown as the input file in Figure 2, CMS EXECs were developed to delete everything on each line to the right of two blank spaces and from the second comma on to the right. The manual's identifier was then edited to the end of the remaining contents, as shown in Figure 3. Thus the page number references are replaced by the manual's identifier.

After the manual identifiers have replaced the page number references, all of the input files are merged into a single file using the CMS COPY command with the APPEND option. The merged file is then processed by the same index format program that is used to format the indexes in our books.

The formatted SCRIPT file is then imbedded into another file that contains the front and back matter for the master index.

Input File Example:

```
Abel, Cain  &I0401
Adam, descendants  &I0501
Adam, Eden, garden, sent out  &I0322
Adam, Eden, sent out  &I0322
Cain, Abel  &I0401
Cain, descendants  &I0417
creation, story  &I0101
descendants, Adam  &I0501
descendants, Cain  &I0417
disobedience, man  &I0201
```

Resulting SCRIPT File:

```
.TB 3 6;.NJ
.OF 0;.OF 1;Abel, Cain  &I0401
.OF 0;.OF 1;Adam
.OF 0;.OF 4; descendants  &I0501
.OF 0;.OF 4; Eden
.OF 0;.OF 7;  garden, sent out  &I0322
.OF 0;.OF 7;  sent out  &I0322
.SK 2
.OF 0;.OF 1;Cain
.OF 0;.OF 4; Abel  &I0401
.OF 0;.OF 4; descendants  &I0417
.OF 0;.OF 1;creation, story  &I0101
.SK 2
.OF 0;.OF 1;descendents
.OF 0;.OF 4; Adam  &I0501
.OF 0;.OF 1;disobedience, man  &I0201
```

Resulting Documentation:

```
Abel, Cain  5
Adam
   descendants  6
   Eden
      garden, sent out  5
      sent out  5

Cain
   Abel  5
   descendants  6
creation, story  1

descendants
   Adam  6
   Cain  6
disobedience, man  4
```

Figure 2. File For Developing Indexes

Edited Master Index File:

```
Abel, Cain  Genesis
Adam, descendants  Genesis
Adam, Eden  Genesis
Adam, Eden  Genesis
Cain, descendants  Genesis
Cain, Abel  Genesis
creation, story  Genesis
descendents, Adam  Genesis
descendents, Cain  Genesis
disobedience, man  Genesis
```

Resulting SCRIPT File:

```
.TB 3 6;.NJ
.OF 0;.OF 1;Abel, Cain  Genesis
.OF 0;.OF 1;Adam
.OF 0;.OF 4;descendants  Genesis
.OF 0;.OF 4;Eden  Genesis
.SK 2
.OF 0;.OF 1;Cain
.OF 0;.OF 4;Abel  Genesis
.OF 0;.OF 4;descendants  Genesis
.OF 0;.OF 1;creation, story  Genesis
.SK 2
.OF 0;.OF 1;descendents
.OF 0;.OF 4;Adam  Genesis
.OF 0;.OF 4;Cain  Genesis
.OF 0;.OF 1;disobedience, man  Genesis
```

Resulting Documents:

```
Abel, Cain  Genesis
Adam
   descendants  Genesis
   Eden  Genesis

Cain
   Abel  Genesis
   descendants  Genesis
creation, story  Genesis

descendents
   Adam  Genesis
   Cain  Genesis
disobedience, man  Genesis
```

Figure 3. Edited File For Master Index

SCRIPT/VS PROCESS

We have not used the following process; however, it is being used by another programming location. SCRIPT/VS has an index facility where index entries may be developed using the SCRIPT/VS put index command (.pi).

Having entered the index entry in the text stream of the SCRIPT file, duplicate the index entry text line, changing the duplicated put index command (.pi) to the SCRIPT/VS write-to-file command (.wf). The file produced by the write-to-file command would then be processed in the manner as the input files are currently processed.

PROCESS

Because the organization is spread over a number of geographic locations, it was necessary for the central organization to establish a date with the programming locations on which index files would be transmit to the central organization. The files that the remote programming locations transmit to the central organization already have the correct manual identifiers and have been reduced to having only two levels per index entry.

The central organization merges the location index files into a single file and then uses the index format program to produce the SCRIPT file that will be inserted into the master index.

Since the entire process is programmed, it takes the central organization only a few days to produce and check the master index. There always seems to be some quirk in the input or process that is discovered in each release. However, one week after the closing date for all other manuals supporting a release of the application program, the locations transmit their indexes to the central organization. The central organization needs about one week to prepare the reproduction copy of the master index for printing.

OUR CHALLENGE

Our challenge now is not the producing of the master index. What we have accomplished is working successfully and to the satisfaction of our customers.

What needs to be done now, and it is an ongoing process, is to improve the indexes in the individual manuals so that references that are quite acceptable in the context of a manual will give the user of the master index a better understanding of the index entry.

ABOUT THE AUTHOR

Les J. Kizer is a Staff writer with IBM in Poughkeepsie, a senior member of STC, and a past STC chapter chairman, joining STC in 1961. During his career in teaching and publications, he has written training manuals, been an engineering writer, written and edited customer proposals, supervised programming project development documentation, written and planned IBM operating system documentation, and currently plans for, edits, and writes for documentation and education for the IBM Engineering Design System (EDS) in Poughkeepsie, New York. He has served on the Corporate Terminology Committee and on the Corporate Indexing Guidelines Committee.

THE BARE ESSENTIALS—
SETTING UP A SMALL PUBLICATIONS DEPARTMENT

Miriam Klammer
Racal-Vadic, Inc.

Miriam Klammer
Racal-Vadic, Inc.
Technical Publications Manager
222 Caspian Dr.
Sunnyvale, CA 94086
(408) 744-0810

The basic organization and the beginning elements of a technical publications department are vital to its functioning and future success as a support group to a company or to independent users. Those who are relatively new to the publications field may be faced with a myriad of questions and considerations with no clear direction in which to proceed. Total organization is a key to early success.

Racal-Vadic is a fast growing manufacturer of modems and peripheral telecommunications equipment in California's Silicon Valley. The publications department is presently comprised of seven people who successfully cover the basic functions discussed in this paper by following a work plan and "wearing many hats."

INTRODUCTION

Publications departments are organized and administered in various ways depending on function, size, type of work, and the political structure of the company. There are, however, some basic elements which are common to most publications departments. These include ORGANIZATION, a basic breakdown of the functions of the department; BASIC ELEMENTS, which include people, equipment, and resources; and ESSENTIAL TASKS, which include management functions such as establishing procedures, responsibilities, and schedules.

ORGANIZATION

The basic functions in a publications department are:

- Writing and Editing —includes gathering source material; setting formats and standards; generating rough text; and editing for consistency, grammar, and conformance to standards.
- Production —includes typing (on a typewriter or word processor); generating illustrations and charts; taking photographs and creating halftones if required; typesetting; designing layouts and pasting up; and preparing camera-ready art for printing.
- Printing and Distribution —includes reproducing and binding the copy in any number of possible ways; and distributing the finished product to the end user(s).

In the very smallest of departments, all of these functions can be carried out by outside vendors with only a coordinator or manager in-house to tie it all together. However, this is usually not cost effective and control over the final product is minimal.

In a majority of medium to large size companies, all of the functions are carried out in-house except perhaps printing,which is often done by outside vendors due to equipment and expertise required. A typical basic functional organizational structure is shown in Figure 1. Some functions may be carried out by the same person(s) (e.g., writing and editing, writing and word processing, or graphic arts and typesetting).

BASIC ELEMENTS

Because a small or new publications department is just that, hiring the "right" people, purchasing the "right" equipment, and having the "right" resources available is very important.

People

Obviously, the initial staff of a department does not allow for trainees; there is no one to perform the training. The philosophy "We are small, just getting started, we will learn" won't get the department

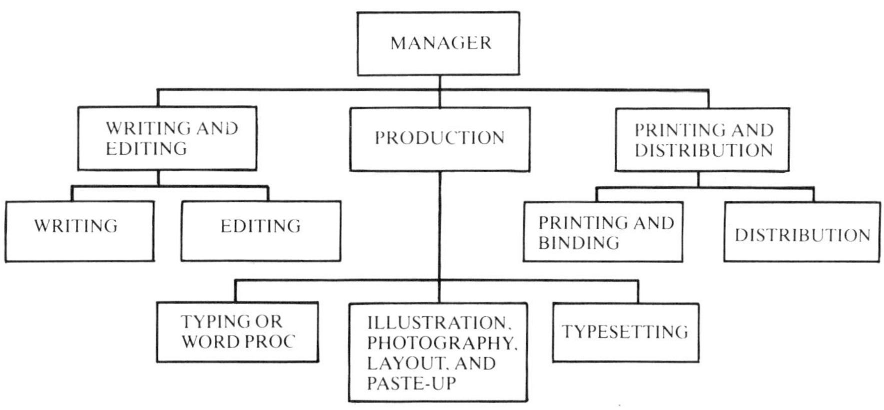

Figure 1. Typical Organizational Structure

off the ground. Many a new manager makes this mistake and ends up vending the work out in spite of an in-house staff.

A writer should have demonstrated organizational and writing skills but should, above all else, have the ability to analyze the subject being written about (e.g., if an instruction manual is to be written on a piece of equipment, a writer should have the technical knowledge and ability to analyze the equipment and determine what information and the level of that information needed by the user). Likewise, a word processor operator should not be a novice to the trade since he/she will need to handle all areas of that function. The graphic artist who is initially hired has the greatest variety of responsibilities to ensure efficient and quality production of camera ready art. Qualifications for this person should include demonstrated proficiency in layout and design, ability to spec type (if typesetting is to be used), knowledge of printing processes and photography, and good organizational skills. No novice could successfully carry out all of these tasks. An editor should have an interest and "love" for grammar and, like a writer, should have a demonstrated ability to organize thoughts on paper and write clearly and concisely. But he/she should also have a high level of coordinating ability. An editor in a small department will undoubtedly spend more time keeping track and chasing down all the pieces that will eventually emerge as a finished document than on actual marking up and proofing.

Equipment

The functions that are handled in-house will determine what equipment is needed. Some is more basic than others.

With the pace of business what it is today, the need for word processing is almost a must. There are numerous agencies popping up all over which will provide this service, the major drawback being lack of control, turnaround time, and getting the material back and forth. It has been my experience that even the smallest publications department will benefit greatly from a word processing system of some kind, even if it is a single terminal and printer. A word processor has become a fundamental workhorse. How publications departments functioned without has now become a mystery.

Racal-Vadic provides word processing terminals and printers to the technical writers and editor. The technical writers have developed boilerplate on diskettes from which they can create new documents of particular types and "massage" their own text, saving the muscle strain of handwriting. This has proved a tremendous success, saved time, and provided a typewritten copy for review with fewer steps. By wearing several hats, the writers and editor have made it possible to get by with just one additional word processor operator to incorporate changes, prepare final copy, and handle miscellaneous word processing projects.

The equipment needed for a production function is more varied but can still be stripped down to the bare essentials. Besides such small but important items as rapidograph pens, parallel rules, electric erasers, and layout paper, some basic larger items which make quality production possible are:

- drafting table
- light table
- flat files
- waxer
- paper cutter
- stat camera (this is to the production effort what a word processor is to the writing/editing effort—a workhorse)
- Space (production takes a lot of room)
- Plumbing (vital for camera work; extremely convenient for production efforts in general)

Resources

The third basic element in a publications department, equally as important as people and equipment, is resources. Resources can include a variety of outside vendors needed to provide service not available in-house or to serve as a back-up. These include word processing, typesetting, graphic arts, printing, distribution, and temporary help agencies. Even if many of these functions are handled in-house, many a crisis can be averted by maintaining a good rapport with a back-up source.

Resources also include contact people and source documents without which it is difficult to generate a usable technical document. Contact people are usually those within the company who have special technical knowledge. Source documents from both inside the company and out provide much needed and often necessary background that can save time, money, and frustration. They also provide stimulation which influences enthusiasm and inevitably results in better finished goods.

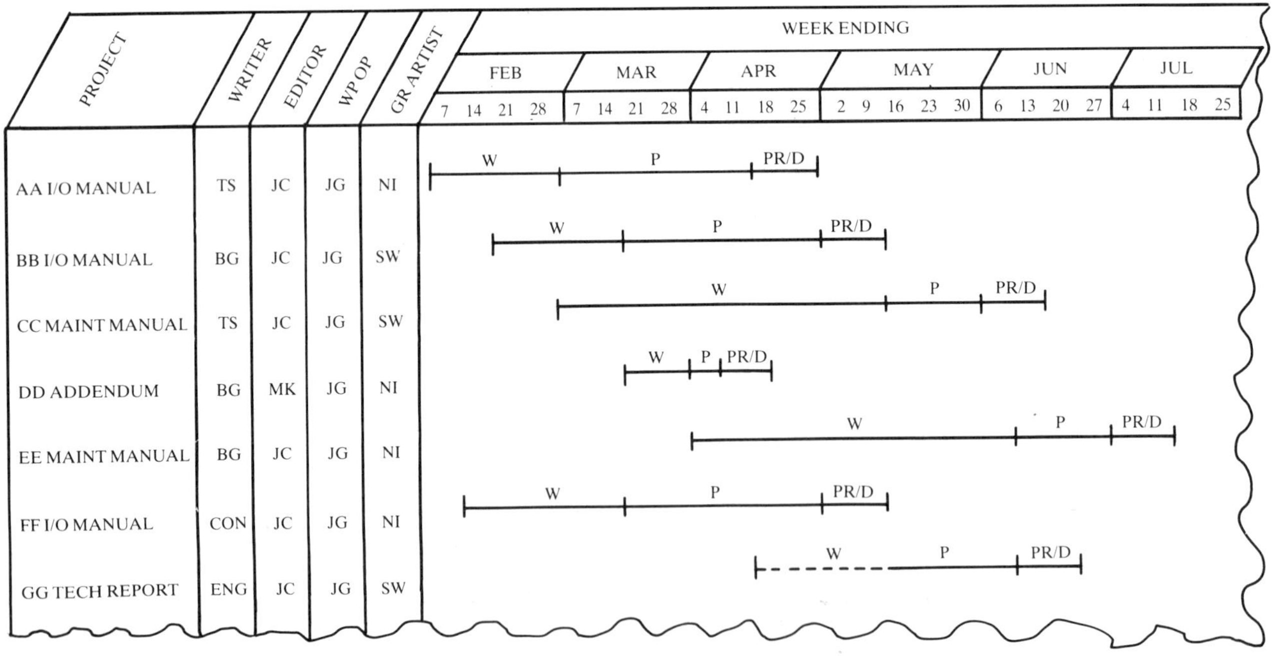

Figure 2. Publications Schedule

Another fairly important resource for everyone in a publications department is a well maintained control file. This may include a writer's draft, reviewers' copies, typeset dummy, and a final printed control copy where need for future changes can be noted.

Ensuring that these resources are available is a responsibility carried out by an efficient manager. It involves a bit of public relations, a lot of phone calls, and keeping a constant eye out for publications and seminars that would be beneficial to the whole effort.

ESSENTIAL TASKS

Some essential tasks which fall primarily to the manager of a publications department include scheduling; setting up basic procedures and a plan for work flow; defining responsibilities; and eliciting support and cooperation from support sources, management, and users.

Scheduling

The starting point in producing technical documents is knowing what needs to be produced, who and what is needed for the effort, how long it will take, and when it is due. A request and a deadline are usually handed down by someone outside the department. But the rest is up to the manager. Scheduling should involve the people who will be doing the work; they are the best estimators.

An overall schedule can be worked out on a grid with as much detail as necessary. One example is shown in Figure 2. It is important to remember that vendor turnaround time is variable. For this reason and basic belief in Murphy's Law, it is very important to include contingency time in the schedule.

Procedures and Work Flow Plan

It is advisable to set up some plan of orderly attack even if it is informal and there is plenty of communication between the few people in a small department. The less haphazard the effort is, the more chance for catching errors, achieving quality, and meeting the deadline.

Figure 3 shows a general work flow plan that has proved usable in several small publications departments. The plan can be revised and expanded as the department grows but getting the initial function moving is a primary management function.

Defining Responsibilities

Writing job descriptions and defining specific responsibilities can seem tedious and very peripheral. But effort expended once or twice a year can be well worth it. Job descriptions can help in hiring the right people, serve as justification to management for additional people and equipment, prevent slip-ups in attending to detail, give employees a sense of responsibility and a clear-cut understanding of what they should contribute, and help eliminate employee squabbles.

Job descriptions and defining responsibilities need not be inflexible. They can provide guidelines, which may be tailored to use the talents and interests of employees. E.g., a technical writer who is particularly talented in diagraming and drafting can produce good quality preliminary art for review purposes allowing the art department more time for other tasks; a word processor operator who particularly enjoys people contact can take over administrative functions such as dealing with Purchasing and Accounting, distribution of documentation inhouse, and hunting down needed information.

Eliciting Support

The manager's function is to keep problems away from those doing the work. A good way to do this is to achieve good credibility and rapport with upper management and support people. There are many ways to work at this, which are commonly covered in management courses. Some ways for a publications manager are:

- Produce a very good document and make it visible. One success breeds a lot of support.
- Keep upper management and other support people informed by status reports and copies of correspondence. Many people oversimplify because they are unfamiliar with the process of producing publications. If they are subtly educated to all the many aspects, they tend to be more appreciative.
- Occasionally do favors for other departments. They will remember and return a favor.

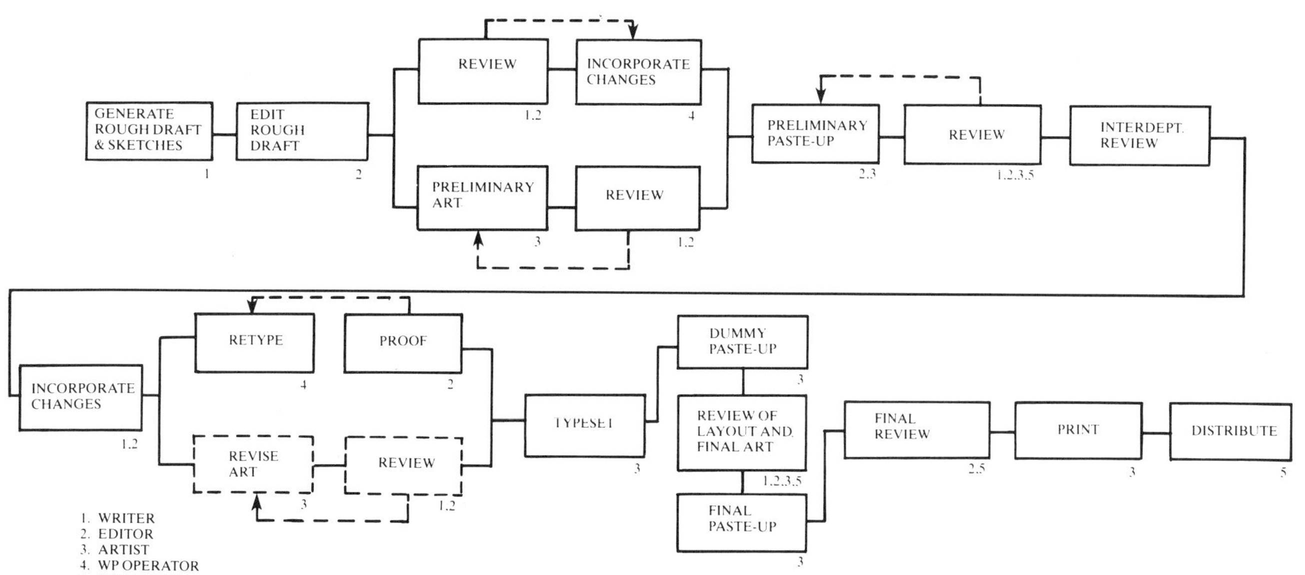

Figure 3. Publications Work Flow

SUMMARY OF DO'S AND DON'T'S

Although any one of the subjects touched upon could be developed much further, I have attempted to outline and simplify in an attempt to provide basic guidelines and direction to those who find themselves facing the challenge of starting from scratch in setting up a publications department.

Following are some do's and don't's learned through trial and error.

- **Do** hire people who can and want to "wear" many hats; it saves time, increases motivation, and expands growth potential.
- **Don't** hire trainees to start a department; there will be no one to train them but you and, unless you can perform every function and make time stand still, you and your department will be in trouble.
- **Don't** hire technical writers who are 90% English and 10% technical; you may get a well worded document that can't be used.
- **Don't** hire an editor who wants only to mark up and proof; an abundance of red marks will not help you find the foldouts.
- **Don't** depend on a sole source for any outside function; the V.P. won't believe the sign maker got the flu.
- **Do** develop a good rapport with the technical "experts;" their input could make or break your document.
- **Do** keep a petty cash fund for emergency supplies; waiting three weeks to order amberlith through the usual route could drastically affect your schedule.
- **Don't** underestimate production time; it involves much more than cutting and pasting.
- **Don't** expect reviewers to return documents on time; you are not their first priority.
- **Do** check bluelines carefully; you may be sorry for missing a transposed part number.
- **Do** make sure equipment service is readily available; a broken printer can render a word processor useless.
- **Don't** let the Purchasing Department buy your printing; you may get a technical manual on newsprint.
- **Don't** expect any document to come back from the printer without a single mistake; you will always be disappointed.
- **Do** make and use checklists for every phase of production; you may hold up printing because no one took the photographs.
- **Don't** make promises when the chances are slim; you will acquire many angry users.
- **Do** expect writers to get a mental block at times; you cannot force creativity as you can productivity.
- **Do** spend time daily getting organized; at least you'll know when the rough times are about to hit.

THE OFFICE OF THE FUTURE

Charles Martin
Samaritan Health Service

Charles Martin
Samaritan Health Service

1410 North 3rd Street
Phoenix, Arizona 85002
(602) 257-4126

The equipment and the know-how for the office of the future is available today. All that remains is the committment by industry to provide cost-effective equipment and the marketing strategy for implementation.

It will not happen

TOMORROW

but

it must happen.

You give little thought to the office lights and your personnel computer link being activated by your body temperature as you enter the room. This is normal.

At the touch of your TD and SCHED keyboard buttons, a display screen rises from your desk. Today's meetings scroll slowly up the screen. The RESEARCH message blinks in the upper corner above the 10:00 a.m. scheduled meeting. You recall entering that code when the meeting was scheduled last week.

After you move the cursor next to the meeting topic, depress the DATA SRCH button to recall associated data from the computer cross-reference file.

The first batch of data appears as the meeting notice is reduced and moves to the upper left quadrant of your screen.

A quick scan of the system highlited key data elements in each document provides nothing needed. Depress CONT, and the displayed data is refiled as a new batch of cross referenced data is displayed.

You mentally note that several of the highlited data elements are pertinent to your meeting. These are easy to save. When you move the cursor to the first position of each selected element and depress the SAVE button, the data is duplicated immediately below your meeting schedule. After you've finished, the CONT button files all data and brings forth new data.

When the last computer stored batch is displayed and you have made your selections, the pulsating FICHE code in the lower right corner of your screen alerts you to additional references.

There may be pertinent data in these files so you depress the SCAN button to display the topic, author and date of data filed on FICHE. Three of these seem appropriate and you key the SELECT code. While some unknown distant file clerk extracts the selected FICHE and inserts them in the Optical Coupler, you quickly review saved data elements.

Even before you've finished, the FILE SRCH lite on your display screen indicates your requested data is ready for review. A simple press of the DISPLAY button screens each FICHE document. A quick search shows only three items of interest in the second document. A simple move of the cursor and depressing the SELECT button posts this information to your meeting reference file.

Whey you have completed your selections, key the RELEASE button so the FICHE can be returned to file.

All the data you need for the meeting is displayed. Now it is only a matter of mentally filtering this information and arranging it for your presentation.

It is time to dictate your thoughts.

When you depress the DICTATE key, your meeting notice and all references are neatly aligned on the left half of your screen. The soft background music in the office is suddenly stilled.

Although your personalized computer voice recognition dictionary contains only 2500 words, it should be adequate. Your spoken comments are automatically displayed on the right half of the screen. An occasional word pulsates to indicate that the computer did not understand. You continue dictating.

With the dictation finished, the SPELL button automatically moves the cursor to the first pulsating word not understood by the computer. A few correcting keystrokes and the SRCH button moves to the next word. Corrections are automatically filed in your dictionary.

Additions or deletions are simple. At the touch of a key, words disappear or spaces open for dictation of additional thoughts. The product is molded. You are satisfied.

Depressing the PRINT, FILE, and EXECUTE buttons in turn provides a hard copy for your meeting and a computer stored file copy. However, the computer file is not accepted. You forgot a key element. You must key in the file retention and category codes that determine individual access to your document and establish retention time and type of storage. Once done, your hard copy is quickly produced by the adjacent laser print.

Is all this a dream? No, they exist today. The electronic typewriter has been around for many years. Energy Electronics makes a Control Lite system that actuates electric switches when a person enters the room. Display screens, the CRT (cathodo ray tube) are in wide use in the word processing field. Voice actuated recording devices are not new. They have even been used by Presidents.

Voice recognition devices, although in the infancy stage and quite expensive, are the key to the future for many business executives. These are all available. Why then, are they not used?

I believe the answer lies in the lack of marketing expertise. Not just the selling. That is the end result. However, in the development of a strategy of use for the purchaser, the total system must be integrated into a sales package that includes a complete training and necessary retraining in every aspect of the office environment of the future. The equipment exists - marketing capability does not.

APPLICATION OF WORD PROCESSING TO ENGINEERING DOCUMENTATION

Clinton McGirr
General Electric Company

Clinton J. McGirr

General Electric Co.
*Neutron Devices Department**
P. O. Box 11508
St. Petersburg, FL 33733
(813) 541-8378

Technical Writer

Word processing has reduced the time required to produce engineering documents at the General Electric Neutron Devices Department. Production time reductions are 10 percent for original typing, 35 percent for document revisions and 90 percent for repetitive typing. Total man hours saved annually are 1,155. The system consists of four work stations, a master central processing unit and two daisy printers. One work station is tied into a large scale computer.

INTRODUCTION

If engineering documents are subject to one thing, it's change. Another characteristic of engineering documentation is repetition. Formats are standardized and specifications frequently repeat one another. At the General Electric Neutron Devices Department (GEND), the Engineering Documentation group generates and maintains product drawings, equipment calibration and operation procedures, material and process specifications, change orders, and a variety of technical publications. All of these documents require frequent revisions and updates. Formats are similar from one document to another; headings, paragraphs and sections are repeated from one specification to another.

FEASIBILITY STUDY

When serious consideration was given in 1979 to purchasing the latest word processing equipment, a feasibility study was completed. GEND is a Department of Energy contractor and, to begin that study, other DOE contractors with WP equipment were first asked for recommendations. Their answers, plus extensive literature research, indicated that WP equipment would probably not significantly reduce operating time for original typing; however, considerable time would be saved in making revisions and in repetitive typing chores. Evaluation of the Engineering Documentation workload indicated that the equipment needed for the initial conversion stage would consist of two work stations, a shared central processor and a printer (Figure 1). One of the work stations would be connected to the large scale in-house computer.

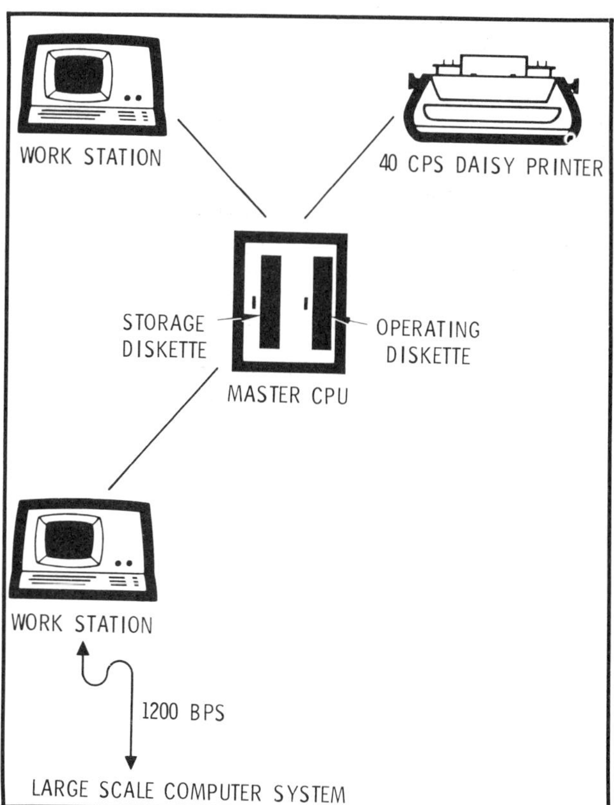

Figure 1

PURCHASE

Specifications were then prepared and bids requested from manufacturers. In evaluating the equipment, cost was one of the most important factors, of course; but there were others. Proven reliability was a must. Another critical feature was the availability of maintenance service: If the equipment is down, how long before the service technician appears? When Purchasing finally placed the order, it went to a manufacturer with an area maintenance facility within a half hour's drive of the GEND plant.

Another factor considered in equipment purchase was operator training. The vendor whose equipment was selected also has a local training facility, with an experienced instructor available to visit the plant upon request during the operators' on-the-job learning period.

SITE PREPARATION

Considerable thought had to be given to the site where the equipment was to be located. A word processing system, with its magnetic medium on which the data is recorded and its sensitive electronic circuitry, is sensitive to various types of electromagnetic interference (EMI). For example, it's possible to degauss sections of a magnetic disk simply by operating a floor polisher near the disk. Therefore, the disks must be carefully handled and stored to protect them from magnetic fields. Other EMI that might cause equipment malfunction can originate in the electrical circuit supplying power to the equipment. For that reason, a power line at GEND was designated for exclusive use of the word processing system. This designated line eliminated the possibility of other office equipment or electrical devices within the plant generating EMI that could travel along the line and affect the system operation. The designated line also helped ensure current and voltage stability for the equipment, another critical factor for satisfactory operation.

Fortunately, the temperature and humidity in the selected working area at GEND are controlled and well within the manufacturer's requirements.

ADVANTAGES

The advantages of the word processing system are many. Note in Figure 1 that a work station is connected by telecommunications to an in-house, large scale computer system. The purpose of this tie-in is to enable engineering data to be sent to, and retrieved from, a large control computer. Prior to the WP acquisition, this function was accomplished using punched paper tapes. These tapes were hand carried to the computer area where they were re-read for transmission to other DOE contractors. Use of the word processor has eliminated the $1800/year rental cost of the paper tape teletype.

The word processing equipment increased efficiency in several ways. One rather unique application involves the preparation of parts lists for engineering drawings. Previously these lists were computer maintained under a separate identity from their related graphic drawings and were printed on conventional computer printers.

The superior quality and range of type fonts for the word processing printer prompted combining the parts list and graphic drawings into one document, this was done by downloading the parts list data to the word processing printer and printing on adhesive backed material using a large type font. Overall savings in document handling, microfilming, etc., from this step are $25,000 annually.

Another piece of equipment excessed by the advent of WP was the Magnetic Tape Electric Typewriter (MTST). This mag tape typewriter was used to prepare equipment operation and calibration procedures. These procedures are now stored on diskettes, eliminating the bulky tapes and the more awkward, slower method of entering changes with the MTST. The information on the tapes had to be transferred to diskettes, and this service was performed by the word processor manufacturer.

Incorporating the new WP equipment into the facility posed minor problems in the beginning. For example, one problem encountered in connecting the word processor to the large computer was that some characters typed into the word processor keyboard would not print out on the computer printer. A percent sign keyed into the WP was recognized by the computer as a pound sign. Likewise with a question mark and an equal sign. However, once such differences were fully defined, the problem was solved by eliminating characters unacceptable to the computer when transmitting engineering documents.

The two-station system was installed in March 1979. It worked so well that plans were soon made to expand the system to four work stations with a 2000-page operating disk capability.

SYSTEM EXPANSION

In order to justify expansion of the two-station system, a study was made to determine actual savings with word processing. The study was designed to compare the time required to prepare typical engineering documents by word processing equipment with the time required by equipment previously used.

One of the documents used in the study was a Final Change Order (FCO). This change order is highly formatted, and is usually composed of two pages. Figure 2 illustrates the first page of a typical FCO.

The study showed that preparation of this page by an ASR 35 teletypewriter (the previous method) was 8.6 minutes. Use of the word processor cut the time to 5.8 minutes. The reason for this 32 percent saving in time is the rigid document formatting. The basic format can be created, stored in a glossary and recalled with two keystrokes when a new document is to be prepared. Figure 3 shows an FCO created by adding applicable data to a format recalled from the glossary.

Another interesting result of this study was obtained by timing the preparation of a Product Acceptance Specification (PS). This multi-page document defines a product's engineering acceptance specifications. For this phase of the study, an extensive revision was made to an existing PS, first with an electric typewriter and then with the word processor. This revision required complete retyping by both methods. The difference in typing time was small. The electric typewriter required 2 hours, 50 minutes, the word processor required 2 hours, 38 minutes--hardly a significant difference in original document preparation.

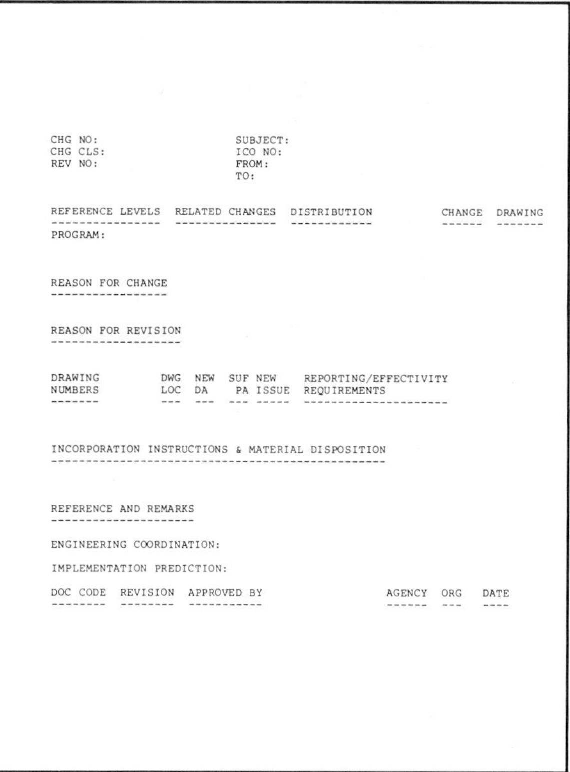

```
CHG NO:                        SUBJECT:
CHG CLS:                       ICO NO:
REV NO:                        FROM:
                               TO:

REFERENCE LEVELS  RELATED CHANGES  DISTRIBUTION        CHANGE  DRAWING
----------------  ---------------  ------------        ------  -------
PROGRAM:

REASON FOR CHANGE
-----------------

REASON FOR REVISION
-------------------

DRAWING          DWG  NEW  SUF NEW    REPORTING/EFFECTIVITY
NUMBERS          LOC  DA    PA ISSUE  REQUIREMENTS
-------          ---  ---  --- -----  ---------------------

INCORPORATION INSTRUCTIONS & MATERIAL DISPOSITION
-------------------------------------------------

REFERENCE AND REMARKS
---------------------

ENGINEERING COORDINATION:

IMPLEMENTATION PREDICTION:

DOC CODE  REVISION  APPROVED BY                  AGENCY  ORG  DATE
--------  --------  -----------                  ------  ---  ----
```

Figure 2

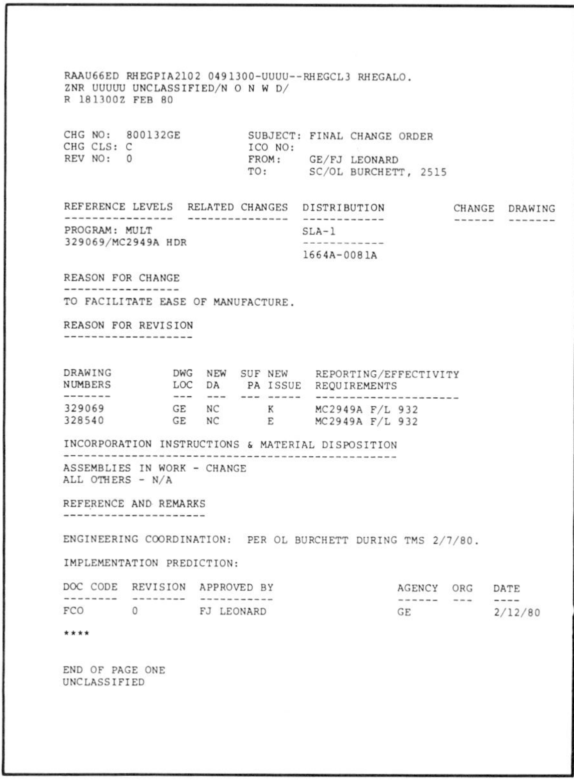

```
RAAU66ED RHEGPIA2102 0491300-UUUU--RHEGCL3 RHEGALO.
ZNR UUUUU UNCLASSIFIED/N O N W D/
R 181300Z FEB 80

CHG NO:  800132GE              SUBJECT: FINAL CHANGE ORDER
CHG CLS: C                     ICO NO:
REV NO:  0                     FROM:    GE/FJ LEONARD
                               TO:      SC/OL BURCHETT, 2515

REFERENCE LEVELS  RELATED CHANGES  DISTRIBUTION        CHANGE  DRAWING
----------------  ---------------  ------------        ------  -------
PROGRAM: MULT                      SLA-1
329069/MC2949A HDR                 ------------
                                   1664A-0081A

REASON FOR CHANGE
-----------------
TO FACILITATE EASE OF MANUFACTURE.

REASON FOR REVISION
-------------------

DRAWING          DWG  NEW  SUF NEW    REPORTING/EFFECTIVITY
NUMBERS          LOC  DA    PA ISSUE  REQUIREMENTS
-------          ---  ---  --- -----  ---------------------
329069           GE   NC       K      MC2949A F/L 932
328540           GE   NC       E      MC2949A F/L 932

INCORPORATION INSTRUCTIONS & MATERIAL DISPOSITION
-------------------------------------------------
ASSEMBLIES IN WORK - CHANGE
ALL OTHERS - N/A

REFERENCE AND REMARKS
---------------------

ENGINEERING COORDINATION:  PER OL BURCHETT DURING TMS 2/7/80.

IMPLEMENTATION PREDICTION:

DOC CODE  REVISION  APPROVED BY                  AGENCY  ORG  DATE
--------  --------  -----------                  ------  ---  ----
FCO       0         FJ LEONARD                   GE           2/12/80

****

END OF PAGE ONE
UNCLASSIFIED
```

Figure 3

However, this same Product Specification was then marked up with a typical, extensive revision. The typing difference for this revision was significant: the electric typewriter - 1 hour, 40 minutes; the word processor - 38 minutes. The reason for the 62 percent reduction in typing time is that an extensive revision with an electric typewriter requires retyping many paragraphs, followed by cutting and pasting of the text to rearrange the copy. But with the word processor, this same rearranging of text is done with just a few keystrokes, making extensive revisions far less time consuming. In fact, experience has shown that text revision is an area of publication savings that can justify word processing equipment many times over.

Another characteristic of the Product Acceptance Specification is that large portions of the text of one specification can often be adopted for another. To compare the times for this operation, such a modification was made to an existing specification. The electric typewriter required 2 hours and 50 minutes to prepare the the new document. The word processor required only ten minutes--a 94 percent time savings. The word processing equipment was able to recall most of the text from a document previously created and stored. With these minor changes incorporated, the PS was ready to go to the floor within an hour.

The advantage of this feature was illustrated with another type of document, Calibration Procedures. The paragraphs of these procedures may also be copied from one document to another. In this step of the study, a four-page Calibration Procedure was created by recalling and modifying a procedure stored on the word processor magnetic disk, and an identical document was created with an electric typewriter. The resulting preparation times were 11.8 minutes per page with the electric typewriter and two minutes per page with the word processor--an 83 percent savings in time.

The productivity gain from the cumulative data generated from the studies showed the following average time savings:

Original Typing - 10% reduction,
Revisions - 35% reduction,
Repetitive Typing - 90% reduction.

Applying these percentages to the annual work time required to publish engineering documents results in the following annual time savings estimate:

Original Typing - 10% of 3,228 hours = 323 hours,
Revisions - 35% of 1,809 hours = 633 hours,
Repetitive Typing - 90% of 221 hours = 199 hours,
Total hours saved annually = 1,155.

NEW SYSTEM

Results of that study justified the expansion of the existing WP equipment at GEND. The new equipment is composed of four work stations, three storage diskette drives, two printers and a master central processing unit (Figure 4). The new four-station system has a much larger working capacity--2000 pages as compared to the original system's 80 pages. This system can be expanded to use 14 peripherals including work stations, printers, telecommunications and photo composition devices.

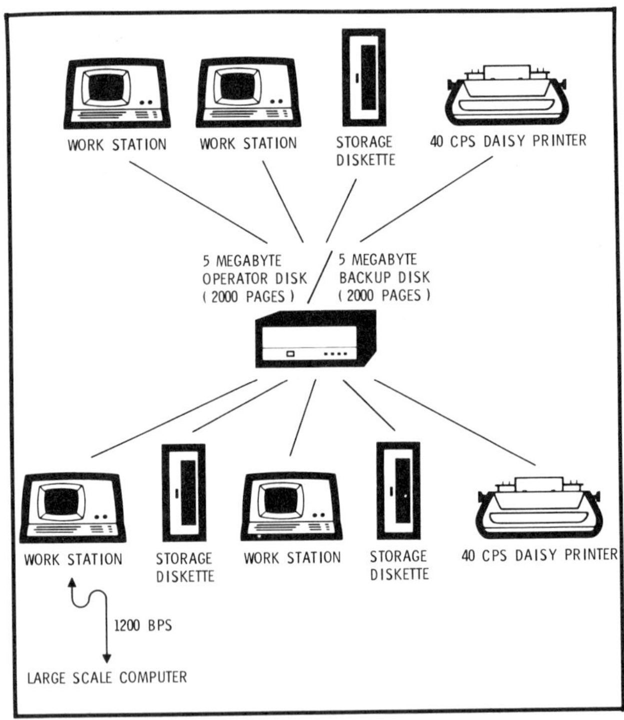

Figure 4

**Operated for the U. S. Department of Energy by General Electric Company Under Contract No. DE-AC04-76DP00656*

WORD PROCESSING FOR TECHNICAL WRITERS AND TEACHERS

Carolyn J. Mullins   Thomas W. West
Information and Computer Services, Indiana University

Carolyn J. Mullins
Indiana University
Staff Assistant
Bryan Hall 7
Bloomington, IN 47405
(812) 336-5465
Complete Writing Guide (Prentice-Hall, 1980); Guide to Writing and Pub. (Wiley-Int. 1977)

Thomas W. West
Indiana University
University Dir., OICS
Bryan Hall 7
Bloomington, IN 47405
(812) 337-1053

At Indiana University in the late 1970s, the demand for word processing, the burden on the university's text processors, and the variety of commercial equipment demanded coordination to meet needs systematically. Now the university's long-range policy fosters compatibility of systems with each other and with the university's computers, enables adequate maintenance, promotes staff mobility, and still meets the needs of technical writers and editors, teachers, students, clerical staff, counselors, and administrators.

ONCE UPON A TIME: A PROBLEM

Johanna McVey, Senior Technical Writer, fidgited as she listened. The topic was modern office technology in general and word processing in particular, and the plans being discussed seemed either short-sighted or irrelevant to her professional needs. Across the table Jim Grayce, who taught business and technical writing, seemed equally disturbed.

The group had been discussing the need for features such as word wrap, global search and replace, simultaneous input and printing, automatic decimal centering, spelling dictionaries, automatic page numbering, math packs (which automatically calculate and total columns of figures, thus preventing errors in either hand calculation or transcription), automatic headers and footers, easy reformatting, and so forth. A secretary from the Admissions Office wanted "merge" functions for joining blocks of letter text with names and addresses in mailing lists.

These features sounded fine, but McVey wanted more. She often worked on long documents and didn't want to have to worry about running out of file space. Because she often worked with Greek and mathematical symbols, she needed super- and subscripts (preferably shown on the screen) and dual-head printers, which accommodate two different printing elements simultaneously. Because she did more revising than drafting, she also wanted automatic marking of texts that have been altered by the most recent revision. Finally, she needed access to data on the main computers, and sometimes had wanted to include computerized data and graphs in documents without retyping. Her friends those who wrote documentation needed automatic paragraph and outline numbering, and automatic tables of contents, lists of tables and illustrations, and indexes.

When the group took a break, she sought out Jim, who turned out to want even more features than she did. "I want electronic mail and a desktop terminal to check and answer mail, send memos to colleagues elsewhere, write notes to students, and make assignments," he said. "I also want to be able to use the same terminal for computer work. And sometimes I want to see a student's transcript, when I write letter of recommendation, for instance. I don't want to call the registrar and wait for a copy to come in the mail--I need it right then."

"Most of all, I want my students to use word processing. Most of them are so hassled by typing that they can't concentrate on learning how to revise properly. When they get jobs, they won't have to worry about typing production copy, why should typing be a bother now? Besides, my most recent graduates tell me they are working on word processors and they do everything, from drafting to printing, by themselves, with no secretarial typing, no printers--nothing. If that's how professionals work, that's how I want my students to learn. After all, if I were teaching science, I'd require my students to learn how to use computers. I think word processors have the same value to technical writers."

"Listen," Johanna said, "We've got to express our needs more forcefully. Most of the people here want stand-alone processors, but many others on campus have needs like ours. We've got to start somewhere . . . "

THE NEEDS

Many of you are probably nodding. These needs seem "far out" to those who want word processing for standard office purposes. Some people with normal office needs think that meeting our needs means not meeting theirs. Such is not the case, although meeting just their needs can sometimes mean not meeting ours.

Professionals in our business require (1) large amounts of file space, (2) easy communication between files, between word processors and computers, and between computer systems, (3) graphics, (4) photocomposition interfaces, and (5) word-processing features such as floating footnotes, column swapping, and easy display and printing of Greek letters and mathematical symbols. (6) Teachers need a system that can be conveniently available for students.

THE ENVIRONMENT AND POSSIBILITIES

Indiana University, an eight-campus system with a total of approximately 79,000 students, is served by an extensive computing network that includes IBM computers for administrative use and IBM, DEC, CDC, and PRIME computers for academic uses. All computers are linked by an elaborate data communication system and form the Indiana University Computing Network.

We considered putting a sophisticated command-based text processor, such as NROFF, on one or more computers, with access through ordinary terminals. Equipping terminals in student clusters with floppy disk archiving stations would enable students to handle their own storage and save disk space on the system. Unfortunately, most text processors were not "user-friendly." They lacked the extensive menus that help operators to use the commercial word processors. Because many university offices have a high rate of staff turnover, training would be very costly.

On the other hand, commercial systems, which make training easier, lacked many of the necessary features and would be hard to make available to students. To make matters worse, the marketplace presented a bewildering variety of brands and combinations of equipment, with little standardization. Thus, choosing the right equipment for an office required some technical knowledge and systematic office study.

OBJECTIVE, POLICIES, AND PLAN

Objective

To provide faculty, administrators, staff, and students with the resources needed to do their work more effectively, the university's long-range objective is to have a common system, for both word processing and text processing, that integrates with the data processing capabilities of the IU Computing Network. Unfortunately, such a system does not now exist in the marketplace and is not likely to become available for the next several years.

Central to the long-range objective is the **policy of technical concurrence**, which requires that units choose software and equipment from an approved ily restrictive but rather to make possible a high quality of long-range, overall service. A major reason for the limited list is the need to facilitate an eventual interface between pre-existing campus word processors through the computing network. A second reason is to make maintenance and repair more efficient and cost effective. A third reason is to facilitate training and job changes by trained operators.

Short-Range Plan

Word processors will become available to administrative and academic users in several ways. (1) Shared-logic systems in several campus locations will serve academic and administrative users who have both data processing and word processing needs. Users who need only word processing will also be served in this fashion. (2) When service cannot be provided from a shared campus site, the university will authorize stand-alone word processors whose software is compatible with that of the shared systems. (3) Microcomputers with approved word-processing software may be acquired by individual faculty for research uses.

Operating Guidelines

For the next two to four years, while awaiting the technological advances needed to integrate all word and data processing, Indiana University will work to achieve as much compatibility and integration as is feasible within the available and everchanging technologies. Specifically, the policies and plans call for:

1. On each campus, strategically locating time-shared or shared-logic word processors to serve several offices.

2. Supporting a limited number of software packages on approved microcomputers for use by individuals and small project groups of faculty and students.

3. Establishing text management capabilities within the IU Computing Network's resources for academic users who need access to both text and data files simultaneously. Currently available processors have adversely affected faculty and students who use the computers for other purposes. The goal, then, is to develop on one or more other computers a fully interactive word processor that can be accessed through the Computing Network.

4. Continuing to install stand-alone word processors in units with self-contained needs.

5. Further limiting the number and types of word and text processing software and hardware to be installed at IU. Currently, Indiana University approves word processors from three commercial vendors.

Initial funding for "ports" into the clusters has been provided in the form of four- or five-year loans from the university, acquired by means of a contract with our office.

All offices in need of word or text processing begin the process by contacting an office consultant, who helps with analysis of need, assessment of available equipment or software, preparation of purchase requests, training of staff, and installation of equipment. In addition, during 1980-81 the consulting staff and purchasing agents sent out a Request for Information to all vendors, inventoried word-processing equipment on all campuses, designed a method for studying the needs of office systems, and prepared a booklet that describes the what, why, and how of word processing (designed for all staff and administrators on all IU campuses).

THE RESULTS

On the Bloomington campus, the first cluster system for administrative users, a Wang OIS 140, was installed in November 1980. At the same time, a WORD-11 system on a DEC computer, intended primarily for academic users, was installed at IUPUI. Both are trial systems. If they satisfy the users, more will be installed. A third trial system is the MUSE word processing software that was installed in October on the Northwest (Gary) campus's PRIME 550 computer. Like many of the commercial systems, MUSE operates by command keys and thus holds the possibility of serving all users well. Indeed, MUSE's main drawback was the lack of some features, such as automatic tables of contents and indexes, that technical writers value.

THE BENEFITS

The Wang system came with "CICS pass-through," which enabled users to reach the administrative data on the IBM computer, and has been judged very friendly to users. For academic users, the system has two drawbacks: it cannot access the academic computers, and the terminals are too expensive to install in unsupervised clusters for students.

The WORD-11 cluster enables academic users to communicate with all university computers, and at least one user department was planning trial access for students on ordinary terminals. Graphics packages were available on the computers. Photocomposition interfaces were on the drawing boards, and with one exception, WORD-11 had the formatting features needed. The exception was easy formatting of equations with on-screen display of Greek, math, and super- and subscripts. Short-term plans to meet this need include use of A. B. Dick stand-alone systems, which handle equations well, and investigation of techniques devised by other users of WORD-11. For the long run, WORD-11's developers plan to add this capability.

Limiting the hardware and software has enabled planning for eventual compatibility. Equally important is the planning we can do to provide "staff compatibility" as, for instance, when staff move from offices with Wang equipment to offices with DEC or A. B. Dick equipment.

Going slowly helps users in all groups to find out what processors and combinations of equipment best meet ther needs <u>before</u> they have invested too much to back off. It does not lock technical writers and teachers or office staffs into systems that don't meet their needs. It is also enabling the university to use its computing expertise to solve communication problems and to study modern office technologies. The DEC system also has one further advantage: DEC computers are popular with our faculty. If WORD-11 proves unsatisfactory, the university can convert the computers to academic use.

CONCLUSION

Indiana University's approach to modern office technology is important to technical writers and teachers in academic institutions because it directs action in a community with diverse needs. However, the approach is generally important because it gives a long-range view of responsible acquisition in this highly technical field.

ON-LINE DOCUMENTS
HAYSTACKS OR BUILDING BLOCKS?

Diana Patterson and Paul Evitts
SYSDOC Dyad Computer Systems Inc.

Diana Patterson
SYSDOC International Inc
213 Albany Avenue
Toronto, Ontario

Paul Evitts
Dyad Computer Systems Inc
80 Bloor Street, W
Toronto, Ontario

Chairman ACM/SIGDOC
former STC chapter chairman, newsletter, &c.

Considerable advances have been made in word processing and photo-composition. But these are only some of the important uses of computers and documentation.

Rather than producing paper documents by computer, we will consider computer documents that are meant to be read from a screen in a paperless office. The computer is used to scan through the material, to format it, to make the connections between one document and another, and to keep the documentation secure. This makes considerable sense when documenting very large systems, such as aircraft, computer, or library systems.

We must consider the special writing techniques and structuring of documents that will be read from the index rather than the table of contents.

On-line documentation is information stored on a computer system and used on that computer system rather than printed out and read in book form. For our purposes it is not especially concerned with the documentation of on-line computer systems, nor the processing of printed pages as in "word processing."

Background

A typical on-line document that has been around for many years does happen to have the computer system itself as its subject; this document is called the HELP command. A set of lines or screensful of information is stored on disks and forms a document that can be examined when a user, or reader, types

HELP

or

HELP subjectname

on his television style terminal. Here is an example:

HELP COST

THE COST COMMAND WILL TOTAL
THE DOLLAR FIGURE FOR THIS
TERMINAL SESSION UP TO THE
SECOND BEFORE YOU TYPED
"COST".

As you can see it is a primitive document, usually available only in upper case. HELP is certainly far from the ideal on-line document, but it is the most common. Many times it is a most unhelpful document because it must be short, and therefore often uses jargon in its descriptions. And inevitably you must ask exactly the right question to get any answer at all. HELP does, however, illustrate several important differences between on-line documents and books.

The average computer screen made today is 23 lines of 80 characters. Only special word processing computer screens have the standard US paper size, 8 1/2 by 11 inches. The 23 lines of 80 characters is very much smaller than that.* Only one page is visible at a time. To look at the next or preceding page requires pushing a button, and therefore waiting for the computer to respond. At peak times on timesharing computers there may be minutes between time of requesting the next page and seeing it; thus, it has always been assumed that only one page is the best size for a HELP document.

HELP commands have little or no structure. The reader types HELP followed by only those commands he wants to know more about. If he types HELP HELP he will get some kind of subject list, but usually this is in no particular order except the order in which the entries were added. Thus the reader selects the order in which he will read the

* At the time of writing, IBM has announced a screen that will display 132 character lines. It does this by shrinking the type size.

material. If he wants to know about copying files, he will only type HELP COPY. The screenful of information that results is all of the document he will read; there will be no beginning, middle and end as in most manuals.

Concerns

Looking at information in the reader's order, rather than the writer's order, creates one of the major challenges for document specialists dealing with on-line documents. It is not a new idea--at least not for the reader! Many readers have been skimming books for years, reading the end first, and the preface last. They have always been considered by the writers, however, as outcasts who were breaking the rules of order and therefore deserved what they got. On-line documents, however, have no rules of order, and their chief use will be by those who need specific information as quickly as possible without the preface, the author's introduction, and so on.

This newly recognized order is what we call "heuristic" rather than 'Hierarchical". It's natural consequence is to play havoc with the background and references that the writer can assume for the reader. The reader is the customer, as we know, and the customer is always right. Therefore, as writers and graphic artists, we must take up the challenges presented by a breakdown in hierarchy.

Within about the next three years networks of computers will be set up in homes and businesses to provide information as a utility, like natural gas, electricity or telephone service. This will introduce a new approach to information that will incorporate the heuristic approach. And as a result of these changes there will be changes in everyone's thinking.

If current trends continue the impact of these information utilities will be very significant. Many, many people will have these new style on-line documents in their homes, and as with computer games and television (the combination of which they will very much resemble) there will be social and psychological influences.

In order to examine the impact on the writer we will design a document system and see how it is used. The system we are using for illustration is not a working one. It is a hypothetical, heuristic approach. However, our system does not differ much from an existing one on the market, and all of the machinery described already exists and could be put together as we describe.

Organization

A manual's organization is usually summarized in the table of contents. It is an overview of the manual's hierarchy. Here is a typical table of contents:

1. Table of Contents
2. Overview
3. Commands
    3.1 ALLOCATE
    3.2 BACKUP
    .
    .
    .
    3.8 CHARGE
        3.8.1 Schedules
        3.8.2 CHARGing to OTHERS
4. Appendices
    4.1 First Appendix
    4.2 Second Appendix
5. Index

The numbers indicate the position and rank in the hierarchy. If a single level list (such as Commands in this example) becomes very long, say several pages, it is hard to see this hierarchy without numbers. If the hierarchy is complex it may be necessary to have volume or chapter tables of contents, as is done in multi-volume reference books.

This hierarchy is based on the writer's assumption that the pages of the manual are ordered, even in a loose leaf binder, and that a naïve reader will read the manual through, in the order in which the pages are bound, at least once. Further references will probably be by index, if there is one, or by the table of contents.

The reader frequently does read an important manual through at least once, because it is the most reliable means of knowing where things are in a poorly indexed, or unreliably indexed work. This is partly a result of the lonely process of reading in which one cannot ask the index a more probing question of one's own concoction.

The organization of an on-line document is dependent on the means in which it is stored and the way it will be searched. The most primitive of the HELP commands used to store text line by line in a sequentially ordered file--meaning one that you search sequentially starting at the top or the bottom. These were usually printed on continuous rolls of paper so that the only way to search it was to start at the beginning and read through until you found something important. Even these computer documents tended to be small.

The kinds of documents that will be optimal for on-line storage will be very large ones that can be broken down into one screen per subject. If the documents are not very large, they will be in large sets to form encyclopedic collections. Either way the information will be worthy of being called a very large collection of data.

Computer technology has been able to set up some very efficient and sophisticated means of searching large and very large data bases. A data base is just a very large collection of data. The term , however, suggests that it has been organized to make searching and sorting the data efficient.

Since our on-line document is a very large collection of data that we will want to index and search, it should be organized as a data base. This organization cannot be solely the choice of an author or a standards committee; it is dependent on the rules established by the particular data base type. But once the data are in the data base's format the cross indexing becomes excellent. A single screen can be reached from nearly all other screens.

The order is precisely fixed, and yet hierarchy is beginning to break down.

In our document system each item in the data base is a screen. Aliases are supplied for all items so that the index is a useful tool even for the non-expert. In order that the index is easy to search, it is layered, so that it has a hierarchy, if only an accidental one. Here is how the layering works:

Highest layer	lower layer	lowest layer
1.	2.1	2.2.1
2.	2.2	2.2.2
3.	2.3	2.2.3
4.	2.4	2.2.4

At any time on the screen, the reader sees only one layer. If he asks for one item to be expanded he will get the next layer down if there is one. When he wants to move more quickly he can move up one layer if there is one.

The index shown in the illustration is only a numbered index, not an alphabetic one. That is because information might be indexed by any order, alphabetic, numeric, or any other that the organizer wishes to impose. This can be by some arbitrary table of contents hierarchy.

Thus the data base organization can theoretically allow the information to be put together in an infinite number of ways. Actually, the way that the data are entered presupposes the order in which the data will be searched most often.

Still, there is nothing sacred about either a table of contents or an index. It is equally possible to order information by the user's job requirements, key words in and out of context, the date of the last screen correction, or by the page numbers of the books referenced by the screens.

Communication

If you are overwhelmed by the possible order of an on-line document, you will be at least whelmed by the communication possibilities.

Our system is based on a small computer. It has a library of its own, but that is reasonably small. The real value of the system will be in its ability to contact the central library held by the utility. This central facility will provide a considerable amount of data which will make the system "economically feasible" such as catalogues of consumer goods that can be purchased through the telephone or cable lines that supply the catalogue. Once the two way communication is established with a computer communications utility, there can be no way to prevent one network reader from talking to another. Electronic mail is already a reality pretty well around the world. The mail service is as much a part of an information utility as sports news and weather will be. A reader should, with all the correct security clearances, be allowed to read documents on any other reader's library. With the correct clearances he will be able to copy portions, change them to his own purposes, mix them with other documents, and so on, using the computer mail or catalogue order communications.

The Impact on the Writer

The writer writing and editing an on-line document has to have the skill of one of those lucky people who makes a living by writing short paragraphs. Screens are not very large, even if the manufacturer will let you squeeze more letters on a page to make you think it holds quite a lot. Readers looking at a screen have just as great a dislike for cramped pages as do manual readers. This means the text must be very brief. If possible, position can be used to replace verbs, e.g.:

COMMAND==>

replaces:

Type the command on the top line.

The first of these invites the reader to do what it says. If you do, the computer will respond, which introduces the intermingling of subject and object, but this is not something we'll discuss further.

Because very few words are used on each screen, the impact of each word should be greater both for reader and writer. Terms must be consistent from screen to screen because of the new importance each word takes on.

Text should be integrated with graphics and tables although they must not be too dense to be legible. Clean, uncluttered drawing can often replace a table. Since everything is backlit on one of these screens, it can become very tiresome doing any prolonged close study. This is especially true of "positive" screens, where the background is white light.

References forward and backward in the text as it was written, now become more important than ever. But now we have less room for footnotes than ever. Our solution to the problem is in the document's machinery. For any piece of text on a screen, another screen can be displayed giving the references for that page. (See the Operation section below.) This does mean, however, that the reference needs to be made for every use of a term, not just the first. Glossary items should have screen references made both ways. Making the connections is a human task, either in whole or in part. It is tedious, but very important.

Although references are available, they should be used as little possible. The information ought to be perfectly clear to as wide an audience as possible. This brings us around to the importance of words. They must be chosen carefully. Whenever jargon can be avoided it should be. Short words are to be preferred to long ones. Be snappy, clear and consistent.

Impact on the Reader

We must assume that the reader will get a minimum search time when he goes looking for information. Whatever that means. But frankly, it probably won't matter to the reader after a while. On-line documents will have machinery that will be simply fun to work on, so that he gets hooked on the game of it even if he could read the whole book faster than retrieve a single item in it.

Not only do these devices replace books, they replace letter writing and driving to the catalogue store. The communication is two ways, which means that readers become writers.

Dialogues can be going on while you are reading so that it need no longer be a private and quiet experience like reading a book or watching television. This is the C B radio of the library set.

Information will be brought into homes and offices over a cable network, many of them the same ones that carry television. The screens of initial models will be home televisions with a keyboard attachment. There will be no avoiding paying for this kind of television, and information will become a commodity as it has never been before. Caveat!

A Touch of Fantasy

So that you can appreciate the impact of on-line documentation on a reader, we will illustrate the system we have been discussing.

You are sitting a a multi-screen console. This is the basic model and has three screens. Each screen is switchable to one of four functions in two modes--no more complicated than the average wrist watch.

In the standard configuration the screens look like this:

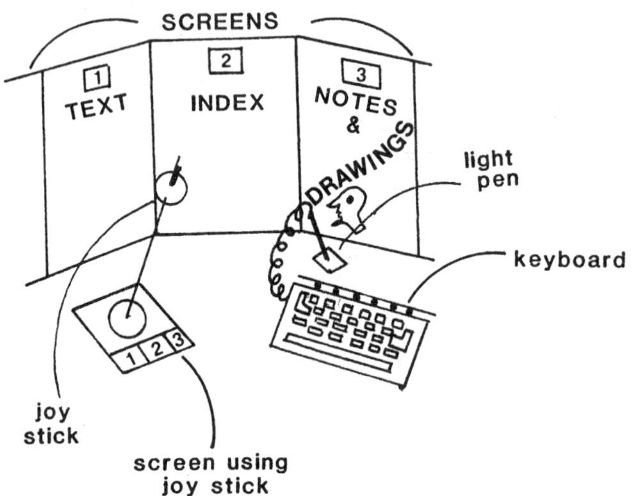

This is Reading Mode. The joy stick is connected to the Index screen. The entries are displayed on a plane with lines between the entries to indicate the density of entries one level down. Using the joy stick you can seem to drive down the index plane, then dive through the upper index level, breaking into the next layer of index, and drive along there for a while.

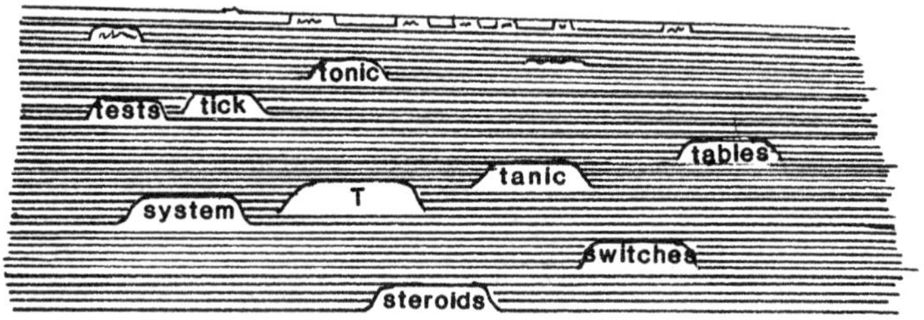

When you select a detailed entry, press the DISPLAY button, the text of the document, which is ideally one screen of information, will appear on the TEXT screen.

Here is a detail index entry for SYSTEM. This is a catchall word, and thus this screen has been designed as a menu to get further information. If the light pen is pressed against the block, the first screen of the document will be displayed. If a more detailed breakdown of the topics under one of these systems is required, the user must look under "2000" or "accounting" in the main index.

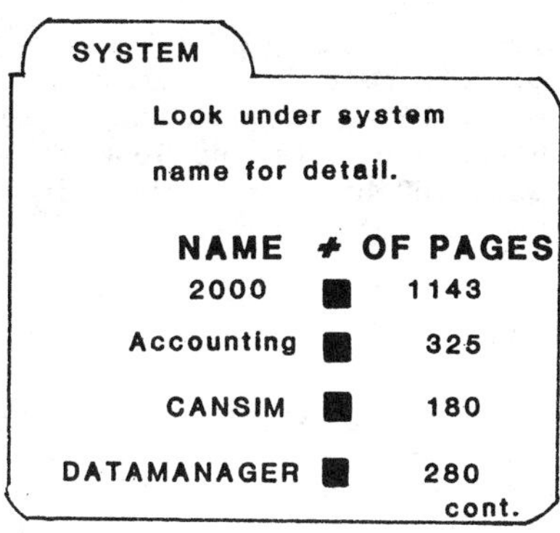

At any time during this process of looking at screens the text can be copied to the notes and drawing screen. Drawings can be added using a light pen.

This next screen would be the text screen for one of the standard on-line reference works in a library. This is from a bibliography of PhD theses. On this menu there are three touch areas. The light pen pressed against one of them will either cause the order screen to appear, the cost calculation screen, or print an abstract on one or more screens.

Conclusion

The system I have shown you is a fantasy, but all of the equipment necessary to make one is available. As any visitor to a computer games parlour will tell you, the system I have described is very tame stuff.

But I repeat, that there are systems like this but without the fancy visual effects, now being sold. Screen by screen documentation is a new means of communication to be explored. On-line documents will be a considerable force in our lives quite soon!

This paper has been only an introduction to the subject. We have already begun writing specifications for such a system, so that when new systems are built they will have some input from the potential writers as well as the potential readers or consumers. There is much work to be done, and we hope that we have got you thinking about the subject so that writer, editors and graphic designers have systems really suited to the task.

PRODUCING COST-EFFECTIVE PUBLICATIONS WITH THE AID OF WP

Charles B. Pinkham
Firman Technical Publications, Inc.

Charles B. Pinkham
Firman Technical Publications, Inc.
Production Manager
95 Church Street
Pembroke, MA 02359
617-826-5148

By integrating the use of word-processing into the creation/production process, you can reduce the cost of producing the mechanicals for a technical publications. All three major production operations are affected: typing or typesetting, illustrating, and layout/paste-up.

INTRODUCTION

The production work involved in a technical publications consists of three major operations: setting the copy (typing or typesetting); producing the line art; and laying-out and pasting-up the copy, the line art, photographs, and the captions; and then dropping-in special items and last-minute corrections.

When a document has been created on a word-processing (WP) system, the work involved in all three of these operations is reduced. The reduced work - and the consequent reduction in cost - are also an incentive to do certain things that might otherwise be neglected, and therefore the publication tends to be a little more complete and a little more effective.

This saving in work and cost is emphasised when WP is integrated into the creation process as it is at Firman Publications (1,2,3). When Firman Publications techniques are used, the creator of a publication (who used to be a "technical writer") has virtually conplete control over the layout. Furthermore, WP functions as a tool that allows the creator to experiment with, and to develop, more effective layouts. If the creator has good visualization and graphic-design skills, and if he/ she works closely with the production manager on the initial design of the publication, then the resulting publication is not only more effective, but also less costly because the production work is greatly reduced.

Thus it is apparent that from a production point of view, WP is a tool for producing more cost-effective publications. Let us look at some of the ways in which WP has this effect in each of the three areas mentioned above.

SETTING THE COPY

The bulk of the publications produced by most publication departments are support publications such as technical manuals, rather than more elaborate promotional publications; for this reason, most copy produced is typed rather than typeset. It is true that typesetting offers such advantages as a choice of typefaces, flexible layout, and right-hand justification; but it is also true that typesetting usually costs $20-$40/page, whereas typing costs perhaps one tenth of that. Furthermore, both typing and typesetting must be proofread very carefully, and extra cost is incurred in making the inevitable corrections.

The bulk of the copy produced by Firman Publications is also support publications; but we never type them, and we rarely typeset them: we mainly use WP output. A WP system provides "letter-quality" output, which is at least as good as any typing quality. WP output also offers most of the advantages of typesetting, including a choice of typefaces, flexible layout, and right-hand justification - yet it is not only less expensive than typesetting, it is also only about half the cost of typing (or about 5% of the cost of typesetting). Furthermore, proofreading and correction of the final copy are not really required because the copy has been stored in the computer throughout the creation process: it has all been proofread during earlier stages of creation - and most of it has been proofread repeatedly.

Let's consider some of the advantages of WP in setting text.

<u>Choice of typefaces</u> Our WP system gives us a choice of typefaces, in 10-pitch, 12-pitch, and proportional spacing. We have about a dozen faces, including the examples shown in Figure 1. Included in these faces are a "Scientific" face that includes extensive math symbols and Greek letters, and an "OCR" (optical character recognition) face that is very attractive in replicating CRT displays such as the example shown in Figure 2.

Letter Gothic:	abcdefghijklmnopqrstuvwxyz ABCDEFGHIJKLMNOPQRSTUVWXYZ 1234567890-=½;',./!@#$%¢&*()_+¼:",.?
Vintage:	abcdefghijklmnopqrstuvwxyz ABCDEFGHIJKLMNOPQRSTUVWXYZ 1234567890-=½;',./!@#$%¢&*()_+¼:",.?
Cubic:	abcdefghijklmnopqrstuvwxyz ABCDEFGHIJKLMNOPQRSTUVWXYZ 1234567890-=½;',./!@#$%¢&*()_+¼:",.?
Titan Italic:	*abcdefghijklmnopqrstuvwxyz ABCDEFGHIJKLMNOPQRSTUVWXYZ* *1234567890-=½;',./!@#$%¢&*()_+¼:",.?*
OCR-A:	abcdefghijklmnopqrstuvwxyz ABCDEFGHIJKLMNOPQRSTUVWXYZ 1234567890-=▮;',./ΥΗ⑀⑁%|&*{}—+▮:",.?
Scientific:	αβψφε>ληι∫κωμν∱ρλθστξ†δχυς ∇|ΨΦ÷<Λ¶}∫§Ω∂√↑ℓΓΘΣ™Ξ∝Δ∞Τ≃ 1234567890 =©πx^^~}√≡↔±{∑®()_+Π°•´~{
SPOKESMAN:	ABCDEFGHIJKLMNOPQRSTUVWXYZ ABCDEFGHIJKLMNOPQRSTUVWXYZ 1234567890-=½;',./!@#$%¢&*()_+¼:",.?

Figure 1. Examples of just a few of the typefaces available on a word-processing system. WP systems can provide 10-pitch, 12-pitch, and proportional spacing.

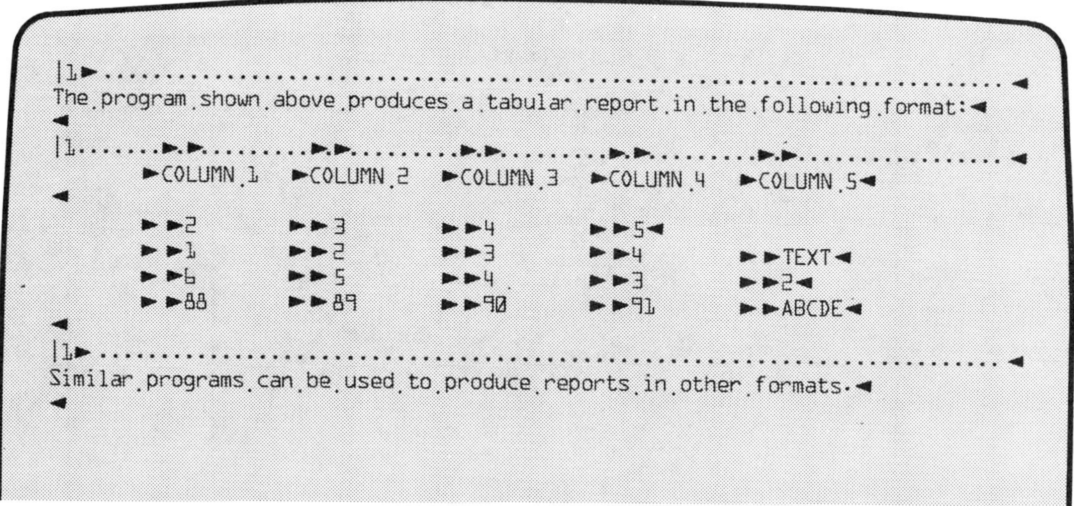

Figure 2. The OCR-A typeface can be used to simulate CRT displays. (This particular CRT display shows the appearance of the table in Figure 4 when it is created on our WP system.)

(It must be admitted, however, that the WP printer will not conveniently mix typefaces during printout. When you wish to mix faces, you have to paste-up. However, the amount of paste-up involved is still well below the level of pasting-up galleys throughout an entire publication.)

Type size It is true that although WP provides a choice of type style, and a small choice of pitch, it does not offer a variety of sizes. However, a process camera can be used to reduce or enlarge the type: the quality is such that you can easily double the size of the printout without significant loss of quality. The title of this paper is an example of considerable enlargement of WP printout. How much does this effect cost? You can photostat a complete page of text for only $3-4; that plus the $1-or-so it costs for the printout is still only slightly higher than the cost of ordinary typing - and still well below the cost of typesetting.

Format flexibility WP gives the writer complete control over the format of the document, and our writers develop the format as they develop the content. (This process is described in detail in the preceding paper by Pat Smith, "Creating Cost-Effective Publications with the Aid of WP".) The WP system can print lines up to 150 characters wide, and can print two columns of copy on a single sheet. This all means that the final printout from the system can be not merely galleys, but a fully formatted page: one or two columns, of any width (or even varying width to allow for artwork), and with or without right-hand justification. A fully formatted page of text requires no layout and paste-up work, which saves a great deal of labor in production.

<u>Headers, footers, and page numbers</u> Our WP system can insert a standard header and/or footer on each page, which is convenient for including such information as the title of the publication and the title of the section. Part of a header or footer can be a page number: the system automatically counts the pages as it prints them, and plugs in a suitable number in each case. Figure 3 shows some examples of headers, footers, and automatic pagination.

PRODUCING LINE ART

The use of WP to create a document can also reduce the work of producing illustrations. In discussing format flexibility above, I pointed out that "WP gives the writer complete control over the format of the document..." This flexibility, together with the instantaneous CRT display, makes WP very useful for designing many common types of technical illustration.

Many technical illustrations are basically organized text. Perhaps the most obvious example of this is a table, such as the one shown in Figure 4: it consists of regular columns of simple text, plus columns of numbers. Less obvious are block diagrams, such as the one shown in Figure 5; and flow diagrams, such as the one shown in Figure 6. All of these illustrations were created on WP: the illustrator's task was limited to ruling the ink lines.

The CRT display in Figure 2 actually shows what appears on the WP screen when the table in Figure 4 is created. At the top of the screen a "format line" shows where tabs have been set for a standard page of text; a second format line sets the tabs for the columns in the table; and a third format line restores the original format. You can imagine how easy it was to create this table; if a particular entry did not fit, the writer simply modified the appropriate format line with a few keystrokes, and the system automatically and instantly reformatted the entire display accordingly.

Even in more complex illustrations WP can be used to provide callouts and captions. The callouts can be set in a suitable typeface, photographically reduced or enlarged to a suitable size. (WP output is more attractive and less expensive than Leroy or dry transfers.) The captions can also be set in a suitable typeface, perhaps different from the body copy so as to set them off as shown by the captions for the illustrations in this paper.

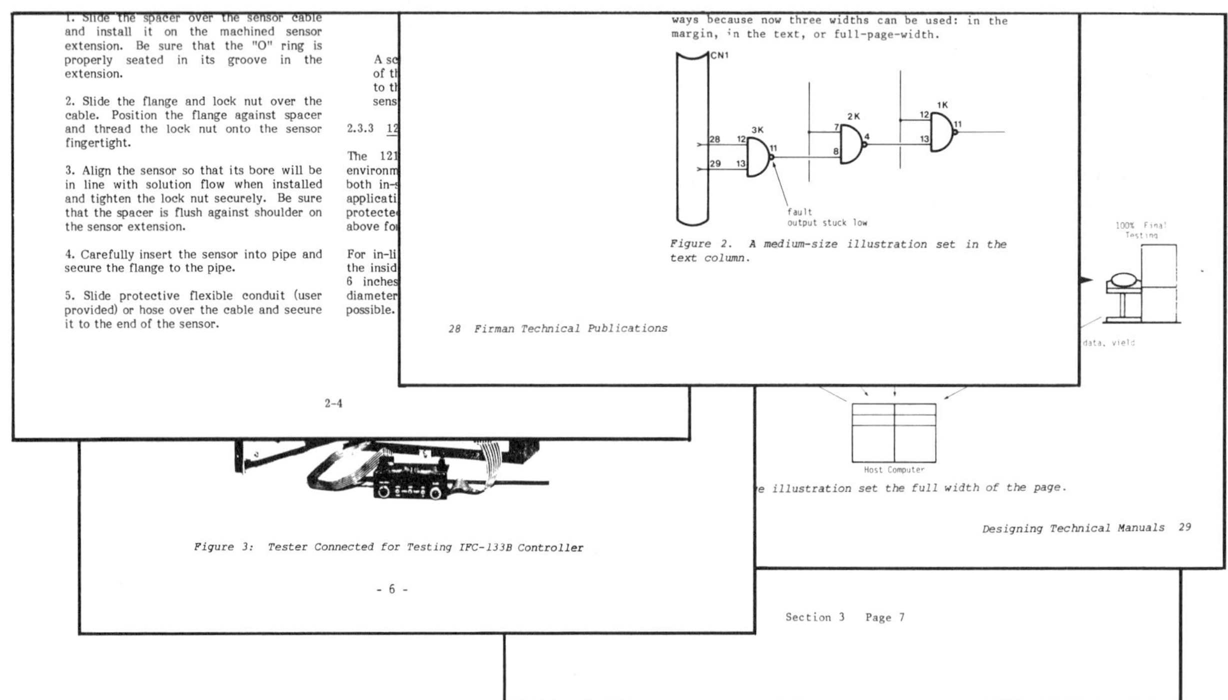

Figure 3. Firman Publications' Wang WP system can be programmed to insert headers, footers, and page numbers automatically, as shown by these examples.

COLUMN 1	COLUMN 2	COLUMN 3	COLUMN 4	COLUMN 5
2	3	4	5	
1	2	3	4	TEXT
6	5	4	3	2
88	89	9Ø	91	ABCDE

Figure 4. Tables like this example are common in technical manuals; a table typically consists of regular columns of simple text and/or numbers. You can format a table directly on the WP system, so all the illustrator has to do is rule the lines around it.

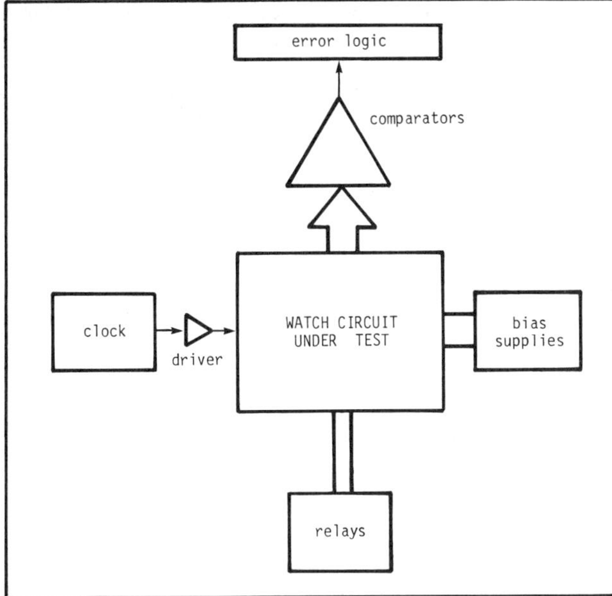

Figure 5. Block diagrams like this example are also common in technical manuals. Simple block diagrams can often be formatted directly on the WP system, so again all the illustrator has to do is rule in the boxes and interconnections.

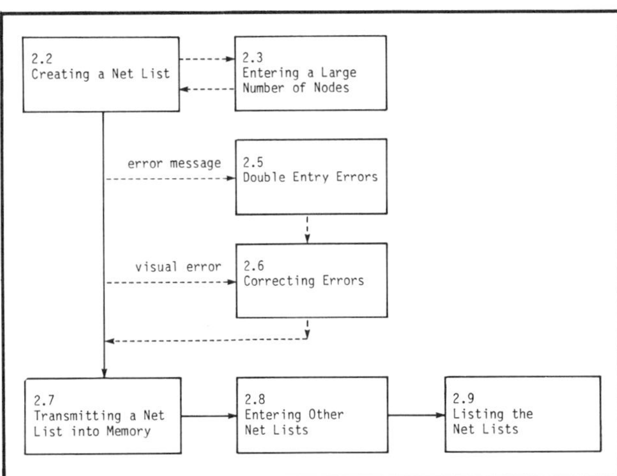

Figure 6. A third common type of diagram in technical manuals is the flow chart. Like block diagrams, flow charts can often be formatted directly on the WP system, so again all the illustrator has to do is rule in the boxes and interconnections.

LAYING-OUT AND PASTING-UP

The use of WP to create a document has a dramatic effect on the work of laying-out and pasting-up. I have already explained at some length the extent to which the writer lays-out the copy as he/she creates the text; and in the preceding paper, Pat Smith has described the process in even more detail. I have also described how WP prints out the final copy in fully formatted pages. It is apparent therefore that my production work does not include "laying-out" in the conventional sense, although I often work with the writer on the original design concept.

When the writer finishes the task of creating a publication, what I get from WP is a stack of pages containing all the body text:
- formatted as required in one or two columns;
- right-hand justified (if required);
- including headers and/or footers; and
- including page numbers.

My job then involves:
- dropping-in figures that have been shot to size;
- dropping-in captions (if they are set in a different typeface from the text);
- cutting-in special type, such as Italicised words, mathematical symbols, and oversized headings.
- cutting-in any last-minute corrections - yes, I'm afraid they are still required occasionally!

This process is illustrated in Figure 7.

The job of pasting-up is quite a simple one therefore - a blessing that most illustrators would be glad of!

OTHER FACTORS

I have indicated that WP provides many of the advantages of typesetting, yet at a much lower cost. There are however certain things that WP cannot do, such as mixing type styles, and providing different sizes of type. These disadvantages are not terribly serious, it turns out, because WP offers some shortcuts to solving them.

One of the many operations of a WP system is copying: with a few keystrokes you can make a copy of a specified string of letters or words, and move that copy to any other location in the document in computer memory. If you wish to Italicise certain words, you can simply set up a page in the back of the document, head it "Italic typeface", and copy all the appropriate words and phrases from the text to that page. When the document is printed out, the computer operator can easily change the typeface before printing-out this page.

In a similar way, all headings can be copied to a separate page at the back of the document: the printout of this page can then be enlarged on a process camera to obtain the oversize headings to be cut-in on the mechanicals.

This principle of copy-to-a-selected-page can also be used to create a table of contents or a list of figures. After the bulk of the work has been done by copying, you can easily adjust the format and add-in the appropriate page numbers to achieve the desired table with minimal effort.

These procedures can be simplied still further by writing a glossary that semi-automatically copies selected text to the selected page at the back of the document: see Ref. 3 for details.

CONCLUSIONS

Producing the mechanicals for a technical publications consists of three major operations:

- setting the copy (typing or typesetting);
- producing the line art; and
- laying-out and pasting-up the copy, the line art, photographs, and the captions; and then dropping-in special items and last-minute corrections.

The amount of production work required in all three operations can be dramatically reduced by the integration of word-processing into the entire creation/production process.

The techniques described in this paper have been in everyday use at Firman Publications for several years now, and they make an important contribution to the improvements in cost-effectiveness that the company has managed to accomplish.

REFERENCES

1) "Managing a Professional-level WP System", A.H.Firman, Proc. 27th ITCC
2) "Making Whoopie: A Professional Writer Learns WP", P.S.Smith, Proc. 27th ITCC
3) "Glossaries: Using the Computer Power of WP", R.B.Bailey, Proc. 27th ITCC

The above three papers are also published in the booklet "Making Whoopie", available from Firman Technical Publications, Inc.

4) "WP as a Tool in Improving the Cost-Effectiveness of Technical Publications", A.H.Firman, Proc. 28th ITCC.
5) "Creating Cost-Effective Publications with the Aid of WP", P.S.Smith, Proc. 28th ITCC.

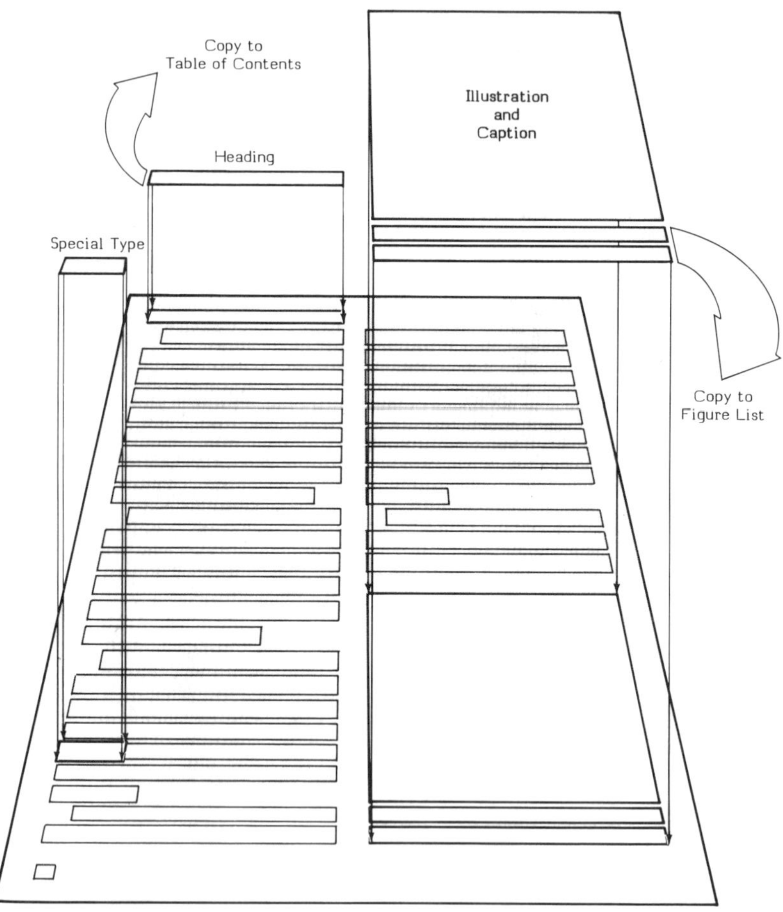

Figure 7. This illustration summarizes my operations in pasting-up a typical manual page: dropping-in figures and captions; cutting-in special type, such as Italicised words, mathematical symbols, and oversized headings; and cutting-in any last-minute corrections. This illustration also indicates how the WP system copies headings and captions to separate pages where they form the basis of tables of contents and figure lists.

OCR -- A SOLUTION OR ANOTHER PROBLEM

August J. Pugliese
Automation Industries, Vitro Laboratories Division

A.J. (Pug) Pugliese
Vitro Laboratories
2345 Statham Blvd.
Oxnard, California 93033

(805) 487-5411

Senior Publications Engineer

Can your office justify capitalization or lease of an $8,000 to $16,000 word processor for a secretary when she types only 20 percent of the time or for a word processing operator when she types redundantly 40 to 60 percent of the time? Most can't! Optical character readers (OCRs) offer an opportunity to solve this problem by placing an input terminal (a $900 typewriter) at every user work station so that they may instantaneously access word and data processing power.

Today, the only limitations to obtaining an OCR are cost and volume of work necessary to justify such a capital expenditure. However, there are problems to be expected and faced before and after installation. If there are problems, why then should you be interested in an OCR and its potential in solving word processing scheduling, turnaround, and workflow problems? To formulate answers to these and other questions, let's address the OCR, its solutions to production problems, and problems to be expected with its implementation. You must decide, as we have, whether it's a solution for you or just another problem. The success and acceptance of an OCR within your office, as well as the success of your word processing (WP) center, may well hang in the balance.

WHAT ARE OPTICAL CHARACTER READERS?

Optical character readers are machines that automatically read and electronically convert written or printed data into machine readable codes for computer or word processing. This conversion is achieved at high speeds and with greater accuracy than methods requiring human participation. Optical character readers are generally being considered to be the best opportunity for saving time and money in an organization requiring fast throughput and rapid turnaround.

OCRs have been available long before today's word processing technology, but past limitations and costs did little to encourage their use. One major problem, now overcome, was the limited number of readable fonts available for scanning. State-of-the-art omni-font reading scanners (like a Kurzweil at ≈ $100,000*) and multiple font reading scanners (such as a CompuScan Alphaword OCR) have now overcome this problem. The Kurzweil scans and recognizes ordinary print in any type or combination of fonts, and in a wide variety of sizes; it converts uniformly or proportionally spaced text to computer compatible form. Affordable machines (like CompuScan's Alphaword, Hendricks, ERM, etc) in the price range of $12,000 to $32,000* or with lease options ranging from $800 to $1,200* per month have also overcome these and other limitations. Along with these lower costs have come the following technological advances:

- Reading or typing errors can be corrected at the OCR and accuracy is up to 99 percent (1 error in 100,000 characters is possible, but up to 3 or more per page are to be expected while users, their secretarial staff, and OCR operators learn basic input procedures)

- Typed text can be read in single or double line spacing. Double spacing is recommended because it reduces errors, increases reading speed, and lowers acquisition costs

- Scanners are software or header sheet programmable to fit desired applications, thereby allowing most tabular and statistical data to be read

- Interfaces are made possible with other word processing and data processing (DP) work stations

*Prices shown are averages compiled from trade publications and 1980 trade show information; they will vary today based on features selected and current vendor price schedules.

● Message transmission via an OCR is made more cost effective than by FAX, Telex, or TWX.

WHY THEN, AN OCR?

When data processing and, later, word processing first began to make an impact on how businesses operated, data processors and word processors were thought of as mysterious, magical machines. The computer room was an isolated entity with specialists, technicians, and reels of magnetic tape and line printers churning out endless ribbons of continuous form paper. The WP center (also an isolated area, with a centralized group of editors, machines, and machine operators) was the domain of specialists of another sort. DP and WP integration was not considered practical, nor feasible. With optical character recognition machines, this integration is now practical.

OCRs have tremendous strength in forcing integration of the total office workforce and in improving productivity of the total office staff. OCRs integrate organizations and improve productivity by eliminating redundant typing , by maintaining user control over initial draft typing, by allowing more efficient use of WP and DP machines as text revision machines, by reducing capital expenditures for more WP machines, and by lowering production costs.

How does this occur? The answer is that OCRs provide the missing link between office secretaries and shared logic, shared media, and distributed WP systems. OCRs link common office typewriters electronically and cost effectively through scanned copy transmitted via hard wires, telephone lines, or over communication networks to WP systems. OCRs make the office network a reality and, because of this, you can look beyond today's support requirements and address the office of tomorrow concept being developed by equipment vendors.

OCRs--THE WORKFLOW

In engineering and technical publication organizations, production needs vary by size of the organization and functions to be performed, but certain production functions are common to all organizations: documents must be generated, filed, retrieved, and communicated; the staff and machines must be efficient and effective. When working with scanner copy, users type input copy using common typing elements (OCR-B, OCR-A, PE, Courier, or others designated for a particular OCR). The author then edits directly on the typed copy by crossing out text to be deleted with a black felt-tip marker pen (the scanner stops at the black) and by writing insertions with a red felt-tip pen (the OCR is blind to red). Marks in black pen are automatically deleted and those in red are ignored. Once scanned, the copy is edited in minutes at a WP system to match format specifications or changes marked in red.

OCRs--ARE THEY THE WP SOLUTION?

The greatest advantage obtained from an OCR is that it helps eliminate the need for high skilled WP operators to keyboard previously typed or recorded documents. Instead, office secretaries or typists become the input keyboarders to capitalize on their familiarity with the needs of the originator and to improve the overall production cycle by eliminating redundant input keyboarding at a costly WP system. This can be proven in your office (as it was in ours) by time-and-motion studies that will establish WP operators spend 50 to 60 percent of their time preparing original text and their remaining time revising these documents and performing normal clerical duties. Using an OCR for input to a word processor for editing and output, can overcome this inefficient approach to document production by maximizing productivity of the input keyboarding and output editing operations.

Before arriving at a decision, consider the following production statistics for double spaced, final copy: typewriter -- 13 pages per day (25 to 40 pages per day rough draft); WP (input and editing) -- 42 pages per day; WP (editing only) --105 pages per day. From these figures, it can be seen that WP production gains up to 60 percent (42 to 105 pages per day) can be obtained by limiting center activity to editing only. These figures are further supported by OCR vendors who imply that one OCR device can handle the full-time output of six Selectric II typewriters, generate as much final copy as six word processors, and cut production costs by one-half or more. Furthermore, OCRs increase productivity still further because of their reading speeds that vary from 200 to 2,000 characters per second (CPS)--depending on the OCR used, the amount of text, and the readability of the text--and because of their ability to access word processor output printer speeds of 45 to 1,800 CPS. This increase in productivity translates into seconds vice minutes per page.

The preceding statistics were based on vendor inputs and an analysis of our normal workload, which was required before any capital expenditures were authorized for an OCR. In our analysis and justification memorandum, we stated that approximately 50 to 60 percent of the text received in the center had already been typed on project by our secretarial staff, or had been previously typed using old Magnetic Tape Selectric Typewriters (MTSTs) whose magnetic media (tape) was not compatible with that of our newer OS-6s (diskettes and magnetic cards). In an effort to eliminate this duplication of effort, and to retrieve previously recorded data (we had already made the transition from MTSTs to IBM Magnetic Card Selectric II and Office System 6 machines), we recommended an OCR. To date, an OCR at our Oxnard Facility and another at our Silver Spring, Maryland Laboratory, have proven to be successful, so much so that our main Laboratatory has ordered a Kurzweil omni-font reader. With Kurzweil, we can open the door to scanning copy produced with any font, and capturing all keystrokes for document changes or revisions. The productivity gains and business potential cannot be overestimated.

We have proven what OCR vendors state "that 10 typists , typing one page each, can keyboard a 20 page change to a proposal in about 7 minutes. Tests have proven that final proposal pages can be prepared in 22 minutes, a fact arrived at by adding the 5 minutes it takes to set up and scan these 20 pages, 10 minutes to proofread and assemble these pages, and the 7 minutes it takes to type the pages. Of course, you recognize this and others tests recommended by OCR vendors are conducted under ideal conditions. What of the everyday tasks where this same change, using WP equipment alone, would normally take, at least, 70 minutes. Such tests resulted in the same overall time expenditures, but production time and difficult scheduling problems are averted. A few salary dollars are also saved in the process, since the original typing can be accomplished using lower-priced secretarial talent. Such tests are proof enough that capital expenditures can be lowered though OCR and that this technology can create a more competitive posture for an office if properly utilized.

Another indication of OCR power in an actual work situation occurred at our Oxnard Facility when two large proposal efforts required 400 pages (200 pages in one proposal and 150 in another) to be typed, edited, and revised in two days. Input text was prepared at various locations within the facility using available typists and text originators as proofreaders. We turned around these 400 pages in two days using two WP operators and one editor proposal coordinator without unduly affecting other scheduled tasks. Because of such successes, we are gaining more user acceptance of the OCR and WP interface, and are expanding OCR use into more technical manual production programs.

When an office becomes more fully oriented to using an OCR; that is, when all users accept the concept and use available OCR readable fonts more readily, and their typists follow simple instructions provided, an office can save literally thousands of keystokes (and hundreds of dollars) daily, and vastly increase production without major capital expenditures for new WP systems. (By the way, we solved the font problem simply by directing our Purchasing Office to replace broken PE fonts or new PE font requirements with OCR readable PE "legal" fonts.) Now, all our office Selectric II typewriters are equipped with OCR readable fonts. We are still selling the concept, but slow, steady progress can only be achieved through deeds--not rhetoric.

OCRs--WHAT ARE THE PROBLEMS?

OCR, WP, and DP systems address identical problems, but from different perspectives; their solutions are not clear-cut, but their problems can be identified as follows:

- The captured text (OCR output) problem
- The bottleneck problem
- The user resistance problem
- The organization (centralization vs decentralization) problem
- The justification problem

THE CAPTURED TEXT (OCR OUTPUT) PROBLEM. One aspect of an OCR that seems to receive little attention is what to do with all its captured text. With an OCR, more work is processed through a WP system. Now, the problem becomes what to do when you don't have the number of skilled operators or WP systems to process this captured text. The obvious answer is that all keystrokes need not be captured. Other answers (if you can't stop or control the flow) are more effective planning, scheduling, or design of a new word processing center system network. These are approaches we, at Vitro Laboratories, have taken. We are still designing the ultimate answer by looking into a state-of-the-art network involving new WP systems, a DEC PDP 11/34 computer, Radio Shack minicomputer terminals, and our IBM computer mainframe.

Another problem created by OCR document creation is printer throughput scheduling, which now (because an OCR can tie up a printer) receives all the extra work along with that of the output of other WP systems without printers. We chose to ignore the problem because of our workflow and availability of other printers when needed. Small centers cannot ignore or solve the problem so easily.

THE BOTTLENECK PROBLEM. The bottleneck (or workflow) problem is well-recognized by all in word processing and is a common experience faced by most WP managers. Solutions are offered that range from expanding the staff and machine count; procuring new, better, and faster word processing equipment; reorganizing for better production and management; or replacing the staff and starting over. These recommendations are, many times, complete with productivity figures, cost analyses, and implementation and system design phase-in plans. The impact of such solutions have been, nevertheless, subject to a continuation of the same solutions--calls for increased production, new equipment, and new personnel. None of these solutions can solve the problem alone.

To overcome the bottleneck problem, CompuScan, the manufacturer of our Alphaword OCR, points out that adding word processors is a costly way to break the resultant workflow backup. Instead, teaming an OCR with the word processing center triples the productive capacity of any system configuration. This is particulary true in large centers, where one OCR unit can accept the output of 20 or more project typists on a daily basis. Again, how does one handle this potential bottleneck without knowing the productivity of your current machines and personnel? All too often, increases in staffing and WP system procurements are presented as answers before production figures are collected.

THE USER RESISTANCE PROBLEM. Many times, user resistance is traceable to miscommunicated equipment capabilities; fear of building a support organization rivaling their own and costing as much, or more, to maintain; or little or no productivity gains once new equipment is installed and running. Because of miscommunicated equipment capabilities and these fears, users resist centralized WP centers; instead, they build their own centralized areas on a smaller scale and term them decentralization. They do this in much the same manner and with confidence they can do a better job. Without concentrating on the publication production problems alone, they soon recognize they have inherited the same problems and have not achieved any better results. So what is the answer? An OCR can help, but only help in resolving or reducing conflicts caused by user and WP center misunderstandings of priorities, organizational roles, and responsibilities.

OCRs offer a solution to the production, training, and turnaround problems without disrupting organizations or physical structures, or without requiring changes that cause personnel and user resistance to its use. Possible solutions to these office problems can now be addressed in the user organizations instead of, as in the past, in the WP center only. With an OCR, it is possible to join user and WP organization forces and still maintain administrative control (in the WP center) over the WP machine and personnel acquisitions, thereby standardizing the product and improving its quality, without adding WP systems that will not be cost-effective if used part time to handle individual user organization needs.

Another source of user resistance is that many word processing managers focus their organization on more efficient delivery of all office tasks, while ignoring individual user tasks that are equally important, such as instantaneous response to typing support needs, administrative support, personal interactions, and budgetary restraints. Typing pools, equipped with only Selectric II typewriters, and specifically devoted to particular needs of a given user organization, are the answer.

In summary, let me state my opinion; the user resistance problem may well boil down to favoring one-on-one relationships regardless of costs, downtime, etc. To the user, visibility and personal contact on a day-to-day basis are number one priorities and bring about the confidence that their support money is being wisely or productively spent; something not always believed when presented by managers of a WP center. An OCR can support this need without eliminating the benefits to be gained from a centralized word processing center.

THE ORGANIZATION, CENTRALIZATION VERSUS DECENTRALIZATION PROBLEM. The organization problems must be addressed. Who will control the organization? Who will operate the OCR? Where will it and the WP systems be housed? Experience has proven that one centralized location for the OCR and a WP system to which it is interfaced is best. What to do about the other WP systems, and the suggested typing pool, will also require decisive action. Whatever is decided, DO CONNECT the OCR to a dual-media word processor so that the magnetic media can be duplicated and the work distributed to several WP systems for editing. This distribution problem may well be solved by WP networks now being promoted in the industry.

OCRs bring about formation of work groups or tiger-team clusters to serve the short-term needs of user organizations. They allow word processing machines to be installed in user organizations, thereby, removing much user resistance to WP growth; to be connected to a computer and word processing network; to be tied together to shared peripherals and local media storage devices; or to be connected to a local or distant data bases.

An OCR works best with a project matrix management structure because a one-time tiger team gets the task accomplished. Accuracy of the text is increased, as the original text is verified at its source. If handled properly, user language is effectively transmitted as the "user intended", not as someone interpreted it. It is also expeditiously transmitted from the user source in the format intended. The secretary preparing the original copy becomes more productive, as only one-time typing is required, and all changes are taken care of at a WP system. Pride in their work, and feelings of contributing to the final product are retained by all. However, what can be done for the WP operator who now feels stuck with changes only? Another problem to solve, but one that can be solved by upgrading WP operator job classifications to editor (like the machine they operate) to capitalize on their production editor and proofreading skills.

Once a project matrix management structure is implemented, word originators are in a supporting role in developing the final document; they can be trained to be writers or editors, and to make extensive changes to a draft in a productive fashion. However, a word of caution. Approach this solution with a proper training program or with a strong editorial support staff.

Whatever the problems, there are still tremendous sociological advantages to be gained in using a project matrix management structure for a documentation program. Impact is made at the individual office level and in the WP center. Office secretaries and typists develop typing that forms the basis for the final document; they have a feeling of contributing to the final product. The WP operator, free from the keyboarding task, immediately begins the job of document assembly. All typists become part of the team and work flow. More effective interaction occurs among word originators, their secretaries or typists, and the WP center.

Finally, five other potential problem areas must be considered before selecting an OCR, regardless of whether the decision concerns a centralized or decentralized organization:

- Trade-offs have to be considered: in the area of centralized versus decentralized authority, cost versus performance, telecommunications versus computing or WP costs, and opportunities versus risks. Each organization's environment and business problems (such as overhead restraints) must weigh heavily on the decision process.

- Availability of information at a WP system on a timely basis: the reliability of the selected OCR and its interfacing WP system is an important factor when availability is being considered.

- The potential of the OCR to grow: additional applications and peripheral interfaces are important considerations.

- The provisions for security: this is a major consideration due to the confidential or discrete nature of much of the information to be processed. Again, individual network-based terminals at user work stations will solve such problems.

- Vendor attitude, maintenance policy, and development capabilities for future technologies: the vendor should be prepared to demonstrate their ability to keep abreast of your publication applications.

Perhaps, when all is said and done, cooperation and coordination are the best means for solving the organization, centralization versus decentralization problem. Users and WP personnel alike must recognize and accept that many corporate resources are needed to properly amalgamate people, systems, and users and to make them successful operations: cooperation of all organizations are certainly a requirement; it can only be achieved through day-to-day communication.

THE JUSTIFICATION PROBLEM. As with any office equipment, OCRs must be economically justified. The same analysis techniques used to determine WP needs must be used to support an OCR. A common method is to substitute one for two (or more) based on the analysis of the amount of redundant typing being performed throughout the office. At today's costs, OCRs and WP units are similar in lease and purchase costs. This presents no problem, since one OCR can replace the need for two or more additional WP systems with Selectric II typewriters.

The key to justification is document volume. You need a sufficient volume of documents so that it is less expensive to process them on an OCR than it is to provide the keyboard units and operators to key the data.

Volume by itself, however, is not the complete story. It must be volume you can control. You must be able to control forms design; the printing or typing devices that prepare the copy; and the conditions under which the copy is prepared, such as, the frequency of cleaning the machines, the type of ribbons and fonts used throughout the organization, and the frequency at which the ribbons are replaced. Such control is not difficult in a centralized WP center or where publication drafts are prepared in one location. But, what of the case where documents are prepared by typewriters scattered throughout a building, around a city, or across the country? Control can be extremely difficult. If faced with such a problem, reconcile yourself to the fact that it will take time to educate the masses and to gain the control necessary. Backing at the highest possible management level won't hurt.

How much volume is necessary to justify the effort to seriously look at an OCR? If you have over 5,000 pages generated per day thoughout your office, a low-cost OCR is worth looking into; if 10,000 or more, a Kurzweil or equivalent machine bears investigation. These are general guidelines, of course, and your situation could be an exception.

OCR--ARE THERE BENEFITS?

YES! The benefits of an OCR as an entry system may be summarized as follows:

- The operator, who routinely spends 50 to 60 percent of the time keying in text and the remainder revising documents, can now spend 100 percent of the time revising and editing. (This can open an editorial career path in some organizations.) Every station becomes more productive, so that a four terminal word processing system becomes a 16-terminal system with the addition of OCR.

- Speed is increased vitally. Average operator speed is 60 WPM or 6 CPS, whereas certain OCRs read 200 CPS to 2,000 characters per second. The OCR can accept 25 to 300 pages per hour automatically (one-half that figure is normal), while operating unattended. In comparison, good typists can develop 20 double spaced draft pages per day. Even these figures can be improved if the typists are encouraged to type documents at rough draft speed. Then, after the draft is typed, the typist or originator can review the draft for format, spelling of technical terms, and accurate statistical information. DO LET the WP operators perform the editing (correction) functions only.

- Equipment capital expenditures can be reduced. Every existing typewriter becomes a "communicating" terminal without the expense of communication networks or telecommunications devices. The number of word processors can be reduced as the owned Selectric IIs can produce the input for scanning. Higher productivity at the Selectric IIs is also realized. This efficiency allows management to control staffing levels by eliminating redundant efforts. This, of course, may take the form of reduction of support personnel, or can mean increased support needs or capabilities from the current staff.

- The proven method of one-time keyboarding can be achieved. Errors normally associated with retyping are virtually eliminated as draft copy is prepared once and proofread at the office of its source. Only corrections are retyped, but at a WP system designed for the purpose. The OCR allows management to retain, in large part, the current organization with a smaller WP center or to limit WP center growth to only that necessary to support input editing and processing requirements.

- Typing projects can be subdivided down to page-for-page increments if necessary. OCR allows use of 20 typists for 20 pages. Users are provided with a viable solution to peak and valley conditions. Instead of hiring or training highly skilled and highly paid operators, excellent typists can be developed from lower payed or temporary typists. Overtime requirements are reduced or eliminated.

- The WP operator training dilemma can be somewhat reduced. (I've heard of no available WP training program, or device, that can eliminate this problem.) OCR operator training occurs in-house and in one-half day. Typists providing inputs to the OCR can be furnished with "how to" instructions.

- Nearly error-free repro copy is produced. The CompuScan OCR allows operators to make editorial changes during scanning. This feature provides nearly the same editorial capabilities as those available at a word processor; i.e., normal editorial changes such as deletion of characters, words, paragraphs are possible along with limited global search capabilities.

- Media conversion is practical. OCRs can be used to read text or data stored on magnetic tape, diskettes, or magnetic cards and duplicate these data on a new media of another WP system. Conversions done internally save time and money and help to preserve discrete or classified data. This conversion capability also provides a dual backup media should copy be lost, and can serve as a central respository of documents, thereby, eliminating file copies in the originating offices.

- Electronic message transmission or electronic mail becomes a reality. Transmission speeds and copy quality far exceed those of FAX machines. On one OCR, this operation is controlled through header sheets, which automatically dial up the desired number and transmits the text. The ability to communicate offline is readily available and is highly accurate. The digitizing of text, with positive verification, allows the scanner to accept masses of data for high-speed transfer to other systems. However, all is not roses. Some problems exist with underlining, centering, and tabular data. Some OCRs can be programmed to ignore these problems and continue scanning.

IS AN OCR RIGHT FOR YOU?

It depends! If you are just getting into word processing, definitely not! If, on the other hand, you are expanding or cannot afford further expansion, the answer is yes. But, first, look to your future needs and expected volume of work--not just to today's problems before deciding an OCR is the answer. Just remember, as with any modern technology, the introduction of an OCR into the office can create varied reactions. The same problems as those involved when you introduced word processing prevail: people, procedures, and equipment capital expenditures.

- The people problem can be handled with communication, training, and exposure to an OCR in operation. Explain the uses of an OCR; provide live demonsrations using actual jobs provided by potential users; and, more importantly, explain how an OCR will benefit the typists, WP operators, users, and management.

- The procedures problem is easily handled by using the same input procedures as those established for the WP systems; inputs are still in rough draft format. Handwritten and cut-and-paste copy are discouraged today, but are an expected reality for the near future I'm told. Today, it is even possible to scan Xerox copy with some OCRs, but is also not recommended unless you can verify that the copy was produced with a Xerox 9200, 9400, or better machine.

- The capital expenditures problem can only be reduced by an OCR or by limiting or stopping growth. The WP systems are there and so are the Selectric II typewriters. If anything, this equipment will be used more for the purpose intended, thereby, becoming more cost effective.

Looking to the future, Vitro Laboratories plans to continue to stay abreast of WP technological advances, to determine their applicability, and to gauge their effect on streamlining our capital equipment procurements. We still believe that productivity improvement is not a short-term, one-shot project; it is a continuing evolutionary, permanent management concern. It is now my belief that such OCR based WP centers as ours can offer a low-cost way to modernize office word processing, accounting, FAX, and computer processing capabilities.

PRODUCTIVITY IMPROVEMENT IN TECHNICAL COMMUNICATION

Dr. Carl F. Rohne
MCAUTO Health Services

Dr. Carl F. Rohne
MCAUTO Health Services
Publications Specialist
5775 Campus Parkway
Hazelwood, MO 63043
314/232-6672.

Senior member STC

Productivity improvement is an issue of major concern in American industry today, including the health care industry served by MCAUTO Health Services. This paper briefly outlines MCAUTO's ambitious productivity programs, and focuses on the role played by a variety of technical communicators in that improvement. Several significant operational enhancements, centering on word processing/computer/phototypesetting interfaces, constitute significant productivity contributions by technical communicators, and are, therefore, discussed in detail.

INTRODUCTION

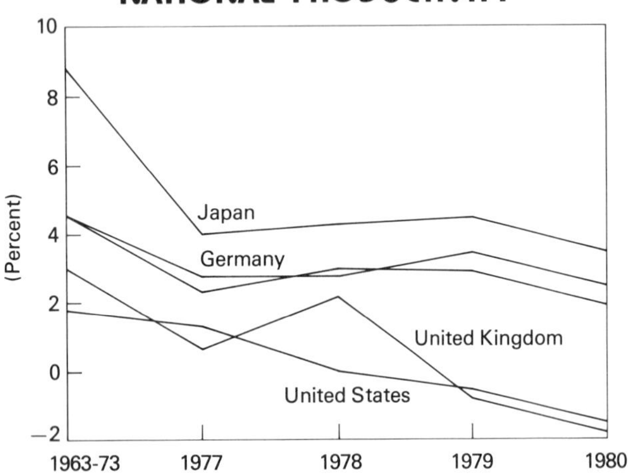

Facing every business and industry in the U.S. today is the issue of IMPROVING PRODUCTIVITY.

At MCAUTO Health Services Division (HSD), we recognize that our clients, the health care industry, are keenly concerned about productivity and quality of service. Health care is a labor-intensive operation, with labor accounting for as much as 60% of the typical hospital's annual budget. Voluntary Effort programs, of course, have been established by the health care industry in an attempt to contain the rising cost of health care in the United States, and to improve the level of patient care per dollar spent. Since our clients are very aware of productivity, we felt that HSD had to make significant efforts to improve its productivity.

A NEW PROGRAM

In 1979, we began a program to establish and develop productivity-oriented programs which would directly benefit not only our own profit line but, more importantly, the quality of service we could provide our clients.

First, we sent members of our corporate staff to seminars held by the American Productivity Center, based in Houston, Texas, and headed by Dr. C. Grayson Jackson. These seminars, and resulting feedback, lead to the establishment of several programs which, in many ways, directly affected technical communicators at MCAUTO.

Health Services first established a productivity council, made up of managers, and chaired by one of our directors who had already had substantial previous experience with productivity improvement programs. Members of the council have developed a series of programs which directly improve the productivity of HSD. These programs revolve around a number of issues, including Supervisory and Management Development programs, Performance Measurement, Quality Circles, a Better Ideas program, a "Most Valuable Performer" program, and others.

Since the Productivity Improvement Program has been in effect, we have experienced significant advances in many areas of productivity and, what makes us feel good, we have been able to pass many of the improvements, especially cost savings, on to our clients, while improving our own bottom line.

OUR TECHNICAL COMMUNICATORS

Technical communicators obviously form a significant bridge within many technical organizations, a point that I have stressed in previous papers for ITCC conventions. With the Productivity Improvement Program we began to involve our technical communicators directly in the process of becoming aware of productivity issues and then of doing something about them.

Through our newly developed "Better Ideas" program, we were offered suggestions by employes (using a special, bright new form), which could directly affect the productivity of our technical communicators. Through the Most Valuable Player and other recognition programs, we rewarded those suggestions. The synergism had begun!

Many of our technical communicators have become so enthused with the ongoing Productivity Improvement Program that they have contributed ideas far outside the normal bounds of technical communication. To this degree, they are developing a view of the company which is less parochial, and less centered purely on technical communication — a view which is far more aware of the profit and loss statement by which we have to exist and by which our jobs are guaranteed. Ideas have ranged from improvements in administrative procedure, to ideas which have been of major consequence to communication.

A NEW INTERFACE

One of the exciting ideas we received was a suggested interface between the Steno Services word processing system in administration, and the main computers located in our Data Center, just two floors below.

Why such an interface? Simply put — to save endless retyping of reports to achieve camera-ready originals, as well as to ensure currency of samples and the ability to rapidly update and change report samples.

In our environment, we use a number of output reports which are produced by our computers for our clients. As communicators, we use them to illustrate (we hope graphically) how our systems work.

In tying the computers to our word processing system, we had to develop an interface which was far more complex than we had thought at the beginning, but which has now given really excellent results. The interface would, for the first time, establish direct operating ties between the Steno Services Redactor word processors and the output generating system of the big IBMs we run. We had to develop new operating procedures, programs to allow those procedures to reach the reports, coordination processes to find the required reports which were on file that day, and methods of protecting client confidentiality (critical in the health care field). We also had to develop a careful list of which clients used which of our systems most meaningfully. Those were the reports we wanted! Perhaps most importantly, the interface required the close cooperation of many personnel from Steno Services, the Data Center staff, and our technical communication areas.

THE USES

As we began to consider the working interface, we explored the possibilities of how many people within the organization could use it. We were astounded at the response.

Marketing Support wanted live reports for marketing visits, which meant that they wanted to custom tailor the reports to an individual hospital prospect. In other words, could we give them the hospital name, city, etc., as a marketing tool? Of course, since the word processors could easily make those kinds of changes on anything that we managed to pull off the IBM operating system.

Could we provide customized reports for the many proposals — over three hundred each year — that we submit? Again, yes.

Could we provide visuals for both marketing and technical training sessions? Of course, since we have developed some fairly sophisticated slide techniques to expand and expound upon the raw output reports. Also, and perhaps most importantly, could our tech writers get the latest live reports for use in our technical documentation? Naturally, since that was the key to improving the quality of our documentation and the productivity of our documentors. We needed the very latest live reports (modified, of course, to protect patient and client names).

BEGINNING DEVELOPMENT

The process of developing the interface between the computers and the word processing system is an interesting one to analyze.

How could we pull output reports off our computers, put them directly onto the magnetic media of the word processors, and then manipulate those reports without any retyping involved? This could save thousands of hours each year of retyping, since previously we took all output reports and retyped them to achieve camera-ready copy.

To look at this problem in more depth for a moment, it required our technical communicators to begin really understanding how the Data Center, filled with very expensive and very complex IBM computers, operated. In addition, communicators were required to develop close contact with members of the Data Center, with whom they did not normally interface.

Achieving that interface took diplomacy, tact, and the willingness to spend some very long hours trying to learn and understand what the special problems of a 24-hour-a-day, seven-day-a-week Data Center might be. In accomplishing that task, our technical communicators were provided with an in-depth understanding of the Data Center to which they would not have been exposed previously.

Since this interface had never been tried, it was of interest to both stenographic services and computer management within the company, which saw real possibilities in improving marketing and documentation turnaround times. Since we had the support of the Productivity Council, and top management, we proceeded to develop the most sophisticated interface we could within the time constraints facing us.

THE PROCESS

We began with meetings among all involved parties to see what the Data Center and the steno pool actually had. This involved conversations involving such terms as "RS232 porting", "storage capacity", "line speed" (Bauds or bits, take your pick), "line drop" problems, and how we could possibly find what we wanted with some 575 hospitals on line with our Data Center each day.

In fact, finding the reports we desired was one of the biggest headaches we had in the interface, since it involved bringing in new groups of our own experts to explain how we could access our own client data base (and not get sued doing it!)

Finally, we got down to the kind of one-on-one relations which proved marvelously beneficial, long lasting, and ultimately productive. We could all see the productivity benefit in trying to reduce the cost of producing the materials we needed from steno services. One of the most fascinating things we encountered as communicators was that we really did not understand the totality of our own large Data Center's operating structure.

Since the Data Center runs every day all day, interesting things occur at 2 and 3 a.m. that were new to us. For example, we had to learn all about month-end scheduling (vital to our clients), as well as what pieces of equipment were used at what times of the day. For example, we could not schedule a laser printer (an IBM 3800) during certain hours, because it was fully occupied with client output reports.

We also went through the usual purchasing problems, including a number of visits to less-than sympathetic corporate types who wondered why we wanted to spend quite so many thousands of dollars. Alas, to be productive, as to makes sales, you must occasionally spend some money, and that could be hard to convey. But again, technical communication, sometimes written, more often verbal, was the key to convincing sceptical purchasing personnel that we really had the need and that the benefits would be well worthwhile.

With equipment in hand, and a prototype demonstration under our belts, we set about the final development of this interface between the word processors and our output reporting system. This now involved running lines between floors, making connections, finding a spare (read cheap) modem, and beginning a new testing program. At first, we were confounded by the results. What had worked so well in tests proved more difficult to accomplish in actual practice.

When we discovered our new technical problems, a few phone calls found us a team of experts from the word processing company, and within a day we and they had our problems solved. Now the process is relatively simple, involving calling the Data Center, a description of the reports that we want (date, time, type, etc.), and a telephone call to Steno Services that says "here we come".

Steno Services then activates the word processor communication link, absorbs the reports onto cassette and transfers them to diskette for viewing on CRT screens. After making any modifications that are required, steno personnel print the reports clean and ready to go to Graphic Arts for incorporation into any medium that is to be the end product.

The reduction in typing time has been enormous (some 90% per report) and has saved all of our technical communicators a great deal of "thumb twiddling" while they wait for the output reports they need to write manuals, prepare presentations, write proposals, or teach a class with new and up-to-date visuals.

The key point is that our technical communicators have actively involved themselves in an invigorating new program — productivity.

THE IDEA SPREADS

Several other communicational ideas have also reflected the growing awareness of writers about a total company-wide approach to communicational techniques. Since we phototypeset all of our technical manuals and promotional materials, using a centralized service of the Graphic Arts department of another division, we began to look at how we could improve that interface in providing top-quality phototypesetting, while at the same time saving a bundle of keyboarding time (between 80 and 90%, in fact).

Because we use an optical character recognition (OCR) system for most entry to the phototypesetting units, we have developed a variety of techniques to utilize our word processing systems (with CRT displays) to expedite the process. Since OCR units hooked to typesetters need special coding, we have the graphic arts experts select type styles, faces, leading, etc., then code our draft manuscripts, which have been prepared on the word processing units and stored on diskettes. The codes, when read by the OCR units, ultimately tell the phototypesetters all the information just mentioned.

The coded draft is returned to the word processing operators, who enter the codes. First, we did it mechanically, then quickly discovered that with a dual disc drive we could store the codes on one disc and play them against the disc storing the manuscript. Again, a significant productivity improvement!

Now all the operator has to do is display lines of text, press a key, select the appropriate phototypesetting code from the other disc, and combine the two into a line of text which is totally unintelligible to any human, but which the phototypesetting unit thinks is akin to Shakespeare!

The amount of time we are saving in typing odd-looking special codes is with practice, averaging a 95% reduction per job. That makes both the word processing and the graphic arts phototypesetting operators very much more efficient, and represented only a realization of what our equipment and a few hours of training could do for our productivity.

HOW SIMPLE CAN WE MAKE IT?

We also have a new health care product which uses its own minicomputers for many purposes, including creating documentation. The challenge here was to interface the output of the new system's printers with the phototypesetting system in graphic arts. A knowledge of what we were doing corporate-wide was invaluable. By virtue of conversations and on-going dialogue with stenographic management, we learned that Steno Services had just acquired their own OCR scanner, a unit that was different from the one in graphic arts that fed the phototypesetting unit.

Now, how could we interface that unit with the phototypesetting unit without spending vast sums on some exotic electronic bridge? The answer was simple, but involved technical communicators who knew all angles of this interface. We bought a daisy-wheel-type element for the minicomputer printers, a type face which could be read by the newly acquired OCR unit in the steno pool. Magic!!

All we have to do now is print copy from the minicomputer-attached printers, scan it through OCR in the steno pool, and put it on diskette (the OCR scanner can do it automatically), then have our steno pool typists add the codes for phototypesetting which we mentioned earlier. With the codes in place, it is a simple matter to read that document through the phototypesetting scanner, hence into the phototypesetting unit. No operator has to rekey a single word, and most of the work can be done easily and routinely by the clerical staff with little intervention from the writers or the graphic arts designers. Again, we feel that we have used our existing technology in a new and exciting way to achieve enhanced productivity, without adding anything to the cost of our existing operation.

NEW AREAS

Our communicators have also been encouraged to become involved in a variety of new areas outside the written word. A whole series of special presentations, on slides and overhead transparencies, has been produced for our field offices, along with accompanying handouts to guide even novice field personnel through technical presentations.

And, of course, we have begun using video tape extensively.

VIDEO

We recently completed a tape on — you guessed it — productivity and quality circles at MCAUTO Health Services. The tape has several different audiences, which, of course, compounded the problems of its production. First, there was our own internal need to explain the quality circle program to our employes and management. Then, we decided that the tape, reproduced in quantity, could be a most effective tool for our health care clients. Naturally, we also wanted to attract the attention and interest of our own top management within the McDonnell Douglas corporation. And finally, one Saturday, during one of our viewings of the many preliminary versions, a director of programming suggested that it might also serve as an ideal tool for recruiting. Voila — we were now up to four audiences, and I'm sure we'll find more!

We produced the tape using the remarkably able help, guidance, and skills of our own McDonnell Douglas in-house facility, a facility akin to the better local TV stations but lacking an antenna. The video experts gave us a crash course in TV production. Then we started taping. We worked without a script (I wouldn't do it again) because we knew that we basically wanted to focus on existing quality circles at HSD. We filmed the circles live (one meets at midnight, which made for some interesting logistics), then began to edit and add. We included a good deal of voice-over narrative, to emphasize and explain the circle process in action, then reviewed the "draft" tape, took copious notes, went back to the typewriters, added more narrative, and created many new graphics. With the graphics done, our tape editors (some of the best communicators I have ever encountered), put the final version of the tape together.

We had created a vehicle to "sell" productivity, and had, simultaneously, learned more ways of being productive ourselves as technical communicators.

NEW CHALLENGES

These new applications represent some very exciting challenges for our personnel, with new techniques to learn, and new methods of communicating just opening up. All of them are geared to the concept that, to remain competitive, we must become more and more productive. That involves using new media and new resources.

We have found many ways to improve our own productivity without sacrificing quality, our concern for our clients, or our concern for our communicators. It has been a fun process, an exciting process, and a process that has been stimulating to those making these productivity improvements come to life. There are many ways in which ALL communicators can become more aware of the productivity problem and do something about it.

For example, productivity measurement tools need further assessment. How does each of us know if he is being productive, beyond a gut feeling that "I'm doing a good job"? Well, as the American Productivity Center says, "If you can measure it, maybe you can manage it". Techniques for measuring output versus input are a fruitful area for new research by both academic and business communicators.

How aware are technical communication managers of productivity? Again, studies have indicated that managers have a ways to go. What are managers doing to avail themselves of the latest studies on productivity? Are managers attending courses on increasing productivity awareness? Are managers discussing the issues of productivity improvement with their employes? How about employe involvement in suggesting innovative ways to "get the job of communicating done"?

All of these opportunities, and many more, will face technical communicators in the next few years. The tools to increased productivity are there — we communicators have to pick them up and learn to use them!

WORD PROCESSING: PLANNING FOR FUTURE COMMUNICATION

Philip M. Rubens

Rensselaer Polytechnic Institute

Philip M. Rubens
R.P.I.
Assoc. Professor, Dept. of Language, Literature, Communication Troy, NY 12180

More than any other single technological feat, the computer promises to alter our communal conception of the communication process. Texts, libraries, and corporate and educational institutes will either change radically or completely disappear. All of these tasks--including archival and textual activities--can be efficiently managed by electronic information systems. Global networks, instant access and feedback, and the implications of the new areas of development will influence our concepts of progress, change, and the very nature of the communication act. Such prospects place an incredible burden on both industry's adoption and use of electronic information handling and the faculty who train future communicators who will work in this strangely different (and, in many ways, better) communication area.

The most dominant feature of contemporary society is the omnipresence of the computer. Communication, at every level, will be influenced by this technology. In this paper, I would like to explore the place of computerized text creation in its relationship to other similar developments, to the future of communication, and to both industry and education.

AN HISTORICAL OVERVIEW

When I sat down to write this paper, I took pen in hand, prepared an outline, and wrote a rough draft. I like to write on scrap paper (the backs of old memos) with a fine-point, felt tip pen (always black ink). All that, despite two typewriters and access to word processing equipment. The important issue here, however, is not my personal inclinations in writing habits; instead, my habits point out the fact that little has changed in the communication process during the past 5,000 years. I took pen to paper and inscribed a series of standard symbols on its surface with a visible, and somewhat permanent, medium.

Writing, the first major advance in communication storage and "global" transmission, created some specific language behaviors. Alphabets, words, grammatical units, and punctuation came into existence. Some degree of uniformity evolved--regional at first, then national. Writing, however, was still somehow imbued with an element of magic. Though generally created (as texts) by scribes who were merely human recording machines, writing was still the province of the educated and powerful who controlled the content of writing. As a consequence, early texts are written in Latin and are primarily concerned with state affairs, religion, business, or "science." The general public is effectively excluded from participation in such works.

Inadvertently or otherwise these early texts create much of the "technology" that underlies information transfer up to the present. If one is to record events, then some very specific elements must be present: a recording surface (clay tablet, paper, film, magnetic tape, electronic switches); a recording instrument (reed, quill, pen, type, typewriter, negatives, electronic impulses); a recording medium (clay, paper, film, circuits); and a set of appropriate symbols (alphabets, punctuation, math and science notations, binary codes). One wonders how anyone ever thought to put all these elements together. Fortunately, they did.

The second major advance in information handling occurs with the invention of the printing press. Someone sees the mass production process of etchings, adds that to the grape press, and multiple copies of texts become possible. New problems admittedly arise--the mass production of type, inks, and papers; "real" printing problems such as imposition, type handling, paper folding, binding, and the like. But all of these are resolved. In fact, virtually all of this early technology still forms the basis for text handling.

More importantly, however, printing results in a variety of intriguing social changes. Suddenly the text has become an object that can be carried around, stored, retrieved, and in general, read at leisure. No longer does a scientist have to visit or write an authority in a particular discipline to discover the state of the art. Printed texts provide such evidence rapidly and iteratively. New work can move beyond this evidence faster. A concomitant feature of the widespread immediacy of texts is that scientific argument arises. Once observations are detailed in print, they are used and wherever they are found lacking or in error are challenged. Such disagreement advances scientific knowledge.

Another interesting social change is the dissolution of scribal activities (with the exception of secretaries). The author/scientist takes not only direct responsibility for the content and recording of texts but also intervenes actively in the printing process. This becomes especially true of scientists who feel compelled to direct the engraving of specialized graphics--charts, graphs, illustrations, etc.--for their books. Authors also begin to standardize language recording at this time. Paragraphs, sentences, words, punctuation, and the like begin to take form. Print technology makes demands on the language, and the language responds to the demands.

The next major change in information handling is based on the tremendous power and flexibility of electronics. Electronic memory systems of various kinds allow recording and storage of huge amounts of information. A cathode ray tube or similar device supplants paper, and a typewriter keyboard replaces pens (though light pens and electronic scratch pads simulate "pens"). While electronic wizardry (a loaded term) affords new ways to manipulate and generate texts, much of the old print technology still underlies the process: type size and faces remain the same; copy fitting and layout still occur; and printing occurs on presses with ink, type, and paper.

But there are changes here; changes that are at first glance invisible; changes that are far reaching and which influence every level of society. For instance, the library and its repository function are dissolving (much like the profession of scribe did after the introduction of print). Another, more important change, is the sense of immediacy created by electronic text production. Anyone, anywhere on the globe (at least at this writing), who has access to a terminal and a method of receiving-transmitting has the potential to plug into a computer network. The science fiction scenarios of electronic libraries and the McLuhanish global villages are here. With such capabilities the possibility of change and the rapidity of change, the potential for new knowledge, has taken a quantum leap.

THE COMPUTER AND TEXT MANIPULATION

In much the same way as printing, the computer offers a combination of techniques to record, store, and manipulate texts. However, because these techniques are based on electronics, they are potentially invisible in contrast to the physical nature of printed texts. This quality of electronic texts offers some intriguing advantages and problems for both reader and writer.

Recording texts has once again become a closed task; that is, a select community holds the key to the recording process. Much like early text creation, which utilized Latin or other cabalistic languages, the computer's control of texts is accomplished through transparent electronic controls which are not universally understood. While one can argue that these control measures are accessible in public texts, that is not entirely true. Local "dialects," security codes, and specialized control languages deter accessibility. A more likely candidate for creating a Tower of Babel, a plethora of unknown and indecipherable tongues, than the computer has not occurred in some time.

This sense of inaccessibility creates incredible tension for both the public and potential users. There is a general feeling that a class of shamen, of priests has arisen whose task it is to preen the new god. I do not make this claim lightly. Computer consultants and teachers are known for their impatience in dealing with novitiates. Computer symbology also contributes to this feeling of alienation. Circuits are engraved on host-like wafers and split apart to spread the word of the "original." My own school has forwarded the mythos by installing their computing center in a chapel where stained glass saints still bless our programs and the chirp of electronic switches echoes from the vaulted ceiling. When one assesses the sense of awe surrounding computers, there is little wonder that an aura of fear still surrounds them.

Once computers have recorded our data, it must (or can) be stored. This process is only slightly more open than recording data (or texts). Most text handling (in its widest sense) relies on personal catalogs and codes to place information in usable files and to maintain that information in a secure fashion, i.e. free from accidental or intentional destruction, change, or perusal. Both of these areas--cataloging and security--are highly subjective ventures within the constraints of the operating system. For instance, one system may demand that a file be cataloged with three characters (numbers, letters, symbols, blanks, or a combination of these) followed by a specialized symbol (asterisk, ampersand, etc.) followed by an additional character set and supporting code. So your file might be: ******* or PMR*TO*. Such cabalistic scratches both retrieve files and act as security for them. It is little wonder one is confused by computer text handling.

Electronic media, however, have tremendous storage capacity. It is this feature that makes computer text handling so attractive. The storage space for text copy, particularly texts that are subject to occasional or constant revision (instructions, reports, financial data, etc.), is reduced from several (or many) file cabinets to a handful of floppy disks. In addition, this material becomes accessible for editorial change and printing with little lead time. Another aspect of this storage feature is that it alters our conception of library functions and bibliographic/database usage. Once widespread electronic text manipulation becomes common, then libraries will tend to dissolve. Information access will/may occur directly from an electronic device (CRT or the like), and paperless information systems will supplant our current print technology. A concomitant event will be the rise of bibliographic/database systems. Careful key word catalogs of governmental, scientific, business, and academic information will create universal databases accessible from electronic networks. Once again, change and the rate of change will accelerate.

Recording and storing texts and the implications of those elements of electronic information handling will depend on our conceptions and use of the methods for text creation. Basically, we can look at electronic text creation as a combination of media and methods that allows nearly simultaneous text creation, manipulation (change and formatting), recording, and storage. Like handwriting or typing there is an element of immediacy to electronic text creation. Electronics, however, has the added advantage of allowing the writer/editor to change and re-format texts without erasures, cross-outs, interlining, squiggly arrows, or the like. Cut-and-paste can also be done electronically to move entire sections of text and present a "finished" text almost as fast as commands can be typed. Just as printing altered our writing habits, the computer text processor will/can create the impression that editing is possible.

Electronic text creation will also radically change our information technologies. If printed volumes cease to exist, will permanent or semi-permanent texts evolve into micro-forms or electrostatic, xerographic hard copy on home printers (on paper, plastic, film)? Will transmission occur over phone lines, microwaves, television? How will these technologies change? What will a "newspaper" look like? Who will know what Caslon type is? Will it matter? Will our reading habits change? What will "texts" look like? Another change in our views about texts will involve multiple authorship--as in scientific papers and technical reports--and the speed of information transfer. Texts in progress, even globally, will be team efforts which have largely been argued and agreed upon. The background to these texts will remain essentially invisible, perhaps even electronically destroyed, but the end results will represent scientific advance. The same will be true of any kind of collaborative text.

It is also anticipated that much electronic text creation will rely on the powerful computational and analytic abilities of the computer to control texts in a variety of ways. Readability and legibility programs will examine texts throughout the writing process and will offer relevant advice on textual problems. Specialized programs for audience analysis, diction, spelling, word frequency, sentence complexity, and the like will be implemented in virtually all text handling processes. While such programs will offer advice (and on occasion demand compliance as in the case of legal terms), the writer/editor will still be the arbiter of taste and understanding.

Text formatting, a final area of electronic text creation, will also radically alter our perception of texts. Composition and layout, typography, and copy fitting will all be integral steps in the writing process. At present, these steps all rely on extant print technology; that is, descriptions of texts utilize methods in current use for text production. Type faces and sizes, layout, copy-fitting, and the like are handled in the same manner either "manually" or electronically.

Some interesting changes have already occurred in text formatting as a result of electronic capabilities. First, texts have become global. Writing groups in one region can and do transmit texts to other regions. Second, the strong screen orientation of electronic texts has begun to minimize graphics. An editor establishes the validity and accuracy of a page of text on a CRT and immediately turns that text into camera-ready copy. If graphics have not been considered at this stage, then they are left out; they simply complicate text production. What the consequences of this trend will be--particularly for scientific and technical material--no one can guess. A final change concerns the accessibility of type in an electronic system. It is entirely possible and quite easy to use any type in an electronic system. As a consequence, an editor could use a different type for each line in a text. Some examples of this pot-pourri have already emerged; they look deceptively like late nineteenth century magazines. How will the reader react to the demands of such texts? How will these frequent changes in type influence readability? legibility? What about computer type? Electronic text formatting, then, has a number of potential hazards that will demand both investigation and resolution.

CRYSTAL BALL GAZING

Until several months ago, I was convinced that scenarios for electronic text creation were too optimistic. I had visited or spoken to a number of research facilities involved in electronic research, and everyone of them seemed to be on the cutting edge. However, they also all were stalled. Had we gone as far as we could? Yet, as I reflected on what I had seen, it became increasingly apparent that there was always an undercurrent in each facility. There were gaps, cracks, interstices, and silences that I began to fill in. Now I've gone 180 degrees and believe taht the electronic future is nearer than ever. Let me share with you the scenario in its most basic form.

Electronic libraries and databases will (and are) arising with considerable rapidity. Highly specialized, these information systems will provide access to enormous volumes of material, including speculative works-in-progress. This access will create "invisible colleges" which postulate, examine, refute, and generate both new information and disciplines. The needs of these colleges and others will require the assistance of professional information borkers who understand electronics as well as databases. Once these elements are in place, change will take another quantum leap.

Electronic innovation will reinvent the "pen" as an electronic device (many versions now exist) which will have vast textual emendation capabilities. Specific programming languages will allow faster text creation and more efficient handling of elements within the text--graphics, type, etc. Massive textual analysis will indeed occur and exert a tremendous influence over our conception of writing. The entire electronic process will also influence our writing habits and forms; grammatical units will dissolve, and a writing style closer to programming styles will emerge in which modules (sub-routines) become an important element. Finally, the physical characteristics of texts will indeed (and unfortunately for bibliophiles) change. Printed texts will continue to exist but in highly altered form. All in all, our world will be different; it will be smaller, rapidly changing, and dominated by a need for information.

IMPLICATIONS FOR INDUSTRY AND EDUCATION

During the early days of Christian missionary work, one of the church fathers in England wrote back to the Pope complaining that his pagan flock persisted in painting their bodies blue and worshipping trees; he wanted advice. The Pope's terse reply was: "Keep whatever you can use." That advice informs my own personal philosophy of education and, as a consequence, what I see as the importance of electronics in text processing and creation.

For science, business, and industry, the speed of text creation and dissemination has the potential to advance knowledge, save lives, and create profits. It also has the capacity to save resources. Since work styles will change (much work can be done at home), large corporate centers will no longer be required nor will huge transportation systems or facilities. Paper and print facilities as well as traditional information and learning centers will also dissolve. Large editorial staffs will become decentralized as will teachers. Society, in fact, may become rural; large corporate work centers in inner cities will become obsolete as will college campuses. Social changes may also occur; ethnic, racial, sexual, and physical backgrounds will be unimportant in an essentially invisible work force. Working and education can become 24 hour activities, and attention to duties can be predicated on personal inclination or corporate need. Free time as well as social and enrichment activities will enjoy similar freedom.

More immediately, however, we must begin to explore the consequences of such changes for our current educational process. What do we, in training scientific and technical communicators, need to do to prepare our students for such changes? Should we investigate the implications and consequences of such changes in terms of their influence on language? language development? ethics? scientific advance? technology? I wish I had an answer for any one of these questions; I don't. I do know, though, that we need to be aware of these issues, to make our students aware of them, and to prepare these students for a significantly different future in writing.

AUTOMATED COMPOSITION FOR THE SMALL-BUDGET PUBLICATIONS DEPARTMENT

or

Making a Small Equipment Investment Pay Off

by Irwin (Irv) Sachs
Miles-Samuelson, Inc.

Irwin (Irv) Sachs

Vice-President,
Sales and Marketing

Miles-Samuelson, Inc.
15 East 26th Street
New York, NY 10010
(212) 689-4554

More times than I care to remember, we've said, "if only we had that old document in some kind of storage medium, would we be able to produce this one quickly and economically!" We've also discussed the advantages of expanding our capabilities in order to provide services we had not been able to furnish previously. Within the last two years, we've made the move into both phototypesetting and word processing at relatively low cost and with greatly expanded services to provide for our clients. This paper deals with our search for, and ultimate choice of, hardware.

I. THE EARLY YEARS

For almost 30 years, my company, a publications subcontractor with over 100 employees dedicated to producing technical documentation of all kinds, produced the final output copy of our manuals using "cold-composition" technique of either IBM Executive Bold 1 or 2 proportional spacing typewriters or IBM stand-alone composers. When a client insisted on typeset copy output, we subcontracted the material.

While the composer provided a crisp typeface look with a small but suitable range of sizes and faces, there was no acceptable means of storage for updates and preparation of similar manuals. Had IBM chosen to introduce an Electronic Selectric Composer with floppy disk rather than mag card storage, we would probably not have pursued the phototypeset portion of our automated composition system search. Figure 1 shows several of the strike-on or cold-composition faces and their corresponding phototype equivalents. Differences in size should be ignored because the phototypesetter can change the size of an entire document with a single keystroke.

When my company decided it was time to automate our operation, I was tasked with researching "what was out there," and what would be best for us.

As subcontractors, our decision on hardware acquisition was slightly more complicated than yours might be. In addition to choosing systems that would be cost-effective, easy to learn, and readily expandable, we wanted to purchase equipment that would interface well with present and future customer requirements. In addition, we wanted to be able to provide typesetting services for our advertising agency division.

II. PHOTOTYPESETTER

The Requirements

In our type of application, which generally deals with large technical manuals produced in short time frames, it immediately became obvious that we would need at least two input stations, although we could get along with one typesetter unit. In the event of a breakdown, our operators could continue inputting copy to disc storage until the typesetter was repaired.

IBM Executive	ABCDEFGHIJ.7890
Phototype Century Textbook	ABCDEFGHIJ.7890
IBM Composer Univers	ABCDEFGHIJKLM.7890
Phototype Univers	ABCDEFGHIJKLM.7890
IBM Composer Press Roman	ABCDEFGHIJKLM.7890
Phototype English Times	ABCDEFGHIJKLM.7890
IBM Composer Theme	ABCDEFGHIJKLM.7890
Phototype Oracle	ABCDEFGHIJKLM.7890

Note: Various phototypesetting manufacturers have their own names for many equivalent typefaces.

Figure 1. A Look at Some Strike-On Typefaces and Phototype Equivalents

VIP

The creative director in our internal advertising agency division (J.B. Rundle) generally used VIP photocomposition for most of his typeset requirements. The type was purchased in galley form from various typesetters in the area. Because my thinking ran along the lines of helping pay for the equipment by doing a great deal of our own advertising copy, my first reaction was to investigate the VIP hardware. Ultimately, several factors led me away from this equipment. The price was very high and the users I spoke to needed a complete backup system because of high repair incidence.

Itek

The Itek Quadritek was a nicely priced unit but was not geared to cost-effective expansion, being a standalone unit.

Alphatype CRS

I considered the Alphatype phototypesetter driven by the now merged Key Corporation's front end keyboard unit because of the ease with which we could train our typists. The quality of the output was very high, but the single sheet feeding and darkroom requirements were not right for us.

CompEdit

The AM CompEdit has a lot going for it, the best features being the 12-inch paper size suitable for preparing large tabular matter, and the many point sizes available. At the time of our choice, AM did not have an inexpensive off-line terminal available.

Compugraphic Editwriter

The Editwriter advantages were that several potential customers already had this equipment and there were two styles of off-line terminals available: a non-counting model, 1750, and a 2750, which was identical to the basic Editwriter but did not have the typesetting unit.

The Selection

Initially, we purchased the Compugraphic 7700 and 1750 units and very quickly realized that the 1750 was a mistake because the 7700 operator had to manipulate all of the 1750 output copy before typesetting. This nullified much of the advantage of the second unit. Compugraphic allowed us to trade up to the 2750 terminal and the benefits were immediate. The actual setting of type on the 7700 did not prevent our operators from simultaneously inputting new material on both units.

Processor

A cost item that is easily overlooked in a phototypeset system is the processor, which develops the final output product. The multitude of styles and types on the market basically come down to a choice among three families of units:

- Plumbing requirements vs standalone unit
- Darkroom vs daylight unit
- Stabilization vs RC type

We required an RC (resin-coated paper) processor because we needed a permanent, nonfading or yellowing output that our customers could store. Although the RC processor had a higher initial cost than the stabilization (nonpermanent) type, the RC paper cost was lower.

Our daylight unit consideration tied in with the choice of phototypesetter. We did not want to provide the required space for a darkroom.

Although a processor permanently connected to a plumbing system is very cost-effective in the long run, our final choice was based on other factors. While RC standalone processors sold in the range of $4,000 to $6000, we decided to buy a VGC "Pos One" Model CPS 520 daylight camera for about the same price. This camera unit includes a daylight RC processor at no additional cost.

Cost Summary

Model 7700 Editwriter (with options)	$21,000
Model 2750 Terminal	8,500
Processor/Camera	5,000
Total Expense	$34,500

III. WORD PROCESSOR

Selection

The variety of word processors on the market can make the selection a mind-boggling experience. When three or more input terminals are required, a shared-logic computer becomes worthwhile, because each additional terminal is generally priced in the $3000 range. If one or two terminals is sufficient, standalone word processors become the cost-effective choice, as a single line printer can easily support two units. Because most of us require high volume outputs, an alternative choice can be considered (refer to Section IV).

We chose the AMText 425 word processors largely because several of our customers had this system, and we have been well satisfied with this choice. The units are dependable, efficient and easy to learn.

Cost Summary

AMText 425 Terminal (each)	$10,000
Line Printer	5,000

NOTE: One printer can easily support two terminals.

IV. OPTICAL CHARACTER READER (OCR)

The Requirements

Very few people have a requirement for an OCR scanner. What they have is a problem, as we had, in that we were limited in how much automated equipment we could afford to purchase, and we needed much more output than a single word processor could provide.

Scanning The Scanners

I looked at an enormous variety of OCR scanners over a period of a year, carrying with me to each of the demonstrations several pages of copy I had prepared using each manufacturer's recommended typeface. My copy was based on a typical manual that included centered notes and cautions, indented procedural steps, and single- and double-spaced lines of copy.

The Magic Phrase

In speaking to manufacturer's representatives, I learned the magic phrase, "our unit captures the keystrokes". After running copy through each scanner, I learned that some captured quite well, while others filled the screen with question marks or underscores, to alert the operator to a possible error. Some machines questioned correct characters and produced wrong characters without question. I even looked at a scanner that had to be taught to read copy.

Narrowing the Choice

After discarding the units that could not "read" a page the same way twice, I discovered that there was more to this game than capturing the keystrokes. Most of the scanners were "dumb" in that they couldn't recognize an indented line from a flush left line. I visited several of my customers who had operating OCR-word processing systems to see how they handled this problem. Two of them showed me their setups and indicated that they only entered text copy with the scanner. All copy was typed flush left with an OCR element, and after the scanner read the copy and displayed it on the word processor screen, the operator manually edited the indented and centered lines to properly position the copy.

Selection

After deciding that scanners reading OCR-B typeface as a class seemed to be more accurate than others, I decided to rent a Context (Burroughs) low-priced unit ($15,000) for a 30-day trial, with an option to buy and apply the rental charge against the purchase. The scanner read the copy with outstanding accuracy, whether single- or double-spaced, and it correctly read copy that had stripped-in or white-paint corrections. Despite various attempts at a fix, however, we ultimately had to give up on this unit because of spacing errors in tabular and indented copy. After finally trying an ECRM scanner (their low-priced unit marketed by AM International) with our AM word processor, we discovered the perfect marriage. The quality was excellent and the scanner essentially "took a picture" of the copy, preserving all margins and indents. Because the line printer of our word processor zips along at a rapid 55 characters per second or 550 words a minute, the productivity gains from our new system were outstanding.

Cost Summary

ECRM OCR Scanner	$18,000
Communication adapter to AMText 425 word processor	1,500

V. INTERFACE

The Need

Because they were produced by the same manufacturer, we did not require an interface device to adapt our OCR scanner to the word processor. Our phototypesetter, being a Compugraphic rather than a CompEdit (AM International product), was not compatible.

Selection

We contacted Shaffstall Corporation as they were beginning to advertise a Compugraphic Editwriter interface device. They were cooperative in setting up a trial run and we bought their EDI-7700 "black box." This unit is now linked to both our OCR scanner and word processor. (See Figure 2.) The unit produces a phototypeset disc off-line, which is of great benefit as it does not tie up a phototype input device. The drawback on this type of unit is that only keystrokes are captured. With varying size and proportionally spaced characters, a centered and indented line has no meaning in conversion from a word processor to a phototypesetter.

Cost Summary

EDI-7700 Interface Box	$8,000

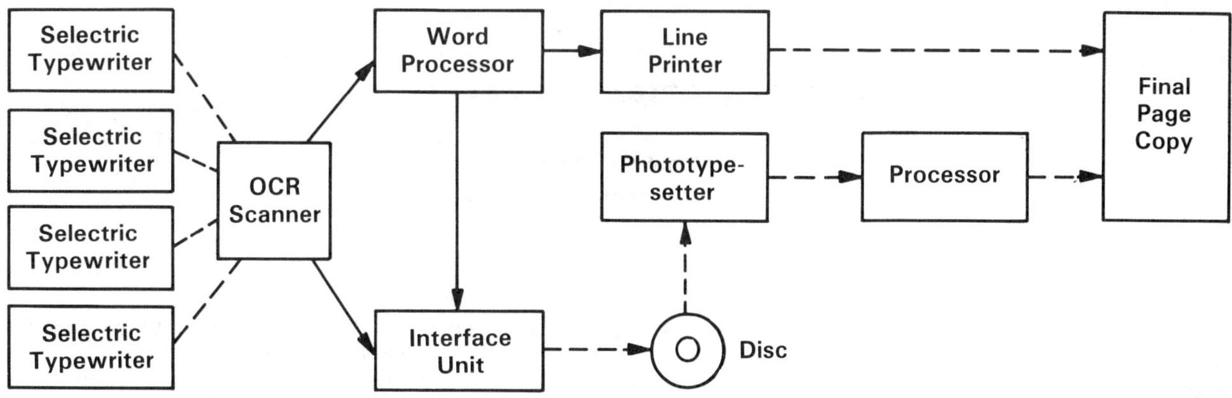

Figure 2. OCR, Word Processor and Phototypesetter Interface Arrangement

VI. WHAT DOES IT ALL MEAN?

Expenses

What we've done is spend about $75,000 for our automation process to date, ($90,000 if we consider our word processor setup in an out-of-town office). We are contemplating some additional telephone interface modems and/or couplers to link our equipment for further expanded capabilities.

The Benefits

Our word processors, OCR scanner, and phototypesetters are operating pretty much on a full-time basis. While we are not running three shifts as many commercial type shops do, we are satisfied that the increased capabilities we now offer to our clients in effectively-priced, readily-available range of typefaces and sizes, efficient preparation of new documents similar to those in permanent disc storage, and ease and efficiency of revising existing documents, have justified our expenditures.

Your Choices

You may be holding back because of rapidly changing technologies. While it is true that more advanced hardware and software is being developed every day, I feel that particularly at the low-priced end of the automated composition picture, a proper selection of equipment today will not be regretted tomorrow. You will, instead look back at your productivity increase and feel justified in the dollars and time spent on this subject.

VII. A FINAL NOTE ON DUPLICATING/ PRINTING SYSTEMS

No discussion of an automated office can be complete without some mention of a copying and/or printing system. While I would guess there's not a single office that doesn't contain at least a single copy duplicator (Xerox, IBM, etc.), the kind of unit I am thinking of is the type that efficiently produces and collates two-sided copies of documents in the range of 20-200 sets.

The methods generally used include printing offset copies using metal plates, offset copies using paper or plastic plates, and Xerographic copies from a high-speed unit such as the Xerox 9400.

For those offices that have small quantity requirements (less than 50,000 copies monthly, for example) or spend less than $1000 per month on outside printing requirements, there would be little reason to consider alternatives.

When it became clear to my company that our 8-1/2 x 11 copy requirements were increasing greatly, we invested in a Xerox 9400 copier, both to produce our own copies and to provide a copying service to outside clients.

We find that our copy quality comes very close to that of an offset page, and where few if any photos and foldout pages are involved, we can demonstrate extremely competitive pricing when compared to commercial print shops. Being able to control our work flow without the need of any outside services has proven of great benefit to us. Whether a Xerox type system, paper-plate or full-blown photo-offset setup is right for you is based on individual needs, and is too complex a problem to explore here; it is my intent only to indicate the range of equipment available, and that can be selected to complete your "office of the future", today.

DOCUMENTATION DATA BASES

Geoffrey James Sickler

Honeywell Information Systems, Inc.

Geoffrey James Sickler

Honeywell Information Systems, Inc.
Senior Engineer/Software Writer
5250 W. Century Blvd.
Los Angeles, CA 90045
(213) 649-6870 x324

Many of the communications techniques discussed at this conference will be obsolete in ten years.

The nature of technology in our society is moving away from standardization and mass production. In the computer industry, for example, standardized software is giving way to "unbundled" systems, where the customer selects features and functionalities that fit his individual requirements.[1] Recent breakthroughs in programmable robots will soon make manufacturing one-of-a-kind items as easy as mass production was in the past.[2]

The demand for technical information about new products will never decrease, but now the requirement for pinpoint accuracy is combined with a desire for elasticity of content. Because a customized product cannot be accurately described by a standardized document, the technology and medium through which technical communication takes place must adapt to this new environment.

I recently contacted a major software development firm whose technical writers were still writing drafts in longhand and producing documents on manual typewriters. The situation was typical of the computer industry...new releases every few months with a never-ending demand for updated documentation. In an attempt to keep up, the department installed a word-processing system. They discovered, much to their surprise, that the new system was only marginally quicker than the old method. They remained the whipping boys when problems occurred at customer sites from simple lack of information.

This failure of word processing to significantly increase the response time for new documentation was a result of the environment moving too fast for conventional communications methodology. The product releases were too closely spaced, and because of the competitiveness in the high technology arena, management could not afford to delay releases to wait for documentation. It was a simple case of the message outgrowing the medium.

Whether documents are created and maintained on word processing systems or by hand-typing, certain time factors remain constant. The document must be written, reviewed, printed, and distributed. After distribution it must be supported, usually via clumsy update pages. The time constraints of publication, distribution, and support are constant, regardless of the speed of the typing procedure. The end product remains an essentially monolithic structure, ideal for a describing a standard, mass-produced item, but highly inappropriate for a customized product that is likely to change significantly in the next few months. (See Figures 1 and 2.)

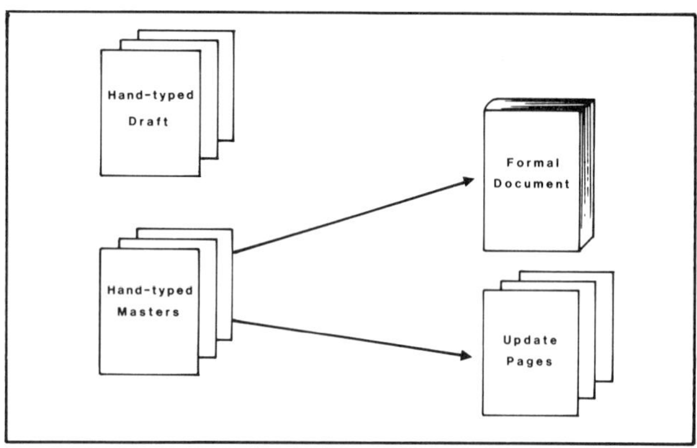

Figure 1. Without Word Processing

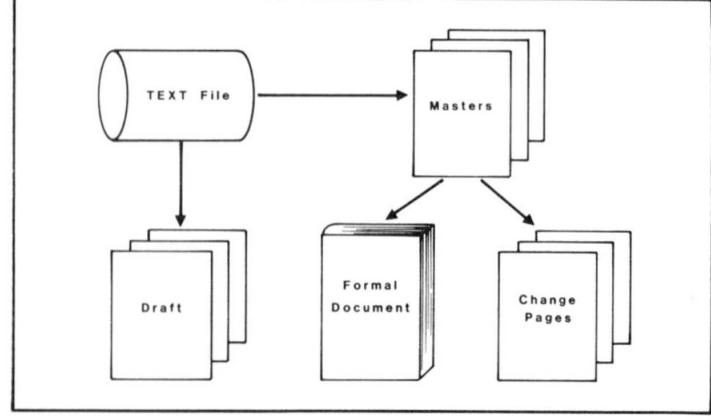

Figure 2. With Word Processing

This snowballing demand for accurate, customized information about an evolving technology makes the model for technical communication -- the published book or journal -- obsolete. A new model must be devised, one as fast and elastic as the high technology that it documents.[3] Until recently such an alternative was simply nonexistent. However, the impact of deflation in the computer industry, as well as improvements in the usability of software, have made data bases available for the storage and retrieval of technical information.

A data base is by its nature fluid and can be dynamically updated as quickly as a product can be modified. Like a published book, a data base is a formalized way of arranging information, but differs in that the structure of that information is not determined at the time of its creation, but at the time of its consumption. The reader, not the writer, decides what he wants to read. Additionally, if there is a specific requirement for published documentation or for sets of customized documentation, the information in the data base can be crystalized into a multitude of different forms and formats -- without requiring reverification of the source material. (See Figure 3.)

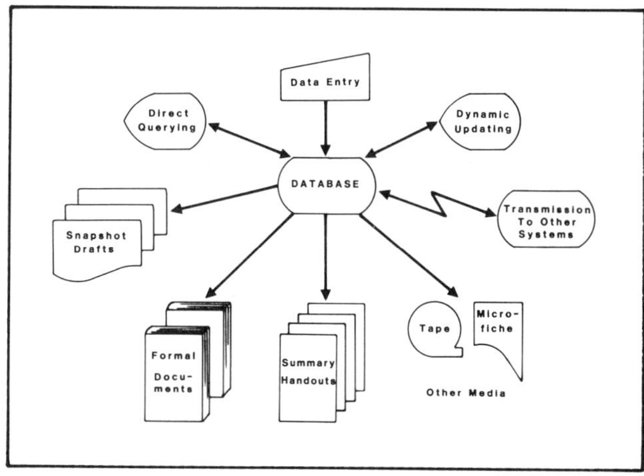

Figure 3. Beyond Word Processing

Databases are not really complex or esoteric.[4] Imagine a set of items that need to be documented. In a book or manual, these might be chapters A, B, and C, with subchapters A.1, A.2, B.1, B.2, etc. In a data base, these items are pictured as a two-dimensional grid. Just as in a published book, a person can read from "cover to cover", perusing each element in a sequential order. (See Figure 4.) However, with a data base, the capability exists to display specific levels of detail without intervening text; this is called associative or parallel access. (See Figure 5.) Further, with a data base, the reader is free to randomly wander from item to item, in effect restructuring the information according to whim. (See Figure 6.)

Thus, rather than a fixed arrangement of contents, as with a manual or a book, a data base contains the potentiality for combining and recombining pieces of available information. A reader could get the same effect by copying an existing book, then cutting and pasting to create a new book...but this would hardly be cost-effective.

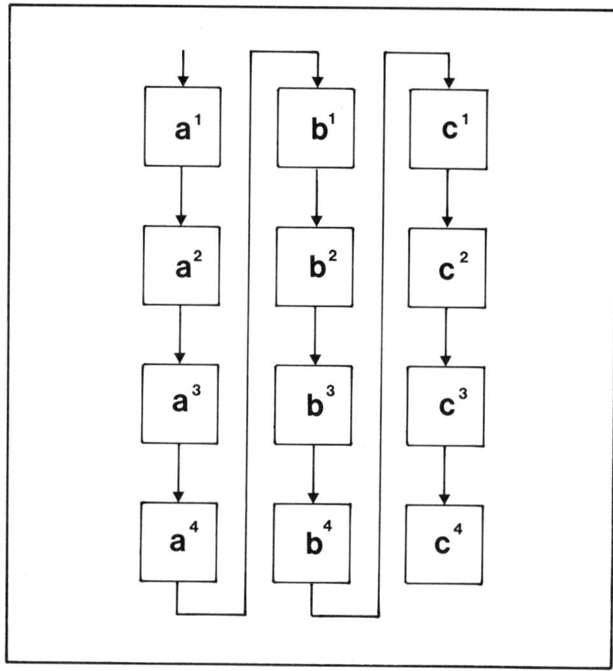

Figure 4. Sequential Access

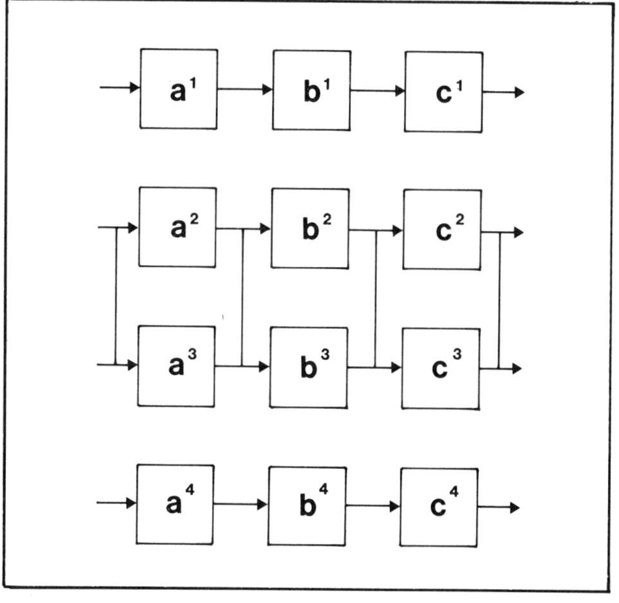

Figure 5. Associative Access

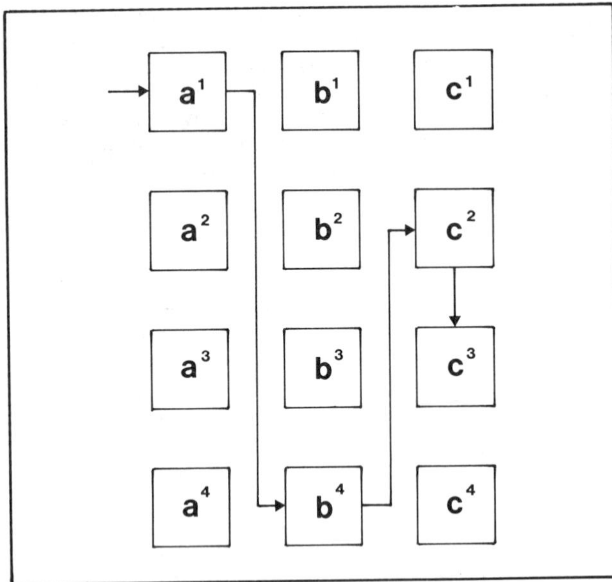

Figure 6. Random Access

Documentation data bases are better explained by specific example. In a typical writer's library are a set of reference books, similarly structured, containing a variety of related information. The bookshelves might sport a Thesaurus,a Webster's, perhaps an Oxford English Dictionary, three encyclopediae (scientific, general, and condensed). Maybe you have a spelling dictionary is in your briefcase. All this information could have been combined into a single volume -- but we'd need a forklift to turn the pages. Yet the sum total would be smaller, because there would be no need for duplication of information, such as correct spelling, pronunciation, etc.

If all this information were stored on a data base, we would be able to retrieve specific items or sets of items on our computer terminals. Using the data base as a documentation tool, we could command the computer to select and print any subset of the total information base, thus automatically compiling dictionaries and encyclopediae of different types and levels of detail. More importantly, we could keep the information current with relative ease. Should a government agency decree that "octopus" shall now be spelled "octopod", we would merely make a single change to the data base, and every subsequent document generated from that data base would have the new accepted spelling.

In the high pressure world of computer documentation, this power and flexibility is essential because major functionality changes so quickly that traditional methods of documentation are swamped.

At Honeywell's Los Angeles Development Center, I designed and implemented a data base for the documentation of the new CP-6 computer operating system. Information is entered into the data base by specially trained production personnel.5 Each item of information, each command, each language element, is stored in a controlled area of the computer's storage. This information is kept in a state of accuracy through an electronic updating interchange system.

From our data base, we create "snapshot" drafts that document the momentary state of development of any one product, subproduct, or set of products. At major release time, we define an appropriate set of documents, and print camera-ready masters. Appendices are generated from the same material as the main body of the text -- nobody has to waste time updating both detailed and summary descriptions. This methodology is an extremely efficient and cost-effective means to produce a large number of documents of different levels of content, in a very small amount of time.

To make the dissemination of information even more speedy, we transfer pieces of the data base directly to our customer's systems. This information is then accessed from an on-line terminal or printed out as hard-copy documents at the customer's site. The customer can modify the data base to reflect any changes or additions specific to the individual computer installation. (See Figure 7.)

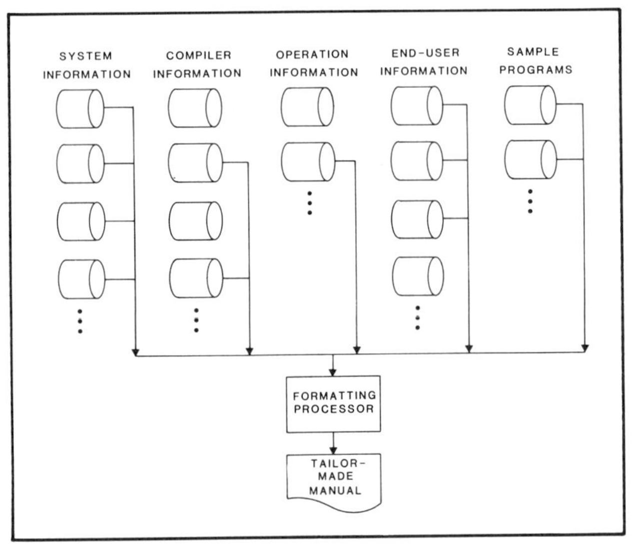

Figure 7. Customer Generation of Tailored Manuals

Since our primary customers are computer programmers, the on-line querying capability has been well received. Why should a programmer be forced to thumb through a technical manual when his terminal is connected to a computer capable of displaying information in seconds? In the past, computer vendors have added their manuals to the system as a set of files that could be printed at the terminal. Yet offering a file of a manual or specification is neither efficient nor useful, because the structure of the documentation remains linear and monolithic. It is actually easier to flip through a hard-copy manual than to wander through pages of terminal output.

This problem does not exist with a well-planned documentation data base. In the case of the CP-6 system, the data base (called a HELP facility) is structured into a hierarchic tree. Items of information are gathered into associated groups, which are summarized by overview messages. At the root are a set of central messages that provide the gateways to various groupings of text. This structure provides a set of natural "paths" through which the documentation can be accessed, yet provides immediate access to any specific item. (See Figure 8.) The HELP facility is interfaced to the error reporting routines, so that the system offers timely advice by printing the correct format of the erroneous entry.

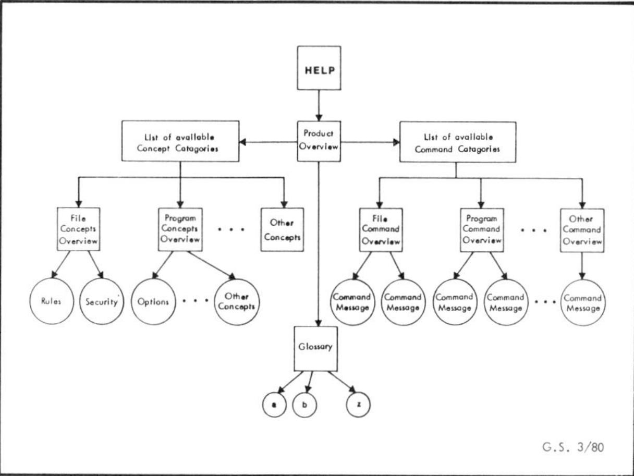

Figure 8. Structure for CP-6 Documentation Data Base

With a complete HELP facility, a programmer works in an environment that is self-contained -- a hard-copy manual becomes a luxury rather than a necessity. More importantly, we can dynamically update the customer's data base to reflect changes in the system as problems are fixed and new features are added.[6]

As a spinoff, we offer our customers the capability to build their own documentation data bases. Because a documentation data base is easy to create, even a non-programmer can build and maintain personal on-line querying systems. For example, our local personnel assistant uses a mini-HELP facility to keep track of everybody's phone extensions and room numbers. Data base storage and retrieval tools are already changing the ways that people communicate with one another.

To summarize, the advantages of a documentation data base over more traditional methods of technical communication are:

- Each item of information is documented in one place; no duplication of effort is required when an item of information is to be included in more than one publication. This also minimizes the effort required to translate the documentation in a different language.

- A variety of different hard-copy documents, at different levels of detail, can be produced from a single data base.

- The data base can be queried from an on-line terminal, presenting immediate information gratification.

- The data base can be maintained in a state of day-to-day or even hour-to-hour accuracy, without cumbersome addenda and change pages.

Some of the disadvantages are:

- The form and scope of the data base must be appropriate to the information it contains; this is much more difficult that merely outlining a technical manual.

- Retraining of both writing and production personnel is required; this is a new methodology and many of the old "that's the way we've always done it" rules are null and void.

- Most importantly, a data base assumes that a computer is available for use by the documentation staff -- not to mention the relatively complex software that makes data base storage and retrieval easy-to-use.

As computers become more prevalent in our society, the data base approach to documentation will become more than a specialized application -- it will become a common or even the accepted means of technical communication. New hardware advances, particularly in the area of color graphics terminals, will make creativity the only limitation to technical communication.

Futurists already predict the "death" of published book.[7] As computer memory grows cheaper and the cost of paper skyrockets, it becomes more cost-effective to leave technical information in soft-copy data bases. The proliferation of microcomputers will hasten this process. Home computers are already being configured with large-scale computers to form information networks, creating a demand for usable information on many different subjects. A documentation data base is the only effective means of storing and retrieving technical information for distribution through a computer network.

The future will bring a demand for greater amounts of information distributed at greater speeds. This is not to say that published documents will disappear completely; but they will be become less important as a primary means of information dissemination. The ability to communicate ideas clearly and effectively will always be much in demand. The growth of a new communication technology will only increase the demand, as a world hungry for information moves into an exciting and complex future.

FOOTNOTES

[1]Toffler, Alvin; The Third Wave; William Morrow, N.Y.; 1980; Chapter 15, "Beyond Mass Production".

[2]See Leopold Froehlich; "Robots to the Rescue"; Datamation; January, 1981, for information about the current state of robotics.

[3]A good overview of various attempts to streamline and improve the documentation process can be found in Improving the Dissemination of Scientific and Technical Information published by the Capital Systems Group of the National Science Foundation.

[4]A not unreadable presentation of elementary data base theory: Alfonso F. Cardenas; Data Base Management Systems; Allyn and Bacon, Boston; 1979.

[5]See "Shared Resources Text Processing" by James C. Carr (published in last year's proceedings) for an overview of our production facilities.

[6]See "Computer Controlled Document Updates" by Linda Socol (published elsewhere in this year's proceedings) for a description of part of our dynamic updating process.

[7]Christopher Evans; The Micro Millenium; Viking, N.Y.; 1979; Chapter 8, "The Death of the Printed Word".

CREATING COST-EFFECTIVE PUBLICATIONS WITH THE AID OF WP

Patricia S. Smith
Firman Technical Publications, Inc.

Patricia S. Smith
Firman Technical Publications, Inc.
Account Executive
95 Church Street
Pembroke, MA 02359
617-826-5148

The integration of word-processing into the creation/production process gives a professional communicator complete control over the format of a publication as well as its content. This control allows a communicator to arrange the formats in the most cost-effective way possible, and also to experiment with new, more effective formats without a major cost penalty.

INTRODUCTION

In the two years since I joined Firman Technical Publications, Inc., I have produced documents in several format types (Figure 1) with the aid of WP:

- single page-width column
- dual column
- mixed column
- text mapped

At Firman Publications we believe that format is a very important part of the publication, that the medium really is the message. We believe that effective organization is the essence of communication. We believe that unless the material is attractively laid out and the reader can find the information he needs, there is no point in publishing, because the reader has other alternatives such as refusing to buy the product on the one hand, or pestering your personnel for information on the other.

We also believe that format has a major effect on costs. We believe that changing format for a publication can be horrendously expensive given traditional methods. We believe that moving material inside an established format can be difficult and expensive given traditional methods. We believe that experimenting with formats to achieve improved communication is prohibitive given traditional methods.

Nevertheless, we are committed to format as an important aspect of communication, not just as a cosmetic additive, but as a fundamental condition of technical communication.

We can *afford* to believe all this because we possess an incredibly powerful computer-based word processing system that allows us both to achieve effective, attractive layouts, and to reduce the costs of changing formats, moving material, and experimenting with formats.

CONFORMING TO OR CREATING FORMATS

When our client has an established format we adapt to it. We can produce manuals in just about any layout, using all the client's established formatting criteria in design.

Often, however, we take on a client whose format has been, up until now, "engineering standard": full page-width, typed, and photocopied. For these clients we design new formats, using a variety of typefaces, to give them a new and distinctive "image", and to improve their customers' ability to locate and use information.

In adapting to an existing format, or in designing a new format, the problems are always the same:

- how to make sure everything fits;
- how to make sure the format works, to cover all the requirements of the information;
- how to change a format after a document has been designed to meet new design criteria;
- how to reorganize a large block of information involving several pages - or a whole section - within an existing format;
- how to rearrange information inside a page.

Given the choice between content and format, most publications departments I know of give first consideration to content, rather than the limitations the format will place upon the content. We on the other hand design a format for a new client, or plan to meet an existing format, before we even look at the content involved.

When we meet with a client to discuss a publication, we take many samples of existing publications with us. Most of these samples are from our own work for other clients. We also take publications samples from other sources if we can use them to give the client new ideas about the format his publication could take. If we can get agreement from the client about the "look" he wants for his publications, we can do a more complete job of planning it.

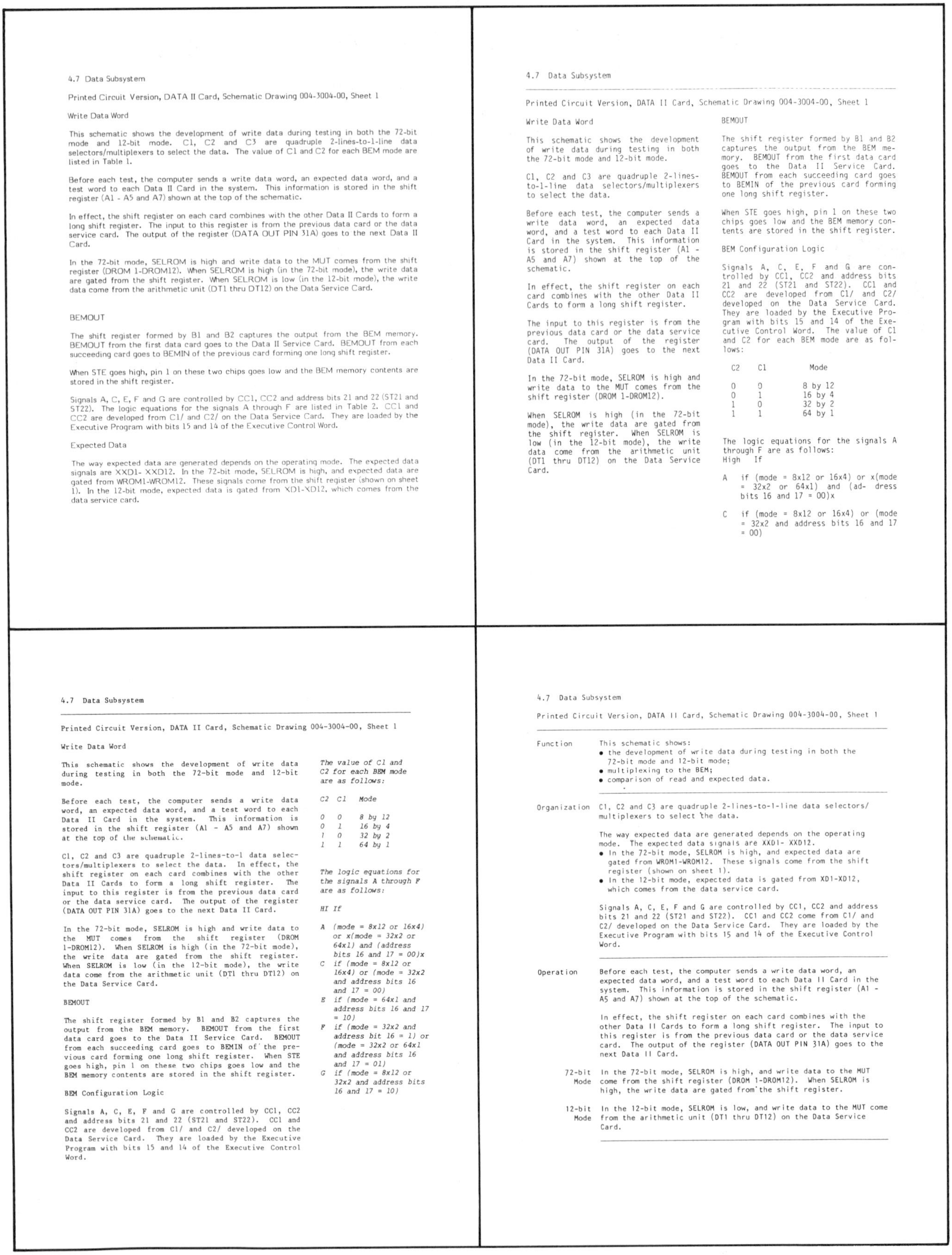

4.7 Data Subsystem

Printed Circuit Version, DATA II Card, Schematic Drawing 004-3004-00, Sheet 1

Write Data Word

This schematic shows the development of write data during testing in both the 72-bit mode and 12-bit mode. C1, C2 and C3 are quadruple 2-lines-to-1-line data selectors/multiplexers to select the data. The value of C1 and C2 for each BEM mode are listed in Table 1.

Before each test, the computer sends a write data word, an expected data word, and a test word to each Data II Card in the system. This information is stored in the shift register (A1 - A5 and A7) shown at the top of the schematic.

In effect, the shift register on each card combines with the other Data II Cards to form a long shift register. The input to this register is from the previous data card or the data service card. The output of the register (DATA OUT PIN 31A) goes to the next Data II Card.

In the 72-bit mode, SELROM is high and write data to the MUT comes from the shift register (DROM 1-DROM12). When SELROM is high (in the 72-bit mode), the write data are gated from the shift register. When SELROM is low (in the 12-bit mode), the write data come from the arithmetic unit (DT1 thru DT12) on the Data Service Card.

BEMOUT

The shift register formed by B1 and B2 captures the output from the BEM memory. BEMOUT from the first data card goes to the Data II Service Card. BEMOUT from each succeeding card goes to BEMIN of the previous card forming one long shift register.

When STE goes high, pin 1 on these two chips goes low and the BEM memory contents are stored in the shift register.

Signals A, C, E, F and G are controlled by CC1, CC2 and address bits 21 and 22 (ST21 and ST22). The logic equations for the signals A through F are listed in Table 2. CC1 and CC2 are developed from C1/ and C2/ on the Data Service Card. They are loaded by the Executive Program with bits 15 and 14 of the Executive Control Word.

Expected Data

The way expected data are generated depends on the operating mode. The expected data signals are XXD1- XXD12. In the 72-bit mode, SELROM is high, and expected data are gated from WROM1-WROM12. These signals come from the shift register (shown on sheet 1). In the 12-bit mode, expected data is gated from XD1-XD12, which comes from the data service card.

4.7 Data Subsystem

Printed Circuit Version, DATA II Card, Schematic Drawing 004-3004-00, Sheet 1

Write Data Word

This schematic shows the development of write data during testing in both the 72-bit mode and 12-bit mode.

C1, C2 and C3 are quadruple 2-lines-to-1-line data selectors/multiplexers to select the data.

Before each test, the computer sends a write data word, an expected data word, and a test word to each Data II Card in the system. This information is stored in the shift register (A1 - A5 and A7) shown at the top of the schematic.

In effect, the shift register on each card combines with the other Data II Cards to form a long shift register.

The input to this register is from the previous data card or the data service card. The output of the register (DATA OUT PIN 31A) goes to the next Data II Card.

In the 72-bit mode, SELROM is high and write data to the MUT comes from the shift register (DROM 1-DROM12).

When SELROM is high (in the 72-bit mode), the write data are gated from the shift register. When SELROM is low (in the 12-bit mode), the write data come from the arithmetic unit (DT1 thru DT12) on the Data Service Card.

BEMOUT

The shift register formed by B1 and B2 captures the output from the BEM memory. BEMOUT from the first data card goes to the Data II Service Card. BEMOUT from each succeeding card goes to BEMIN of the previous card forming one long shift register.

When STE goes high, pin 1 on these two chips goes low and the BEM memory contents are stored in the shift register.

BEM Configuration Logic

Signals A, C, E, F and G are controlled by CC1, CC2 and address bits 21 and 22 (ST21 and ST22). CC1 and CC2 are developed from C1/ and C2/ developed on the Data Service Card. They are loaded by the Executive Program with bits 15 and 14 of the Executive Control Word. The value of C1 and C2 for each BEM mode are as follows:

C2	C1	Mode
0	0	8 by 12
0	1	16 by 4
1	0	32 by 2
1	1	64 by 1

The logic equations for the signals A through F are as follows:

High If

A if (mode = 8x12 or 16x4) or x(mode = 32x2 or 64x1) and (ad- dress bits 16 and 17 = 00)x

C if (mode = 8x12 or 16x4) or (mode = 32x2 and address bits 16 and 17 = 00)

4.7 Data Subsystem

Printed Circuit Version, DATA II Card, Schematic Drawing 004-3004-00, Sheet 1

Write Data Word

This schematic shows the development of write data during testing in both the 72-bit mode and 12-bit mode.

Before each test, the computer sends a write data word, an expected data word, and a test word to each Data II Card in the system. This information is stored in the shift register (A1 - A5 and A7) shown at the top of the schematic.

C1, C2 and C3 are quadruple 2-lines-to-1 data selectors/multiplexers to select the data. In effect, the shift register on each card combines with the other Data II Cards to form a long shift register. The input to this register is from the previous data card or the data service card. The output of the register (DATA OUT PIN 31A) goes to the next Data II Card.

In the 72-bit mode, SELROM is high and write data to the MUT comes from the shift register (DROM 1-DROM12). When SELROM is high (in the 72-bit mode), the write data are gated from the shift register. When SELROM is low (in the 12-bit mode), the write data come from the arithmetic unit (DT1 thru DT12) on the Data Service Card.

BEMOUT

The shift register formed by B1 and B2 captures the output from the BEM memory. BEMOUT from the first data card goes to the Data II Service Card. BEMOUT from each succeeding card goes to BEMIN of the previous card forming one long shift register. When STE goes high, pin 1 on these two chips goes low and the BEM memory contents are stored in the shift register.

BEM Configuration Logic

Signals A, C, E, F and G are controlled by CC1, CC2 and address bits 21 and 22 (ST21 and ST22). CC1 and CC2 are developed from C1/ and C2/ developed on the Data Service Card. They are loaded by the Executive Program with bits 15 and 14 of the Executive Control Word.

The value of C1 and C2 for each BEM mode are as follows:

C2	*C1*	*Mode*
0	*0*	*8 by 12*
0	*1*	*16 by 4*
1	*0*	*32 by 2*
1	*1*	*64 by 1*

The logic equations for the signals A through F are as follows:

HI If

A (mode = 8x12 or 16x4) or x(mode = 32x2 or 64x1) and (address bits 16 and 17 = 00)x

C if (mode = 8x12 or 16x4) or (mode = 32x2 and address bits 16 and 17 = 00)

E if (mode = 64x1 and address bits 16 and 17 = 10)

F if (mode = 32x2 and address bit 16 = 1) or (mode = 32x2 or 64x1 and address bits 16 and 17 = 01)

G if (mode = 8x12 or 32x2 and address bits 16 and 17 = 10)

4.7 Data Subsystem

Printed Circuit Version, DATA II Card, Schematic Drawing 004-3004-00, Sheet 1

Function

This schematic shows:

- the development of write data during testing in both the 72-bit mode and 12-bit mode;
- multiplexing to the BEM;
- comparison of read and expected data.

Organization

C1, C2 and C3 are quadruple 2-lines-to-1-line data selectors/multiplexers to select the data.

The way expected data are generated depends on the operating mode. The expected data signals are XXD1- XXD12.

- In the 72-bit mode, SELROM is high, and expected data are gated from WROM1-WROM12. These signals come from the shift register (shown on sheet 1).
- In the 12-bit mode, expected data is gated from XD1-XD12, which comes from the data service card.

Signals A, C, E, F and G are controlled by CC1, CC2 and address bits 21 and 22 (ST21 and ST22). CC1 and CC2 come from C1/ and C2/ developed on the Data Service Card. They are loaded by the Executive Program with bits 15 and 14 of the Executive Control Word.

Operation

Before each test, the computer sends a write data word, an expected data word, and a test word to each Data II Card in the system. This information is stored in the shift register (A1 - A5 and A7) shown at the top of the schematic.

In effect, the shift register on each card combines with the other Data II Cards to form a long shift register. The input to this register is from the previous data card or the data service card. The output of the register (DATA OUT PIN 31A) goes to the next Data II Card.

72-bit Mode In the 72-bit mode, SELROM is high, and write data to the MUT come from the shift register (DROM 1-DROM12). When SELROM is high, the write data are gated from the shift register.

12-bit Mode In the 12-bit mode, SELROM is low, and write data to the MUT come from the arithmetic unit (DT1 thru DT12) on the Data Service Card.

Figure 1. Examples of four different page layouts: single page-width column, dual-column, mixed-column, and text mapped.

Once we have established a proposed format for a document, we move immediately to WP to develop the outline. Outlining a publication is an essential step in the estimating process because it results in agreement about the amount and kinds of *content* to be included in the publication.

But the outline alone is not enough. Most of our estimates are based on page/hour rates. Thus, the essential next step is to work out the number of pages the publication *as outlined* is likely to take. We do this by using a page layout sheet and the format we have decided upon. Working from the outline, we assign pages to the cover, front matter, and then the sections, one by one.

The format is essential for two reasons:

- the format helps us assign space, and
- the end product, the completed layout sheet, *shows* the client the way we visualize the publication.

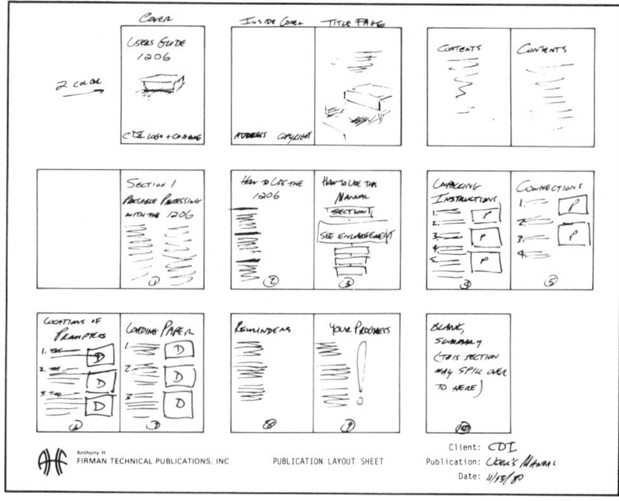

Figure 2. An example "thumbnail" page layout sheet, prepared as part of an estimate.

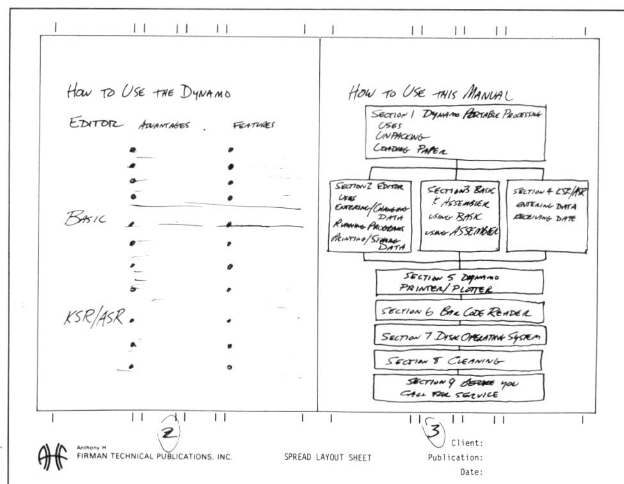

Figure 3. An example "comprehensive" spread layout sheet, used to plan layout details.

At this time we also plan for the illustrations, so we can immediately begin to meet the inevitable challenges in that area before the publication is ever committed to WP.

THE PROBLEM AND THE OPPORTUNITY

When our client is sophisticated, he already understands the importance of page layout and design both in terms of readability for his customers (and consequently the education of his customers) and in terms of company image. Our sophisticated clients appreciate the "look" we can give them by giving them well designed publications. They examine the layout sheets carefully. They worry about how format, words, and illustrations are going to fit together. They demand that we make an effort to achieve format effectiveness.

When our client couldn't care less, we put demands on ourselves that the client does not, *because* every publication we produce is headed for our own company portfolio, and is a new chance to experiment with the medium.

In either case we are faced with a PROBLEM: the need to wrestle with baulky, bulky material in order to develop a readable publication. And in either case we are faced with an OPPORTUNITY: the chance to make an extra effort, to find out more about designing publications, and to get superior results.

In either case the effect is the same. We make as many decisions as we can at the layout stage, and we plan for contingencies:

- how to put related information on one page or facing pages;
- how to add or delete information;
- how to add or delete art;
- how to achieve a format change if one is required;
- how to move material from one page to another if necessary.

COPYFITTING

These problems manifest themselves in many related ways. At the page layout stage, we find ourselves already borrowing trouble in a sense, deciding in advance what we will do to the agreed-upon format to make it work. We allow ourselves to borrow trouble up front because of the power of our word-processor, a system that allows us to count and organize pages, plan space for illustrations, copy, move, insert, and delete information, and to change formats instantly. Traditional publications groups, using clever or not so clever typists, spend a lot of time hoping that such problems will never arise - and a lot more time being frustrated when they do. On the other hand, we know they arise. Sometimes it may even seem that we invite them. But the professional communicators at Firman Publications never put the layout of a page into the hands of a typist, however clever, because we don't have to. We keep control of the publication in our own hands.

For example, one of our strong biases is in favor of placing related information on facing page spreads and contiguous spreads. It is a requirement we make of ourselves, even when the client does not know he wants it. Some formats, such as dual-column, offer more room for information than others, such as mixed column. That means that within a page formatted in dual column there is considerably more flexibility than in a mixed column layout.

So we plan ahead of time to fudge - sometimes with column width, sometimes with the locations of tabs, sometimes with the location and size of an illustration - in order to pack a pair of pages with all the information on a topic, and to set up the next topic for the same kind of treatment.

As another example, adding or deleting information plays havoc with a format. Pages spill over or shrink to insignificance. The material that once took two pages now takes three, or one, or seven. So we plan ahead of time to redesign on WP, using page-length indicators, locations of illustrations, and redesign of tables and paragraphs to make room.

It may seem that we have gotten ahead of ourselves. One cannot redesign a page until one has a page to redesign. One cannot move material until one has material to move. With WP and a competent proposal procedure, one not only can, but does.

Given the format choice - mixed-column, dual-column, full-page, or text-mapped - we can sketch the format on layout sheets.

The page layout for each section is used to estimate the cost of that section. The number of pages, multiplied by the complexity of the page (the amount of background material and client staff available for input and review, the complexity of the subject, the complexity of the language problems, etc.), multiplied by our hourly rate for creation, gives us the cost to create that section. Our estimates always contain a buffer - time to rearrange, time we steal from the other common editing operations because WP performs them automatically.

Similarly, the number of illustrations, multiplied by their complexity, multiplied by our hourly rate for production, gives us the cost to create the illustrations. We also estimate the need for paste-up and producing the WP output as separate line items within each section.

Thus, when a client wants to change the specifications during the life of a project, we have already planned for his changes in a sense, because we can estimate the amount of material to be added, deleted, or moved and what the relative cost is. And we can tell him, in terms of WP operation, just how we intend to make these changes.

THE FORMAT AS PART OF THE DATABASE

The result is that the format becomes as important as the data, and it becomes part of the data base, to be used, copied, and manipulated just as the data itself.

Our WP system provides us with the ability to establish and control formats and layouts, and to determine the feasibility of producing a certain document in a specified format.

When we submit an estimate, it always contains the page layout sheets and sample pages. If the client has not actually settled on the format, we may give him several alternative page layouts using the same text to look at, showing him differences in column-width, typeface, arrangement of material, and location of illustrations.

During the actual creation of a document, we constantly review the concept in overall terms, as well as the way the concept relates to individual pages, to determine whether or not we are formatting effectively, whether the pages fit, and whether the material presented is appropriate for the divisions it is in.

We also educate our clients in the process, so that during the final review of a publication the client often has learned enough to make his own educated assessment of the result. So sophisticated have some of our clients become that they can assess the effect of moving half a sentence or enlarging an illustration on the appearance of a page, and can estimate the number of pages affected if material is moved from one section of a chapter to another.

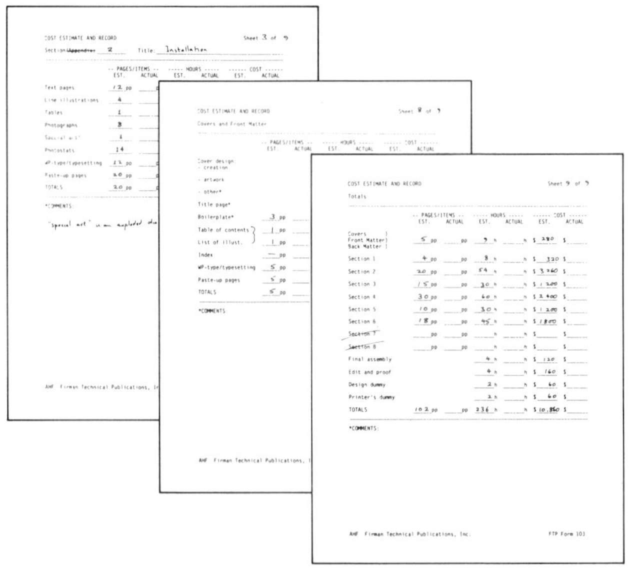
COST ESTIMATE AND RECORD — Sheet 9 of 9

Totals

	PAGES/ITEMS EST.	PAGES/ITEMS ACTUAL	HOURS EST.	HOURS ACTUAL	COST EST.	COST ACTUAL
Covers / Front Matter / Back Matter	5 pp	pp	7 h	h	$ 280	$
Section 1	4 pp	pp	8 h	h	$ 320	$
Section 2	20 pp	pp	54 h	h	$ 3 260	$
Section 3	15 pp	pp	30 h	h	$ 1 200	$
Section 4	30 pp	pp	60 h	h	$ 2 400	$
Section 5	10 pp	pp	30 h	h	$ 1 200	$
Section 6	18 pp	pp	45 h	h	$ 1800	$
~~Section 7~~	pp	pp	h	h	$	$
~~Section 8~~	pp	pp	h	h	$	$
Final assembly			4 h	h	$ 120	$
Edit and proof			4 h	h	$ 160	$
Design dummy			2 h	h	$ 60	$
Printer's dummy			2 h	h	$ 60	$
TOTALS	102 pp	pp	236 h	h	$ 10,860	$

*COMMENTS:

AHF Firman Technical Publications, Inc. — FTP Form 103

Figure 4. Worksheets for estimating the cost of creating and producing a publication. Separate worksheets are used for each section or appendix, for covers and front matter, and for computing the totals.

Using WP for format, then, has several advantages for the communicator:

- Company- or customer-mandated changes of format do not throw us into panics: WP makes such changes relatively easy.
- We can copy a successful "look" easily.
- We allow ourselves the *privilege* of constantly reviewing a format to ensure its ability to communicate.
- Total reorganizations of publications can be accomplished with reasonable ease.

The mechanism for such suicidal missions is the software embodied in our WP system, a Wang Laboratories System 25. Unlike some other word processors we have studied, we have a number of options. "Making Whoopie - A Professional Writer Learns WP" (Ref. 1) spelled out many of the options available on this system.

In essence, the Wang system provides a number of command keys which, combined with the memory capacity of the system, allow some very sophisticated ways of reassembling data. In addition, the Wang system is user-programmable through "glossaries" (Ref. 2). Programmability allows the user to assemble virtually any combination of typewriter and command functions into an automatic procedure that can be activated by two keystrokes.

The functions directly related to formatting include:

- instant formatting at the top of the document;
- the ability to copy or a move a format within a document, or between documents;
- the ability to generate formats from glossary entries;
- the ability to edit formats by means of glossary entries;
- automatic hyphenation and justification.

We also have the ability to search and replace information, to copy and move information, to insert or delete material, and to reorganize page-breaks at will. In addition, we can reorganize tables, for instance to move columns of information or to rearrange information. We can also develop the space on-screen to contain an illustration.

We have said that one of the important reasons for putting WP in the hands of the writer is to allow for the capture of ideas, content, substance. The other reason for putting WP in the hands of the writer is to put control of the document's design in his or her hands. We have found that the cost of capturing content is considerably less than the cost of capturing form. Making WP reduces the cost of capturing the format, and makes "Whoopie" even more profitable.

REFERENCES

1) "Making Whoopie: A Professional Writer Learns WP", P.S.Smith, Proceedings of the 27th ITCC.
2) "Glossaries: Using the Computer Power of WP", R.B.Bailey, Proceedings of the 27th ITCC.

Both of these papers are also published in the booklet "Making Whoopie", available from Firman Technical Publications, Inc.

COMPUTER CONTROLLED DOCUMENT UPDATE

Linda R. Socol

Honeywell Information Systems, Inc.

Linda R. Socol

Honeywell Information Systems, Inc.
Technical Writer
5250 W. Century Blvd.
Los Angeles, CA 90045
(213) 649-6870 x325

INTRODUCTION

Throughout the development of data processing a great deal of attention has been paid to patching and correcting of software, while software documentation has remained static. Everyone agrees that documentation is important but still changes and corrections are slow and laborious through addendums, change pages or rewrites every year or two -- or even longer. In the meantime, the manual holders just suffer along. This method of communicating manual corrections and updates is no longer adequate for today's high speed, electronic communication. In response to this need we have developed a new system for disseminating document updates to Honeywell's Control Program-Six (CP-6) operating system users.

The CP-6 system is new to the marketplace, having only been released by Honeywell for unrestricted marketing to potential new users in January 1981. Prior to January, the CP-6 system was available only to a limited user base. The CP-6 system was developed as a system progression for the former Scientific Data Systems/Xerox Data Systems users after aquisition by Honeywell Information Systems, Inc. in 1975. The CP-6 system is a large-scale, multi-use, distributed processing system designed to perform on Honeywell computer mainframes.

CENTRAL STARLOG DATABASE

Our documentation department supports the software development group which designed a central database called STARBASE to collect customer complaints about the system software and software documentation. This database is governed by an I-D-S/II-type processor (Integrated Data Store/II) called STARLOG. Together they form an interactive on-line information management system which runs on a Honeywell mainframe using the shared resources of the CP-6 operating system. Using STARBASE/STARLOG, we have developed a system for "patching" manuals and transmitting the patches in the form of errata using a computer to computer transmission technique.

For those who do not live your lives with a computer, patching is a term used to describe the correcting of program problems or errors. In software terms, a patch is a correction to the internal coding of the CP-6 operating system. In documentation terms, it refers to the correcting of text in the documentation database. In this paper, I will use the terms patch and errata interchangeably.

Through STARLOG, users may log on to a central support computer from their site and enter a problem report or an improvement suggestion into STARBASE. Our users are a diverse group, including the entire customer base, as well as the software developers themselves. The developers are frequently using and commenting on our manuals before the customers ever see them.

Each individual problem report is called a STAR (System Technical Action Request). Disposition of the STAR is accomplished via notes that are appended to the individual STAR. The notes come in various types and often contain several exchanges of dialogue before a resolution is reached. The sophistication of the STARLOG processor allows the user, through a wide variety of select and report command combinations, to search the STARBASE, sort and then report on almost any category of STARS. Sorting can be based on STAR number, origin date, subject, current status, severity, customer name, assigned programmer, etc.

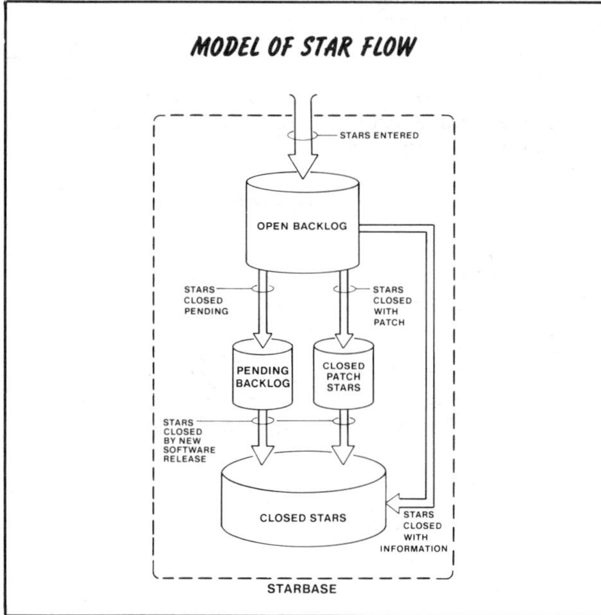

Figure 1

STAR CLOSURES

When a resolution has been reached, a note is appended to the STAR that closes it in one of the following ways (see Figure 1):

CLOSED PATCH

The STAR is closed with an errata/patch to the manual. A disposition note is appended to the STAR containing a manual update or the location of an update.

CLOSED PENDING

The STAR is closed pending the next update to the manual. A disposition note may give the name of the release or an approximate release date.

CLOSED

The STAR is closed by:

- responding to a query with information
- general release of software or manual containing a fix
- the request of the submitter
- a type A or D note containing further information, as in the case of an improvement request that will not be implemented

CLOSED DUPLICATE

The STAR is closed as a duplicate of some other STAR. There will be a note of type R or D attached to the STAR giving duplicate STAR number.

OPEN

The STAR remains open and visible to everyone. Some STARS elicit dialogue and discussion and may remain open indefinitely.

DOCUMENTATION STARS

If the STAR pertains to a manual, it is assigned to documentation where a preassigned technical writer receives it for response. The writer has a specific time frame in which to respond to the STAR depending upon the severity assigned to the problem by the originator. Documentation STARS are generally of lower severity then software STARS. For instance, Severity 1 usually indicates that the user's system is down and requires immediate response. However, most documentation STARS are Severity 4 or 5 (5 being the lowest other than "D", a discussion STAR).

STARS are usually assigned to documentation in one of two ways:

1. When there is no question that the documentation is at fault, e.g., an error or missprint, the STAR is assigned directly to documentation for resolution.

2. If a software correction affects the documentation, then the STAR is reassigned after the software correction has been determined.

ERRATA PATCH & SOURCE CORRECTION

If an errata/patch is determined to be the resolution to the STAR, the patch is prepared by the assigned writer. A note is appended to the STAR telling the user where the patch is located.

Almost all of our manuals are stored on disk in the CP-6 system. The correction is made directly to the manual source file and then the affected portion of text is copied into a special read-only account name :ERRATA. Each errata file pertains to a specific manual page or range of pages and is identified by manual number, manual version, section number and page number, followed by the account name for location, e.g., CE40-01-05-050.:ERRATA.

PATCH DISTRIBUTION

Document errata distribution to users outside of the Los Angeles Development Center is the same as the procedure for CP-6 system patches. All patches, whether documentation or software, are transmitted by a process we call "BEAMing" to the user site, computer to computer, over communication lines on a weekly basis (see Figure 2).

Once the user site receives the errata transmission, they can distribute the individual errata as they see fit, for example:

1. They can copy the errata to a line printer and then distribute them internally to manual holders, or

2. Set up their own :ERRATA account on their system and let each manual holder copy only the patches they need.

There is an added benefit to keeping the errata files on the CP-6 system in that CP-6 File Management does all the sorting and filing, keeps all existing patches in alpha-numeric order, and eliminates paper, paperwork and administration. Plus, the user can easily locate one or all patches by searching for the manual number, version, etc., and not get any unneeded patches. Internal L.A.D.C. users have access to the :ERRATA account and can read or copy patches directly as needed.

CONCLUSION

The system described above is an evolutionary step forward in the area of document maintenance. Our user community benefits and we (software documentation) benefit in several ways:

1. The user can obtain the latest available information at any time, without interminable waiting.

2. The documentation database (see Figure 3) contains all known corrections at any one time, which gives the writer a head start at rewrite time and saves a great deal of production typing time.

3. Printed software documentation can be available in more than one form, i.e., via line printer copies or magnetic tape, as well as the traditional hardcopy book.

4. Cost is lowered through the elimination of printing and distributing of addendums or change pages during the life of the printed manual.

5. We have also developed on-line documentation in the form of "HELP" messages which are extracted from the same document database.

We have been using computer controlled document maintenance for slightly less than one year. Not all of our CP-6 users are on the system and we are still working to educate them on the merits of the STARLOG system. However, even during this startup period, inputs through the STARLOG system have outnumbered inputs via the time-worn system -- tear-out comment sheets at the back of each printed manual -- by a ratio of 20 to 1; and the greater majority of the STARLOG problems and suggested improvements have been valid and worthwhile. While these numbers may suggest we have created more work for ourselves, we believe that the open feedback we receive enables us to steadily improve the quality and timeliness of our software documentation.

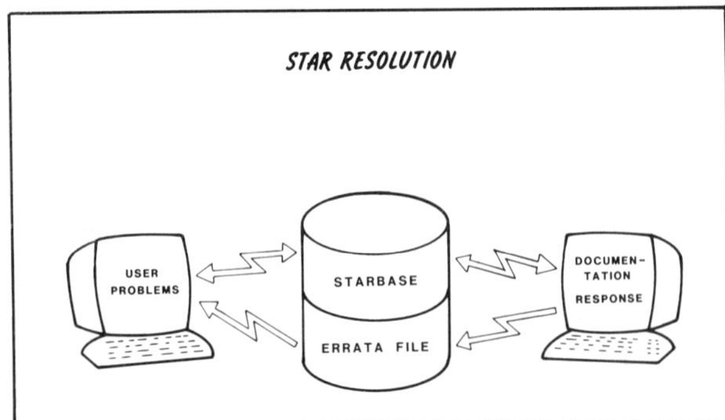

Figure 2

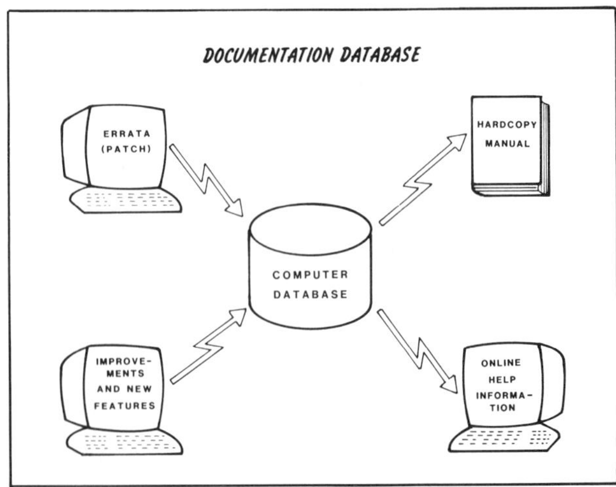

Figure 3

VIDEOTEX WITHOUT VIDEO:
Electronic Delivery Systems for Technical Communications

Eric Somers
University of Wisconsin / Stevens Point

Eric Somers
University of Wisconsin
Stevens Point, WI 54481

Assistant Professor
Department of Communication
(715) 346-3409

Eric Somers is a widely recognized author/lecturer on videotex and electronic graphics. He has also worked as an advertising agency creative director and a corporate communications manager.

Although many technical communications departments may just now be considering the acquistion of word processing equipment, this paper suggests that technical communicators look beyond the more limited "word processors" to videotex systems, comprehensive systems for electronic composition, storage, distribution and display of page materials.

But there is a danger that electronic text delivery systems may become extensions of television communication, not of print, and may become infused with values not conducive to effective and objective learning.

The structural communicative values of both televison and print are analyzed and the characteristics of videotex systems well suited to technical communication are described.

Word processing is dead.

For those just now considering the purchase of word processing equipment, that assertion might seem puzzling or frightening. Isn't word processing a key part of the "office of tomorrow?"

The answer lies in a broader understanding of the future impact of the new information technologies on business communication, and in the realization that the "office of tomorrow" is more than an office based on new electronic devices, it is an office with a new information environment.

A Merging of Disciplines

The traditional concept of "word processing" is derived from an office concept in which words are a distinct entity handled by separate personnel and equipment. If a manager wants to prepare a technical bulletin, he or she may ask for help from a writer — a wordsmith — in creating the message. The writer in turn may prepare a draft of the document on a typewriter and may ultimately use the services of other word specialists, such as editors, typists, translators, etc.

If drawings, photography, or other artwork are needed in the document, special personnel and equipment will be used to perform the required tasks. When the message has been designed, the bulletin will be passed on to others who will take care of having it printed and of distributing the completed copies to the intended receivers.

Contrast the process just described with that of making a telephone call. Usually the acts of designing a telephone message and of sending it are done unconsciously and rapidly by the person making the call. The communication process is at all times under the direct control of the caller (though the wise telephone communicator will always be conscious of the needs and responses of the listener in order to insure effective communication).

Traditional word processing seeks to automate some of the tasks associated with preparing written words. Most products called "word processors" have little or nothing to do with picture creation or with distributing messages.

But the future — the near future — is much more promising. Many people are concerned with the creation of a whole new medium of page communication in which the author can be his own editor, art director, printer, and publisher. The new medium is called "videotex" and represents a merging of the disciplines of communication and computer science.

Videotex is often talked about in terms of mass communication into the home. Expressions like "electronic newspaper," and "a library in your TV set" are used in the popular press. Of course, big money may ultimately be made in these large scale enterprises, but they will develop slowly and initially be very costly to their owners.

But technical communicators can use the new technology to reach small specialized audiences quickly and cheaply. In fact there are at least two good reasons why technical communication is an especially good application of videotex in its formative years:

1. Technical people are often more willing, even anxious, than the public at large to try out new communications systems, and are more tolerant of breakdowns and limitations associated with pioneering efforts.
2. Since technical communicators often have small and very well defined audiences for their products, problems of standardization and hardware acquisition may be minimized.

Videotex systems are integrated information systems for creating, storing, transmitting, and receiving *page* information. Although many people like to talk about videotex systems as electronic *text* systems, I think that term is far too narrow. Most systems have been designed with color graphics capability and videotex demonstrations are heavily graphics oriented. So the more general term "page" seems best suited to identify the kind of information communicated.

Pages by Wire or Radio

There are two types of videotex systems. *Broadcast* videotex systems, also called "teletext" systems, use radio waves for transmission of the information pages. These are usually multiplexed onto existing radio and television channels using parts of the spectrum unused in conventional radio and television broadcasting. This makes teletext transmission very economical. But the mass distribution of radio and television really makes teletext technology more suited for consumer mass market communications (the "electronic newspaper") than for more specialized technical communications.

Technical communicators should probably direct their attention to *interactive* videotex systems, in which pages of information are distributed over wires, such as telephone or cable television lines. Although the standard telephone service is often used, data can be sent more cheaply over packet switched high speed services such as Telenet and Tymnet (which generally serve only moderate-to-large communities).

Since videotex systems are still in a developmental stage, communications managers are in a position now to influence the development of videotex systems along lines most useful in meeting their communicative needs. And by understanding the powers and limitations of the new medium early, costly mistakes in hardware investment can be avoided.

These mistakes were often not avoided in the early days of instructional television. A number of corporations and schools made huge investments in hardware that ultimately proved incompatible with the communicative objectives the equipment was designed to meet.

Is Videotex Print or TV?

A key question, and the one that inspired the title of this paper, is whether videotex should be considered a print form or a television medium. Gilbert Seldes, who was involved with the television industry in its infancy, once expressed the view that television as an entertainment medium might have developed along entirely different lines had the motion picture industry, rather than the radio broadcasting industry, had seen the medium's economic potential and developed networks of stations. [5]

At this early stage in the development of viewdata in the United States both print and broadcasting interests are involved in videotex development. But we shall see that the communicative strengths of these two media lie in entirely different realms, and a videotex service developed to serve one industry might not be suited to the other. Before the technology becomes fixed in one specific form it would be wise for all communicators to consider what communicative values would make the medium most effective.

Before discussing actual or proposed videotex systems, let us first look at the way information is stored and communicated in the broadcast media (especially television) and in the print media.

Perhaps the most important question concerns control of the to be communicated. In television and radio the information provider has a great deal of control over the flow of information which is presented to listener/viewer in serial form. The viewer can choose to watch a TV program for example, but he cannot easily pick out small parts of the program that interest him and look only at those parts.

Even if the television program is recorded on videotape and the viewer has his own player, the process of searching for information on a reel of tape containing information is very crude and difficult. By contrast, a person can quickly scan a printed document, or even jump right into a desired portion by use of an index, but the video viewer is pretty much confined to viewing the information as the producer intended it to be seen. This seriously weakens television as a medium of education since there is a reasonable body of information to suggest that learning progresses most rapidly and constructively when the learner can control the information being learned.

Behavioral and Abstract Information

Another key difference between television and print is related to the *type* of information best communicated. Television is a medium of human behavior, while print is a medium of abstract ideas represented in symbolic form.

The information on a television screen tends to be personal in nature. That is, the so-called "objective" content cannot easily be separated from the personality of the person doing the communicating. This is especially evident in fiction television where the TV role of an actor, say the role of a doctor, becomes so merged with the personal characteristics of the man, that viewers ask the actor for medical advice.

In the area of "pure" information on television it can be noted that news reporters are sometimes evaluated by test audiences who, with skin sensors attached, watch the reporters' stories . The purpose is to measure positive physiological responses to each reporter's personal characteristics apart from the content of the stories he/she is reporting.

Thus television communication seems to be a kind of human interaction in which the focus is on people, not ideas. Edmund Carpenter, a Canadian researcher who has studied communication in many cultures, has commented that "Educational TV loves to do profiles that 'let the viewers see the man,' that is, deal with his loves and eccentricities, but ignore his genius, which may belong to a medium ill suited for TV presentation." [1]

Of course the personal nature of television makes it an ideal medium for the presentation of human behavior, whether in the performing arts or in "actuality" broadcasts of various personages. Indeed the success of shows such as The Johnny Carson Show can be attributed to the fact that the viewer seems to participate in a conversation with famous personalities.

Marshall McLuhan and others have noted that getting information through participation and group experience is a "tribal" form of communication much like that practiced in pre-literate societies. Since one must be a member of a group to receive information, the flow and control the information will be heavily influenced by group dynamics and the values held commonly by leaders of the group. Such an arrangement does not encourage radical new ideas that may threaten group values or detract from the power held by leaders. It promotes a "yes man" conformity and political maneuvering much as can be found in the "tribal" board rooms of many modern corporations.

If television is a medium of personality then print is the medium of "genius" and ideas. The use of alphabetic symbolic coding (of language) serves to remove the reception of information from an experience of interaction and involvement and places it in one of isolated perception and reflection.

The writer can communicate information to the reader, yet neither has to conform to the social standards of each other, or even to the social standards of society at large. Marshall McLuhan has suggested that it is no accident that human invention and creativity have expanded and flourished in western civilization in direct proportion to the availability, to individual members of society, of written or printed information. [3]

Information Definition

There is yet another way in which television information can be contrasted with print. Television messages are fuzzy and indistinct. This is due in part to the very nature of the cathode ray tube and to the way television fragments a picture into a single moving dot that serially transmits each frame to our brain for reconstruction. "Fuzziness" is also a characteristic of highly personalized communication utilizing spoken words and non-verbal communication forms. The receiver of the information "fills in" missing elements by participation in the tribal communication process. Thus two lovers talking on the telephone can talk in incomplete phrases and in sighs and breaths and feel that a great deal of communication is taking place.

But those same lovers deciding to keep in touch by letter soon discover that the written language is highly defined and each must spend a considerable amount of time composing a love message to the other in order for the experience to be meaningful. Print communication seems to require both visual clarity (hence the use of typewriters and other regularized type fonts in situations where communication accuracy must be very good) and clarity of verbal expression.

It is easy to see from the discussion so far why propagandists, including advertisers and politicians, find television so powerful. The information provider is in control of the information flow; viewers must participate on a personal emotional level in order to receive the information; and the medium works best when "objective" logic and clarity are supplanted by an engaging personal magnetism. Television, or any other medium with these characteristics, is ideal for controlling people.

But if the goal of a communicator is not to control people, but to stimulate them to let their own ideas and imaginations soar, then television cannot be effective. The tribal involvement and group dynamic associated with television do not encourage a free marketplace of ideas. So the true scholar, the educator, and the poet find more power in the medium of print.

What does all of this have to do with videotex? Is it not obvious, one might ask, that videotex is a form of print communication, though electronically transmitted, and might be used to supplant print for certain types of applications?

A closer look will reveal that in terms of the form of display, the control of the information flow, and the degree of user involvement in a group dynamic, many videotex systems bear a much closer resemblance to television than to print.

True, the information is presented in an alphabetic form, but does anyone really find the television screen a satisfying display device in comparison to the printed page? The low definition of the television screen, though well suited to "cool," involving, personal forms of communication, is frustratingly incompatible with the high definition of the printed word.

Gordon Thompson, of the Institute for Research on Public Policy in Montreal, feels that this single issue, that of an appropriate visual display, could jeopardize the future of the new videotex medium:

> Although this new medium could use aspects of television technology, particularly in the beginning, it must be separate enough so as not to confuse the user. It is not television. [6]

He later adds:

> Early confusion with television, in order to defray costs, could jeopardize the whole medium. [6]

The question is not entirely theoretical. Already some unions are protesting the fatigue of using some kinds of video terminals, especially color displays, and there are some doctors who feel the terminals used in newsrooms to prepare stories may cause cataracts. For whatever reason, "a library in your TV set" should not be the goal of videotex. A display medium more compatible with print needs to be found.

Of course printing terminals could be used. But these devices certainly would not save paper (an often stated advantage of videotex services, especially when used for electronic mail, news, and other quickly discarded communications) and would require, for longer documents, that the receiver be his/her own bookbinder. Probably a better solution could be found in a high definition volatile display.

Gas plasma devices are being used for some terminals. One of the most intriguing, the VuePoint by General Digital Corporation, features a small flat screen that stands upright on a table or desk. Besides having a very high definition alphabetic display, the screen surface is touch sensitive so that a selection from a displayed index, for example, can be made simply by pointing to the desired listing.

This direct tactile interaction, coupled with a sharp display, gives that kind of terminal a much greater kinship to the printed page, than does a TV display. But these values are carried still further in experimental liquid crystal displays. These screens can display pictures, drawings or text on a small, flat, high definition screen, but have the added advantage of being viewed by reflected light, like a piece of paper, rather than being luminescent.

Future technology may hold further promise in the area of videotex displays. Computers have already been used to generate holograms — high resolution three dimensional displays — and holograms have been produced on reuseable media. With considerable refinement of technique and an increase in processing power of microprocessor chips, it is not inconceivable that we may someday see a terminal producing three dimensional color graphics and high resolution text viewed by reflected light on a flat screen.

Information Control

If technology may save us in the area of providing a suitable display for videotex, it is probably much more meaningful to look at the element of information control. And here I wish to examine two kinds of control, the control the viewer has over the information received, and the control the information author has in preparing the information for transmission.

We noted earlier the value of print in giving the reader control over the information received. Yet many existing videotex systems are structured along the lines of participatory media, such as television.

It is taken almost as an axiom by most computer technicians that two way communication is stronger than one way. And they are right if, like the propagandist, their goal is to control people by involving them in information retrieval by group interaction.

But however benign the intent, group interaction must always involve some kind of control by the leader (or owner) of the information structure and even by other users.

In their book, *The Network Nation,* Starr Roxanne Hiltz and Murray Turoff have made quite a thorough analysis of the group dynamics of computer conferencing, a communications form they advocate.[2] On a more informal level, one can point to discussions in the popular press about the "problem" of frank sexual information and objectionable language that occasionally has appeared on timeshare computer information networks. And even when there is an attempt to avoid any form of regulation of information, the very structure of interactive communications systems tends to create "channels" that favor some forms of information over others. Two way communication can never be free from the sanctioning effect of group interaction.

A still more serious problem with centralized viewdata systems is a result of the fact that information is requested by the user on a piece-by-piece basis. Besides the obvious dangers of censorship and monitoring of the kind of information received by an individual, we must consider the problem of delays. No timesharing system is fast when many people request information at the same time.

If broadcast teletext seems to offer a solution, it can be seen that this one way service is often structured to serve as little more than a kind of text television. Most teletext services transmit a small number of pages, perhaps 200-700, on a continuously recycling basis. When you request a page on your terminal, it waits until the page number you want is transmitted — a wait that can be up to 25 seconds on the British Ceefax system, for example — then "grabs" it and puts it on your screen.

Although in the short run, and in terms of the limited communicative goals of most technical publishing, we can probably live with the interactive nature of viewdata, in the more distant future I am suggesting that videotex would do well to copy the structure of the book, in which a large block of randomly accessible data is presented to the reader which he/she can search, reorganize, or destroy at will.

The key to this approach, of course, is to give the videotex user considerable processing and storage power in the display terminal. It has been suggested that videodisks or even more exotic high density storage media be used for publishing and the information be physically delivered to the reader.

Another similar approach is that proposed by the Datacast project which proposes to broadcast thousands of information pages in a non-repeating format via signals multiplexed onto FM radio station broadcasts, and let each would-be reader use a personal computer to automatically retrieve the information categories desired for future reference.

In the area of technical communications this distributed approach makes sense, even when used with wired transmission, since information users are apt to own capable computer systems and be knowledgeable about their use. A single phone call, for example, could cause an instruction manual to be placed in a customer's computer for constant reference (which could later be printed out in hard copy form, if needed). Since the reader would not have to come on-line to the information provider's computer every time he/she sought information, the burden on the provider's computer system would be considerably lessened.

And, of course, a significant benefit of this all-electronic form of publishing, which can handle text, photographs, and drawings, is that updates can be made instantly and old editions already in the field can be corrected via a simple phone call.

User Control

Another important aspect of control is the control the author of the information has over its final form. Early in this paper I remarked on the difference in effort and personnel involvement between making a telephone call and publishing a printed document.

Although the print medium imposes a structure that may always require more thought and skill of presentation than simple speech (though speech input into computers is already a limited success!), the fact remains that often many other people "process" an author's ideas, and possibly distort or otherwise modify them, before they are realized in the final document. How often have you printed something only to discover after it was distributed that a key line was accidentally dropped in typesetting or a chart was miscaptioned?

Electronic word processing, as the term is generally used now, at best eliminates one or two typists, and their potential errors, in the process of document preparation. But the technology of videotex allows for the possible creation of a complete document, *including all graphics materials,*by the original author. Just as the typewriter has eliminated the need for a writer to have legible handwriting, graphics systems now in existence (with limited powers) or being developed (with much more power) will allow individuals with very poor art ability to draw high quality graphs, charts, schematic diagrams, three dimensional representations, etc.

Ted Nelson, a brilliant maverick in the computer science field, emphasizes the need for text/graphics systems that are easily and directly controlled by users who are not computer technicians. He likens an ideal system to an arcade video game:

> The truly interactive system, as in the arcades, needs no carriage returns; each user action creates an instant response — and may not echo what you typed. [4]

He also compares the highly interactive system to a movie in which the user can control all of the action:

> The interactive system, I think, may best be thought of as a new kind of movie; but a movie that is about something you wish to affect. And it is the imagining of this interactive movie that is the important design task. Never mind the nuts and bolts — what's the dream? [4]

Nelson has modeled various system concepts including a system that contains a "word processor, scheduling system, graphics package, bookkeeping package, typesetting and layout programs, etc." [4] The complete system is operated by only four basic controls that can be learned in minutes, not by dozens of control commands as is the case with many commercial "word processors" that can only produce typewriting.

This is the future of videotex and of the "office of tomorrow." It is a future in which publishing of page materials can be done completely by a single author; in which distribution can be fast and cheap using radio waves, high density portable media, or telephone lines; and one in which the reader, not the sender, controls the information flow and uses a display designed to promote easy reading, not cause fatigue.

This is a far cry from what passes for "word processing" today. And indeed many of the innovations I have described are either not yet on the market, or exist only in expensive pioneering versions that might not make economic sense in meeting the needs of technical communications departments.

But I urge all technical communicators to remain flexible and open to the kind of interactive electronic text communications described above. And beware of huge investments in dedicated "word processing" packages that are in reality expensive glorified typewriters.

In general, it is better to look at text and graphics systems designed around common general purpose mini and micro computers. These usually allow for expansion and updating more easily that dedicated hardware designed to perform a single text processing task.

If one believes in the kind of user oriented comprehensive electronic page communications system described in this article, then these views should be expressed to word processing salespeople and other industry representatives. The equipment will be produced when professional communicators let it be known that there is a market.

Summary

The title of this article is based on the premise that videotex commumications systems offer great great benefits to technical communicators, but the new systems must be designed to promote the values commonly associated with print communications, rather than the values associated with the best known form of electronic communication, television.

In addition, we have seen that videotex offers a total systems approach to page production of text and graphics; high speed, low cost transmission; and user controlled display, editing, and storage of information received.

Videotex may not develop along the lines this article has advocated, but the field is new and the technology has not yet been standardized. There is still time for potential users, technical communicators, to influence the development of these systems along lines that would provide the greatest communicative utility.

References

[1]
Carpenter, Edmund, *Oh What a Blow That Phantom Gave Me*, Holt, Rinehart and Winston, (New York), 1972.

[2]
Hiltz, Starr Roxanne, and Turoff, Murray, *The Network Nation*, Addison-Wesley Publishing Co.,(Reading, Mass.), 1978.

[3]
McLuhan, Marshall, *The Gutenberg Galaxy*, University of Toronto Press, 1962.

[4]
Nelson, Ted, "Interactive Systems and the Design of Virtuality," *Creative Computing*, Vol.6, .No.11 and Vol.6 No.12 (two parts), November and December, 1980.

[5]
Seldes, Gilbert, *Television USA:The First Thirteen Seasons* Museum of Modern Art (New York), 1962.

[6]
Thompson, Gordon, *Memo From Mercury: Information Technology is Different*, Institute for Research on Public Policy (Montreal), 1979.

PROPER SITE PLANNING AND PREPARATION: ESSENTIAL FOR OPTIMUM WORD PROCESSING OPERATIONS

H. Allen Thomas

Honeywell Information Systems, Inc.
Small Systems and Terminals Division

H. Allen Thomas
Honeywell Info Systems, Inc.
Sr. Technical Writer
300 Concord Rd.
Billerica, MA 01821
(617) 671-3135

B.S. Liberal Arts & Mng't; 1972, Northeastern Univ.

STC; Sr. Member, Boston Admin Council, BosChap

This paper discusses, from an engineering point of view, the criteria required to integrate a new word processing system into your total technical publications operation. It covers general electrical requirements including proper power connections and grounding; the effects of equipment noise on operators and other personnel; the importance of proper lighting, both natural and artificial, and suggestions for reducing glare from display terminals. This paper addresses the problems arising from the buildup of static electricity, and gives suggestions for reducing electrostatic discharge. Optimum environmental recommendations are given for operation of your system in environmentally uncontrolled areas, along with suggestions for storage of supplies and security of your data base. Proper site planning may increase your installation costs, but it will be justified in terms of high operator efficiency and equipment reliability.

In the current market, word processing systems are available in basically three types; standalone, shared-logic and interactive systems. Furthermore, these systems may be configured as; single stations, clusters, or combinations of single stations and clusters. The recommendations discussed here are generally appropriate for any type of system and configuration. Proper site planning and preparation basically involves careful site selection, proper allowance for electrical power requirements, examination of recommended environmental conditions and effective work area layout. Investments and efforts afforded at this stage of installation will yield greater efficiency and productivity later on.

SITE SELECTION

Selecting a site for your word processing (hereafter called WP) system is the most important single factor in planning and preparing your installation. Because sites and configurations are unique, and WP objectives are varied, tradeoffs are to be expected. However, in any installation, operator safety must always be of prime consideration. When selecting the site for your WP system, consider also:

- Impact of local zoning regulations, insurance costs, and possible disruption to your present operations,
- Accessibility to the site for delivery of equipment and supplies,
- Adequacy of both floor space and storage space for both current and anticipated future needs.
- Examination of any configuration rules imposed upon the system by your vendor,
- Proper electrical power source of sufficient capacity (with allowance for expansion) is required for all equipment at all locations,
- Examination of facility to determine if site can meet recommended environmental conditions. Determination of requirements for air conditioning, proper lighting, and noise abatement, if necessary,
- Review of planned communications requirements for capacity and quality of both internal and external voice and system networking.

Much is said here of planning for expansion in the future. The trend over recent years shows that a significant number of word processing operations are growing rapidly. This, coupled with the fact that many WP systems are available with upgrading and expansion capabilities, should prompt you into planning or allowing for the economies of additional wiring, air conditioning, communications lines and floor space now rather than waiting for the need to arise.

ELECTRICAL POWER

Primary Power Source

The ac power source should have sufficient capacity to handle any loads that are likely to be imposed by expansion of the system in the future. The ac source should be independent of other non-word processing system loads such as air conditioning equipment, lighting, convenience outlets, or other office equipment which may cause disturbances.

Often, a WP vendor will recommend that all interconnected devices in the system regardless of location, are powered from the same side of the same electric utility power transformer. If isolation is required (and this is often likely with large WP systems), a separate transformer and a service drop may be necessary. They must also be on the same side of the same utility power transformer. In addition, the branches from the transformer should be dedicated strictly to WP system devices. This ensures that each device in your WP system is grounded at the same physical point, and minimizes possibilities of electrical noise and other disturbances from feeding back through primary power and data lines. Failure to provide a common-point ground may result in degraded performance of all the equipment on the line.

Power Flucuations

Many problems with system operations are caused by fluctuations in the ac power system. Disturbances on the commercial power line such as electrical noise, power interruptions, and lightning may enter the system and cause equipment outages. Most of the disturbances, however, are generated within the building itself. Your plant engineer and electric utility representative must identify these disturbances and take necessary corrective action.

Convenience Outlets

All WP work areas should have adequate 120-volt ac convenience outlets for test equipment, vacuum cleaners, floor buffers etc. Circuit capacities are usually 15 to 20 amperes. Convenience outlets should be on a feeder line which is separate from the CP to prevent electrical noise and other disturbances.

Grounding

When providing ac power receptacles or connectors, it is often the user's responsibility to ensure that an equipment ground is provided along with the circuit conductors supplying power to the equipment. This ground conductor must be at least the same size as the circuit conductors, and must be bonded securely to the building ground conductor.

Ac Neutral

The ac neutral must not be confused with the protective ground (equipment frame). Damage to the equipment may result if an ac neutral is connected to the frame of any unit or protective ground.

Lightning Protection

Whenever power is brought into your building from an overhead service drop, if not required, it is advisable to have lightning protection at the service entrance of the building. This protects not only your equipment from damage, but minimizes the loss of your working data base from the disturbance. This protection is usually necessary in areas where electrical storms are frequent.

Static Electricity

Static electricity is generated when two dissimilar materials in close proximity are moved relative to each other. Electrostatic discharge (ESD) is a low-current discharge from one surface onto a third surface. This can affect WP operations; from simple inconveniences such as printer paper sticking to itself, to much more serious results such as data loss, or damage to sensitive components in the system devices. The situation becomes more critical during the winter heating months when the relative humidity is low. Use of high resistance floor surfaces or carpets and other materials such as plastic seat covers add to the problem in environmentally uncontrolled work areas.

Static discharge problems are a possibility whenever any WP equipment is installed in uncontrolled environment work areas, and no manufacturer has thus far made equipment that is unaffected by the problem. Several methods are available that can minimize the problem.

Antistatic floor surfaces such as the gridded-raised floor tile found in computer rooms or low-pile antistatic carpeting come in a variety of patterns and colors. They have the desirable attributes of conventional floor surfaces and carpeting. Antistatic floor cleaners, polishes and sprays prolong the effectiveness of the floors or carpets, but they should never be used as a sole protection from ESD. One of the most effective means of minimizing ESD is adequate humidity. This is not very practical for large open areas unless a central humidification system is installed.

Electrical Interferences

In some situations, your proposed site may experience electromagnetic interference or radio frequency interference. These interferences could result from nearby radio-frequency sources, industrial equipment and power distribution lines. Magnetic sources include electric floor heaters, control panels, transformers, rotating machinery such as dentist drills and power distribution lines.

ENVIRONMENTAL CONSIDERATIONS

Before locating your installation, set up a planning review to assess the environment. Determine well in advance your special installation or product needs to ensure normal system operation and product protection.

Temperature/Humidity Requirements

WP systems operate reliably within the manufacturer's recommended temperature and humidity ranges. The optimum site operating conditions for most systems are a range of dry bulb temperature of 68 to 78 degrees Farenheit and relative humidity range of 40 to 60 percent. Although most equipment operates well individually outside of these ranges, they are provided for integrated system operation. The farther away from these ranges that the equipment is operated, the likelihood of static and other operational problems increase.

If it is determined that air conditioning is necessary, keep in mind that vendor specifications usually give heat gain in British Thermal Units per hour (Btu/hr) or in Kilocalories per hour (Kcal/hr). These values are for equipment only. When calculating for total air conditioning capacity requirements, do not overlook other sources of heat such as sun loading, personnel, and non-WP equipment.

Lighting

Proper lighting, both natural and artificial, is essential for efficient and productive operation. Offices should have light-colored walls with white ceilings to allow for light reflection instead of light absorption. Artificial lighting should be compatible with natural lighting. Light intensity should be uniform at desk level and range from 40 to 80 foot-candles .

It is desirable to place the terminals so that the terminals and operators are essentially parallel to the longitudinal dimension of the fluorescent lamps. The terminals should also be placed so that neither the operator, nor the screen of the terminal directly faces the windows. Shading, such as venetian blinds, drapes or tinted glass may be used to reduce glare. In any case, direct sunlight should be avoided.

Noise

Although some disk drives and diskette drives produce a constant hum or whine, the noise generated by these devices is not normally objectionable. The location of impact-type printers requires greater attention, however, because they tend to be noisier than other devices in the system. Your word processing personnel usually become accustomed to the noise generated by the printers, but other people in adjacent office areas may find the printer noise objectionable. Many WP vendors offer quality sound-deadening enclosures for the printers, and should be able to recommend to you enclosures that are particularly suited for your printers. Noise from non-impact printers is generally not objectionable.

FLOOR AREA REQUIREMENTS

The floor area required for your WP system layout is determined by the complement of equipment. It is also affected by the length-to-width ratio of space available and by any columns, air conditioning units or other obstructions that may be present. When planning the system layout, allow sufficient maintenance and working access to the equipment, and for storage of media and supplies.

You should provide a storage area for disk cartridges, diskettes, printer paper and other printer accessories, spare parts and manuals. These items should be housed in fire-resistant metal containers or cabinets. For the most efficient operation, storage facilities should be located within or adjacent to the WP work area.

To ensure logical workflow in your WP operations, check the guidelines below:

- The heart of the WP center is the central processor. It should be readily accessible by your office supervisory personnel.
- Aisles leading to and from your CP subsystem should be at least 36 inches wide to allow access to the ancillary equipment, media, supplies and maintenance equipment.
- Free standing equipment such as the cartridge module disks or storage module units should be arranged according to frequency of use.

- Locate your printers according to functional requirements keeping in mind the noise problems that they may create.

- Be sure to check the configuration rules required by your vendor about cable distances for locating terminals from your central processor; and if networking is involved, to fully understand the communications techniques, communications equipment and protocols required.

- Give particular attention to minimizing the buildup of electrostatic charge.

Check with your WP system vendor marketing representative to approve the floor plan jointly.

Operator Workstations

Because the bulk of work on your WP system is performed at terminals and printers, operator comfort, efficiency and safety are paramount. Besides proper lighting, the physical structure of each work station is of high importance.

Many vendors recommend that the keyboard of a terminal should not be any higher than 26 or 27 inches above the floor. Experience has shown that placing a keyboard on standard office desks and tables is too high. This tends to fatigue an operator quite rapidly. The terminal should also have sufficient slack in cabling to allow the operator to turn the terminal to the left or right, and forward or back for visual comfort and to reduce glare. Ideally, terminals should be equipped with a separable keyboard with a cable that allows the operator to move the keyboard and/or the terminal independently for visual comfort. The table should also have sufficient width to allow the operator to place working manuscripts or editing material to the side of the terminal.

The chair at a terminal should allow the operator to be comfortable for extended periods of time, should be of a swivel type, adjustable in height and backrest. Remember also, that in the contemporary office environment, men as well as women spend considerable amounts of time at an operator workstation. Any operator should be able to keep both feet flat on the floor.

Printers

Printers should be placed according to functional group. However, as discussed earlier, the noise generated by printers should be considered when placing the printers. Besides requiring service clearance, all printers using form feed paper require space in the front of the printer for supply paper, and in the rear for takeup or printed paper.

SAFETY

Safety is a important factor in planning and preparing your word processing systems. This includes having a safe operation for your personnel, and protection of your capital equipment and data base.

Personnel

Most WP equipment offered by reputable manufacturers has been designed to operate without safety hazards provided reasonable safety precautions have been observed. In addition, it is incumbent upon you and your installation personnel to observe and implement all of the procedures and requirements of local, state and national electrical and fire codes as they apply to your installation.

Electrical shock hazards exist at any time that equipment is opened up or has covers removed. Activities that require covers to be removed or electronic components to be exposed are best left to authorized and qualified personnel.

Media and Printed Data Security

Do not overlook the serious matter of protecting your word processing center security. Many WP systems provide their customers with a high level means of security to prevent unauthorized access to the data base and use of the equipment. Proper storage of disk cartridges and diskettes as well as printed material are equally important to prevent compromising the security system. In addition, a locked security system also minimizes the possibility of physical damage to the media. At the very least, extend your existing building security system to include any word processing centers.

CONCLUSION

Many ideas have been expressed here in these few pages. Some of them are not thought about when discussing your word processing objectives with vendors or when attending demonstration sessions. These discussions often focus upon the features of one system vs. another. It is the hope of this author through this paper, that some of the unseen extras that may be required of the user can become more visible. With proper site planning and preparation, these ideas should lead the user to greater word processing operations productivity.

TYPOGRAPHICAL TERMINOLOGY

T. R. Walker
Oak Ridge National Laboratory

T. R. Walker

Oak Ridge National Laboratory
Oak Ridge, TN 37830
(615) 574-6963

Supervisor,
Composition Section

Every craft develops its own language and printers are no exception. Technical writers and editors, in daily contact with printers and compositors, sometimes become confused by the language used. Today's compositors use high-speed, computerized phototypesetters instead of type cases and Linotype machines but many of the words they use were developed during the early days of printing when all type was set by hand from a typecase. Although the words, as used today, still have the same basic meaning, the editor or writer has no way of knowing the meaning unless they have a background in printing or typesetting. A listing of the most commonly used words and their origins and definitions will help clear some of the confusion. The inability to communicate with printers and typesetters frequently leads to misunderstandings and the production of work that does not meet the required standards.

An explanation of families, branches, and fonts of type with some suggestions on how to identify and match typefaces will also aid the editor and writer in the performance of his duties.

Typefaces have developed and evolved for hundreds of years. The history of this evolution into the forms of type used today helps to instill an appreciation of the art involved in designing a typeface.

The language of the printer is constantly changing but the heritage of the past remains with us.

Welcome to the world of typesetting. When I began my career in printing more than 30 years ago, most typesetting was done by hand; that is, each character was removed individually from a typecase, assembled in a compositor's stick, and transferred to a chase, which was locked by means of furniture, reglets, and quoins. The chase rested either on an imposing stone or on a form truck, also called a "turtle," which was on wheels. It was on the "turtle" that the locked-up type was rolled to the press for printing.

Many of the words that I have just used are probably unfamiliar to most of you—unless, of course, you have had several years of experience in the printing industry. To the printer, however, these words were as familiar and complementary as the combination of ham and eggs. A chase is a metal frame that was used to hold the type in place on the press bed. When placed inside the frame, the type was surrounded by wooden blocks called furniture and smaller strips of wood called reglets. Metal quoins—sliding metal wedges that were adjusted by turning a large key—were then locked tightly to hold the form in place. Pages of type were composed originally on tables with heavy stone tops—hence, the name stone. All these words were a part of the printer's typographical terminology, which for many years remained unchanged.

Then came the Linotype machine, which accomplished more rapidly the work done formerly by the hand compositor. Actually, it is not a typesetting machine, because individual pieces of type are not used in any of its operations. Instead, the process consists of assembling brass matrices, or letter molds, from the magazine into a line interspersed with spacebands, then sending the line to the casting mechanism, where the face is cast on the top of a bar of metal. This produces a finished line, or slug, accurately trimmed for height and thickness and ready for use by the printer. With the exception of the assembling of the line by the operator and the raising of the assembling elevator, all actions of the machine are entirely automatic. By the 1960s, more than 100,000 linecasting machines had been distributed throughout the world, with matrices for more than 900 languages and dialects. Large headings were still handset, but all else had changed.

For years the Mergenthaler Company's Linotype machine, together with linecasting machines built by other manufacturers, dominated the printing industry both here and abroad. Then came the computer. Today, gone are almost all the Linotype and other linecasting machines, and in their place reigns the new lord of printing—the computer. The air is filled with new typographical terms such as cursor, configuration, and display resolution. Indeed, computerized phototypesetting is no longer a dream of the future. The "future" is now.

All these changes in the printing industry have resulted not in a replacement of previous typographical terms at each step of the way, but in an interweaving and pyramiding of both old and new technical terms. The result has been a breakdown in communication between those steeped in the terminology of the past and those just becoming acquainted with the typesetting practices of the present. Without a bridge to span the changing currents in communication, it is little wonder that some frustration has been the result. Let's cross that bridge together and, at the same time, take a look at the world of tomorrow.

Although almost all the words that were used years ago still have basically the same general meaning, no clue is found in their spelling that would unravel their usage today;

Oak Ridge National Laboratory, Oak Ridge, Tennessee, is operated by Union Carbide Corporation's Nuclear Division under contract no. W-7405-eng-26 for the U.S. Department of Energy.

for example, current usage of the word "quad" is not at all like that of the past. In the early days when type was set by hand, quads referred to the blank elements that were used to fill in at the end of a line of type. All lines had to be the same length; otherwise, when a page of type was picked up, the short lines would fall out. Each line had to be tightly justified. Thin spaces and quads of 1- and 3-em widths were used. Today, when a compositor says that a heading will be typeset quad center and that all the text will be quadded left to achieve a ragged-right margin, he is still speaking of spacing, but pieces of type metal are not used to achieve the desired result. Instead, a line of type is positioned, usually automatically, in a certain place on a page.

Suppose a compositor receives a manuscript to be typeset. The instructions are: "Typeset in Times Roman, 10-point type, with a 28-pica line length and 14-point leading. Indent all paragraphs 2 ems, and justify all text. Set all heads bold and centered. Keep the galleys under 12 inches in length." Do you understand these instructions? Several of you probably know exactly what is going to happen to the type, but do you know why those particular words were used? Let's examine the origins and meanings of some of the more commonly used typographical terms.

What is a type font, such as 10-point Times Roman? A font is a complete assortment of characters—letters, numbers, punctuation marks—that makes up a branch of a family of type. Later I will discuss, in depth, type classifications and families.

The next instruction was to use 10-point type. All type is measured in points. One point equals about 1/72 of an inch, which is the smallest unit used in the point system. Line lengths are always designated in picas; 1 pica is equal to 12 points. The manufacturers of type in the early days had no typographic standards, no uniform measurements. Type from one manufacturer was not compatible with that from another. Typesetting was almost complete chaos. In 1764, a French printer devised the point system. His standard point was not the same as that used today, but his was a major step forward in clearing the confusion. However, printers in the United States were slow to accept change, and it was over 100 years before the point system came into this country. In 1878, a major type-manufacturing company lost all its equipment in a fire. When they started rebuilding, they adopted the point system. By 1887, the United States Type Founders' Association had adopted 0.01384 of an inch as a "point." This is still the standard.

We have established the typeface to be used, the type size, and the length of the line. The next instruction specifies 14-point leading. Whether type was composed by hand or typeset on a Linotype machine, it was necessary to place some form of spacing between the lines of type. Thicker pieces, 6 points or more in thickness, are known as slugs. These leads and slugs are cut to standard lengths and stored in racks. The thickness of the lead, combined with the number of leads used between lines, is known as leading. This term is still used to indicate the amount of space between lines. Fourteen-point leading means that the distance from the base of one line to the base of the next line is 14 points. Ten-point type on 14-point leading means that 4 points of white space exist between the lines.

"Indent paragraphs 2 ems." What is an em? This is a square of the type size being used, such as 6-point, 10-point, 18-point ems, etc. You may hear someone use pica and em interchangeably, but they are incorrect unless they are talking about 12-point type. The term "em" originated at a time when the lowercase letter "m" was cast on a body with dimensions equal to the type width. This is not always the case today. Remember, an em is always as wide as it is deep. We speak of em, en, and thin spaces. An en is one-half the width of an em. All arabic numerals are almost always the same width as ens; thin spaces equal the widths of periods and commas—also designed, of course, to aid in typesetting tabular work. Two thin spaces will equal an en in most fonts.

"Keep the galleys under 12 inches in length." What are galleys? Galley is a word that is heard often. A galley originally was a metal pan, which was usually 20 to 24 inches long and varied in width from perhaps 10 picas wide to the width of a newspaper page. Made of steel, they looked like shallow bread pans—open on one end so that type could be pushed directly out into the page form. Type was placed in a galley, and magnets or heavy weights were placed at the end of the column of type to hold it erect. The galley of type was then placed on a proof press. The type was inked, and a proof copy was printed. A galley proof simply distinguished that proof from one that had been made on a press. The printed copy has since acquired the name of the equipment that was used to produce it.

Capital and small letters are often referred to as uppercase and lowercase characters. Handset type is stored in trays called cases. A case has a number of compartments of different sizes; the larger compartments contain frequently used characters, such as the letter "e." The capital, or uppercase, letters are located in the right side; the lowercase, in the left. Capitals are in alphabetical order; lowercase are not. Originally, two cases called "news" were placed on a stand one above the other—capital letters on top, small letters on the bottom. Thus the terms uppercase and lowercase came into being.

Linotype, Intertype, and Monotype linecasters use molten metal to set type. When the phototypesetting process came into use, the type was simply not "hot" any longer; hence, the term "cold."

This brings us to today's computerized phototypesetter and, along with it, more new words. A unit is a measurement smaller than a point. Units can range from 6 to 9 per em to 400 or more per em.

Earlier, I referred to fonts of type. Identifying a particular face of type can be difficult. Several faces may appear to be the same, but closer examination will reveal differences. Study of the capital letters "T" and "A" and the lowercase letters "g," "e," "r," and "t" will help in this identification. Observe the lengths of the ascenders and the descenders and the sweep of the serifs. What is a serif? This is a short crossline at the end of the stroke of a roman letter.

Many of the editors at the Oak Ridge National Laboratory (ORNL) aid in the design and preparation of brochures and other specialty publications. To reflect the degree of professionalism requisite at ORNL, the editor must have a broad knowledge of typefaces. Similarly, wherever you are employed, the experienced compositor will probably recommend certain fonts of type for your publication, but will you really understand what he has recommended?

It is important that you know how type is organized. Type may be divided into (1) groups, or races, (2) families, (3) fonts, and (4) series. Type groups, or races, are distinguished by both structural form and the historical development of the face. These groups can be divided into Text Roman, Gothic, Script (Cursive), and Novelty. Other groups and subdivisions are sometimes listed, but these five basically cover the spectrum.

Text, the first group, originated in Germany and is often a variation of Gutenberg type. It has heavy, angular strokes. William Caxton, in the 1400s in England, designed a type copied in part from the German. The text face, called Old English Text, is the present-day form. It may be seen in church literature and wedding invitations.

Roman faces evolved from Text faces. Garamond, Caslon, Bodoni, Baskerville, Century—all are examples of Roman type. (The upright version of a face is also called "roman" by printers to distinguish it from the italic face.) Roman is subdivided into Old Style (Caslon), Modern (Bodoni), and Transitional (Baskerville).

Gothic faces are skeletal, very plain, and machinelike. Easy to design and manufacture, they are often called Sans-serif, Block, or Contemporary. Gothic typefaces make good headlines but are not recommended for text. Typeset in small print, it looks very monotonous and uninviting. Examples are Univers and Megaron.

Scripts and *cursives* are typefaces that resemble handwriting—they are designed so that the characters join together. Coronet is a typical Script typeface.

Novelty includes typefaces such as Broadway, P. T. Barnum, Typewriter, or almost any other face that does not fit the above-mentioned categories.

"Family" is descriptive of all the variations within one typeface, for example, the Century family. Whether one selects Century Bold, Century Italic, or Century Condensed, the basic design stays the same.

As I discussed earlier, a font refers to an assortment of alphabetical letters, arabic numerals, and punctuation marks—all of which make up a branch of a family.

"Series" depicts the range of sizes available in a certain typeface, for example, from 6- to 92-point type.

Today's world of typesetting is far different from what it was 30 years ago. While we continue to build on the basic and solid foundation of the past, new technologies, new words, and new ideas propel us into the future. I hope that I have helped you to understand better the language of compositors.

TOTAL CONCEPT IN TECHNICAL WRITING

Donald L. Wright
The Foxboro Company

Donald L. Wright
The Foxboro Company
Senior Software Technical Writer
38 Neponset Avenue
Foxboro, MA. 02035
(617) 543-8750 Ext 3425

The technical writer is a prime key to the documentation process. By directly using a word processor, the writer can complete the job more efficiently. The writer must do more than "just his part," if the job is to be done well. The end user sees not only the technical content, but other important factors such as format and style of presentation. If your users find material difficult to use, your efforts do not inform them no matter how much time and cost goes into the preparation. Since word processing gives the writer an effective controlling tool in writing, you must use it to the fullest extent to control the total job. The writer writes, designs the style, performs much of the editing, controls the format, and even prepares the photocomposed "camera ready copy" by effectively using a word processor. Certain concepts in most word processing systems allow effective production of a better job with less effort. Some of the features a writer can use aid in the draft preparation, editing, and layout of the publication. The technical writer plans the entire job while making the word processor do the majority of work. The writer visualizes the page on the computer screen before starting the final phase. Thus the writer "calls the shots" first and prepares a more usable and effective publication. In addition the writer has the satisfaction of completing the entire job without needless delay. Word processing has come of age, and now the professional technical writer must start using it effectively.

Word processing is practically limitless when used properly. Some people use only a small part of what word processing equipment is designed to do and they create more work for themselves. Word processing companies have generally oriented their equipment to either one of two fields, the secretarial typist, or programmer. As a result of these two extremes, some companies use their word processing just like an ordinary typewriter. It is time for the technical writers to speak up for improvements which will make their writing more productive. Word processing is not a glorified typewriter, but a highly powerful, and relatively easy tool to use in the complete writing effort. It must be accessible to every working technical writer.

ROUGH DRAFT

When I prepare material on the CRT, I try to think of the total job.

- Initially, I use word processing to input original thoughts and ideas onto the CRT.
- Then I organize, outline and manipulate the structure at a time when the content is easy to handle. This may begin with a lot of scattered ideas from the engineering, training, and marketing departments.
- Then I use keywords effectively since these abbreviated thoughts appear on one CRT screen. I develop these ideas at a later time.

The use of a word processing tool called "merge" in the "WANG System" and indexing in the IBM ATMS-II system allows one to tie in the specific outline item to developed paragraphs and print with an organized format. This concept is very similar to a computer subroutine which branches out to the developed ideas and then returns to the main program or, in this sense, the outline.

One of the major benefits of this outline approach is that one can see very early in the writing effort if the contents covers the required subject clearly. It reduces the large wasted effort of a false start in writing new material and better guarantees the writer staying on the intended subject.

EDITING THE ROUGH DRAFT ON THE CRT

Once you complete the rough draft on the CRT, use another computer facility to edit the writing before sending it to anyone. I edit my document using one of the most powerful features on most word processors called search. Technical writers can use this to improve the grammar, structure, and readability of what they write. For example, use search to check for commonly misspelled words. Some more expensive systems have a data dictionary which checks spelling for you. But on most systems one can input the 100 most often misspelled words and let the computer do the work of searching for them. Another feature of search looks for jargon, deadwood, or abstract phrases such as the small number of possible examples shown below:

accordingly
aforementioned
duly
enhance
first and foremost
few and far between
furthermore
in close proximity
minimum quantity
provided that
the manner in which
with regard to, etc.

These searches may be as extensive as required. There are many publications written today on clear writing which list such phrases to avoid. What better place to put them than into your computer and let it do the work for you. I also use search to help reduce excessive use of the passive voice. Passive voice emphasizes the receiver of the action rather than the doer of the action. It is usually signified by the any form of the verb "to be." My list of computer searches include:

be
is
was
were
are
was
being
been
shall
will

Remember to include the space before and after the word so that the computer does not falsely detect words within other words (i.e., d*is*tract). Thus a routine that searches for these words, in effect, searches for the passive voice. The word processor allows you to change the structure of the sentence to the active voice if desired. In most cases active voice shortens the sentence. I am not totally against passive voice, but most technical writers and engineers use it excessively. A suggested use of no more than an average of four or five occurrences of passive voice per page might work best in your documents. This also depends on the size of your page and whatever limitation you wish to set up.

FORMATTING YOUR EDITED DRAFT COPY

Once I substantially complete the draft, I suggest using the computer a step further. For some writers, sending unformatted galleys of text for review may be acceptable, but many times the reviewers want to know how the job will really look. I let the computer do the work for me and at the same time let the computer format the page close to the final product. Documentation production people misuse this area of computer processing greatly. The computer should do all the work including headers, footers, and other page make-up functions. The computer should perform the pasteup operation of two-or three-column work; not a pasteup artist. The computer formatting facilities allow the industrial artist to perform more meaningful talents. In fact, the computer should format your page exactly as the finished product appears. Many of the easier-to-use word processing systems have only a limited capability in this area. But a minimum set of features should allow full page formatting without *any* pasteup operation. The page numbers should sequence on a single, or multilevel numbering scheme with applications of Arabic numbers, Roman numerals, or letters.

When formatting, other important areas are figures and illustrations. The author should send the illustration requests to the industrial artist as soon as completing the art mark-up. Select the size by choosing from a variety of predetermined standard sizes. Decide the location where each figure appears. The actual placement of the white space for the figure can be done by the computer. Some computers actually hold or keep together certain areas such as illustration or table areas. In these systems the author can indicate if the illustration space appears at the top or bottom of the next page. Or the computer can check the next available space (top or bottom) to allow you to do illustration figure placement page layout during the preliminary stages of the document preparation. This helps also in the preparation of tables of information which must appear together. Perhaps the best approach is the use of standard illustration or table sizes. Some convenient and achievable standard sizes are: 1/6, 1/3, 1/2, 3/4, 5/6, and full page depth windows. The width of the illustration equals the full measure of the column. When you request a figure, the computer automatically inserts the figure reference in the text, searches for the next best place to put it, and prompts you to check its decision. When the computer allots the space, it automatically prompts for the illustration title and inserts a counter or variable which later converts to the actual figure number. Using this variable allows you to insert and delete figures and references in the front of a document without affecting the downstream numbers of figures or tables.

At this point you have the draft and first edit and initial page layout complete and ready for the review cycle. During the input of the document many other features give you advanced capabilities. One of these features is the use of imbedded commands in text. The IBM ATMS-II identifies them as text comment lines as signified by the tag "!tcm." In the "Wang", they are called notes and referenced by "!!" or two adjacent exclamation marks. These facilities allow you to perform many functions. One of the most useful of these applications allows communication with editors, typist, phototypsetting people or anyone who reviews the text on the CRT screen. The writer can, for example, ask the editor to check a particular format. The editor can suggest that the author perform certain steps in a specified sequence. These imbedded messages do not necessarily appear in the finished text although they can print as an option. When the author or editors review the text, they can search for notes calling out problems and solve each of them, one at a time. When imbedded notes are used to highlight problems, you can use a prewritten program to remove the notes after resolving the problems.

Another important feature of imbedded commands is to place tags in the text which do not appear on the finished product, but which allow the user to create a table of contents, indexes or glossaries. For example, each time a certain level heading appears, an imbedded tag indicates the heading level for a later search. In the same sense one can create an index by tagging only those words requiring indexing at the final document pagination. This reduces much of the work an author or production person spends in creating complete and accurate indexes. In reference books, this concept helps greatly because the reader does not usually read the book from cover to cover, but only looks for a specific subject. In this case the publication appears no better than its index. The author can also tag glossary entries by imbedded commands which the computer recognizes using a special routine after the author completes the writing effort.

TYPESETTING

If your company has an integrated, word processing-to-typesetting system, you can imbed most of the typesetting commands in the text with tags. Instead of waiting until after a document is written, you can, in certain predetermined instances, allow the computer to insert a standard typesetting variable into the text during creation time. This then translates to a typesetting command via a special table (character set). For example, when you input a first order heading, which must be set in large, bold type, and centered with all capital letters, you would type a two-or three-keycode sequence at the word processing keyboard. This inserts a special variable and later translates, for example, to GENEVA bold in 24 point with 28 point leading. The computer asks you to input the heading while reminding you to enter all caps. Upon completion of the heading, indicate completion by a special key (EXECUTE or ENTER) and again the computer inputs a predetermined sequence of instructions which return the text size and style to the main body copy and set the appropriate leading. Another typesetting example allows you to indicate italics by using a standard program oriented approach for setting up standard type.

Thus in some word processors the typesetting instructions can closely integrate with the word processing. I realize that not all typesetting commands can or should be entered at the time a document is written. However, where requirements of predescribed typesetting codes occur, these methods allow greater publication throughput. Then the phototypesetting person can add the specialized codes to complete the typesetting. And the author feels a greater sense of accomplishment in seeing the finished product sooner.

REVISION CONTROL

Once everyone reviews a document and the revisions start coming back, another computer feature shortens the repetitive reading task. Revision markers allow you to indicate where text changes. In many engineering areas the workers have little time to review the entire document. They appreciate being told where the changes have been made. By using a small vertical line on the left or right margin you indicate the start and end of a change. This saves time in reproofing lengthy material with few changes. In some systems the revision markers equal a single upper or lower case letter or a number. These alphanumerics can represent levels of change, or even people. The levels of change can further be coded on a front page to indicate a particular date or phase. If the revision letter represents people, the user can easily tell who made a change.

OTHER USEFUL FEATURES

Another feature that saves writer time is that of retrieving repetitive text (boilerplating). Repetitive text can appear in most any form: standard notes, cautions, warnings, etc. Thus you can call up repeating groups of text and very easily place them in the body copy. Not only does the time saving appear in the typing stage, but consider that the editors, engineers, etc., do not need to reproof the standardized statements. In addition, the control standards group can standardize all statements of similar nature from a central file and change them to suit the company requirements as necessary.

A CHALLENGE

Document creativity is best performed at the hands of the author, and even page layout can be directed by letting the computer do most of the work. Word processing companies should take a harder look at expanding word processing toward the professional technical writer's needs along with typists, secretaries, and programmers. To the technical writers, I challenge you to use your word processor to the fullest extent and let it do the repetitive tasks for which it has been designed.

E
Education and Research

Stem Managers

Dr. Bertie E. Fearing
Assistant Professor
Department of English
East Carolina University
Greenville, North Carolina 27834

Editor, *Teaching English in the Two-Year College*
Editor, "Recent and Relevant"
Associate Editor, *Technical Communication*

Chair, Bibliography Committee, STC
Chair, Education Committee, Carolina Chapter, STC
Senior Member, Carolina Chapter, STC

Thomas L. Warren
Director, Technical Writing Program
English Department
Oklahoma State University
Stillwater, Oklahoma 74078

Senior Member, STC
Past Chairman, Oklahoma Chapter, STC
Past President, Council for Programs in Technical and Scientific Communication

EDUCATION AND RESEARCH

Education and research are inextricably interwoven in all academic disciplines, with research forming the base upon which the education stands. In technical communication, we are all educators: Our mission is to inform others, to impart knowledge. Yet, we are not all researchers who systematically inquire into our discipline to discover or test or revise the assumptions upon which the discipline rests. We need not be both. All disciplines have not only their practitioners but also their researchers, and both share their knowledge with one another. The Education and Research Stem accomplishes this objective.

- The *Education Component* is eminently practical. Presenters describe courses in technical communication as well as techniques for teaching course content. Courses span the discipline — high school, college, graduate school, in-plant workshops, and seminars teaching not only written and verbal but also visual skills.
- The *Research Component* assesses the current state of research in technical communication and suggests its future direction. Papers focus on current research, needed research, and projected research. This component also shows how research in other disciplines can enhance the growing body of knowledge in technical communication.

KEYNOTE ADDRESS

The Education and Research stem leads off with "The Universality of Thinking," by William A. Mambert. Noted consultant, lecturer, and author, Mambert believes not only that thinking is universal, but that it is basic to effective communication. His address sets the tone for both the theme and the dual role of the stem.

PANELS

Following the keynote address on Thursday are three panels. The first, "Historical Dimensions of Technical Communication," traces the development of technical communication from Chaucer's *Treatise on the Astrolabe* to modern Environmental Impact Statements. The second panel focuses on "Simplicity: It's Not a Simple Matter." The day concludes with "Let's Hear It From Industry: Essential Skills for Technical Communicators," in which three practicing communicators educate educators about "real world" expectations from academe.

Friday begins with a special discussion on "Technical English as a Second Language." Authorities from the US and Europe address this problem. The next panel, sponsored by CPTSC, presents four leading educators on "What Academe Does for the Profession of Technical Communication." The final panel for the day complements with "Technical Training: From High School Through Continuing Education." In between, two panels composed of practicing communicators give advice on "Research Methods for Testing Documents" and "Purity, Jargon, or Creativity in Language."

Saturday is again a blend of research and application. The two panels discuss "In-House Research and Training" and "Language Control: Theory and Practice."

WORKSHOPS AND POSTER SESSIONS

But panels are not all the Education and Research Stem has to offer. Running concurrently with the panels are five outstanding workshops and three new poster sessions:

- "Fundamentals of Information Structure"
- "Educating Women Into Management"
- "How Humans Read: Psychological Theory and Writing"
- "TECH COMM — An Interpersonal Communication Game"
- "The Rhetoric of Graphs and Charts"
- "Salaries in Technical Communication"
- "The Analysis of Narrative in Technical Communication"
- "On-Line Source Material: A Step Toward Better Publications"

This year, as in years past, the Education and Research Stem has something for everyone's intellectual palate. The stem is truly an education for educators and researchers.

NINETEENTH CENTURY AMERICAN HOUSE PATTERN BOOKS
A RHETORICAL ANALYSIS

Deborah C. and William D. Andrews

Drexel Univ. / Philadelphia College of Textiles and Science

Deborah C. Andrews
Drexel University
Director of Tech. Writing
Philadelphia PA 19104

coauthor, Technical Writing: Principles and Forms

In 19th Century America, a widely felt need for information on house-building resulted in a sizable body of literature that addressed homes as both fact and symbol and developed a form and a language for conveying architectural information to a broad middle-class audience. One of the most popular of the house pattern books was A.J. Downing's The Architecture of Country Houses (1850). Downing was an effective popularizer of architectural information whose language dealt simply and effectively with both the technical details of building, plumbing, ventilation, heating, and the cultural and symbolic aspects of home in a period of great change and growing concern for developing an American architecture.

INTRODUCTION

In 19th Century America, a widely felt need for information on house-building resulted in a sizable body of literature that addressed homes as both fact and symbol and developed a form and a language for conveying architectural information to a broad audience. This presentation outlines a rhetorical approach to this literature. First we look at the context of American home literature and some statistics on publication to show its significance. Then we look closely at one leading house pattern book: Andrew Jackson Downing's The Architecture of Country Houses. First published in 1850, it had sold 16,000 copies by 1865 and had been reprinted 9 times by 1866.

THE AMERICAN HOME

"He and his eighteenth-century, troglodytic Boston," remarks Henry Adams in a much-quoted passage from the Education,"were suddenly cut apart--separated forever--in act if not in sentiment, by the opening of the Boston and Albany Railroad; the appearance of the first Cunard steamers in the bay; and the telegraphic messages which carried from Baltimore to Washington the news that Henry Clay and James K. Polk were nominated for the Presidency (1)."

The expansion and increasing presence of technology; the increase in size and diversity of population, especially through immigration; the geographic extension of the country and new forms of transportation and communication that could link city and country, section and section; these changes indeed required adjustments in individual lives. One institution particularly affected was the home, which served potentially conflicting roles as absorber of the new in science and technology and refuge, first against the weather and then against unwanted social changes. An increased general level of affluence as well as advances in engineering like the balloon frame and new materials of construction made home ownership more broadly possible. Within the home, technology was accepted as both product and way of thinking. Developments in plumbing made possible the placement of a bath and toilet inside even modest homes by the end of the century. The ability to bring running water into the kitchen as well as bath, improved means of drainage, and more reliable means of waste disposal contributed significantly to making life in the house both healthier and more pleasant. Moreover, the systematic approach to running factories and laboratories gradually was applied to the efficient running of a household.

At the same time, the house was seen as a refuge against change, a barrier, in Downing's terms, against "vice, immorality,and bad habits." A "domestic feeling . . . at once purifies the heart and binds us more closely to our fellow beings (2)." The conflict between technology and values that readers in the 1980s may see in the role of home was not apparent, however, to writers and readers in the mid-19th Century. In part, fact and symbol were held together through a transcendental concept of architecture which, for Downing, was inspired by a close and engaging reading of Ralph Waldo Emerson. For many authors, the material fact of home design readily took on symbolic significance. Homes served symbolic functions.

They became, at one level, the focus of the development of a particularly American architecture against remembered forms from England, a country from whom we had recently won political independence. They also served to distinguish their owners from one another in the developing pattern of individualism and privatism in the country. They fostered a cherishing of the family as an institution, and especially the predominance of the woman who became "queen of home." One authority sees the design of homes as reflecting reform movements in the same way as temperance and suffrage struggles (3).

READERS AND WRITERS OF HOME LITERATURE

Such changes in the conception of the home and of the best way to design and build it were codified in popular magazines and in the architectural books we are concerned with here. These books addressed what we can broadly call the middle class. For their housing needs, the upper classes had access to architects who prepared designs for individual clients. The lower classes lived often in rental properties and could not afford to build. But the middle class, a large and literate group, turned to books to discover what they needed to know about building. This group reflected what one writer, Bruce Price, called "that American trait which inspires every man, no matter how subordinate his position in the business world, to assert his individuality and independence by owning a home which is the outgrowth of his special tastes and needs (4)." The books replaced memory and oral tradition, the first sources of architectural information used by early settlers.

The books were written by builders, architects, landscape gardeners, physicians, ministers, and a phrenologist (Orson Fowler). In the period we are discussing (roughly 1830-1900), architecture itself was being designed as a profession regulated by educational and training requirements. (The American Institute of Architects was founded in 1857.) Many of the writers, while they might also practice architecture, turned more to writing than building.

FROM BUILDERS' GUIDES TO HOUSE PATTERN BOOKS

Early architectural publications in America were, not surprisingly, reprints of British works, the first in 1775. The first American architectural publication was Asher Benjamin's Country Builder's Assistant, in 1797. Others, by such authors as Owen Biddle and Minard Lafever, went through several editions from 1800-1830. These "builders' guides" consist mainly of plates of the orders and of facades of buildings. They also provide advice on reproducing ornamentation for a readership of amateur carpenter-builders.

In the 1830s, a new form of publication appeared: house pattern books. These show interior plans and give less emphasis to facade and ornament. Moreover, they tend to include more textual than graphic material. In house pattern books, writers consider the context of the architecture more than just the physical dimensions of the buildings they describe in text and views. Most such works begin with introductions, ranging from several-page prefaces to forty- and fifty-page essays, which place domestic architecture in social, aesthetic, and historical perspective by relating building art to social trends, advancing generalized theories of art, and surveying styles and forms as they dominate periods and cultures. Following such contextual introductions, the books then take up special problems of building, commenting on matters like site, materials, and methods of construction. Specific plans are then presented in drawings, usually accompanied by textual discussion of the relative advantages and disadvantages of each and the special requirements one should consider in selecting a plan. As technological innovations came to be applied in house building, house pattern books included appended chapters on plumbing, heating, and ventilation.

After about 1840, house pattern books replaced the builders' guides as the favored means of conveying information about domestic architecture. Complemented throughout the 19th Century by discussions of architecture in periodicals--discussions often authored by writers of successful house pattern books--these works remained well into the 20th Century the chief source of written advice about home design and home construction. They were the forerunners of currently available handyman's (and handywoman's) guides and various whole earth catalogues. It is impossible, of course, to ascertain precisely how many such works were published, but a count of surviving house pattern books reveals that at least 160 works appeared between 1830 and 1900, excluding American reprints of foreign publications and second and later editions of American ones. A great many of these 160 or so works went into 8 or more later editions, a further sign of their popularity. In addition to market demand, publication of these books depended on new techniques in printing that allowed inexpensive reproduction of drawings and economical bindings so that books could be cheaply reproduced and priced.

The chronological distribution of house pattern books is enlightening. Divided according to the decade in which they were first published, the works are distributed as follows:

decade of the	
1830s	4
1840s	14
1850s	31
1860s	13
1870s	39
1880s	40
1890s	20

It is probably true, as Henry Russell Hitchcock observes, that "the output of the publishers in this field parallels rather closely the curve of building production (5)." The periods of 1840 to 1860 and 1870 to 1890 were expansive ones, when economic growth made possible increased home production; the fact that only 13 books were published in the 1860s shows rather clearly the decline in building that resulted from the Civil War. The decrease in numbers at the end of the century may reflect an increase in the role of professional architects and builders in designing individual residences.

House pattern books have two explicit purposes, both consistent with their authors' perceptions of the popular desire for architectural help. By providing detailed plans for houses, with appropriate commentary on how to execute them, they attempted to satisfy what was seen as an urgent need for practical advice. And by providing written context for the plans, the authors tried to educate their audience in the historical, social, and artistic implications of home building. Written in the layman's language--avoiding technical jargon and all but the simplest mathematics--the books were directed toward a literate audience but one untrained in the special techniques of building. These two goals suggest that the shift in popularity from builders' guides to house pattern books reflects not merely a chronological development but a class-oriented one. The architectural writer appealed directly to the middle class, which came to constitute the primary audience for all forms of publication. The oral, or "folk" mode of conveying information about building continued, of course; but among the middle class, increasingly suburban in location, the published book became the chief source of information about domestic building.

DOWNING'S ARCHITECTURE OF COUNTRY HOUSES

As interest in home restoration and alternatives to oil home heating increases today, several of the authors of 19th Century house pattern books are being rediscovered to suit a new generation. Some names may thus be familiar: Gervase Wheeler, George E. Woodward, Bicknell and Comstock, Orson Fowler, George Palliser, Calvert Vaux. But the most significant author was Andrew Jackson Downing. Let's conclude this brief presentation with a look at his work.

A.J. Downing's formal education ended at age 16 when he joined his brother in running their father's nursery in Newburgh, New York. He later married into a wealthy and prominent family but continued in the nursery business until 1847, when he began devoting himself to house design, a practice which culminated in 1851 in a partnership with the British architect Calvert Vaux. Downing made few actual designs for individual homeowners; instead, his real career was that of an architectural man of letters, a career cut short when he was drowned at age 37 in a freakish steamboating accident. Downing wrote several works on both horticulture and architecture and edited The Horticulturalist. Frederika Bremer, the Swedish traveler and writer, remarked that Downing's books were "to be found every where, and nobody, whether rich or poor, builds a house or lays out a garden without consulting Downing's work. Every young couple who sets up housekeeping buys them (6)." Downing's Swedish contemporary may exaggerate some, especially about the appeal to the "poor," but the works were indeed popular. The best modern student of Downing notes that his writings were "by far the most popular and influential writings of their kind ever published by an American (7)."

Downing was less of an innovator than a popularizer. Architect friends, like A.J. Davis and Richard Upjohn, provided the finished drawings. Much of the text, too, was inspired by J.C. Loudon's Encyclopaedia of Cottage, Farm, and Villa Architecture and Furniture, first published in London in 1833. But Downing simplified Loudon's work, directed it toward an American audience, and infused his book with his own views on taste in America and the factual and symbolic meaning of houses.

Country Houses begins with a section "On the real meaning of architecture" and ends with "Warming and ventilation." In between, Downing discusses cottages, farm-houses, and villas. For each type of housing, he indicated first "What it should be" and then presents designs (in drawings) and miscellaneous details. The presentation includes advice on chimneys and fireplaces, roof ornaments, patterns for decorative shingles, woodwork staining, cheap varnish, durable oil paint, gutters, ventilators, management of swine, furniture selection, and interior decoration. Drawings--floor plans, elevations, sectional details, views of furnishings--are used throughout.

Downing's purpose in the book is both to inform and to uplift. Practical advice on managing swine and stuccoing walls is wrapped in a moral lesson. Throughout, the tone is didactic. But the language and approach are straightforward. He begins the preface, "There are three excellent reasons why my countrymen should have good houses." Each following paragraph cites one: 1. a good house is a "powerful means of civilization"; 2. "the individual home has a great social value for a people"; 3. "There is a moral influence in a country home." All these are tied to a particularly American expression of taste and character. Because of the importance of home, then, home design should not be left to unconscious clumsiness, but to informed and elevated taste.The information may be difficult.But "I have endeavored to explain the whole subject in so familiar a manner, as to interest all classes of readers who can find any thing interesting in the beauty, convenience, or fitness of a house in the country," notes Downing.

Downing's designs reflect his preference for the Gothic revival style as opposed to the Greek precedents evoked earlier in the century. He found the Greek glaring, unsuited to individual expression of differences or to the American climate which, to Downing, should encourage porches, halls, other intermediate spaces between outdoors and indoors, between public and private areas of the house."In adapting any style for imitation our preference should be guided not only by the intrinsic beauty which we see in a particular style, but its appropriateness to our uses. This will generally be indicated by the climate, the site, or situation, and the wants of the family who are to inhabit it (2)." The house is a statement of individual character; the Greek style imposed a certain regimentation and lack of individualism. Downing spoke out against excessive ornamentation as well (although by our current stripped-down standards his bargeboards and ornamented chimneys look highly decorative). He warned against ostentatious display in housing, the excesses that would come after his death in the mansions of the robber barons. These he would have found decidedly un-American.

The text expressing these ideas is long--484 pages.But it is liberally sectioned for ease of access, laced with frequent headings, composed of short paragraphs, usually begun by a generalization then supported with evidence, as in the following:

> The great advantage which grained wood-work has over that which is simply painted white or any plain neutral tint is, that it is so easily kept clean. The surface of painted wood is always somewhat rough, and catches dirt readily, and white-lead (or other light shades of which it is the base) always oxidizes or changes color, more or less. The grained surface, on the contrary, being made smooth by varnishing, does not readily become soiled, and when it does, a moment's application of a damp cloth will make all clean and bright:--while, if the same surface were painted only, it would require frequent and most vigorous scrubbing by the house-maid, to restore it to its original condition. Every one who has made a trial of grained or stained and varnished wood-work, will agree with us that it is great economy of time and labor in housekeeping, while the addition to the cost of plain painting is very trifling. (p. 367)

Many other examples of effective technical exposition could be cited. In some places, Downing's language swoops with ease from fact to philosophy, even within one sentence:

> We greatly prefer the vertical to the horizontal boarding not only because it is more durable, but because it has an expression of strength and truthfulness which the other has not. (p. 51)

Downing, then, approaches the home as a whole in his work, compiling a kind of whole house catalog to direct the reader toward clean woodwork and clean beauty.

SIGNIFICANCE OF HOUSE PATTERN BOOKS

In the mid- and late-19th Century in America, house pattern books like The Architecture of Country Houses were addressed to a broad middle class audience for whom they provided information about building and maintaining houses at a time when such activity was expanding throughout the country. They also provided lessons in taste and values as they were developing in a particularly American way with special focus on the individual and his family. Houses were an index to the material progress of the country. They were also an index to its social progress. Houses both reflected the character of the individuals living within--and shaped it.

Because of the importance of houses--and contributing to it--a literature of the American home grew and flourished. Although early builders' guides addressed only the physical aspects, later house pattern books saw homes in a rich symbolic context and provided a form and a language for conveying architectural information that significantly shaped architectural publishing in America and, in general, the transfer of technical information to a nontechnical audience.

REFERENCES

1. Adams, Henry. The Education of Henry Adams. Houghton Mifflin: Boston, 1961, p. 5.
2. Cottage Residences.4th ed. Wiley: New York, 1853.
3. Clark, Clifford E. Jr. "Domestic Architecture as an Index to Social History: The Romantic Revival and the Cult of Domesticity in America, 1840-1870," Journal of Interdisciplinary History. 7:1 (Summer 1976), 33-56.
4. Price quoted in Russell Sturgis et al. Homes in City and Country. Scribner's: New York, 1893, p. 98.
5. American Architectural Books. Univ. of Minnesota Press: Minneapolis, 1962.
6. The Homes of the New World: Impressions of America. 2 vols. Harper: New York, 1854. I, 46.
7. Tatum, George B. "Introduction" to The Architecture of Country Houses. DaCapo Press: New York, 1968, p. vi.

COMMUNITY COLLEGE PROGRAMS: PREPARING THE ENTRY-LEVEL TECHNICAL WRITER

Dave Bloomstrand
ROCK VALLEY COLLEGE

Dave Bloomstrand
Rock Valley College
Professor of English
3301 N. Mulford Road
Rockford, IL 61101
(815) 654-4398

Instructor, A. A. S.
in Technical Writing
Member, Chicago Chapter
STC, ATTW, CPTSC

THE COMMUNICATIONS DIVISION, ROCK VALLEY COLLEGE, OFFERS BOTH THE ASSOCIATE OF APPLIED SCIENCE AND A CERTIFICATE IN TECHNICAL WRITING. THE DEGREE CURRICULUM INCLUDES THREE COMPONENTS: A TECHNICAL COMMUNICATIONS CORE, A TECHNICAL OR BUSINESS AREA CONCENTRATION, AND GENERAL COLLEGE REQUIREMENTS. THE PROGRAM REFLECTS A CO-OPERATIVE EFFORT BY LOCAL INDUSTRY AND THE COLLEGE THROUGH THE TECHNICAL COMMUNICATIONS ADVISORY COMMITTEE.

INTRODUCTION

The Rock Valley College technical writing curriculum is representative of the emerging two-year programs for training entry-level writers.

Developed from an earlier single elective course in technical writing, the Rock Valley program was in direct response to the needs of local industry. A survey of technical writing needs and opportunities in fifteen local industries was undertaken in 1974-75. The degree and certificate programs were developed by the Technical Communications Advisory Committee representing both industry and the college.

RATIONALE

The community college attempts to provide both job-entry training and skill improvement or retraining. Three major types of technical writing students might be identified:

• Technical writing degree students are typically older than the average community college students, and very likely have work experience. Because of employment or family responsibilities, these students are likely to be attending part-time, requiring perhaps as long as four years to complete degrees.

• Certificate students very likely hold college degrees. Promotion or better employment opportunities motivate these students.

• Students completing electives, particularly for technology majors, make up at least half of the class enrollments in technical writing. These students very likely take only one course as a "reinforcement" for their majors, and frequently enroll on the recommendation of faculty advisors or in some instances, their employers.

CURRICULUM

The two-year technical writing graduate is necessarily limited in preparation. The compression of writer training, a technical subject area, and the usual college requirements into a two-year curriculum imposes some obvious restrictions. At Rock Valley College, we attempt to provide a balance of three areas of study:

- The technical communications core
- A technical or business concentration
- General college requirements and electives

TECHNICAL COMMUNICATIONS COURSES

Beyond the required first-year composition courses, technical writing majors complete English 110, Introductory Technical Writing, and English 210, Technical Writing.

English 110 serves both technical writing majors and other technology majors as an elective. Attention is focused on "in-house" technical reports, with particular attention to laboratory and research reports, technical definition, and process writing.

English 210 stresses external communication: a technical article, a short assembly or maintenance manual "written to specs," a formal report to customer or client, and a sales proposal or brochure are likely assignments. Attention is given to readability standards, graphics, and general editing practices in the field.

English 220 is a technical writing internship or field placement. Minimum internship responsibility is twelve hours per week for fifteen weeks, and some full-time, paid internships are available.

In addition to the above courses, majors complete at least three other elective courses, selecting from public speaking, introduction to linguistics, newswriting or feature writing, and creative writing.

TECHNICAL AND GENERAL COURSES

Technical writing majors are required to complete at least twenty semester hours in no more than two of the following subject areas: automotive service technology, building construction technology, commercial pilot, data processing, electronics, engineering, industrial engineering, machine design, marketing and advertising, occupational safety and health, personal and public safety, and quality assurance technology.

General requirements include six semester hours in mathematics or computer science, general and industrial psychology, and such electives as design, audio-visual media, blueprint reading, and drawing interpretation.

PROGRAM ASSESSMENT

The community college has some unique opportunities to prepare technical writers, and it has some obvious limitations as well.

Community college students in technical writing represent a broad range of backgrounds and abilities, from begin- "writers" with no technical background whatsoever, to those with engineering degrees. Many students are sampling different areas in search of a "right" career.

Technical support courses are readily available in the community colleges, but technical writing students sometimes find enrollment priorities go to the majors in the technical areas, or that technology courses are offered in very strict sequences or times. For some of our technical writing majors, mastery of both writing skills and technical subject areas is too demanding, and it is this need for competence in the two areas that also accounts for the attrition of technical writing majors.

In addition, technical writing starting salaries are below those of comparable entry-level jobs in electronics or data processing. Technical writing majors frequently switch to these higher-paying major fields, or in other cases, do take jobs outside of technical writing. At present, industry is just not paying as much for employees with communication skills and general writer training as for those with pure technical backgrounds.

One further concern remains. At Rock Valley, technical writing is a terminal rather than college-transfer program. By this fact, we cannot accept the four-year degree student in the two year terminal program, and this seriously restricts our enrollment of technical writing majors who seek the bachelor's degree after two years in the community college.

THE TWO-YEAR PROGRAM GRADUATE

Students represent a broad range of backgrounds and abilities. Programs or majors are not strictly departmentalized, and co-ordination of programs between business, technology, and communications is relatively easy.

Ties with local business and industry are close, with direct relationships established through advisory committees and through student internship placement and supervision.

The two-year technical writing graduates are necessarily limited, however, in preparation. Graduates of a two-year program in technical writing are prepared to assume positions in business and industry which require in-house laboratory and investigative report writing. Customer relations and technical sales correspondence are within their range of ability and preparation. Operations manuals and repair and overhaul procedural manuals are a major part of writers' preparation. Limitations are most notably in such areas as graphics and production, although some community colleges are now offering these courses, in sales and publicity writing, and in formal proposal writing.

FIELD EVALUATING DOCUMENTS: WHAT CAN WE REALLY EXPECT?

E. Scott Brendel
Bell Laboratories

E. Scott Brendel
Bell Laboratories
Whippany, New Jersey 07981
(201) – 386-6568

Technical Editor,
Technical Publication Department

B.S., M.S. Rensselaer Polytechnic Institute

Bell Labs recently conducted a formal evaluation of some training and reference documents using computerized readability formulas, the cloze procedure, opinion surveys, and discussions with trial users. The evaluation told us as much about the techniques as it did about the documents.

INTRODUCTION

The idea of field evaluating technical documents is not a new one. However, too often it is dismissed as being too difficult or too expensive. This does not have to be the case.

The Technical Publication Department at Bell Labs has long been interested in techniques for measuring the effectiveness of documentation. Recently, we conducted a formal evaluation of some new software training and reference documents. The evaluation methodology included readability formulas, the cloze procedure, user opinion surveys, and observing subjects using the documents. (It was based in part on a field evaluation guide developed at Bell Labs. [1]) The methodology is a relatively straightforward approach which directly involves typical users.

The formal evaluation gave us some important insights into the testing techniques, as well as the documents being evaluated. This paper will describe the techniques we used, what we learned about them, and what they can tell us about our documents.

READABILITY FORMULAS

Readability formulas are well-known tools of our profession that have been available for over 50 years. At Bell Labs, readability scores can be calculated by computer. (2) We used programmed versions of the Flesch Reading Ease formula (3), the Kincaid formula (4), and the Dale-Chall formula (5) to evaluate our documents.

The Flesch, Kincaid, and Dale-Chall readability formulas assess text difficulty in large part by measuring average word and/or sentence length. The formulas assume that text containing long words and long sentences is more difficult to understand than text with shorter words and shorter sentences. By comparing readability scores to the average educational level of a document's audience, we can predict whether the document will be difficult for that audience to read.

The Dale-Chall formula considers an additional factor that others do not: word familiarity. The Dale-Chall formula compares the text sample against a list of 3,000 familiar words. Unfamiliar words (those not on the list) are counted and factored into the readability rating of the document.

Using Readability Formulas to Evaluate Documents

Computerized readability formulas are easy to use and take little time to apply. A writer or editor inputs 100-word text samples selected at random and lets the program do the work (for text stored on-line, there is one less step). Scores are reported as education level ranges. Figure 1 shows a sample printout. Because they are so easy to use, readability scores can be run on a document at any point in its preparation. We encourage our writers and editors to test their first draft as they prepare it.

```
TEXT 1 - READABILITY ANALYSIS

  Flesch Score  =    34.72 (Reading Ease Score)
                      This score relates to a
                      grade level of 13-16
  Kincaid RGL   =    12.47 (Reading Grade Level)
  Dale-Chall    =     9.46 (   Formula Score    )
                      This score relates to a
                      a grade level of 13-15

     **** FORMULA VARIABLES ****

  SYL =    1.84 (word length in syllables)
  LET =    5.21 (word length in letters)

  SNL =   16.32 (sentence length in words)
  NSN =    6.13 (sentences per 100 words)
```

Figure 1: Typical print-out from Bell Labs' computerized readability program. The text of this paper was used as the sample.

The program can also print out the unfamiliar words in the text samples. This allows writers to review their choice of language and can expose any penchant for jargon or acronyms.

Other considerations, however, limit the usefulness of readability formulas:

- Readability formulas cannot measure a myriad of factors which can affect a reader's understanding. This includes a document's organization, the level of abstraction of ideas, and an audience's familiarity with subject, among many others.

- These formulas do not adequately consider an audience's familiarity with a technical vocabulary (generally longer words that are not on the Dale list). (Something else to keep in mind about the Dale-Chall formula is that the study was conducted in 1948; the original list of 3,000 familiar words has not been updated since then. As a result, some very familiar words [such as "television"] do not appear on the list and are scored as unfamiliar.)

- If the educational level of an audience cannot be accurately estimated, readability scores have little value.

- It is not practical to revise a document strictly according to readability results alone. To do so would mean shortening sentences and using shorter words; this by itself does not guarantee that a document will be easier to read. For instance, readability formulas cannot diagnose problems caused by poor organization.

Readability measures, therefore, can only help predict when an audience *may* have problems with a document. We use them only as a very general guide or preliminary check of work.

CLOZE PROCEDURE

The cloze procedure is a readability measure which was first developed in the early 1950s. (6) In its most typical form, every fifth word of a sample passage of text is deleted and replaced with a blank; the subject is then asked to fill in the blanks by guessing the missing word using clues from the surrounding text. Figure 2 shows an example of a cloze test.

The cloze procedure's value stems from its constant consideration of context. It accounts for the background of test subjects. It reflects the influence on understanding of subtle elements, including the organization of content. "Unlike many [readability] formulae, it is not fooled by long, easy sentences or short, hard words." (7)

Although the cloze procedure was originally developed as a readability measure, researchers now feel that it successfully measures how well a reader understands text. Bormuth (8) correlated cloze scores with oral comprehension test scores. Rankin and Culhane (9) corroborated his study. Figure 3 tabulates equivalent scores for cloze tests and multiple-choice tests. (9) This gives us a context in which to interpret cloze test results.

Multiple-choice Scores	Cloze Scores (Bormuth)	Cloze Scores (Rankin & Culhane)
50	19	10
55	23	15
60	27	22
65	31	28
70	35	35
75	38	41
80	42	48
85	46	54
90	50	61
95	53	67
100	57	74

Figure 3: Equivalent cloze and multiple-choice percentage scores. (From Rankin and Culhane, 1968 [9])

Using the Cloze Procedure to Evaluate Documents

Many features of the cloze procedure make it an excellent tool for evaluating documents.

Bell Labs recently __________ a formal evaluation of __________ training and reference documents __________ computerized readability formulas, the __________ procedure, opinion surveys, and __________ with trial users. The __________ told us as much __________ the techniques as it __________ about the documents.

Figure 2: An example of a cloze test with every fifth word deleted. The abstract of this paper was used.

- Cloze tests are easy to construct, administer, score, and interpret. (We are trying to make it even easier by computerizing the test construction and scoring steps.)

- Cloze tests can account for many contextual factors that readability formulas cannot; this includes specialized vocabularies known to an audience.

- Unlike multiple-choice comprehension tests, it is virtually impossible to introduce bias into a cloze test when constructing or scoring it.

- Whereas readability scores are compared to an *estimate* of audience educational level, cloze tests directly assess a document's ability to communicate by using members of that audience.

Interestingly, one of the greatest virtues of the technique may be its greatest disadvantage. Representative document users should be used for cloze testing; borrowing potential test subjects was the largest single problem that we faced.

OPINION SURVEYS

Opinion surveys may be one of the most useful and generally applicable tools for evaluating documents. We used written opinion surveys to gather subjective comments about the documents' content, organization, artwork, writing, physical characteristics, and overall effectiveness.

They had several advantages:

- Like the cloze procedure, opinion surveys put us in direct touch with the people we write for.

- Opinion surveys are easy to use: by modifying one basic questionnaire, you can develop a tool for gathering specific information about any document. (10)

- Written opinion surveys allow subjects to comment privately; this frequently turns up very candid opinions and suggestions.

Although the opinion surveys gave valuable information, we discovered two problems:

- Some people diligently completed the surveys even though they had not used the document in question. This was a problem with the questionnaire that almost went undetected.

- Most people preferred talking about the documents rather than writing about them. As a result, some comments from the questionnaires were sketchy.

To counter these problems, follow-up interviews could be used to verify the information from opinion surveys and identify the context in which those opinions were offered.

ON-SITE OBSERVATION

An unusual feature of this field evaluation was that we were present while the subjects used the training documents being evaluated. We encouraged them to immediately bring to our attention anything about the documents that they liked or disliked and to mark any sections that gave them problems. This was particularly valuable in identifying specific parts of the documents that were difficult to understand or that otherwise needed further work.

It is difficult to place a value on our direct contact with our audience, but it was important. We received spontaneous comments and suggestions that otherwise might have gone unmentioned; by dealing directly with typical users, we verified some aspects of our original audience analysis (regarding educational background) and modified others (relating to job skills); we saw the job environment first-hand; all intangibles, but all so important.

WHAT CAN WE REALLY EXPECT?

Our formal field evaluation told us as much about the techniques we used as about the documents we evaluated. We found that this combination of techniques addresses many of our needs and concerns:

- *Early warning check:* applied to early drafts, readability formulas can flag potential problems while they can easily be changed.

- *Suitability for an audience:* readability formulas can estimate whether we have written to our audience; the cloze procedure more directly measures if that audience can understand what we have written.

- *Direct response from the target audience:* the cloze procedure, opinion survey, and discussions with typical users allow us to evaluate whether our work is truly user-oriented.

- *Specific, as well as general, evaluation:* the cloze procedure and readability formulas are general indices of a document's suitability; opinion surveys and discussions with users can identify specific problems.

- *Suggestions for improvements:* dealing directly with representative users can present a new and more pragmatic perspective on what is needed in our documents.

CONCLUSION

The testing techniques discussed in this paper have their respective advantages and limitations. In combination, they can help us evaluate a document according to certain parameters and then suggest both general and specific ways for improving it. However, this system of methods is by no means complete; in fact, what we have now is a mere framework. We still require basic techniques capable of directly and accurately assessing documents; we also need techniques which specifically identify what must be done to improve those documents.

Since the bottom line is monitoring the successful transfer of information, the development of new techniques may be outside the province of technical writers and editors; fundamental issues involving cognition, learning, and communication are at the heart of the problem. However, we should be vocal in guiding the basic research so that the needs of our profession are not overlooked.

As intimidating as the long-range requirements seem, we do have some tools to work with. They are crude, to be sure; but their continued use and refinement will eventually make more powerful and sophisticated tools possible.

REFERENCES

(1) K. M. Gibson, "Field Testing Technical Documents: One Department's Approach," *Proceedings of the 26th ITCC,* May 1979.

(2) The original programs were prepared by E. U. Coke and M. E. Koether and later modified by W. G. Repsher (all of Bell Labs).

(3) R. Flesch, "A New Readability Yardstick," *Journal of Applied Psychology,* Volume 32, #3, June 1948.

(4) J. P. Kincaid, R. P. Fishburne, R. L. Rogers, and B. S. Chissom, "Derivation of New Readability Formulas (Automated Readability Index, Fog Count, and Flesch Reading Ease Formula) for Navy Enlisted Personnel," Naval Training Command Research Branch Report 8-75, February 1975.

(5) E. Dale and J. S. Chall, "A Formula for Predicting Readability," *Educational Research Bulletin,* January 21, 1948.

(6) W. L. Taylor, " 'Cloze Procedure': A New Tool For Measuring Readability," *Journalism Quarterly,* Volume 30, Fall 1953.

(7) E. F. Rankin, "Grade Level Interpretation of Cloze Readability Scores," from F. P. Greene (ed.), *Reading: The Right to Participate,* (20th Yearbook of the National Reading Conference), National Reading Conference, Milwaukee, 1971.

(8) J. R. Bormuth, "Cloze Test Readability: Criterion Reference Scores," *Journal of Educational Measurement,* Volume 5, #3, Fall 1968.

(9) E. F. Rankin and J. W. Culhane, "Comparable Cloze and Multiple-Choice Comprehension Test Scores," *Journal of Reading,* Volume 13, December 1969.

(10) K. M. Gibson, Bell Laboratories internal communication.

EDUCATING THE COMMUNICATOR AS COMMUNICATOR IN A FOUR-YEAR SCHOOL

William O. Coggin, Ph.D.
Coordinator, Technical Writing Programs
English Department
Bowling Green State University

Expanding technologies create a need for people who are able to communicate these technologies. Four-year colleges and universities are responding to this need by establishing technical communication programs. These four-year programs should have as their goals the training of communicators who are communicators. To achieve this goal the programs need three primary components: 1) requirements for study in a technical or scientific area; 2) requirements for study in all communication media, rather than only in print; and, 3) opportunities for the students to gain practical experience in the field.

Change in science and technology is creating an increasing need for people who can communicate in these areas. The result is that institutions across the country, especially four-year colleges and universities, are rushing to include technical communication programs in their current offerings. Of the new programs, many will meet with success, while others will not. The reasons for success or failure will, of course, be various, but those programs which establish a clear philosophy of training communicators as communicators will, I believe, be most successful. At the same time, departments seeking to develop technical communication programs must also recognize the occupational and practical orientation of the programs and respond to the needs of potential employers of their graduates.

The philosophy, or orientation, of technical communication programs is developed from contact with a number of sources, among them representatives of business and industry. The people in business and industry with whom I'm in contact are consistent in telling me that the people whom they need for thier communications divisions must have a solid background in science and technology. Moreover, these representatives are primarily interested in people with training in the particular scientific or technical fields with which their concerns are involved. Thus, a computer firm wants a technical communicator who is also a computer scientist.

Technical communication programs at four-year schools must, obviously, respond to the need of business and industry by training people in scientific and technical areas. Yet, they must also train them, and primarily so, in communication, for the graduates of technical communication programs will become communicators, not scientists or technologists. And, since technical communication programs at four-year schools must be practically and occupationally oriented, they must provide the students with practical experience in communications so frequently requested by employers.

Thus, technical communication programs at four-year schools need to recognize that their purpose is to provide students with training which consists of three components; 1) scientific or technical training; 2) training in various communication media; and, 3) opportunities for practical application of their communication skills. The following paper discusses the technical communication programs at one four-year university.

The technical communication programs at Bowling Green State University are designed to respond to the requirement that they have the ability to recognize the proper medium for a particular communication need, and the requirement for practical experience in technical communication.

To respond to the need for students to have training in at least one technical or scientific area, the BGSU Technical Communication Programs require students to pursue a cognate, or concentration, in an area of science or technology. The cognate is comprised of the numbers of courses required to satisfy the requirements of a minor. Each student designs a cognate in consultation with me and a professor from the other department. The cognate may be in any area other than one of the humanities, but I encourage restricting the cognate by discussing with the student the virtues of pursuing cognates in, for example, computer science, physics, environmental studies, marketing, advertising, journalism, and geology. The purpose of the cognate area is not to give the students thorough, sophisicated knowledge of the cognate, although by the time students complete 5 or 6 courses beyond the general survey courses in the area, they are conversant in it. Instead, the cognate is designed to give them the background which they will need to be able to learn more on the job.

In addition to the courses required for the cognate, undergraduate students are required to complete at least two courses in computer science; graduate students are required to complete at least one graduate level computer science course. Also, on the undergraduate level, students are required to take courses in such areas as electronics, design technology, and blueprint-reading. Graduate students take one course which is designed to present the background and basics of photography, printing, film and television, and to provide experience in the technology of visual communications. Through these allied courses the students learn more about a number of areas which will help them become more effective communicators.

Although the cognate and allied areas are essential to the training of the technical communicator, the core of the program, again, must be in communication. For that reason we provide two components in the actual communications experience: audio-visual communications and written communications. On the graduate level, we require one course which gives the students the opportunity to increase their expertise in the areas of radio, television, film and photography. Through the course the students are provided experience in the technology of visual communications.

The purpose in requiring courses in communications media other than print is, again, not to produce experts in the use of those media. Rather, it is to give the students knowledge of concepts and terminology in each area. These courses allow students to develop a sense of what can be done with the media so that in later experiences they can identify the best medium or combination of media for the communication; they allow our graduates to talk with experts in those fields about integration or adaptation of the media to a particular need.

The remainder of the course work is comprised of training in research and bibliography, and in writing. Through the research course the students learn the process of systematic investigation of problems, both experimental and through the library. They are introduced to computer information retrieval and other documentation sources. Essentially, they gain an awareness of the complexities of the flow of information in technical and scientific fields.

In the introductory technical writing course, the students are exposed to the theories of writing and to a number of kinds of technical writing; for example, they write instructions, descriptions, proposals, and reports. In this class, students are also given their first opportunity to get actual experience in technical writing, an opportunity which is repeated in each technical writing course they take. For each class, each student is required to research and write a major project.

The major projects provide the opportunities for students to combine the classroom experience with practical experience. Many universities require that students do a project in their major fields, with help from consultants in those academic departments. We go one step further: our students select topics in their cognate areas, and they may select academic consultants from the areas of their cognates, but they may also find consultants from off-campus sources. The projects the students do may be reports, but they may also be booklets, pamphlets, brochures, proposals, chapters of books, or articles for publication. The primary requirement is that whatever they produce must have a practical use by someone. Except in extraordinary circumstances, projects which will be completed and filed away are not acceptable. So far, students have produced teaching booklets, pamphlets, brochures, newsletters, instructional and operational manuals, and program descriptions, as well as articles for presentation and publication, and teaching films. On campus, many instructors from the Technology Department are using booklets in their classrooms which were produced by our students; the Medical College of Toledo provides many of the patients treated there with home care information booklets written by technical communication students; most of the various campus organizations use information pamphlets written and produced by our students; and, one of our students is presenting a paper at this conference.

Therefore, the students complete projects which receive criticism from a number of different sources, providing them the opportunity to see how audiences react to their work. Moreover, the students must go through the entire process of finding a project, researching it, and producing the final product. All projects are professionally printed and bound or produced (some are audio-visual productions).

The major project is a good introduction to the actual communications experience; yet, even this might be considered too academic by potential employers. For that reason, each student is also required to do an internship with a business or industrial concern in a publications group. During the internship (3 to 6 months), the students are paid for their work and are expected to produce just as the permanent employees produce. The internship at Bowling Green is an essential and organic part of the student's education for the profession. Both the participating business and the students are well aware of this fact before the internship begins.

Educating the technical communicator in a four-year college or university requires that the program be established with a philopophy consistent with the primary goals of such a program; those goals should be to educate the technical communicator as *communicator* primarily, but also to give each the technical experience to be a *technical* communicator. Three components assure the proper training of the technical communicator in the four-year school: training in at least one technical or scientific area; emphasis in technical writing, but training in several communication media; and practical experience, both through course work and through internships. These experiences should achieve two important goals for technical communicators: first, providing technical communicators with the capabilities to become productive in the area of technical communication, thus enhancing the relationship between academia and the business and industrial world; and, second, providing graduates with an understanding of their positions in the business and industrial world.

Attachment 1

Graduate Program in Technical Communication -- Bowling Green State University

Required communication courses

Technical writing
Research in Language/Literature (bibliography)
Internship in Technical Writing
Technical Writing Practicum
Graduate Writer's Workshop: Non-fiction

Cognate area

Any five courses -- at least three in the cognate area itself, with up to two electives -- at the graduate level, selected with the approval of the English Department Graduate Advisor in consultation with the Graduate Advisor in the cognate area.

Other required courses

Topical Computer Applications
Visual Communications for Business and Industry

USING EXPERIMENTAL PSYCHOLOGY IN TECHNICAL WRITING

Morris Dean
International Business Machines Corporation

Morris Dean

IBM Programming Publications
555 Bailey Avenue
San Jose, CA 95150

Teaches, edits user publications

B.A., Philosophy, Yale, 1964

Publications: "Techniques of Creative Technical Writing and Editing," *Proceedings, 23rd ITCC,* 1976; "Invoking the Muse of Technical Writing," *Technical Communication,* 4th Quarter 1973 —won the journal's "Award for Distinguished Technical Communication"

The psychology of learning is the foundation for the human factors of technical publications. This paper describes that foundation, whose five pillars are the principles of psychological set, pattern formation (Gestalt), short-term memory span, meaningfulness, and reinforcement. These principles define information ease of use by forming a model of the way the reader's mind works; the reader can better understand, learn, and remember information that is presented to fit the model.

In their classic style manual, *The Elements of Style,* Strunk and White refer only once to reader psychology—when they give the following argument (in favor of putting statements in positive form):

> Consciously or unconsciously the reader is dissatisfied with being told only what is not; he wants to be told what is. Hence, as a rule, it is better to express even a negative in positive form.[1]

The psychology here may be weak, but the *method* of appealing to psychology for support is not.

An instructor needs such support even more than an author does—for responding face-to-face in the classroom. To develop a course on "Writing for Reader Understanding," I wanted to know the psychological principles underlying effective communication. So I turned to books on the psychology of perception, cognition, learning, memory, and problem solving.[2] I found there such a coherent explanation for *why* good writing is good that I decided to use psychology as the very foundation of the course.

In this paper I construct a psychological model of the reader that includes:

The reader's psychological set
The mind's pattern-forming tendencies
The span of short-term memory
What is meaningful to the reader
The mind's need for reinforcement

By explaining how readers perceive, understand, learn, and remember, the model shows us how to present information for effective communication.

HOW IS THE READER SET?

The principle of psychological set is the readiness of an organism to make a particular response, the disposition of readers to think or act in a particular way.[3] Readers' set is determined by their prior experience and by their needs. *Prior experience* obviously refers to the past—how has their past determined their set? What has happened to them before they encounter your writing? How much experience, if any—and what kind—have they had with the subject? Less obviously, *needs* refer to the future—how does their desired future determine their set? What has to happen as a result of their reading your writing, in order for their needs to be satisfied?

If you fail to take account of readers' set—the set that exists when they first come to your writing, before they've read a word of it—their response may be different from what you hope for. You may get puzzlement over technical details you thought would enlighten them; hunger from omitted details you left out to avoid stuffing them; irritation over humor you provided to entertain them.

When what you say—or how you say it—doesn't match your readers' set, there's *interference.* Some is unavoidable; you can't know perfectly your audience's past or desired future. But what you write *becomes* their past; it will influence their set for what they read next. Interference here is avoidable:

- If you've used one term for an idea, they'll expect you to continue to use it; when you slip in an "elegant variation," they'll wonder whether you're referring to the same idea, or to something else.

- When you use an -ing verb for the first three items in a list, they'll expect the same in the fourth item; switching to a -tion noun causes interference.

- By their prominence, headings, labels, captions, and lead-in sentences determine the reader's response to what follows. Meaningful ones will either lead or mislead, depending on whether they're consistent with what follows. Garbled ones fail to lead at all, but they too set the reader up—to grumble and to expect the worst.

- When an illustration must read from right to left, or from bottom to top, say so! By alerting your readers, you establish a momentary set to replace their well-established disposition to read from left to right, top to bottom.

The most important need of the reader is to know how to *do* something, to accomplish a necessary task; that's the main reason people read technical literature. Where I work, we say our readers are "task-oriented"—and we recognize that our publications have to be task-oriented too. We must thoroughly understand the particular tasks we're writing about. But, in general, we know that when readers are dominated by the need to do a task:

- They consider the technical manual to be a necessary evil —they'd like to get the job done without it, but they know they can't.
- They want to find quickly the specific information they seek; they want immediate relevance to what they have to do—any preliminary information provided must be essential.
- They expect information to be presented in the order and in the form it's needed for doing the task.

When what you say and how you say it match the reader's set, there'll be no interference in the communication.

WHAT PATTERN WILL THE READER FORM?

Gestalt psychology says that the mind tends to seek meaningful patterns—in perception, in understanding and learning, in remembering. It says that people learn best when they can relate new information to something they already know: "A person who has learned with understanding learns new tasks much more readily than a person who learned by rote."[4]

The mind seeks one meaningful pattern ("Gestalt") at a time. For example, there's more than one orientation for the box in this figure, but you can see only one of them at a time:

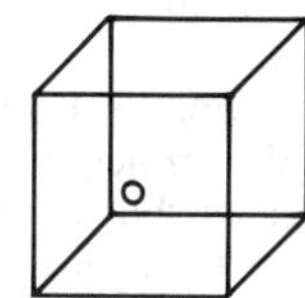

The pattern is called the *figure;* all else in the field is the *ground.* The figure-ground principle has implications for the purely visual aspects of page layout. Headings, highlighted phrases in text, and other devices that visually emphasize particular information on a page fail when overused, because the distinction between figure and ground is lost. When everything is important, nothing is important.

The mind seeks a *good Gestalt*—a pattern that is regular (predictable, harmonious), stable (unambiguous—the box above isn't stable), simple (having the smallest practical number of parts), and complete (accounting for as much of the information as possible). Gestalt psychologists formulated laws of organization by which the mind forms a good Gestalt.

One of those laws is implied in this statement from a book by Ernst Jacobi (who often appeals, implicitly, to Gestalt psychology): "The major reason anything is important to anybody is the existence of a problem, and the best way for a writer to stimulate his reader's interest is by defining and stating his problem."[5] This is the *law of closure*—the mind seeks to complete something that is incomplete. It adds missing parts to a spatial configuration: "n v rth l s." It seeks a solution to close out a problem. The tension aroused by incompleteness is the source of the "hook" in writing of many kinds.

The *law of similarity* underlies what Jacobi says about parallel construction:

> Mixed constructions—also called nonparallel or shifted constructions—are among the most common faults that block understanding. Our minds are trained to expect parallel construction. When our expectations are disappointed, we tend to become confused.[6]

Similar things, or similarly presented things, tend to be perceived as parts of a single pattern. We cause conflict when we present similar or related things in dissimilar or unrelated ways; we're giving contradictory cues for forming a pattern.

The introductory part of a passage is crucial for the reader's pattern formation, as in this example from Jacobi:

> The method of conveying information by starting out with a conclusion is called the deductive method. It is the method used in everyday learning and thinking, and it should be used by every writer who cares about reaching his reader's mind.[7]

Given the conclusion, readers can proceed by relating the details to it. But given the details first, they'll have to look for something to relate them to. Finding no pattern will frustrate them; finding a wrong one will mislead. Given an outline of the whole that follows, readers are less likely to get lost in the parts, especially if we tell them occasionally, "You are here (in relation to the whole)."

Readers seek patterns; that's the way the mind works. Writers can either cooperate with the fact, or resist it. The communicator's objective should be to supply cues for forming helpful patterns—and withhold cues for forming misleading ones.

WILL THE READER'S SHORT-TERM MEMORY BURST?

Short-term memory is what we use when we hold an unfamiliar telephone number temporarily in mind while we dial it, or when we listen to a speaker, or read a sentence. The capacity of short-term memory is 7±2 items.[8] This fact has numerous implications for presenting information: sentence length, number of items to include on a chart, number of parts into which to divide a comprehensible, memorable whole—and the number of psychological principles to teach in a class, or talk about in a paper at the ITCC.

The "length" of a sentence is determined, I believe, by its demands on the reader's short-term memory. Says Jacobi: "Long sentences may become troublesome not because they contain too many words but because they contain too many ideas."[9] The difficulty of defining "idea" may be one reason that exponents of short sentences for readable writing (such as Rudolf Flesch) recommend we count words—we can *do* that.

But the following pair of sentences demonstrates the over-simplicity of word counting. One sentence has three times as many words as the other, yet I've found that people can recite them about equally well:

> The white elephant with the green umbrella in its trunk trotted happily along the rock wall of the small village in the Indian jungle.
>
> Eight small, rough, silently bounding green cubes disappeared.

Clearly, the elephant sentence breaks up into phrases. People form one image (one "idea") of "the white elephant," or even of "the white elephant with the green umbrella in its trunk." Obviously ideas can be made up of numerous smaller ideas. In fact, people who need to hear these sentences several times use the repetitions to build up larger and larger units (i.e., to "recode" them), until the number of units is within the capacity of short-term memory. And people who can repeat either sentence the first time don't necessarily have a greater capacity; they may be better recoders–better able to picture, for example.

Common sentence patterns (such as subject-verb-object) require less recoding, and thus are easier to remember, than unusual constructions. The green-cubes sentence is backhanded–the mind has to pigeonhole all the modifiers until the subject is revealed.

Of course, we're concerned in reading, not with remembering sentences verbatim, but, as Jacobi says, with the sentences not "becoming troublesome." But I think that exceeding the span of short-term memory causes most of the trouble; to understand a sentence, readers must hold all of it in short-term memory.

HOW MUCH MEANING WILL THE READER FIND?

To demonstrate the capacity of short-term memory, I entertain my students with a few "uncontrolled experiments."[10] I have them recite sentences and simple lists of varying length and content. For example, compare these lists:

kol, naz, san, fis, cav, yeg
truth, finance, charm, work, health, division
ball, flower, mirror, package, hair, rock
car, wheel, road, map, station, gas

The lists are short enough that most people can recite each of them after one hearing. But as items are added, there comes a point where one hearing is insufficient. That point comes later in subsequent lists–as I move from nonsense syllables to abstract words, to concrete words, to related concrete words.

These exercises demonstrate that a long list of familiar, concrete, or related items is as easy to remember as a shorter list of unfamiliar, abstract, or unrelated items. Or, generally, the more meaningful the material, the easier it is to remember–and to understand and learn. This, of course, is a central theme of Gestalt psychology.

Information is meaningful when it is familiar, it can be pictured, and its parts are related:

- You improve the chance that information is *familiar* to your audience when you use the vocabulary they use; when you minimize new terminology and define it in terms of the old; when you explain new ideas in terms of ideas they already understand.

- You help your audience *picture* information when you use concrete language–language that appeals to the senses. Show an actual picture, if possible; if not possible, draw a picture in words–give examples, provide an analogy to a situation the audience can imagine.

- You show your audience the *relationship* of the parts when you select them, first, because they really are related, then organize them to reveal the relationship in a single Gestalt.

Information is simple when the number of items presented doesn't exceed the span of short-term memory, and all the items are meaningful. More than about seven items–even familiar ones that can be pictured–will initially swamp the reader's mind. There are too many to form a Gestalt. You assist the reader by recoding the items into a smaller number of groups.[11] Recoding *must* be done–either by you or by the reader–before the reader can comprehend the whole.

The greater the familiarity, concreteness, and relatedness of information, the more readily readers can do the recoding themselves:

- What is *familiar* is easily associated with information in long-term memory, and the association can be used in recoding. Such association is possible with nonsense syllables only to the extent that momentary sense can be imagined for them.

- What is *picturable* can be represented pictorially, and a mental picture can economically protray several objects (though not a thousand)–a white elephant holding a green umbrella in its trunk.

- The *relationship* among items evidently related can be used in recoding–car, wheel, road . . . "going for a drive."

When readers can easily recode information into few enough units for short-term memory, the writer has probably been simple enough.

We've been considering *cognitive* meaningfulness–the reader's being able to understand and process information. But meaningfulness also includes being interesting, motivating the reader. That, in fact, is the theme of Jacobi's book: "Clarity isn't enough . . . to be effective your communication has to be interesting."[12] But presenting relevant information lucidly is the first thing writers should do to keep their readers awake and alert.

IS THE READER REINFORCED?

Perhaps programmed-instruction (PI) texts are writing's most thorough application of the principle of reinforcement. PIs exploit the idea that the reader should immediately respond to instruction, and that correct responses should be acknowledged and incorrect responses remedied.

But in what ways can we use reinforcement in more standard technical writing to improve understanding, learning, and memory? Reinforcement includes:

- Repetition–a certain amount is essential to mastering new material. The consolidation of new information into long-term memory takes time–minutes at least.[13] The writer can't economically provide all the repetition actually required (anyway, readers wouldn't stand for it), but the classical structure of introduction-body-summary gives the writer a chance to repeat main ideas at least three times.

- Verification–checking whether the material has been understood or interpreted correctly.[14] Verification ultimately depends on *doing something,* whose results the reader can observe and judge for success or failure. The writer should judiciously indicate things for readers to do and ways for them to judge results.

- Appealing to more than one sense.[15] Showing reinforces telling; an example reinforces a definition; a side-view reinforces a view from the front; a motor action reinforces passive cognition.

There's a limit to how much reinforcement we can give. Readers are responsible for reviewing the material, for doing an exercise, for imagining alternatives. But we can help by organizing the information so it's easy to review, by providing examples to serve as exercises, by indicating some alternatives.

SUMMARY SKETCH OF READER PSYCHOLOGY

Psychology gives us a coherent set of principles that explain generally why good writing is good, and that can guide our attempts to write well. From these principles we can shape a model of the reader:

- Readers have a mental set, determined by prior experience and future needs. There's interference when what we say, or how we say it, doesn't match that set.
- Readers must form one meaningful pattern at a time. Our job is to ensure they form the pattern we intend.
- Readers have a short-term memory span of 7±2 items. We should recode complex information into mindsized portions.
- Readers find more meaning in the familiar, the picturable, the related. The more meaningful we make information, the easier it is for readers to recode it for themselves.
- Readers require reinforcement. We can help them review the material, verify their interpretation, and see information in alternative ways.

REFERENCES AND NOTES

1. Strunk, W., and White; E. B., *Elements of Style,* Macmillan, New York, 2nd edition, 1972, p. 14.
2. A book about problem solving that is particularly useful for communication is: McKim, Robert H., *Experiences in Visual Thinking,* Brooks/Cole, Monterey, CA, 1972.
3. Krech, David, and Crutchfield, Richard S., *Elements of Psychology,* Alfred A. Knopf, New York, 1962, p. 96.
4. Hilgard, Ernest R., and Bower, Gordon H., *Theories of Learning,* Appleton-Century-Crofts, New York, 1966, p. 256.
5. Jacobi, Ernst, *Writing at Work: Dos, Don'ts, and How Tos,* Hayden, Rochelle Park, NJ, 1976, p. 39.
6. Jacobi, p. 75.
7. Jacobi, p. 47.
8. Miller, George A., *Psychology of Communication,* Penguin, Baltimore, 1967, p. 32.
9. Jacobi, p. 96.
10. Many such experiments are described in: Stevens, S. S. (ed.), *Handbook of Experimental Psychology,* Wiley, New York, 1951.
11. George Miller relates this principles to short-term memory in: "Information and Memory," *Scientific American,* W. H. Freeman, San Francisco, 1956. This essay appears as the first in Miller's book.
12. Jacobi, preface.
13. Hergenhahn, B. R., *An Introduction to Theories of Learning,* Prentice-Hall, Englewood Cliffs, NJ, 1976, p. 329.
14. Hilgard, p. 559.
15. Weinland, James D., *How to Improve Your Memory,* Barnes and Noble, New York, 1957, p. 75.

The JUNGLE OUT THERE Has What the IVORY TOWER Needs ... And Vice Versa:
Sharing Skills and Knowledge for the Successful Teaching of Technical Communication

Robert B. DiGiovanni
Environmental Research Institute of Michigan
&
The University of Michigan

Robert B. DiGiovanni
Director, Communication and Information Services
Environmental Research Institute of Michigan
and
Adjunct Assistant Professor
The University of Michigan

P.O. Box 8618
Ann Arbor, MI 48107

These days the world of higher education has something in common with that of industry, business, and government: in increasing numbers both worlds are offering courses in technical communication so that students and practicing professionals in science, engineering, and other technical fields will be able to communicate their increasingly complex knowledge more effectively and efficiently.

Too often, however, these courses are not as satisfying or successful as they could be. A major problem is that neither world is capable of providing a total teaching and learning situation that really works.

The purpose of this extended abstract is to explore this problem in a general way and to propose some possible solutions to it.

Some obvious major "ingredients" are necessary for success in any formal teaching and learning situation. The teachers must have (1) the incentive and (2) the skills to teach well, and the learners must have (3) the incentive and (4) the skills to learn and perform well. When technical communication is the subject, however, yet another element is needed: (5) a real knowledge of the workworld environment.

The following matrix shows the presence or absence of these five major "ingredients" in technical communication courses as they normally are offered in the two worlds mentioned above.

	higher education classroom	industry/ business/government classroom	
staff teachers	have (1) have (2) lack (5)	have (1) lack (2) have (5)	in-house trainers
students	have (3) have (4) lack (5)	lack (3) have (4) have (5)	participants
total situation	has (1) has (2) has (3) has (4) lacks (5)	has (1) lacks (2) lacks (3) has (4) has (5)	total situation

As can be seen from the matrix, both staff teachers and in-house trainers have the incentive to teach well. Furthermore, both students and participants have the skills to learn and perform well.

But both worlds exhibit significant failings in providing a total teaching and learning situation that really works. Although staff teachers have the skills to teach well and their students the incentive to learn and perform well, neither has a real knowledge of the workworld environment. On the other hand, while in-house trainers and participants both have a real knowledge of the workworld environment, the former often lack the skills to teach well and the latter lack the incentive to learn and perform well.

Some possible solutions to these failings are offered below. Although they may be much more difficult to implement in practice than to expound in theory, they do indicate directions in which both worlds should move if they are truly interested in increasing the effectiveness of their instruction.

In the world of higher education, staff teachers could gain a real knowledge of the workworld environment by doing part-time or freelance work, or, even better, by taking part in work-study/coop programs with a neighboring industry, business, or government agency. Students could gain this knowledge through part-time or summer employment or, better still, through participation in similar work-study/coop programs.

As for the world of industry/business/government, in-house trainers could obtain teaching skills through formal schooling, particularly courses in education theory and methods, and — though this is highly unorthodox — through participation in teach-study/coop programs with a neighboring college or university. Finally, participants could gain the incentive to learn and perform well as the result of strong management support of and encouragement for in-house training, including a system to evaluate the performance of participants and report the evaluation to supervisors and other managers.

This last solution is wholly an internal matter that can be handled on an individual basis. But the other solutions may ideally be implemented through a joint venture between a college or university and a nearby industry/business/government agency, as indicated in the following flowchart.

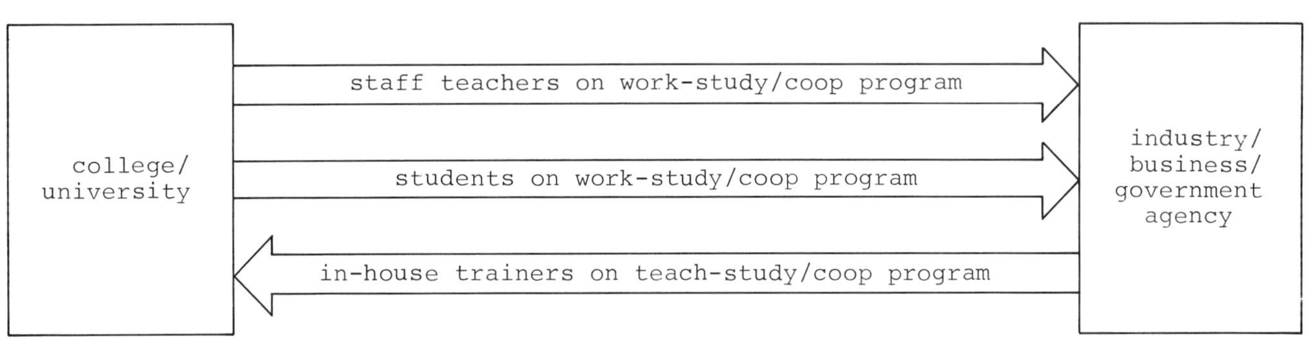

This sharing of skills and knowledge between the JUNGLE OUT THERE and the IVORY TOWER could, indeed, bring about the fully efficient and effective teaching of technical communication in both worlds.

NAVAL TRAINING TECHNIQUES FOR EVER CHANGING NEEDS

DENNIS DUKE AND RALPH LEE
NAVAL TRAINING EQUIPMENT CENTER

Dennis Duke
Education Specialist

Ralph Lee
Training Specialist

Training Acquisition Branch
Naval Training Equipment Center
Orlando, Florida 32813

A survey of research suggests that there exists a growing problem with training today's young people to accommodate various tasks necessary for job performance. Critics often fault training methodologies. Training, though, is not the problem. Rather, it is the solution to the problem.

This paper will investigate the problems currently being experienced by the U.S. Armed Forces on the arena of building a competent combat-ready defense force. Interesting to note; many of the areas of concern within the armed forces are parallel problems equally shared by many business and industrial organizations, as well as by higher educational institutions.

Although many of these problems do not have simple, surefire answers, they do have solutions. These solutions require the implementation of various innovative training techniques. These training techniques provide effective technical instruction which will increase the interest and motivation of the students' direct interaction. Additionally, these techniques are extremely cost-effective because they utilize existing training materials. They also can be self-paced and exportable - an important asset in our fast-paced and demanding training environment.

These novel training techniques require unique technical writing expertise. They involve the ability to visualize not only the major components, but also the peripherals which the trainee encounters.

This paper will illustrate this unique form of technical writing which is currently being employed by the Department of Defense. It will also offer suggestions business and industry may utilize as in-house training vehicles. Recommendations for inclusion into college English courses will also be provided because an examination of the current trends in contractor-provided training reveal that this type of training will provide the necessary catalyst required to meet the ever-changing educational requirements of a twenty-first century society.

The department of the Navy has made significant advances in the development and utilization of innovative methodologies within the past five years. Nonetheless, the Navy still recognizes that current technological developments are vastly outpacing educational systems approaches. This problem is exemplified by comparing the complex weapons systems acquired within the last five years to the academic skills of the individuals who graduated during the same time period. Studies reveal that a wide gap is evident between the skills possessed by the high school graduates and their ability to deal with the complex weapons systems. This hypothetical gap is evident not only in the military but is prominent in business and industry as well as technical schools and universities. (Duke, 1978)

There exists many variables that help perpetuate this dichotomy. It is not the intention of this paper to discuss or examine the myriad of why or how these educational problems exist but, rather, to highlight the fact that there exists a critical lag of education to technology. This hypothetical gap is increasing daily. We, as professional instructional developers and technical writers, must accept the fact that innovation in training development is tantamount to future success in technical education and new inroads must be developed to "lessen the gap." It is imperative that we silence our complaints and reserve our opinions as to whom is to blame for this dilemma.

Instead, we must concentrate our efforts toward lessening this hypothetical gap between educational systems and technological development. What needs to be done to accomplish this? There are many approaches that can be assumed. This paper will focus on the role of technical writer/instructional systems designer and the impact he has in lessening or eliminating the gap.

The first thing we, as professionals, must face is the realization that the theater which utilizes the talents of training developers and technical writers requires much more than the mere interpretation of technical facts or outlines of technical lesson plans. In order for a technical writer/developer to succeed in today's competitive market, he must be able to offer much more than the ability to interpret specific engineering data. He must be able to undertake a rather thorough analysis of the situation with which he is tasked. This not only involves his ability to interpret technical details about his product but, also, his ability to effectively analyze his audience.

In the Navy our audience varies from basic recruits to sophisticated nuclear technicians. The technical writer/developer must be able to alter his style to complement the wide variety of intellectual ability and experience of the people who will be using his material. He must consider the value systems of the cadre. This is one variable which is often neglected and may be one of the biggest reasons for the onset of the hypothetical gap between technological advances and educational systems. If we examine the interests (values) of today's student population, we discover many interesting anomalies from previous generations. High school graduates of the 80's witnessed a war being fought on the other side of the world and a man walking on the moon with no time delay. If they so desired, they could talk to anyone in the world on a whim by simply dialing a telephone. They could also relocate themselves thousands of miles in a few hours by boarding a commercial aircraft. One must realize and accept the fact that today's young generation is used to the convenience of having things immediately available. They are also accustomed to having information presented in a concise and entertaining manner. They are "turned off" by irrelevance and nonprofessionalism. Irrespective of the presence of these new values, we, as technical professionals, must strive to accomplish our objectives which are to present information relative to the operation and maintenance of various technical equipment.

Ironically, we have the same objectives as did our counterparts of a few years ago, yet we will fail in our effort if we continue to try to accomplish them via the same means employed in the past. This may be illustrated by using the analogy of the auto mechanic who is hired to "tune up" two different automobiles - one a 1965 model and the other a new 1981 model. His objective is the same for both - to tune them up. His methodology, however, will differ. The 1965 model will require manually changing points, plugs, and condensor, etc., while the 1981 model will require the utilization of sophisticated engine analyzers, etc., to calibrate the electronic ignition. Just as the mechanic was tasked to perform the same objective of "tuning up" the autos, so too are we tasked to perform the same objectives as our counterparts - to present technical information. The mechanic, out of necessity, has altered his approach in order to cope with Detroit's new products; we must adjust our style accordingly. This paper will illustrate two methods of information presentation which the Navy is using in order to cope with anticipated changes occurring in the 1980's and beyond.

As referred to earlier, today's high school graduates were raised with media. They are used to seeing news "as it happens." If this news is not presented in a professional manner, the public does not hesitate to voice their dissatisfaction. We hear criticisms such as, "Oh, that show put me to sleep." or "Boy, was that TV show terrible." Unfortunately, the training community must cope with harsher critics. We are graded not only in the interest factor but also in the retention factor. The Navy believes that in order to meet these high demands for training excellence it must use the most innovative technology available today. In addition to the traditional methods such as technical manuals, programmed instruction, platform teaching, etc., the Navy is using advanced technology such as videodiscs and interactive television as well as simulation and complementing scenarios in order to reduce the aesthetic distance of the trainee and thus lessen the gap between technology and training. Employment of these new media requires additional demands of technical writers and developers. In order to stay ahead in the competitive market, the technical writer must become more versatile. This paper will examine the media being used and elucidate upon the additional requirements needed by the writer developer in order to produce these training packages for the 80's.

Interactive video is one technique currently being used by the Navy in information presentation. Interactive video is a fusion of two powerful and proven instructional communication technologies - the computer and the electronic media. More specifically, interactive video, as referred to in this paper, refers to the use of micro computers with random access videotape and/or videodisc. The concept used for technical information presentation is actually quite simple. Computers are adept at pacing and sequencing information, branching to different points in a program, and responding to individual student's answers - but computers cannot present real-life images with which the students will be faced. Video can effectively add this dimension. Interactive video utilizes television, a passive medium which is accepted and enjoyed by today's young populace but, as the title implies, adds the variable of interaction. It brings to an essentially passive medium the important psychological elements of identification, attention, and reaction.

The basic design of interactive video consists of a regular television program interspersed with stopping points which often contain situations which require viewer response. Based upon the way he responds to these questions, the program branches to a particular tract. No matter how the student responds, he will be required to become involved. This involvement, combines with an element of uncertainty (What will happen if I do this?), brings about an identification with the lesson. The student perceives himself as being a part of the on-going technical situation. The element of realism is increased. The objective of information transfer, or learning, or whatever you want to call it, is well on the way to being accomplished. It becomes evident that the method of accomplishing the objective involves much more than the traditional technical manual approach. This extra effort is needed today in order to effectively present the highly sophisticated information to our audience. This is where the technical writer/designer fits in. Rather than just continuing in the narrow spectrum of interpreting engineering data in the preparation of traditional technical manuals, the writer/designer must be versatile and be able to do several other tasks including the design of technical training packages and interpreting and understanding engineering data.

Initially, he must be able to interpret or undertake a task analysis which involves analyzing the task, the skill level required, priority of the task in a training schema, and environmental impact, etc., to develop some type of objective strategy for presenting his messages. This involves determining purposes or objectives for the information presentation. He must define his reason for pursuing a particular route. What is the information intended to accomplish? Is it initial exposure to new equipment? Is it correction of existing procedures?

In addition to his objectives, the strategy is influenced by an audience analysis which the writer/designer must be able to undertake and individually interpret. This involves determining such things as who will benefit from the information? What is their intellectual ability? Reading ability? Previous experience or exposure to the information? Visual literacy, etc.?

Secondly, the technical writer/designer must be able to develop a topical outline for the information. Obviously, this requires an understanding of the material, but the technical writer need not be an expert; however, he must know enough about the specific area in order to effectively communicate with the subject matter expert with whom he will be working. Together they must be able to make determinations as to the depth of information coverage and the degree of importance of various topics to be presented. He must be able to anticipate difficulty levels which may pose problems to cognitive transfer as well as define what remedial avenues will be provided for these problem areas. His outline should also indicate test points and the criteria with which they will be evaluated. Lastly, the outline should indicate what media (if any) such as workbooks, reference TM's, etc., are to be used and where they are to be used. This paper elaborates upon the media selection available for presentation of an interactive systems approach.

The technical developer's next step is to transform his outline into some type of workable format. Interactive television (ITV) requires him to flow-chart his (information) system. This is actually the "blueprint" of his design. In order to accomplish this, the technical writer/developer must be able to "visualize" how the receiver of his information may act. Interactive video requires a thorough understanding of television scriptwriting combined with creativity. By its very nature ITV implies interaction, i.e., electronic pencil, written workbook response, typewriter type keyboard, etc., are determined by the developer. He must present the information as clearly and concisely as possible so that minimal confusion is experienced by the viewer upon initial exposure to the information. In addition, he must also write remedial loops. This is for the viewer who did not comprehend the initial presentation, as evidenced by his undesirable interactive response, to be reexposed to the information. A different approach is needed here so that the viewer can perceive the information from another perspective.

Interactive television can be complemented by rather simple, inexpensive media such as printed workbooks and specially produced video taped programs or it can employ rather elaborate computer programs and videodiscs. It is desirable for the modern technical writer/developer to have some knowledge about computer programming operations but it is not necessary; nonetheless, he should be able to explain and direct an experienced computer programmer as to what he wants to accomplish in the way of branching. The amount of realism and involvement desired in information presentation (training) is determined by objectives as well as cost.

Another area of training available to the technical writer/developer, which is somewhat similar to ITV but much more sophisticated, is that of actual simulation. Simulation, for purposes of this paper, is defined as a medium that allows for the introduction of specific phenomena which are likely to be experienced in actual performance. Actual simulation totally involves the participant in a "simulated" event. An example may be that of an individual who is receiving training in an aircraft simulator. He may experience the phenomenon of the actual motion of that simulated aircraft as well as any preprogrammed visual stimuli; i.e., images he may face when taking off, landing, banking, etc., that the instructor may wish to introduce. The result is the creation of high physical fidelity.

Simulation is the answer to a number of environmental, social, educational and economic goals. Because of this, simulation is used today in many areas of military training. Due to cost factors and impracticability, simulation systems are often a prerequisite to hands-on training to insure that the full training expectancies and learning transfer are accommodated in situations where we cannot allow initial learning exposure to occur at the detriment of a multi-million dollar piece of equipment or a priceless life.

There are many advantages to using this concept such as reduced training costs, minimal development risk, and a maximum education training end product. Early simulation systems lacked the modularity required for low-cost operation and did not provide the realistic and exciting stimulus required for current state-of-the-art training. Today's modern technical writer/developer faces the challenge of providing immediate reinforcement and stimulus response as well as perceptual and cognitive learning via a medium which is as realistic and exciting as possible. Actual simulation provides the technical writer/developer with the opportunity to combine the psychomotor phase of the educational hierarchy to the cognitive and affective domains which are offered by instructional television.

A simulator is a teaching machine and a strong positive aspect of teaching machines is the ability of the instructor to control the rate and selection of stimulus to the student. Simulators are designed so that the participant involvement becomes so total that the psycho/ physical senses perceive actual conditions and the cognate and physiological dimensions recognize and react accordingly. The result is that maximum training transfer is accomplished during these situational encounters. Participant reaction, when exposed to these simulated conditions, is positive in most respects when the technical writer/developer professionally designs the system with a realistic and creative approach. As was mentioned earlier, this approach causes maximum motivation and productivity in our target audience which is accustomed to immediate reinforcement in a concise, realistic and professional manner.

The technical writer/developer must use an "outline" of some hypothesized chain of events in order to stimulate the situational training environment. This "outline", referred to as a scenario, is much more detailed than the regular topical outline mentioned earlier and is somewhat similar to a television script. In a scenario the rate of presentation is controlled by the system operator (instructor) or automatic controlling mechanism. It is the task of the technical writer/developer to write scenarios which contain a wide variety of stimulus alternatives which can be modified to meet the needs of a particular participant and/or situation. Instantaneous feedback and reinforcement allow for program modification as dictated by student response. The technical writer/developer must become familiar with the intricacies of the particular simulated system as well as provide realistic and challenging developments.

The Navy maximizes simulation systems due to the nature of the training it undertakes. It is cost-prohibitive and often dangerous to provide initial training at the expense of actual equipment; i.e., teaching a new pilot to fly a tactical aircraft. Significant savings in terms of training dollars have been realized when these simulated systems are used to complement training. As a result of this, plans to use more of this type of training in the military is imminent. This paper does not purport that ITV or simulation or any other methodology used for information presentation is the ultimate panacea. There are disadvantages to using these complex systems. One of these, for example, is the student's psychological realization that he is using a system and, therefore, can afford to assume an attitude of recklessness.

The fact then becomes evident that technical writers/developers, in order to stay competitive and insure themselves a stable position in today's marketplace, must become versatile. This involves acquiring specific skills, in addition to writing, which are necessary to the subject area in which they will be working. Educational developers must be aware of the gap created by the advanced technology being produced and the relative intellectual decline of the target population addressed by the technical writer. Training for these individuals must begin at the undergraduate level. It is no longer adequate to produce graduates well versed in the construction and usage of the English language alone. Course requirements must undergo a process of change. English education majors must supplement their basic language mastery with courses in the technical aspects of the area in which they plan to specialize. For example, if a person desires employment in the electronics industry; i.e., writing electronics material, it is imperative that he have a fairly strong background in electronics. This is absolutely necessary for effective communication with subject matter experts.

In addition, this paper has surfaced the realization that today's technical information presentation is relying more and more upon new technology. This includes, but is not limited to, the areas of computer science and interactive systems. Speculating about future trends, one might strongly recommend that the individual interested in technical writing/training development acquire specific talents pertaining to the areas of television production and computer science. He must also have a background in basic education theory; that is, he must understand what behavioral objectives are and what constitutes criterion referenced instruction, etc. Many corporations, as well as government, require new employees to play a "dual role" of training specialist as well as traditional technical writer; therefore, an individual well versed in both areas has a definite advantage.

We, as technical writers or instructional systems developers, can take steps toward closing and possibly eliminating the hypothetical gap of technology versus education referred to in this paper by encouraging college administrators to implement changes in the curriculum of aspiring technical system designers. When this occurs, a new generation of individuals who are well versed in many facets of training systems will be produced. It is these individuals who will, by effectively assuming a dual role of technical writer/training developer, provide the necessary impetus to close the gap.

Being outpaced by technology is a trend that must be curtailed. It is not beyond conjecture that the hypothetical gap would reach such proportions that we would have to slow the technical development rate and begin what we might best term reverse technological development. Our training future lies in the creation of a new generation of technical/training development specialists who must cope with harnessing the hypothetical gap and must bring it under control by introducing new training methodology.

Duke, Dennis, "Training for Modern Battlefield," videotape produced at WFG-TV, U.S. Army Signal Center and Ft. Gordon, Ft. Gordon, GA, for presentation at National Security Industrial Association Conference, Nov. 1978.

TEACHING TECHNICAL ENGLISH TO SWEDISH ENGINEERING STUDENTS

Gerd Eng
Chalmers University of Technology
S-41296 Göteborg, Sweden

Gerd Eng
Chalmers Univ. of Tech.
Language Consultant
S-41296 Göteborg
Sweden

Graduate students at Chalmers University of Technology have been taught technical reporting in English in a workshop environment. The course consisted of three parts: English grammar and vocabulary, technical writing, and oral presentation. The teaching was done in a small group with individual instruction, and the students wrote and revised a number of assignments. Some specific problems for Swedish students have been identified. At the end of the course the students' communication skills in English were clearly improved.

INTRODUCTION

Communicating information is an essential part of every engineer's or scientist's work. In Sweden, much of the technical and scientific writing is in English in order to be understood by an international audience. However, at the technical universities there are no required courses in English or communication. Formerly, university students had already acquired good skills in these areas in the course of their primary and secondary education, but the Swedish educational system has undergone far-reaching changes, and this is no longer the case. Consequently, the new graduates of the engineering schools to a great extent lack these important skills, and many of the graduate students find it difficult to present the results of their research. Chalmers University of Technology has recognized this problem and offers a course in technical reporting in English to its graduate students.

In this paper I will first discuss some of the problems I have noted when testing the students at the beginning of the course. Then I will describe the course and report on some of the improvements I have found.

LANGUAGE AND COMMUNICATION PROBLEMS

At the first meeting the students are asked to write a brief description of their work and to introduce themselves orally to the class to give me an opportunity to assess their language and communication skills. The results, of course, vary depending on their interest in language and writing and on amount of experience. However, I have found that the majority of the students have problems with subject and verb agreement in English, since the Swedish language uses the same form for all persons. Dangling infinitives are common as are long and complicated sentences. Furthermore, punctuation rules are different from those the students are used to. But above all, spelling creates difficulties. Swedish is a fairly phonetic language, and the students are confused by the variety of spellings of certain sounds in English. Of course, switching from their native language to a foreign language on command in a classroom constitutes stress for many students.

COURSE DESCRIPTION

The course comprises classroom instruction consisting of two-hour meetings twice a week for seven weeks, homework, and individual counseling in connection with the return of corrected papers. The class is limited to twelve students. No grade is given, but 75% attendance and completion of all assignments as well as a final exam are required for credit.

English Language Training

The engineering students at Chalmers have taken English in primary and secondary school, but little emphasis has been placed on writing English correctly. In addition, most engineering students are not language oriented. While studying at Chalmers, they have not been required to take any classes in English, and therefore a review of basic English sentence construction is in most cases needed. About half of the course time is spent on the language training, and as a textbook we have used R. Dodge: How to Read and Write in College,[1] a book used by some American universities in their subject A, freshman composition classes. In addition, we read, analyze, and discuss texts to improve language fluency. A written final exam tests their knowledge of English sentence construction at the end of the course.

The importance of clear, concise, and complete writing in technical communication is pointed out, and weekly written assignments are required.

Technical Reporting

In this part of the course, the principles of technical reporting are taught. The instruction is based on learning-by-doing, and at the end of the quarter, each student is expected to have produced a technical article written according to the requirements for a paper in an international technical journal. Furthermore, all students make a ten-minute oral presentation based on his or her own research.

As textbooks are used M. O'Connor and F. P. Woodford: Writing Scientific Papers in English,[2] as well as handbooks from professional societies.[3-5]

At each meeting a particular aspect of technical reporting is discussed such as the organization of a paper, the elements of an informative abstract, requirements on a good title, effective figures and tables, mathematical English, or steps in preparing oral presentations, including poster presentations.

Articles taken from recent issues of technical journals are used as examples and are analyzed in class. The organization, clarity of language, etc. are discussed.

As mentioned earlier, a written assignment is handed in every week. These assignments consist of parts of a technical paper. The instructor corrects the papers and hands them back to the students at special counseling sessions. Five hours a week are set aside for this activity in addition to the classroom instruction. The students then revise each part as indicated by the instructor.

The reason for having the students rewrite their assignments is that writing frequency alone does not seem to improve the writing very much.[6] A limited number of writing assignments combined with teacher-student conferences and followed by revision have, however, given positive results.

Instruction is also given in how to prepare and give a technical talk, including the use of visual aids. Then we simulate a technical conference, where all the students give a technical presentation, ask questions, and comment on each other's talks. For this part of the course, we use a video tape recorder, so that everybody has an opportunity to see his or her own performance.

RESULTS AND DISCUSSION

The results of the written final exam, the technical paper produced, and the oral presentation provide the basis for an evaluation of the course.

At the end of the course I have found that the sentence-construction skills of the students have improved considerably compared with the initial test. However, the spelling results are less positive.

The final version of the technical paper handed in at the end of the course shows that most students master the concepts of formula writing. However, many still seem to have difficulties in emphasizing the significant part of their own research work in their writing. This is probably not really a writing problem but rather indicates a lack of research experience.

The oral presentations generally go well. The training in speaking English obtained during the course as well as advice on how to present a technical paper gives them confidence. The video tape provides valuable feedback on their performance and makes them aware of mistakes such as reading from the manuscript or talking to the blackboard rather than the audience as well as annoying mannerisms.

CONCLUSION

The course has been very well received by the students, although most of them have complained about the amount of work it required. Their proficiency in English has been raised, and it has made them realize that it is possible by following a formula and by training to write an acceptable technical paper without possessing any special literary talents.

A course like this one is naturally only a "first-aid" measure and has to be followed up with practice to reinforce the learning obtained and to improve analytic as well as linguistic skills.

REFERENCES

1. R. Dodge, How to Read and Write in College. New York, Harper & Row, 1970.

2. M. O'Connor & F. P. Woodford, Writing Scientific Papers in English. Amsterdam, Elsevier, 1976.

3. American Chemical Society, Handbook for Authors. Washington, D.C., 1978.

4. American Institute of Physics, Style Manual, 3rd ed. New York, 1978.

5. Institute of Electrical and Electronics Engineers, Information for IEEE Authors. New York, 1965.

6. J. Stusrud Held & M. Goodman Burns, "Research in Writing: Implications for Future Technical Writing Courses," Paper presented at FORUM '80 - The Second International INTECOM Conf. on Technical Communication, 24-27 August, 1980, Lillehammer, Norway.

ACKNOWLEDGMENTS

I would like to thank Chalmers University of Technology for giving me the opportunity to set up and teach this class and to the students, who have inspired me in my work.

SAVE MONEY, GET RESULTS: PRODUCE YOUR OWN ANTHOLOGY

William E. Evans and Raymond N. MacKenzie
Kansas State University

William E. Evans

Director of Technical Writing

Raymond N. MacKenzie

Instructor

English Dept.
Kansas State University
Manhattan, Ks. 66506
913-532-6716

Academics are well aware of time and money limitations on preparing supplemental examples for use in technical writing courses. Our response to those limitations has been an anthology of assignments, editorial and grading criteria, and--most important--student examples to supplement our text's professional models. We have succeeded in saving time and money for the staff and department while at the same time improving our students' attitudes, critical skills, and writing effectiveness.

I

Before discussing the anthology that we devised for our technical writing course, I would like to set the stage for its evolution. Our course, Written Communications for Engineers, was instituted three years ago as a required course for juniors and seniors in response to employers' complaints about writing deficiencies in recent graduates.

Initially, two graduate students taught the three sections offered in the first year. It was clear that enrollment in the course would increase, and the English department saw the need for development, coordination, and supervision of the course. I accepted the task two years ago. While enrollment increased to eight sections per semester, faculty available to teach the course was severely limited because of other course commitments and an apparent reluctance to teach a course they felt they knew little about. Two regular faculty members and two graduate students currently handle the load. The department and the dean's office seem to feel some ambivalence, reflected in part by a reluctance to commit limited financial resources to research and development for the course.

The College of Engineering, on the other hand, obviously wants a skills course that will better prepare their students for "real-world" writing requirements. They recognize the need for improvement, but they have been reluctant, to this point, to share the financial burden of providing useful supplementary classroom materials that could be distributed through photocopied or duplicated examples.

The shortage of financial resources, secretarial help, and number of staff members having the time to produce supplementary illustrative examples of technical writing, coupled with the observation that students were not responding well to the "professional" examples used in our basic text caused us to compile an anthology that employs student examples exclusively. The anthology has been a success for us in two important ways: 1) it frees the staff and department from the time and expense it had cost to type and duplicate handouts, and 2) it provides the students with "models" that they respond to more readily and critically than they have to "professional models."

Before the anthology was constructed, we spent substantial time duplicating lists of editorial criteria, sample topics, and examples of student work from previous semesters for the current groups. That also involved costs of photocopying or duplicating paper. We considered having that material duplicated through the college's Xerox center, but neither the department nor the staff wished to be committed to "selling" such a collection to the students directly (though that is a common practice in some other departments). Still, the idea of distributing all the supplementary course material in one collection was appealing.

We found the solution by compiling an anthology. Our technical writing staff constructed it, and the department and part of my research money paid for the manuscript typing. We then turned it over to the university's printing department. They photocopied the manuscript and turned it over to the university bookstore for sale. We receive no money from the sales, but we are now relieved of much of the preparation time and all of the expense. While the students incurred additional cost ($4.25 for a 136 page paper bound book), their expenditure was minimal in contrast to other text costs; they also had the semester's work outlined and demonstrated in a single volume.

Our course requires eight writing assignments for the semester. These assignments are keyed to Technical Writing by Mills and Walter. The first few assignments deal with their "basic techniques": definition, description of a process, and classification. For each of these assignments, the anthology supplements the text by providing: an extended statement of the assignment, additional sample topics, two or three examples of papers on these techniques written by students from previous semesters, and an editor's checklist with the grading criteria. The checklist and grading criteria present a series of specific questions aimed at calling attention to the special considerations appropriate to the several papers and to more general questions concerning such elements as audience adaptation, organization, effective and efficient diction, and general format attractiveness.

The four remaining written assignments are perceived as more real-world oriented. The first is a letter of introduction and a resume; the second is a combined memo/informal report/letter project; the third is a progress report; and the fourth is a formal report.

The memo/informal report/letter assignment has been one of our most successful in stimulating student interest. We create a situation in which the student is assumed to work for a company that has received a customer complaint. First, the student assumes the role of supervisor; in that role, he writes a letter to a problem-solving engineer within the company specifying the complaint, asking for an informal report detailing the solution (addressed to the supervisor and higher management personnel), and requesting a letter to the complaining customer specifying the problem's solution and the company's role, if any, in sharing responsibility for the remedy. This assignment broadens student awareness of multiple audiences and multiple rhetorical purposes associated with a single problem. Many students have written about problems they have had first-hand experience with during summer or part-time employment. The anthology provides sample topics and two complete student examples. Since the course text has no example specifically tailored to our simulation, the anthology is especially useful.

The major assignment is a formal report with full apparatus. That assignment is preceded by a progress report on that project which serves a two-fold function: 1) it requires students to be familiar with this kind of report, and 2) it requires them to begin work on the longer paper prior to the last minute. Three samples of progress reports and two complete formal reports are reproduced in the anthology.

Having presented the initial motivation for producing the anthology--primarily time and money considerations--I would like to address some other problems that such an anthology might alleviate for other teachers in different situations.

So far as I am aware, there is no commercial technical writing reader or anthology that extensively or exclusively uses student writing for models. The few that are available (for example, Bowen and Mazzeo's Writing about Science, Harty's Strategies for Business and Technical Writing, and Sparrow and Cunningham's The Practical Craft) draw from published professionals or scholars, focus on teaching techniques, or show students that management personnel in business and industry value good writing. Each of these serves a purpose, but not the one we try to address. Furthermore, our discussions with publishers suggest that no such reader is likely to be forthcoming for some obvious reasons. Publishers need volume sales to make a profit. Because "technical writing" courses vary so widely in their makeup and writing levels--from special groups (such as our engineers) to all-encompassing groups from the basic sciences to agriculture to computer science to business and from freshmen to seniors--it is nearly impossible to target a market.

One of the greatest benefits of creating your own anthology is the opportunity to adapt it to widely diverse student interests. The selected examples can be narrowly or widely distributed by discipline and by writing level. If the anthology is produced by a photocopy process, it can also be revised easily to meet changing student constituencies. If the basic course requirements remain constant, the anthology can be adapted easily to similar courses. In the fall semester, 1981, we will offer a new course entitled "Written Communication for the Sciences." While it will be directed to juniors and seniors, the audience will be far more diverse than our course for engineers. Initially we will have no student models available for a comparable book, but by the end of the first semester, we expect to be able to prepare one for that course using papers from the first group. If the new anthology does not prove to be adequate, it can be revised during the following summer.

A final consideration for me is the matter of convenience. Because the enrollment in our engineering course has grown so rapidly and because we have had to recruit staff from the graduate assistant, temporary instructor, and instructor ranks, the anthology has become a useful supplementary training tool for new teachers. Not only does the anthology provide them with a fully developed syllabus, it also provides a standard by which teachers new to the course may judge their students' work. Whether this assumption is valid depends, of course, on the accuracy with which we have selected good examples for the norm.

In short, there are several potential benefits accruing from making your own anthology. I have tried to focus on pragmatic concerns in teaching and administrating technical writing courses. We turn next to more detailed consideration of psychological and theoretical benefits for the students.

II

There is one potentially serious theoretical problem in using an anthology such as this; I wish to address that problem now. Everyone associated with the discipline, from students to teachers to practicing professionals, agrees that if there is one thing a course in technical writing should be, it is practical. In the university there are some who fear to use that adjective 'vocational,' but as a matter of fact our aim in a tech writing course is to prepare students to do the kind of writing they will be asked to do on the job. With a clientele such as ours at Kansas State the imperative is even clearer. These students expect that we will give them practice in writing that is immediately and directly applicable to the demands their jobs will make on them. If the opposite of "vocational" is "academic," than an academic course is not at all what either they or we want.

Given this emphasis on the 'real world,' then, one might question the value of using student papers as models. Shouldn't the students be given models from industry, from government, from the military? If we have students imitating other students' work, aren't we encouraging a kind of inbreeding, and helping to maintain the gap between the university and industry?

I don't think so. In the first place, the teachers can make sure that the assignments they make are, as nearly as possible, real-world assignments, rather than essays, themes, or other writing types which only rehash the standard composition course. Reference has already been made to our memo-informal report-letter of adjustment assignment, for which many students choose as their subject a situation that actually occurred in their summer of part-time employment. The major assignment for the course is a full-dress formal report. Here, the student is encouraged to use the report as an opportunity to write up and formalize a project he or she is engaged in in one of the engineering courses. In doing field or lab work and then having to organize and formally present the results of that work, the student is doing exactly what will be required on the job. Other assignments include a progress report, a letter of application, and a resume--not at all mere academic exercises.

An equally important justification for the use of student models is the psychological boost they can give a class. Most tech writing texts include samples of professional writing. This, of course, is a sensible inclusion, as it lets students see the principles they have been learning in practice. But if professional writing is all the student sees, a number of negative results are possible.

First, most of our students seem to see themselves primarily as students, not as near-professionals. Perhaps because they see the end of college life approaching, many seem to relish it all the more and draw a thick line between themselves and practicing engineers. As a result, many simply tune out when confronted by examples of professional writing. One of the most arduous tasks I have faced in the course is leading a discussion over a professional report. The first time I tried it, I was rather amazed at how utterly bored the class was with what was, to me, an excellent and stimulating piece of writing.

When I pressed the class, I uncovered a number of interesting reactions. I should not have been surprised, but I was, to find several saying that they found the report difficult to follow because of its highly technical content; many felt that they didn't know--or shouldn't be expected to know--enough to follow the report. Other students indicated that they found the report dull because it was not what they had to be concerned with: they had to write papers to pass this course, not write <u>real</u> reports. The industrial report, as far as they were concerned, just wasn't of practical interest to them. Underlying all this, I believe, was a feeling on the part of many that they simply couldn't do as well as the professional writer had, and as a result were not interested. Whether this is a laudable attitude in a student, I won't offer an opinion; but it was a very real attitude, and a very real problem.

Using student models has produced a dramatic contrast. It is not merely that better class discussions result--though indeed they do: students seem to have read the reports much more closely, and questions and comments come up easily and naturally when we discuss one. Much more important is the upturn in the students' level of interest in what is the subject of the course--effective writing. This heightened interest grows out of an immediate pragmatic sense: "here's how a student last semester met this assignment. What can it tell me about how I should proceed?" I also have found very interesting an occasional note of competitiveness: several have indicated they felt they could do a better job, both in the writing itself and in the technical content. I believe the students are now trying harder and taking their tasks more seriously, and I am now seeing some better writing than I have in the past--writing which is, though generated by student models, more professional all around.

Another indicator of the anthology's success came on the course evaluations at the end of the semester. The students were asked to comment on the helpfulness of each of our texts, and to evaluate them as either "very helpful," "helpful," or "not helpful." Over half found the anthology "very helpful," and the remainder said it was "helpful." Some commented that the book helped them to think of topics, and some said that it was the most useful guide in understanding the assignments. The closest thing to a negative response came from a student who said this: "It was helpful, but most of the papers in it were not all that good. Surely you can find better examples to include." In my view, that is a very positive response, for that student had spent some time reading closely, comparing, and evaluating. When we can get our students that closely involved in thinking about writing, we are well on the way to making them into first-rate writers.

CONCLUSION

A student anthology, then, can help take the burden off overworked teachers and secretaries and a strained departmental budget. It is true that the costs are passed on to the student, but those costs are minimal, and student response indicates that the anthology is perceived as a worthwhile investment. While some might argue that the student should be using only professional models rather than those written by other students, the boosts in class morale and in student interest in the course are benefits of some importance. Most importantly, though, the student models result in better writing--an end which surely justifies the means.

THE TECHNICAL WRITING CONSORTIUM: TWO-YEAR —— FOUR-YEAR COOPERATION

Joanna M. Freeman
Pittsburg State University

Dr. Joanna M. Freeman
Pittsburg State University
Coordinator, Technical Writing
Pittsburg, KS 66762
Basic Technical Writing and Business Writing, Iowa State U. Press
Technical Writing Papers:
International Institute
for Community Colleges,
National Conference on
Developmental Studies

Membership in a technical writing consortium is advantageous both to the local university and the neighboring community colleges because the consortium guarantees accreditation of the community college courses and increases enrollment in the advanced technical writing courses and internship programs of the university. This paper describes the textbook, syllabus, daily schedule and writing assignments suitable for member schools. The continuing success of the informal consortium and of the formally structured consortium——federally or privately funded——depends primarily on the leadership of the director, the commitment of the administrations and the number of participating schools.

Dynamic technical writing programs, cooperatively planned and administered by a consortium, can benefit both the local university and the neighboring community colleges. Such programs establish close professional ties among the faculty of the participating schools and provide a needed community service as they revitalize English departments by attracting students from all departments on campus and from the community itself.

RECIPROCAL BENEFITS

An area wide technical writing consortium can be advantageous to the community colleges because it sets the standards that will guarantee the transfer of community college credits to the university. Sharing the expense of guest speakers and field trips with other schools in the consortium is also beneficial. Certainly, the prestige that comes with course accreditation, with well—known speakers and with shared teaching materials is desirable at the community college level.

There are also advantages for the four—year university that forms a consortium with community colleges. Such cooperation is a valuable recruitment tool; cooperation instead of competition attracts community college transfer students to the university. These transfer students can bolster enrollment in the advanced technical writing courses and in the internship programs. Further, community college faculty can develop new job opportunities in their respective areas for university interns.

INITIAL IMPLEMENTATION

One school, often the university, must take the initiative in forming the consortium. A technical writing coordinator who has contacts with area community colleges can meet with the community college technical writing instructors on an informal level to discuss the needs of the students enrolled in each school. Catalog descriptions of existing technical writing courses can be studied. Once course prerequisites, the level of instruction, and the objectives are discussed and agreed upon, then the detailed work of the technical writing consortium begins.

Using the same textbook will draw participating schools closer together, and selection of that textbook is a key decision. It must include specimen writing from all areas of the job market and must have a wide range of writing assignments suitable for the community college students enrolled in two—year technical and industrial programs, for non—traditional students who may be employed full time in the community and may be enrolled in only one course as a special student, and for the baccalaureate degree candidates who will continue in a graduate school program. The most suitable textbooks are those which include specimen writing from technicians and professionals actually working in their respective fields.

When all schools in the consortium use the same textbook, they can also agree on the content of the introductory technical writing course. The consortium can plan a common syllabus and day by day schedule in rather general terms, stipulating the number and types of papers to be written.

Then instructors in each school can individualize the syllabus with specific writing assignments appropriate for their own students. The sophomore level technical writing course for the community colleges requires the most flexibility in assignments because it serves the needs of so many different types of students. The universty technical writing instructors, either working alone or with others in the consortium, can schedule writing assignments that will prepare their students to handle successfully the report writing which will be required of them in the lower management positions they will fill after graduation. Their topics will reflect the additional training they have had in their major areas. Other writing assignments can develop the skills required in graduate school programs.

Students at every level and in every technical field and profession will be writing mostly for other knowledgeable professionals and must produce writing that tends to be more informative than persuasive. Early in the semester the various methods of exposition can be taught (technical definition, technical description, explanation of a process, instructions and analysis), some in a context—free framework. Later in the semester, several of these methods of exposition can be combined to produce feasibility, progress and formal reports, each pertaining to the students' major areas.

In addition to textbook selection, syllabus writing, and assignment planning, members of the technical writing consortium can also in their annual or semi—annual meetings articulate grading philosophies, exchange ideas on classroom activities and information on guest speakers and field trips to government and industrial technical writing installations. Such an idea exchange provides valuable input both at the community college and university level.

THE INFORMAL CONSORTIUM

The informal consortium usually has a rotating directorship funded by the director's own school for a designated period. This type of consortium seems to be preferable for schools with smaller enrollments in sparsely populated areas. A rotating directorship allows each school to preserve its own autonomy and encourages the participating schools eventually to share the operating expenses, each director's own school providing the funding and services necessary for the specified period. Often the director performs his duties on an overload basis in addition to his regular teaching duties and committee work.

THE FORMAL CONSORTIUM

Establishment of an informal consortium is usually the first step toward a formally structured consortium with a permanent director funded through federal, state, or private grants. Federal or private funding can be secured to develop the cooperative program already underway as long as each school listed in the application shows its commitment to the consortium both on the faculty and administrative levels.

Information about federal grants available for the funding of technical writing consortia can be found in the **CATALOG OF DOMESTIC ASSISTANCE**, in the **IRS ANNUAL REPORT FORMS—AR 990**, and in an April issue of the **FEDERAL REGISTER**. Information about state funding can be requested from each secretary of state. A list of private foundations interested in funding education projects can be found in the **FOUNDATION DIRECTORY, FOUNDATION GRANTS INDEX**, and the **ANNUAL REGISTER OF GRANT SUPPORT**. They include Ford Foundation, Rockerfeller Foundation, Lilly Endowment, W.K. Kellogg Foundation,Kresge Foundation, Exxon Education Foundation and the General Electric Foundation. A computer search can provide the specifics of individual projects, amount of the grants and duration of the funding period.

There are specific steps in writing a successful proposal. After sources of funding have been located and information requested, each funding source must be analyzed and studied carefully. Contacts on the informal level must be developed. Books that would be helpful in preparing a grant proposal would include Krathwohl's **HOW TO PREPARE A RESEARCH PROPOSAL** and Lefferts' **GETTING A GRANT**. In the written proposal, most foundations ask for a brief introducton that includes a biography of the applicant and of the personnel involved, a statement of the existing problem, methods of solving the problem, evaluation of the probable effectiveness of the proposed project, and a budget. Usually the abstract is written last, and then several follow—up letters are written before the funding decision is announced.

Funding can be used to hire a full—time or half—time director, to provide an office and secretarial help, and to offer a more highly structured program and comprehensive services which might include the establishment of a central technical writing library. These libraries can house textbooks written specifically for community college students enrolled in two—year technical and industrial programs, textbooks for university students training for lower level management positions and graduate schools, and textbooks that contain such a wide range of writing assignments and specimen writing that they are suitable for students at both the community college and university level. Some texts are so comprehensive that they can be used in both introductory and advanced courses. In the last decade several collections of essays representative of scholarship in the technical writing field have become available.

THE SOUTHEAST KANSAS TECHNICAL WRITING CONSORTIUM

The technical writing consortium in southeast Kansas is very loosely structured. This informal type of consortium was formed at the instigation of Pittsburg State University to relieve the pressures of overcrowding in introductory technical writing courses. All our sections of technical writing were being filled each semester; additional degree programs were requiring technical writing each year. Because we did not have the funding to offer as many sections as were needed, we decided that encouraging sophomore level technical writing courses at area community colleges would decrease enrollment in our introductory courses and perhaps increase enrollment in our advanced technical writing courses. We also believed that it would be better to assist the community colleges in developing acceptable introductory courses instead of allowing them to develop courses that might be basically remedial English courses or extensions of freshman English and therefore perhaps not eligible for transfer credit.

Three area community colleges were invited to participate in a very informal type of consortium; two accepted. As coordinator of the technical writing program at Pitt State, I met first with the English faculty of each school. We discussed the goals and objectives of the introductory Pitt State technical writing course, the needs of the sophomore level students enrolled at their institutions, and finally the advantages of the consortium for each participating school.

At subsequent group meetings we agreed that our overall objective should be to train our students to write the reports required of them in their first years on the job. We agreed that we could best reach our objective by insisting on a prerequisite of two semesters of freshman English or the equivalent. We agreed that our students should write several short reports during the class period to simulate writing under pressure on the job and should write a longer, formal report requiring some library skills and some training in their major areas.The short reports should be revised.

Each school published approximately the same course description in its catalog:

Introduction to Technical Writing. 3 hours. Practice in writing technical descriptions and definitions, instructions, explanations of processes, progress reports, lab reports, formal reports, and basic business correspondence with emphasis on audience analysis, organization, style and format. Prerequisites: English Composition 101 and 102.

Our next step was to select a textbook which included the specimen writing and writing assignments appropriate for our three types of students: **BASIC TECHNICAL AND BUSINESS WRITING** published by Iowa State University Press in 1979.

Using this text, we wrote a syllabus and day by day schedule which included reading assignments, eight short reports (seven of which were to be written in class) and the formal report. Article—report, memo, letter and blank forms were to be used. The syllabus was written in rather general terms:

February 9—Chapter 4 continued (explanations of processes)
February 11—Paper 3
February 13—Revision of Paper 3

Instructors in each participating school individualized the syllabus with specific writing assignments appropriate for their own students.

The following writing assignments could be used for students enrolled in a two—year associate degree program:

1. Explain a basic testing procedure used in your major area.
2. Analyze career opportunities in your chosen field.
3. Propose a solution to some current problem at your school: parking, vandalism, loss of library books etc.
4. Write a letter of inquiry to a local company. You need this information for the formal report you are writing in technical

writing.

5. Write a formal report on the techniques used in electric arc welding.
6. Assume that you are the president of a campus organization. Write a memo informing members of your group that they must come to a two—hour study session each week.

Since non—traditional students may be employed full—time in the community, the report forms used by their employers can be incorporated into their assignments. The following writing assignments would be appropriate for them:

1. Write instructions telling a new employee how toperform some process.
2. Write a memo to your supervisor recommending the purchase of a piece of equipment to be used on the job.
3. It is the end of the fiscal year. Write a follow—up letter to your charge account customers who have done businesswith you during the last year.
4. An ecology spokesman has written a letter in which he has accused the company you work for of air pollution. He has challenged you to a public discussion of the problem to be broadcast over a local TV station. Answer the letter.
5. Write four expanded definitions of a tool or piece of equipment you use in your work, each one suitable for a different audience: (1) president of the company, (2) workers in your department, (3) third graders and (4) foreign visitors.
6. Propose that the Company Beautification Committee take a specific action to improve a certain area of the grounds.

University students who are enrolled in four—year degree programs and who have had additional training in their major areas would find the following assignments profitable:

1. Select a technical periodical to which you might reasonably submit an article. Write a detailed analysis of that periodical.
2. Examine the papers you have already written in this course. Select one that would be suitable for publication in a technical or semitechnical periodical that you have examined. Rework the paper so that it meets the standards of the periodical.
3. Write a feasibility study determining whether or not existing equipment should be repaired or replaced.
4. Write a formal report in which you explain the various treatments for the emotional problems related to advanced alcoholism and evaluate their effectiveness.
5. Analyze the advantages of no—fault divorce.
6. Assume that you are a first year teacher. Write an unsolicited proposal to your school board suggesting that a certain course be added to the curriculum.

Some writing assignments are appropriate for all students:

1. Write (1) a thorough, professional resume which effectively emphasizes your unique strengths as a job applicant and (2) a covering letter of application which highlights and supplements the resume. Apply to an actual company, organization, institution or agency to the appropriate person by name. Apply for a specific job for which you are qualified based upon your education, interests, and experiences.
2. Assume that you have written an unsolicited letter of application for employment to a company, but you have not received a reply. Write a follow—up letter.

In the future, members of the consortium plan to discuss grading philosophies and to arrange public evening seminars in which community leaders explain and illustrate the types of writing that they do in their work. They hope to schedule training seminars in addition to the planning meetings.

Relationships between the English faculties of the member schools have improved dramatically. And another area community college is now interested in participating in the consortium.

THE SUCCESSFUL ON—GOING CONSORTIUM

The success of both formal and informal technical writing consortia depends to a great extent on the leadership of the designated director, the commitment of the administration of each participating school and the number of schools involved.

The director must have the dedication to accept leadership responsibilities of the informal consortium on an overload basis, or he must have release time to supervise mailouts, newsletters, and organize field trips to area businesses. Full—time directors of formally structured consortia can schedule all the necessary seminars, develop a file of internship opportunities available in the area, visit interns on the job, and publicize current job opportunities.

The administration of each institution must be firmly committed to the technical writing consortium. The administration must offer encouragement to the involved faculty in terms of release time for seminars and public recognition. If they do not actively support the technical writing consortium, then it will fail. Usually faculty members cannot by themselves weld together a consortium of area schools.

The number of schools in the consortium often determines the success of the effort. The smaller the number, the more likelihood both of initial and continuing success. The alliance, like everything else, requires the personal touch. Annual meetings and semi—annual seminars are easier to schedule for fewer schools. A smaller group requires the representatives of each school to share ideas on writing assignments, classroom activities, field trips and guest speakers and to make the necessary contacts in the community.

The technical writing consortium——whether loosely organized or very formally structured——can develop an ongoing program vital to the prestige of the local university and the area community colleges and, finally, can provide the expertise that will guarantee its continued success.

THE COMPLEXITY OF BEING SIMPLE

Reeta Garber

EG&G Idaho, Inc.

Reeta Garber
EG&G Idaho, Inc.
Communications Specialist
Box 1625
Idaho Falls, Idaho 83415

208-526-9450

M.A. in English; Vice Chairman of the Eastern Idaho Chapter of the STC.

This paper suggests that editors need to define and clarify the meaning of simplicity, that undue preoccupation with simple writing is unwise, and that simplicity in itself should not be the only goal of an effective piece of technical writing. Many rhetoric handbooks state that artificiality of tone produces pretentious, wordy and awkward writing. However, these texts seldom go beyond such generalities. Editing technical writing demands that editors be aware of the exception to these generalities. For instance, editors should know that polysyllabic words do not always have precise monosyllabic substitutes, that shorter sentences are not necessarily easier to comprehend than longer ones, and that the passive voice is at times preferable to the active voice. The goal of the editor is to make writing functional and useful, not simple-minded.

Writers of exposition generally agree that simplicity in writing is more desirable than elaborate and decorative prose. In technical exposition, simplicity is more than desirable; it is necessary. According to G. H. Lewes, the desire for simplicity has a psychological basis in impatience with superfluity (i.e., intrusive rhetoric). In his Principles of Success in Literature (p. 134), Lewes states that "The first obligation of simplicity is that of using the simplest means to secure the fullest effect. But although the mind instinctively rejects all needless complexity, we shall greatly err if we fail to recognize the fact that what the mind recoils from is not the complexity, but the needlessness."

Recognizing what is "needless" can be a complex matter requiring careful consideration. Many editors, weary of reading mounds of pompous, affected prose, react by changing or eliminating what is not needless. Therefore, one of the more common deficiencies in technical writing today is the indiscriminate oversimplification of written material, regardless of its intended audience. The resultant style, often condoned by technical communicators, is dull, mediocre and just as offensive as the convoluted, prolix style which it purports to correct.

Editors in particular are responsible for this trend toward deadly writing. Most handle many varieties of "technical writing." For instance, many companies produce a variety of written material: manuals and guides that spell out policies, procedures, and instructions for professionals and nonprofessionals; promotional literature advertising scientific and nonscientific programs and campaigns; journal articles and conference papers for experts in technical fields; and scientific and technical reports informing scientists and technicians of new developments. Editors are oftentimes pressured to make all these types of writing simple and clear. Many editors yield to this pressure, disregarding the fact that procedures manuals and scientific reports vary widely in purpose, content and levels of complexity. Manuals are written for a much wider audience than are scientific reports, which are intended for a more limited audience. It is difficult to simplify scientific reports by applying to them such "rules" for simple writing as: 1. avoid polysyllabic words and "make the unfamiliar, familiar;" 2. shorten sentences; 3. use the active voice.

In technical writing, arbitrary shortening of words and sentences may lead to distortions of meaning. Sometimes, short equivalents for polysyllabic words do not exist. Long sentences requiring qualifying statements cannot always be successfully trimmed to the prescribed twenty words per sentence. The use of the active voice does not always enhance clarity. Unfamiliar concepts requiring unfamiliar terminology cannot always be easily rephrased for a lay audience. In this presentation, I intend to discuss in more detail the applicability of these standard "remedies" in technical writing.

In much scientific writing, long words are used because they are specific and economical, and seldom have shorter equivalents in general usage. It is absurd to expect the writer to find substitutes for such precise words as "dendrochronology" and "crystallographic", especially if they are crucial to the meaning of the sentence. "Simplifying" such polysyllabic words under the mistaken assumption that all long, unfamiliar words are jargon, leads to verbosity and inaccuracy. Mark Twain advised journalists of his day to use the right word, not its second cousin.

Obviously, an editor cannot substitute shorter words for many polysyllabic words, because no synonyms exist. Moreover, there is no need for such substitutions, since most technical reports are not written for the lay audience. However, certain editing skills may be used to clarify the dense prose that is sometimes caused by the use of technical terminology. The editor for even a limited audience may distribute technical details throughout the paragraph, rather than concentrating them in one or two sentences, and may separate words or phrases bunched together too compactly.

As a consequence of advances in science and technology, certain descriptive words have of necessity been coined. Some such words have become part of the standard vocabulary (for example, the word "antihistamine"). However, editors are rightfully wary of another type of coinage--that type which changes a noun to a verb out of ineptitude. Such coinages as "containerize", "pallatize" and "prioritize" should be eliminated because they are superfluous.

Editors are oten told that understandable writing is invariably characterized not only by short words, but also by short sentences. However, reading ease does not always depend upon the number of syllables in a given word or upon the length of the sentences in a given paragraph. Much depends upon the audience. Only an expert in the field can read a technical report quickly and with full comprehension. A lay person without the requisite technical knowledge would have a difficult time comprehending the same material no matter how short the sentences might be.

Some editors miss this simple point and insist on shortening as many sentences as they can. These editors also place an inordinate amount of faith in readability indexes. Rudolf Flesch's formula for reading ease and Robert Gunning's "fog index" do not provide definitive measures of readability in technical reports. Obviously, technical reports often have an abundance of unavoidable polysyllabic words, which builds up the alleged "fog factor." Also, long sentences are fairly common in such writing because they are often comprised of statements requiring qualification. Such sentences can give, and give concisely, more information than several short sentences. For example, consider the following sentence:

> Since the primary purpose of the irradiation experiment was to evaluate the effects of irradiation on the ceramic cements and the electrical properties of the small coils prepared with ceramic insulated wire, the experimental details related to the graphite, mica, and other material included in the experiment will be discussed in Appendix B. (53 words)

Even though this sentence could easily be broken into two shorter sentences, it is still understandable and effective. It shows a cause and effect relationship which would be lost if it were to be divided in accordance with the twenty-word limit prescribed by several readability indexes. Undeniably, long and complex sentences are often not as easy to understand as short ones. However, short and excessively simple constructions (subject-verb-object) are rarely able to indicate logical relationships between ideas, and may contribute to a choppy and monotonous style.

Editors often believe that a third essential element of effective writing is the obligatory substitution of the active for the passive voice. Unfortunately, many scientists and engineers tend to overuse the passive voice because thay have been taught since junior high school that good technical exposition is unprejudiced and detached--and that such a viewpoint is only achieved via the self-effacement of the passive voice. Consequently, in the estimation of such writers, they are suiting the style to the content, since technical documents emphasize things, rather than people. The confrontation of hard-line advocates of the passive and active voices not only causes hard feelings, but also produces poor prose.

Some editors, when confronted with overly "passive" prose, attempt to "humanize" it by indiscriminately changing every single passive construction to the active voice--much to the chagrin of the author, and to the detriment of the prose. Many editors (some of whom also teach classes in technical writing) use examples from literature to prove to the writer how strong and effective the active voice can be.

After all, they say, "Hamlet's soliloquy, 'To be or not to be', has twenty-five verbs, only two of which are in passive constructions." What such editors fail to realize is that technical writing contains neither soliloquies nor speeches. It merely reports on experiments and tests conducted and results obtained therefrom. It is not designed to draw attention to the initiators of the experiments.

Since technical exposition is concerned with the action (the experiment, test, etc.) described by the verb, the action itself may legitimately become the subject of the sentence via the passive construction. Consider the following examples of passive constructions:

A small-break experiment was conducted in the reactor core.

The samples' general condition was observed, including their surface texture and cracking.

Resistance and inductance measurements on the irradiation coils were made and compared with the preirradiated data.

The subject in the first example demands a passive construction because the emphasis is on the experiment itself, not on its conductor. In example two, the general condition of the samples is the subject, and in example three the two kinds of measurements are being emphasized.

There is no doubt that the use of the passive voice leads to an impersonal style. It is difficult for an engineer to be personal when he is describing what transpires inside a computer, or for a chemist to be personal when explaining a chemical reaction. But just because a style is impersonal does not mean that it is uninformative. Active voice personal pronouns do not make sentences any more informative, unless the emphasis is on the person, rather than on the action.

I cannot condone overzealous, uncritical editing. Overreaction in any of the above-delineated areas may lead to simplistic writing, reduced to the lowest common denominator. Long words can be effective in the proper context. Too much "simplicity" can also lead to inaccuracy, and is therefore as undesirable as long-winded prose. In summation, editors must avoid what George Orwell in another context called "... merely the substitution of one piece of nonsense for another" (<u>1984</u>, p. 37).

CORRESPONDENCE/EXTENSION EDUCATION-- A PERSPECTIVE TO INDUSTRY

James G. Hamlett
General Electric Co. Syracuse, N.Y. 13221

James G. Hamlett
General Electric Co.
Project Leader
CSP 4-18, P.O. Box 4840
Syracuse, NY 13221
(315) 456-7712
Associate Fellow-STC, Mgmt
Theory & Practice Stem Chmn
27th ITCC, Eng'g Student Adv'r
Syracuse U, Author/Lecturer

Over the years biases and distortions have been thrust upon correspondence/extension (C/E) education. For the major part they were unwarranted. Simply, these distortions occurred through factual ignorance of this field of education. Admittedly, however, some truths existed from a few schools of questionable stature.

Unfortunately insufficient information was presented to the scientific/business community. Thus a communication gap has existed.

This paper addresses these problems to present this education segment in its proper perspective. The author has the unique advantage of graduating from both a major university and accredited C/E schools. Thus he feels qualified in making an unbiased proposal of the facts.

INTRODUCTION

Correspondence/Extension Study Defined

Correspondence/Extension (C/E) Study is the enrollment in a selected course of intended learning from an educational institution specializing in conducting this type of training while utilizing the postal services. This area of education provides a very useful service to a special segment of the learning population. It should be thoroughly understood that correspondence/extension education is not designed or structured to supplant residential centers of postsecondary education.

The traditional college or university is designed primarily for the young resident student (1). As good as they are they do not fully meet the needs of a major section of the population. As evidenced by over 4,000,000 students (2) enrolled from throughout the U.S. and foreign countries, specific needs are not fully addressed. Table 1 below presents these problems. Table 2 presents solutions to the cited problems.

TABLE 1 TYPICAL PROBLEMS OF POTENTIAL C/E STUDENT

NO.	WEIGHTED	PROBLEMS
1.	Critical	College per credit hour cost very high.
2.	Very Important	College location too distant from home necessitating high travel costs.*
3.	Very Important	Cannot take time off from job (if employed) to attend resident college.*
4.	Very Important	Cannot leave family (married) to attend resident college.*
5.	Important	May not be able to stand the pace of college classroom if a large time gap exists between last schooling date and present.
6.	Important	College class usually possesses a very high instructor-to-student ratio when compared to C/E instruction.
7.	Important	Communication from another town with college instructor can be difficult when attempting to reconcile a confusing assignment.
8.	Important	Some employers will reimburse employee for a job-related college course, but not a job-related C/E course due to non-up-to-date knowledge of C/E training quality.
9.	Important	Advanced degree, education generally not available through C/E at this date.
10.	Important	College course enrollment limited to predetermined class size and fixed start date.
11.	Less Important	Some employers do not recognize completed C/E training due to step 8 bias.
12.	Less Important	Resident classes may prevent student from progressing at a rate related to work and home schedules.

*Some C/E schools i.e. ICS Center for Degree Studies School of Engineering, and the School of Business requires a resident laboratory training period.

TABLE 2 TYPICAL CURRICULUM ADVANTAGES FOR POTENTIAL C/E STUDENT

NO.	WEIGHTED	ADVANTAGES
1.	Critical	Lower cost than college tuition.
2.	Very Important	No travel costs (usually) *for course completion.
3.	Very Important	No time off required (usually)* from job to complete training.
4.	Very Important	No time off required (usually)* from family (married) to complete training.
5.	Very Important	Student studies and proceeds at own pace without formal classroom peer group or fixed semester completion date pressures.
6.	Very Important	Several schools specialize in providing complete vocation change training in a minimum time frame.
7.	Very Important	No. of colleges and universities with C/E programs are increasing and expanding this service.
8.	Very Important	Although hundreds may be enrolled in some training classes, professional instructors assigned to your area actually maintain a one-on-one relationship through postal system or toll-free direct-dial system.

PHILOSOPHY

The educational spectrum of C/E runs from Accounting to Yacht Design. Courses are available covering a depth of instruction from a single-subject course of a few weeks to an in-depth pursuit covering approximately three years. These are average times only. Naturally some students complete their training in less or more time as individual requirements dictate. Entire career planning with free consultation is available from most of the larger C/E institutions with a full range of subject courses leading to a career completion, whether new or changed.

Texts used in C/E education are rather unusual. They are written, but tested for maximum student learning efficiency and retention. They also make excellent reference material.

C/E education constantly researches and monitors their curriculm effectiveness and training techniques. In addition to the two-way postal system many employ audio and video cassettes. Toll-free telephone communications are often utilized to permit students to talk to their instructor on an individual basis concerning subject problems. In the larger C/E institutions most instructors, as in academia hold advanced degrees and are recognized as experts in their profession.

Computer-Aided Education is now developed and only the economics of the application of student accessability remains to be resolved to further enhance C/E capability.

The true test of any C/E program is how well the graduate performs as opportunities occur in the business/industrial/scientific/educational world.

*ibid

STUDENT PROFILE

According to a study conducted by the University of Washington in Seattle (3) a profile has been completed of their participants intending to further pursue specialized training through the University of Washington's Continuing Education activities (conferences and institutes, media development, human resources development, independent study, etc.). People looking toward a university to provide their continued education are: above college age, fairly equally divided by decades of age: (30-60 and then tapering off a bit in the 70's), well educated with over 50% having more than one degree, 66% women - 34% men, and all branches of work are spanned with the professions especially well represented. Courses taken are almost equally divided from work related to general interest.

A study similar to the student profile by the University of Washington has been undertaken by the noncollegiate C/E community. However, a sampling (4, 5) has revealed: a wider age spectrum occurs from 18-70 but the average is 25-34 with decade age peaks occurring due to early necessity of marketable skills and midcareer changes. Since there is a larger younger entry, the educational start level is not as high as the collegiate. The ratio of men to women is approximately 75% to 25%. The professions are not quite as highly represented as in above studies again due to the higher youth entry. The courses taken are vocationally oriented.

A key concern of an employer observing the profile of a C/E student employee follows. The typical C/E graduate student employee is: a self-starter, has tenacity of purpose, dependable, responsible, finishes what is started, has strong initiative with a good eye on the business. These proven qualities represent a strong ratio of knowns vs unknowns.

ACCREDITATION

Few teachers, counselors, students or employers have more than a vague idea about C/E education, and even fewer people understand the meaning and purpose of accreditation (6).

Definition

The word accreditation, simply stated (6) for this paper, is C/E "certification by a recognized body that a school has voluntarily undergone a comprehensive study and examination which has demonstrated that the school does in fact perform the functions that it claims: that the school has set educational objectives for students who enroll, and furnishes materials and services that enable students to meet these stated objectives."

This covers all programs, courses, and C/E "endeavors of a school, including degree, nondegree vocational and avocational program." Unlike regional or specialized accrediting agencies, the Accrediting Commission of the National Home Study Council (NHSC) provides these institutions with a single-source national recognition."

Standards

The following Standards and Benefits of Accreditation discussions is furnished through the courtesy and permission of the NHSC.

For fifty-four years the National Home Study Council has been the standard-setting agency for home study schools. The Council has progressively raised its standards. Its accrediting program employs procedures similar to those of other educational accrediting associations (7).

Each accredited school meets the following standards:

- Has a competent faculty.
- Offers educationally sound and up-to-date courses.
- Carefully screens students for admission.
- Provides satisfactory education services.
- Has demonstrated ample student success and satisfaction.
- Advertises its courses truthfully.
- Is financially able to deliver high-quality educational service.

To become accredited each school has made an intensive study of its own operations, opened its doors to a thorough inspection by an outside examining committee, supplied all information required by the Accrediting Commission, and submitted its instructional material for a thorough review by competent subject-matter specialists. The process is repeated every five years.

The Accrediting Commission

The Accrediting Commission establishes educational, ethical and business standards; it examines and evaluates home study schools in terms of these standards, and accredits those which qualify. The Commission is recognized to accredit both private and non-private correspondence institutions in the United States.

Members of the Accrediting Commission

The distinguished staff include:

- Fred F. Harcleroad, Chairman, Professor of Higher Education, University of Arizona
- John F. Thompson, Vice Chairman, President, McGraw-Hill Continuing Education Center
- Gerald O. Allen, Chairman of the Board, Cleveland Institute of Electronics
- G. Dewey Arnold, Regional Managing Partner, Price Waterhouse & Co.
- Jules G. Fleder, President, Westlawn School of Yacht Design
- James E. Godfrey, Director, Truck Marketing Institute
- John D. Phillips, President, National Association of Independent Colleges and Universities
- Harold P. Rodes, President, General Motors Institute (1960-1976)
- Hester L. Turner, President, American Forestry Association

Benefits Of Accreditation

Some of the major benefits of accreditation for C/E schools include:

- Reliance by counselors, employers, educators, governmental officials, and the public on the accredited status of a school as a reliable index of quality.
- An expression of confidence in the policies and procedures of the school by its peers -- a lasting source of pride for the school.
- An external source of stimulation to improve services, program, and staff through periodic evaluations by an outside agency and self-study.
- Assurance of high standards and educational quality through the adherence to established criteria, policies and standards.
- Some recognition of status by certain states under legislation and regulation as well as recognition given by federal, state and local agencies in referring students to accredited schools.
- The benefits and opportunities given to accredited schools by federal law such as student loan acts, and by federal agencies, such as the Veterans Administration.
- The listing of the school and its courses in the Directory of Accredited Home Study Schools.
- The listing of the school along with other higher education institutions in the Council on Postsecondary Accreditation directory, Accredited Institutions on Postsecondary Education, distributed by the American Council on Education.
- The use of the NHSC seal and reference to accreditation by the Accrediting Commission.
- Students qualify for tuition reimbursement under several industry plans requiring enrollment with only accredited schools.
- Eligibility for participation in the academic credit evaluation procedure conducted by the Office on Educational Credit of the American Council on Education.
- It expedites acceptance of school advertising by newspapers, magazines, radio and television stations and other advertising media.
- The opportunity to serve on accrediting examining committees visiting other schools.

CONTINUING EDUCATION—PROBLEMS

Upon graduation from college, or other learning center, the "time value" clock begins its countdown. Industry/business/education wants to see some evidence of keeping up with state-of-the art. Some states require evidence of continued education for maintaining professional license certification (7).

When courses are taken directly at a college evaluating the course through credit hour rating is simple. However, when completing courses through various university Schools of Continuing Education, and professional societies, where intensive short courses are offered, evaluating courses becomes more difficult. This is due to the lack of a course completion standardized industry-wide acceptable guide.

Some course completions either by C/E or special three-four-or five-day intensive short courses conducted by nationally recognized authorities, "big-name" industry experts, or well-known educators- are evidenced by the award of continuing education units (CEU's). The CEU was originally defined as "10 contact hours of participation in an organized continuing-educational activity under responsible sponsorship, capable direction, and qualified instruction" (7). A widely held criticism of the CEU is that it is only an attendance record, and that its quality is only as good as the reputation of the issuing body. Conversely, at New York University students earning CEU's take exams and produce papers as if they were on a degree course. CEU's were not designed to be a credit record toward an academic degree. The question at this time is "what good is it then?" The Institute of Electrical and Electronic Engineers (IEEE) in 1979 tightened standards and offered the continuing education achievement unit (CEAU) (7). This does have the evaluation component since a measureable test is involved and/or project completed (7).

In addition to the IEEE's efforts, the National Society of Professional Engineers (NSPE) has introduced the professional continuing education unit (PCEU). This again requires an evaluation component (7).

The author recommends establishing a committee to initiate a solution acceptable to all concerned, and have an instantly acceptable unified rating system

In spite of this small problem most employers recognize and emphasize the importance of continued education. Companies are generally quick to recognize a good investment by either prepaying or reimbursing employees for continuing education. The rationale is simple since the benefits exceed the modest dollar expenditure. The benefits are efficiently and primarily obtained by the updating of technical personnel skills (7).

An example of short courses offered by the IEEE in 1979 is shown in Table 3. (7) (Note - Due to space limitation contact the address at table top for identification of sponsors, addresses and contact with phone numbers).

TABLE 3 SHORT COURSES OFFERED BY IEEE-- FALL 1979 (7) (Supplied through courtesy of IEEE)

Here are some of the short courses to be given this fall on subjects of interest to IEEE members. You'll find contact addresses and phone numbers at the end of these listings for further information or enrolling. The listings given here were supplied by *Guide to Continuing Education*, a journal published every other month by Technological Advancement Centers, Inc., P.O. Box 891, East Brunswick, N.J. 08816.

Subject and Title (No. of days)	Date Start, 1979	Fee, $	Location	Sponsor
Acoustics				
Architectural Acoustics Design (3)	9/24	275	WI	WIEU
Underwater Acoustics Short Course (5)	9/24		PA	PSCU
Communications				
Applied Data Communications Systems (3)	9/10	450	CA	IPES
Design of Fiber Optic Communication Systems (5)	9/10	450	WI	WIEU
Spread Spectrum Systems and Interference Rejection Techniques (5)	9/17	475	CA	CAEU
Effective Technical Communications (3)	10/3	470	IL	CFPA
Applied Data Communications Systems (3)	10/8	450	DC	IPES
Computer Science				
Frame Analysis: Statics Workshop (2)	9/5	300	MI	SDRC
DC 2 Operations and Software (5)	9/10	425	IA	FSCC
On-Line Systems Audit Controls (3)	9/19	450	DC	IPES
Computers in Manufacturing (3)	9/25		NY	AIES
Frame Analysis: Statics Workshop (2)	9/26	300	MI	SDRC
Finite Element Analysis: Superb Workshop — II (2)	10/3	300	OH	SDRC
DC 2 Operations and Software (5)	10/8	425	IA	FSCC
DC 2 Provox CCAP and BCAP Techniques (5)	10/8	425	IA	FSCC
On-Line Systems Audit Controls (3)	10/11	450	CA	IPES
Frame Analysis: Statics Workshop (2)	10/17	300	OH	SDRC
Finite Element Analysis: Superb Workshop — I (2)	10/24	300	MI	SDRC
Applied Interactive Computer Graphics (5)	10/29	495	MD	MDCU
Data Processing				
Effective Data Center Management (3)	9/10	450	CA	IPES
Systems Workshop (5)	9/24	665	CN	WARE
Effective Data Center Management (3)	10/8	450	NY	IPES
Systems Workshop (5)	10/22	665	CA	WARE
Database				
Database Design Development and Implementation (3)	9/10	450	DC	IPES
Database Design Development and Implementation (3)	10/10	450	CO	IPES
Design				
Design and Analysis of Industrial Experiments (3)	9/14	295	PA	ACSS
Systems Design and Synthesis (2)	9/18		AL	AUEU
System Modeling (2)	9/19	350	CA	SDRC
Design and Drafting Automation (2)	10/8	200	WI	WIEU
Computer Aided Design and Drafting (3)	10/10	470	IL	CFPA
Geometric Dimensioning and Tolerancing (3)	10/15		NC	NCSU
Design Analysis for Long-Life Damage-Tolerant Structures (5)	10/22	500	CA	CAEU
Digital				
Digital Filters (5)	9/17		GA	GAIU
Digital Switching (5)	10/22	500	MD	MDCU
Electronics				
Electronic Noise Control (3)	9/10	495	NJ	CFPA
Fundamentals of Electricity and Electronics (10)	9/10	250	TX	TXPU
Open Pit Electrics (5)	9/17		NV	NVDU
Electronic Noise Control (3)	9/24	465	CA	CFPA
Preventing Vibration Failure in Electronic Equipment (5)	9/24	450	WI	WIEU
Solid State Electronics for Non-Electrical Engineers (3)	9/24	505	CA	CFPA
Laser Beam Information Systems (3)	9/26		IL	CHCU
Digital Switching (5)	10/22	500	MD	MDCU
Plastics Electrical Properties and Applications (3)	10/23	425	NJ	PIAS
Energy				
Commercial/Energy Audits (4)	9/4	400	WI	WIEU
Energy, Efficient Systems (5)	9/10	445	WI	WIEU
Energy Design and Analysis (5)	9/17	445	WI	WIEU
Fossil Fired Large Power Steam Generation (4)	9/17	595	NJ	CFPA
Energy Management in Buildings (2)	9/18	495	NY	NYMC
Economic Analysis for the Energy Industry (5)	9/24	325	OK	OKLU
Energy Conservation in Heating, Ventilating and Air Conditioning Systems (2)	9/26	200	AL	ALEU
Energy Forecasting (5)	10/8	395	WI	WIEU
Energy Management in Buildings (2)	10/9	495	CA	NYMC
Filter				
Kalman Filtering (5)	9/10	500	MD	MDCU
Digital Filters (5)	9/17		GA	GAIU
Instrumentation				
Instrumentation for Environmental Controls (10)	9/10	250	TX	TXPU
Instruments and Safety Controls (10)	9/10	350	TX	TXPU
Laser				
Laser Beam Information Systems (3)	9/26		IL	CHCU
Management				
Quality Management Institute (5)	9/9	425	CT	CTUU
Basic Safety Management (5)	9/10	440	TX	ISAC
Effective Data Center Management (3)	9/10	450	CA	IPES
Engineering and Management Program (6)	9/10	700	CA	CAEU
Finance and Accounting for the Nonfinancial Manager (2)	9/17	395	CA	SCHA
Managing the Industrial Engineering Function (2)	9/18	150	WI	WIEU
Finance and Accounting for the Nonfinancial Manager (2)	9/20	395	NY	SCHA
Effective Visual and Oral Prestation of Engineering Proposals (2)	9/25	145	WI	WIEU
Finance and Accounting for the Nonfinancial Manager (2)	10/1	395	IL	SCHA
Forecasting Techniques for Engineers and Managers (2)	10/3	200	AL	ALEU
Management by Objectives for Engineers (3)	10/3	475	MI	MIMU
Basic Safety Management (5)	10/8	440	TX	ISAC
Effective Data Center Management (3)	10/8	450	NY	IPES
Finance and Accounting for the Nonfinancial Manager (2)	10/8	395	NJ	SCHA
Finance and Accounting for the Nonfinancial Manager (2)	10/15	395	CA	SCHA
Creative Action in Engineering and Science (3)	10/17	275	WI	WIEU

Manufacturing				
Good Manufacturing Practices for the Pharmaceutical Industry (3)	9/5	460	NJ	CFPA
Practical Machining Principles Shop Application (2)	9/11	300	OH	MCDC
Good Manufacturing Practices for Pharmaceutical Supervisors (2)	9/13	320	IL	CFPA
Fundamentals of Production and Inventory Control (2)	9/17	200	AL	ALEU
Computers in Manufacturing (3)	9/25		NY	AIES
Practical Machining Principles for Shop Application (2)	9/25	300	OH	MCDC
GMP's for Pharmaceutical Supervisors (2)	9/27	325	NJ	CFPA
Practical Machining Principles for Shop Application (2)	10/9	300	OH	MCDC
Geometric Dimensioning and Tolerancing (3)	10/15		NC	NCSU
Practical Machining Principles for Shop Application (2)	10/30	300	OH	MCDC
Microprocessors				
Higher End Microprocessors (4)	9/17	595	CA	CFPA
Microprocessors and Minicomputer-Interfacing (6)	9/23	485	VA	ACSS
Microprocessor Programming Workshop (3)	10/4	495	NY	IEEE
High End Microprocessors (4)	10/15	595	NJ	CFPA
Microscopy				
Advanced Polarized Light Microscopy (5)	9/10	475	IL	MCRU
Transmission Electron Microscopy (5)	9/10	475	IL	MCRU
Polymer Microscopy (5)	9/17	475	IL	MCRU
Selected Area Electron Microscopy (5)	9/17	475	IL	MCRU
Particle Manipulation (5)	9/24	475	IL	MCRU
Applied Polarized Light Microscopy (5)	10/1	475	IL	MCRU
Photomicrography (5)	10/8	500	IL	MCRU
Identification of Small Particles (5)	10/15	475	IL	MCRU
Scanning Electron Microscopy and X-ray Microanalysis: Theory & Practice (5)	10/15		NY	NYPU
Advanced Microscopy (5)	10/22	475	IL	MCRU
Microwave				
Microwave Processing for the Food Industry (2)	9/13	325	IL	CFPA
Microwave Processing for the Food Industry (2)	10/18	325	NJ	CFPA
Mini-Microcomputers				
Distributed Minicomputer Networks (3)	9/12	450	DC	IPES
Microprocessors and Minicomputers—Interfacing (6)	9/23	485	VA	ACSS
Minicomputer Systems (3)	9/24	450	DC	IPES

C/E EFFECTIVITY

The following partial list represent only a small number of distinguished alumni of accredited C/E schools (8, 9) and cites the effectiveness of C/E education:

- J.B. Black, Chairman of the Board, Pacific Gas & Electric
- H.A. Bullis, Director, General Mills, Inc.
- Dr. James L. Busey, Professor of Political Science University of Colorado
- J.L. Collyer, Director, B.F. Goodrich Co.
- R.J. Cordiner, Chairman of the Board, General Electric Co.
- A. Crabtree, President, Fraser Companies, Ltd.
- William J. Cunningham, James Hill Chair of Transporation, Harvard University
- David Eisenhower, Father of Dwight D. Eisenhower, former U.S. President
- Philo Farnsworth, Vice President, Capehart-Farnsworth Corporation, Fort Wayne, Indiana
- Ralph E. Flanders, United States Senator from Vermont
- Charles Fox, Deputy Commissioner of the Internal Revenue Service
- E. Gray-Donald, Vice President, Shawingian Water and Power Co.
- Dr. M. Mason Gaffney, Professor of Agricultural Economics, University of Missouri
- John C. Garand, Inventor of Garand Rifle
- Arthur Godfrey, TV and Radio Star
- J.E. Hammond, Certified Public Accountant, President, California Board of Accountancy
- B.C. Heacock, Chairman, Caterpillar Tractor Company
- G.M. Hobart, President Consolidated Paper Corp. Ltd.
- Luther Hodges, Secretary of Commerce
- Charles R. Hook, Chairman, Armco Steel Corp.
- A.W. Hughes, Chairman, J.C. Penney Co., Inc.
- N.B. Jackson, Director, Chemical Bank New York Trust Company
- W. Frank Jones, President, Borden Co., Ltd. of Canada
- Dan Kimball, Former Secretary of the Navy
- Robert G. LeTourneau, Head, R.G. LeTourneau & Company
- F.L. Marge, Chairman Aluminum Company of America
- John McClellan, United States Senator from Arkansas
- Max McGraw, President, McGraw Electric
- W.L. McKnight, Chairman of the Board, Minnesota Mining & Manufacturing Co.
- T.R. McLagan, President, Canada Steamship Lines Ltd.
- Jule P. Miller, Vice President, Pet Milk Co.
- High Milton, II, President, New Mexico College of Agriculture & Mechanical Arts
- Harry W. Morrison, President, Morris-Knudsen Co., (builders of the Grand Coulee Dam)
- Charles W. Nash, Former President, General Motors
- Lyman T. Newell, Admiral USNR Retired
- Ray C. Newhouse, Chief Engineer, Allis-Chalmers Manufacturing Company
- H.C. Osborn, Vice President, Addressograph-Multigraph Corp.
- James E. Otis, Jr., President Dodge Corporation.
- A.L. Penhale, President, Asbestos Corp. Ltd.
- Russell L. Peters, Vice President & Director, Inland Steel Corporation
- B. Earl Puckett, Chairman, Allied Stores Corp.
- C.A. Putnam, Past President, National Association of Manufactures
- Leonard T. Recker, Vice President, John S. Swift & Company
- Eddie Rickenbacker, Chairman, Eastern Air Lines
- Robert M. Rickover, Son of Admiral Rickover, USN, Student at Yale University
- A. H. Schweitert, President, National Industries Traffice League
- Kenneth L. Smith, President, Chicago Stock Exchange
- A.D. Stephenson, Colonel, Communications Branch U.S. Army
- W. Stuart Symington, United States Senator from Missouri
- M.R. Sullivan, Chairman, Pacific Telephone & Telegraph Co.
- E.J. Thomas, Chairman, Goodyear Tire & Rubber Co.
- H.E. Whitaker, Chairman, The Mead Corporation
- D.A. Whittaker, Director, Sherwin-Williams Co. of Canada, Ltd.
- D. Williams, Vice President, The Dow Chemical Company
- Edwin Williams, Former Dean of Graduate School University of Pennsylvania
- Charles E. Wilson, Former President, General Motors
- John H. Winchell, Chairman, Interstate Commerce Commission
- C.S. Young, President, Federal Reserve Bank of Chicago

SUMMARY

This paper has disclosed that correspondence/extension education can be a major contributor to the industry/business/education/scientific community. It can reduce the problem of keeping up with the state-of-the-art from an employee-employer relationship. There is no intent to supplant resident college courses, but only to perform as a proven alternative when limitations are in effect. The limitations: (1) Remote Locations (2) Costs (3) Time (4) Family Restrictions (5) Job Restrictions and (6) Course Availability.

C/E is cost effective from an employer point-of-view. A recommendation of full accreditation of the learning center should be insisted upon. Over 40,000,000 people have completed C/E education since 1900 (2).

The no-cost publication "A Bibliography On Home Study Education" (10) is recommended reading for all those having responsibility of C/E education funding for either employers or course developers.

C/E according to John Wilheim, Staff Director for IEEE Educational Activities, "overcomes one of the biggest continuous education road blocks--minimal time to attend formalized courses. Furthermore, engineers seem to prefer home study. In a 1977 Educational Activities Board survey of 3,000 IEEE members, ranked home study well above other learning situations (one-day courses and audio tape or cassette were ranked second and third)" (7).

Chairman Takeshi Mitaroi of Canon, Inc., Japan, stated that "people will work harder when they know their efforts are recognized. Team work is still very important and essential to any company's success. But, individual achievement must also be encouraged and applauded. These practices have proven themselves over three decades and I see no reason for them to change radically in the future" (11).

REFERENCES

(1) "Center for Degree Studies", Catalog, Vol. 1 pg. 3 Guided Independent Study, ICS, Scranton, PA 18515, 1979

(2) "Careers for the Homebound", President's Committee on Employment of the Handicapped, Washington, DC 20210

(3) L. Hirasawa, Program Director, "Direct Letter to Author", Univ. of Washington, Seattle, WA 98195, 2 Sept. 1980

(4) Long Distance Telephone Communications to 6 C/E schools", 9-11 Sept. 1980

(5) "Report on Current Educational Practice in NHSC Member Institutions", National Home Study Council, 1601-18th Street, NW, Wash., DC 20009 pgs 1-13

(6) NHSC "Home Study School Accreditation", Accreditation Commission of the National Home Study Council, 1601-18th Street, NW Washington, DC 20009, pgs 1-9

(7) C Patton, "The Challenge of Keeping Current, IEEE Spectrum", the Institute of Electrical and Electronic Engineers, 345 E 47th Street, NY, NY 10017 pgs 53-58, Aug. 1979

(8) "Forging Ahead in Business", Alexander Hamilton Inst., 71 W 23rd Street, NY, NY 10010, pgs 20-41 (1961)

(9) "Distinguished Alumni of Accredited Correspondence Schools", Compiled by National Home Study Council, 1601-18th Street, NW, Washington, DC 20009, April 1963

(10) "A Bibliography on Home Study Education", National Home Study Education 1601-18th Street, NW, Washington, DC 20009, 1980

(11) "Japan 80", Fortune Magazine, Time and Life Bldg., Rockefeller Center, NY, NY 80020, Vol. 102 pgs 51-52, 11 Aug. 1980

WE SAY AND WRITE WHAT WE THINK — NOT WHAT WE MEAN

Wm. Jack Hranicky
Computer Sciences Corporation

WM. JACK HRANICKY
COMPUTER SCIENCES CORPORATION
TECHNICAL PUBLICATIONS EDITOR
6565 ARLINGTON BOULEVARD M/C 224
FALLS CHURCH, VIRGINIA 22046

EDITOR/PUBLISHER: POPULAR ARCHAEOLOGY

This paper briefly examines recent communication theories concerning the basic processes of language production; namely, meaning and thinking. From these theories emerges the hypothesis - we say and write what we think - not what we mean. While partly a behavioralistic approach, the argument for the hypothesis is based on structural linguistics and transformational grammar. Thinking and meaning are further equated with two theoretical levels. These are: surface structure and deep structure in language production. Errors or failures in a communication act or situation involve an individual's ability with transformational grammar at the deep-structure level.

•••••

Using an anthropological perspective, human beings, as well as their culture and society, are in a continuous state of change. This change is evolutionary in nature and involves organic adaptive processes that adjust to progressive evolutionary changes. Every socialized human exists in an environment and has the basic faculties to communicate with other human beings within that environment. Human communication has considerable antiquity, and I feel that communication via a language is synonymous with the first tools that man made 2-1/2 million years ago. The late L.S.B. Leakey suggested toolmaking as the principal difference between man and the rest of the animal kingdom. This, of course, has not proven to be the principal criteria for suggesting a difference between man and his mammalian relatives, i.e., Goodall's studies of chimpanzee tool usage. Even those who have suggested language as a separator have failed, as a number of mammals have demonstrated a propinquity to the communicative process, i.e., wolves, crows, and porpoises. Perhaps we are not different from our mammalian kinship and the differences we suggest are only manifested in the egocentricism of humanity. I will suggest at the end of this paper that the basic cognitive processes are the same in the entire mammalian world. The differences among mammals that I will suggest are in memory capacity and the ability to organize behavior according to a time frame or reference. Naturally, there are anatomical differences, but the concern here is basic cognitive processes with reference to human language production.

Language is also in an evolutionary state of continuous change, and our use of it involves two communication concepts. They are: meaning and thinking. While space is too short in this paper to make a major effort at differentiating between meaning and thinking, I will relate the two terms to the hypothesis - we say and write what we think - not what we mean.

In this paper, I will use the concept of meaning in a culturalistic perspective. That is, the meaning of a word or sentence is such which most of the speakers agree a word or sentence to mean. A norm within any linguistic community can be established for sounds, words, or even sentences. For example, if a large sample of people in the U.S. were asked to define the word "chair," a norm for the word could be established. This is merely the Webster Commonality concept, and when a speaker says "chair," most people would know what he has said. The picture becomes complicated when we ask what cognitive process produced the word "chair"? Even within this simple treatise on meaning, it is doubtful that any two people would give the same definition of the word "chair." However, we have an area within language where meaning is extremely consistent. This is mathematics. Two times two (2 x 2 =) does equal the same thing each time it is used, and the rules for producing this product are consistent and do not change. Does phonology or language production, whether verbal or nonverbal, have the same consistency? Before pursuing affirmative answers to this question, a theoretical background and perspective on language production is needed.

For more than fifty years, Structural Linguistics (SL) has been expanding on its empirical foundations and theoretical perspectives concerning language production. American SL, pioneered by Edward Sapir (1) and Leonard Bloomfield, (2) were theories of sound patterns and phonetic patterns, respectively. About twenty-five years ago, Noam Chomsky (3) introduced the Generative Phonology (GP) or transformational grammar approach, and it served as the pivot of theoretical movements within linguistics. Transformational theorists believe that you cannot adequately explain language behavior in traditional ways. First and foremost, GP makes a basic distinction between two levels in language production: a surface level and a deep level. The surface structure is the phonetic level, the actual uttered sentence. The manifest surface sentence is generated from a deep or underlying structure. The deep structure is an abstract entity (or sentence model) in the mind of the speaker. All surface structures involve transformational rules corresponding to deep structures. All sentence constructions with the same meaning, regardless of form, stem from the same deep structure. GP linguists use a transformational model that considers syntax the central component, and the sentence is taken to be the fundamental unit of language (2). All segments of a language are interpreted with reference to the sentence. Moreover, the grammar is organized in such a way that the surface strings produced in language are derived through a series of explicit stages from abstract rules specifying the deep structure (4). Assuming that an English speaker wished to produce this sentence - Mother is running - a transformational grammar would be represented as the process in Figure 1. Transformational grammar is complex and often very abstract. This brief overview is only an attempt to describe the basic procedures of the GP approach.

Recent trends away from GP are traceable to the heightened desire for genuine psychological/biological correlates for the study of language. Additionally, linguistic and other social theories built on a premise of homogeneity have a long history, but apparently not much future. Whether one looks only at linguistic codes themselves or at these codes in their social and cultural contexts, heterogeneity is inescapable (5). The surest prediction one can make about linguistic theories

I. SENTENCE: MOTHER IS RUNNING

II. PHASE STRUCTURE RULES:

S → T NP AUX VP
T → <+ DECLARATIVE><- NEGATIVE>
NP → N
AUX → <+ AUX>
VP → Vb

III. LEXICON

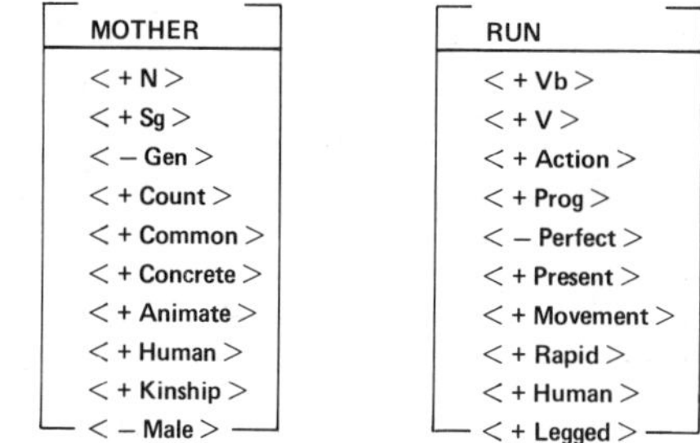

IV. DEEP STRUCTURE REPRESENTATION

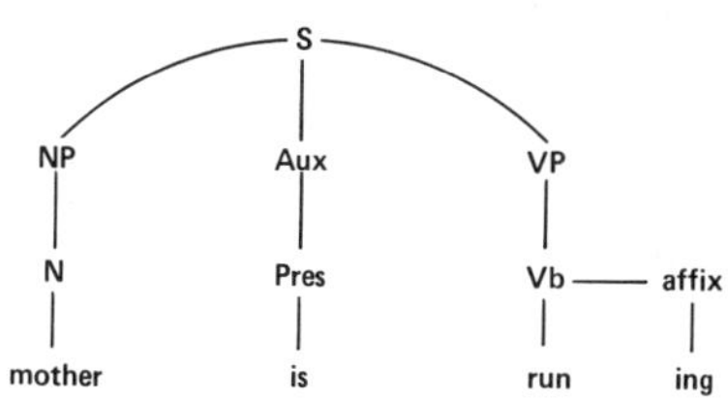

Figure 1. Transformational Grammer Sentence Analysis

is that nothing ever disappears from them. Traditional grammar has not; SL has not; GP will not either. Perhaps a new variety of linguistic theory, Natural Generative Phonology (NGP), will become the dominant theory. It has many of the traits carried over from SL, bases its investigations on a more native phonology, and offers linguistical investigations into the psychological-/biological aspects of language. NGP allows investigations into more areas of language studies; namely, the thought processes of language. This aspect also has a theoretical perspective known as neurolinguistics, but NGP allows an investigator to observe "external world" influences on language production.

A NGP model of language production (see Figure 2) contains surface and deep structure sentence as manifested in the transformational rules, but more importantly, it includes biological and cultural restraints, influences, or controls that are present in the cognitive processes of any individual. These restraints can be overridden in some cases, however, they are a major factor in average speech and written communication. These restraints act as conceptual filters which screen individual behavior and communication acts.

We do know what areas of the brain produce language, but disagreement exists over the extent the two most

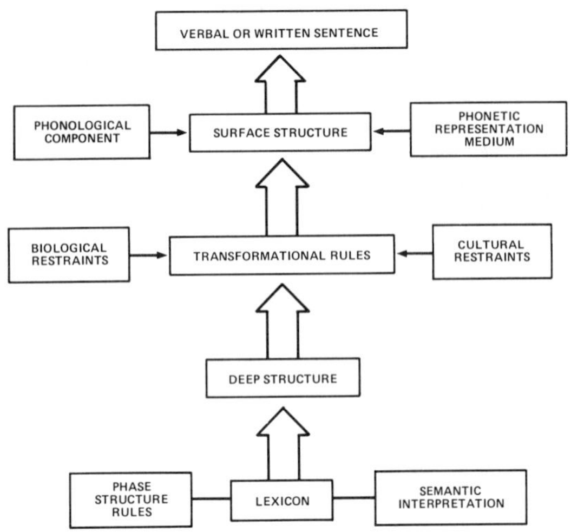

Figure 2. A NGP Model of Language Production

generally recognized "language areas" of the "dominant" hemisphere control. Broca's area in the frontal lobes and Wernicke's area within the tempero-parietal region exist as coherent functional entities tied to specific regions of the cortex.
Controversy remains on the extent to which these areas are specifically linguistic in character versus a manifestation of more general motor and sensory processing systems (6). No consensus has been reached on whether Broca's and Wernicke's areas are true language generation areas or only upper-level speech production areas.

In these language areas, one could propose that each human language makes special use of a set of phonologically related root-forms which, in light of the cultural history of its speakers, possess special core meanings. Such phono-semantics are integral to many lexical units of a language. Access to these lexical units is not understood, but perhaps for a lack of a better term, a synergitic waveform is used that is an internal motor function of the brain. This synergy obtains these units located in the memory and brings them to the surface where they are used for trans-human behavior-communication. This surface structure includes mannerisms, facial expressions, tones-of-voice, and verbal responses.

These lexical units are the cognitive factors of our mental facilities, and we require them at a very early age. These units, like bits to the computer, lie in the deep structure of language production and probably do not show any resemblance in the surface structure treatment. A verbal response to a situation involves a mental-sorting process of the known possibilities and evaluating the best or correct response for the particular circumstances. Ideally, a verbal or written response has only one correct response. Thus, two times two (2 x 2 =) equals four (4) and any other answer involves either: a grammar transformation error or lack of mathematical knowledge. Any individual, given the knowledge, will respond with the same answer to the above problem.
The mental processes are the same; only the language assignment differs.

If we think of the mind as a system of cognitive processes that relies on codes, rules, and physical capacity, we can apply some of the concepts of cybernetics; namely, a systems approach, and a analogy between the mind and the computer (natural versus synthetic intelligence). Consider, for instance, the basic component of the human nervous system, the nerve cell or the neuron, which is to the brain what the transistor is to the computer. The computer can recognize patterns, play chess, translate books, or elucidate even some aspects of the nature of human reasoning by means of highly complicated scanning devices. While the internal circuitry of microprocessors is well understood, the neural circuitry of the brain is still a complete mystery in spite of all the latest neurophysical researches. Jagjit Singh (7) makes a comparison between the mind and the computer by treating the mind as the "black box" containing locations for modus operandi memory functions.

I feel that motor functions of the brain are secondary functions, and the primary functions are the result of basic cognitive responses to a perceived environment or set of circumstances. The actual utterance of words is a motor function, and is the product of the cognitive deep structure. Language is the result of the brain's basic properties and, for lack of a better term, the brain's computer system. The basic capacity or programming of the brain is the same for the entire mammalian world. Each mammal, including humans, begins life by programming information into its memory. This memory is not random in its assignments of locations for data, but is organized into categories for easy access. I suggest four categories for the basic brain organization (see Figure 3).

Briefly, the subjective contains personal concepts; the objective contains the outside world; the auxiliary concepts are nonthreatening items such as play or simple verbal greetings; and the locative concepts are directions, places, and time refer- ences. At one time, I suggested that this latter category separates the mammalian world from mankind in that mankind can plan an activity for tomorrow. However, the African baboon knows areas in his ter-

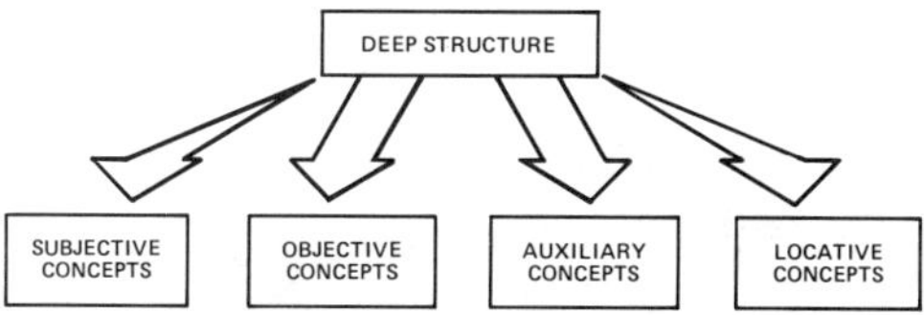

Figure 3. Basic Cognitive Areas of the Brain

ritory where and when certain plants and berries will maturate and arrives at these places in time for a meal. All these concepts are the bases for the deep structure in language production. The brain contains the rules for placing this information in what it perceives is the logical arrangement and brings them to the surface for vocal or nonverbal presen- tations. The differences in surface structures are differences in the physiological bases of each individual. Perhaps these surface structures are uneducated/educated or primitive/modern communicators, but the evaluation is purely a human egocentricism. For example, there is a chimpanzee named Washoe at the University of Oklahoma who has a sign-language vocabulary of nearly 300 words (8). Animals possess complex toolmaking abilities, social structure, and language capabilities, all of which lead us back to the properties of the brain. The language used by mankind's ancestors 2-1/2 million years ago was probably no more than sounds representing objective objects and subjective feelings. It probably lacked surface structure grammar, but contained transformation rules for structuring thought processes; antique stone tools are the artifactual evidence today. For many, this represents primitive forms of language, which has evolved into the modern languages of today. But what about tomorrow: at what point does a primitive language stop and a modern language begin? The question need not be answered, as we must address ourselves to the basic properties and functions pertaining to language production. The point is that we give ourselves away physiologically when we speak because we say and write what we think, not what we mean.

All of us rely on previous conditioning of our memory banks for information to be used in the communication process. Even a simple communication act involves numerous factors, and a failure in any of these factors causes a failure to communicate. These factors include the individual's ability, the world setting that individual has lived in, and the communication channel that is being used (see Figure 4). Sometimes, it is surprising that all these factors actually work, and considering the number of times we engage in a communication act, the frequency of success is quite good. However, we are still a long way from a Utopian form of communication, that is: a successful transmission of all intended data from one individual to another. Like the chimpanzee's primitive language, our language and communication processes of today will be quite primitive tomorrow. Where is the key that will unlock the door of tomorrow? I feel it is our understanding of the mind, its processes, and its functions. Once this door is opened, we will experience the greatest revolution in human history.

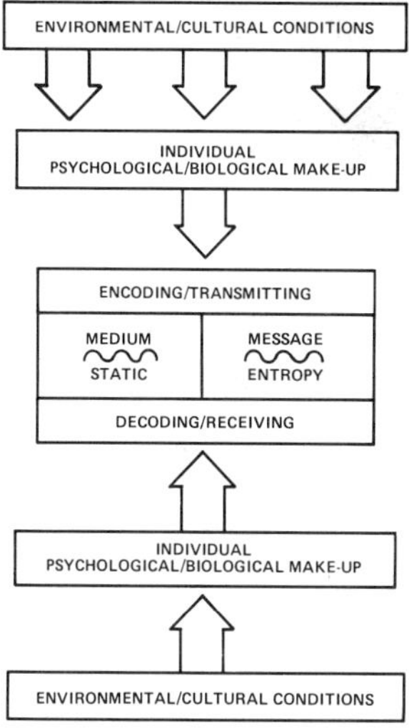

Figure 4. Model of a Communication Process Between Two Individuals

Literature cited:

1. Sapir, E., 1925. Sound patterns in language. Language 1:35–51.
2. Bloomfield, L., 1926. A set of postulates for the science of language. Language 2:153–64.
3. Chomsky, N. A., 1967. Some general properties of phonological rules. Language 43:102–28.
4. Chomsky, N. A., 1965. Aspects of the Theory of Syntax. The M.I.T. Press. Cambridge, MA.
5. Bailey, C. J. N., 1973. Variation and Linguistic Theory. Arlington, VA. Center for Applied Linguistics.
6. Poeck, K. and W. Huber, 1977. To what extent is language a sequential activity? Neuropsychologia 15:359–63.
7. Singh, Jagjit, 1966. Great Ideas in Information Theory, Language, and Cybernetics. Dover Publications, Inc., N.Y.
8. Linden, Eugene, 1974. Apes, Men, and Language. Saturday Review Press/E. P. Dutton & Co., Inc. N.Y.

TEACHING RELATIVE-CLAUSE USAGE TO NON-NATIVE SPEAKERS

Thomas N. Huckin
The University of Michigan

Thomas N. Huckin
University of Michigan
Assistant Professor
Department of Humanities
College of Engineering
Ann Arbor MI 48109
(313) 764-1420

Increasing numbers of foreign students are pursuing technical and scientific studies in the United States, including one or more courses in technical writing. As a consequence, technical writing teachers are finding more and more non-native speakers of English in their classes, many of whom need extra help with English grammar. This paper, containing instructional material on relative-clause usage, illustrates an approach which has been found to appeal to such students. A "deductive" approach, it features general rules for both grammatical and stylistic correctness, examples of good and bad usage, and many exercises.

English is the most widely used language for science and technology in the world today. More and more countries are using it as the medium of scientific and technical instruction, and more and more companies overseas are using it for business communication. At the same time, increasing numbers of foreign students are coming to the United States and using English in their scientific and technical studies. After graduation, many of them stay here and go to work in places where they must write technical reports and give oral presentations in English. Thus, not only should the teaching of English-as-a-Foreign-Language overseas be oriented toward scientific and technical utilization, but the teaching of technical writing here in the United States should include optional special instruction in English for foreign students.

One form of special instruction that the technical writing teacher could provide concerns certain aspects of English grammar that are known to plague non-native speakers. Modal auxiliary verbs, articles, prepositions, relative clauses, and verb tenses, for example, are important features of scientific and technical English and yet are typically problematic for non-native speakers. How does one go about teaching such subjects? This paper is designed to show you, using instruction in relative-clause usage as a model.

Relative clauses are essential to good technical writing. In a study of 44 scientific and technical texts totalling 135,000 words, one researcher found that relative clauses were used at an average rate of 1 per 2.65 sentences (Huddleston 1971). In a similar study of a much smaller corpus, I found the rate to be even higher: 1 per 2.0 sentences. If frequency of use is a good indicator, as I believe it is, relative clauses are clearly important to scientific and technical communication. When used in the "restrictive" sense -- as they most often are in this type of English -- they serve the useful purpose of pinpointing a specific referent for the noun modified, or at least of narrowing down the class of possible referents. For example, if the relative clause (italics) were omitted from the following sentence, the result would constitute an inaccurate statement: "Any celestial body *that emits light waves* may be expected to emit radio waves because both wave forms are part of the same electromagnetic spectrum." The relative clause makes it clear that we are talking not about celestial bodies in general but only about certain kinds of celestial bodies.

Despite the obvious usefulness of relative clauses, however, non-native speakers often avoid using them -- apparently out of a well-founded fear of making grammatical and stylistic mistakes (Mukattash 1978, Gass 1979). Here is an example, from a Japanese graduate engineering student's attempt to discuss the pro's and con's of nuclear energy: "Conventional electricity plants consume huge amounts of oil or coal, the oil or coal reacts with air, whose products include a lot of carbon dioxide, it is not harmful to the human body, but it changes the environment." Why do non-native speakers have such trouble with relative clauses? Mainly because of linguistic differences in the way relative clauses are formed in English and the way they are formed in other languages. In Japanese and many other languages, for example, relative clauses are never headed by a relative pronoun, and they usually precede rather than follow the noun they modify. In other languages, e.g. Arabic, the relative clause follows the noun it modifies (as in English) and is headed by a relative pronoun; but a "duplicate" pronoun will often appear elsewhere in the clause. Thus, Arabic speakers tend to make this kind of mistake in English: "There are four streams which the students can study one of *them*." Still other differences exist in other languages.

The solution to the problem of relative-clause avoidance, therefore, lies in first giving students proper instruction in how to form and use relative clauses correctly. (For most students, this will simply be a review of instruction they've already had sometime in the past.) Once students feel they have control of relative clauses, they are much more likely to use them. In my experience, students in science and technology prefer a "deductive" instructional method featuring an interlocking pattern of explicit rules, clear examples of good and bad usage, and realistic, challenging exercises. The following material exemplifies this type of approach. (NOTE: This material deals solely with restrictive relatives, these being far more numerous in scientific and technical English and having a different function than nonrestrictives. Also the exercise material is condensed here for lack of space.)

USING RELATIVE CLAUSES*

Definition: A relative clause in English is a clause or phrase that follows a noun and serves to narrow the class of possible referents for that noun. English relative clauses typically begin with a relative pronoun (*which, that, who,* etc.), though this can be omitted under certain conditions, as in this example:

"A voltmeter is an instrument *used for measuring the differences of potential between different points of an electric circuit.*"

Rule 1: Put the relative clause directly behind the noun it modifies.

CORRECT: The energy *which is consumed in overcoming friction* is not lost but is converted into heat.
INCORRECT: The energy is not lost *which is consumed in overcoming friction* but is converted into heat.

Exceptions to the rule are allowed in cases where the rule would produce a confusing or extremely unbalanced sentence. In this sentence, for example:

EXCEPTION: A popular large computer system can do computations in one minute *that it would take approximately 50 years to do manually.*

the relative clause modifies the noun *computations* but is separated from it by the phrase *in one minute.* Following the rule in this case would have produced a potentially confusing result: "A popular large computer system can do computations that it would take approximately 50 years to do manually in one minute."

Exercise 1. Check each of the following sentences for proper positioning of the relative clause. Make appropriate changes where necessary.

1. Any celestial body may be expected to emit radio waves that emits light waves because both wave forms are part of the same electromagnetic spectrum.
2. The concepts were known previously which underlie these methods but could not be economically utilized.
3. Today a jet engine develops 40,000 pounds of thrust that weighs 5,000 pounds.

Rule 2: Make sure there is a relative pronoun at the head of the relative clause, optionally preceded by a preposition. (Subject to stylistic variation: see rules 5-7 below.)

CORRECT: A hygrometer is a device *which is used to measure the relative humidity in the atmosphere.*
INCORRECT: A hygrometer is a device *is used to measure the relative humidity in the atmosphere.*

Exercise 2. Check each of the following sentences and see if you can spot grammatical errors involving relative clauses without relative pronouns. Make whatever corrections are necessary.

1. The flow of current through a wire conductor is equal to the number of electrons pass through its cross-sectional area in a particular unit of time.
2. Thermodynamics is the science deals with heat and work and those properties of substances bear a relation to heat and work.
3. All engines operate on the Carnot cycle between two given constant-temperature reservoirs have the same efficiency.

Rule 3: Make sure you don't have a second pronoun duplicating the role of the relative pronoun.

> CORRECT: The difference between research and power reactors is in the kind of uranium *that they use.*
>
> INCORRECT: The difference between research and power reactors is in the kind of uranium *that they use it.*

Many languages of the world (e.g. Arabic, Persian) differ from English in that they do allow a "duplicate" pronoun to appear in the relative clause. But this is because in these languages there is not really a "relative pronoun" as such. Instead, there is only a "relative marker." Let me explain:

A relative pronoun actually performs two functions. One, it serves as a relative marker, signalling the presence of a relative clause. Two, it serves as a pronoun, playing a grammatical role in the relative clause and sharing the same referent as the main-clause noun. This double function can be seen most easily in sentences like the following:

> The punched card concept *which Hollerith devised in the 1880's* turned out to be a major step in the development of the computer.

Since the verb *devise* requires a direct object, we sense that there is a "hole" in this relative clause, immediately after the verb. In effect, the relative pronoun *which* represents the "missing" direct object at the same time as it marks the relative clause. It is as if the sentence went through these "stages of construction":

(1) The punched card concept [*Hollerith devised it in the 1880's*] turned out to be a major step in the development of the computer.

(2) The punched card concept [*which Hollerith devised ___ in the 1880's*] turned out to be a major step in the development of the computer.

Since relative pronouns in English serve not only as relative markers but also as pronouns, there is no need to fill the "hole" with another pronoun; in fact, it would be grammatically wrong to do so:

> INCORRECT: The punched card concept which Hollerith devised *it* in the 1880's turned out to be a major step in the development of the computer.

Exercise 3. Inspect each of the following sentences and cross out any pronouns that duplicate the role of the relative pronoun.

1. A pump is a device that it induces fluid flow through pipes, generally by means of a moving piston.
2. Pulley systems provide a mechanical advantage which it is equal to the number of supporting ropes involved.
3. Such a system would not provide the cooling power that people are now accustomed to it in air-conditioned cars.

STYLISTIC OPTIONS

Once you have firm control of Rules 1-3 above, you can consider using additional rules for stylistic purposes. Unlike Rules 1-3, which should be rigorously followed for the sake of grammatical correctness, these next rules are strictly optional: if you do not feel comfortable using them, do not use them!

Rule 4: You may substitute *that* for *which* if it is not preceded by a preposition.

> CORRECT: The energy *~~which~~ that is consumed in overcoming friction* is not lost but is converted into heat.
>
> INCORRECT: Multi-spectral sensing is a process *in ~~which~~ that the earth's surface is surveyed from an airplane equipped with infrared and ultraviolet sensors.*

Traditionally, the use of *that* as a relative pronoun was thought to be too informal for scientific and technical writing. But times have changed, and now it is fully accepted. Indeed, using *that* instead of *which* generally makes for a smoother, easier-to-pronounce sentence. It is also favored by many writers who feel it makes their writing less stiffly formal.

Exercise 4. Convert *which* to *that* wherever possible in the following sentences.

1. Technology assessment is an opportu-

nity for universities to carry on studies which go beyond the bounds of individual fields of study.
2. An economizer is simply a heat exchanger in which heat is transferred from the products of combustion to the condensate.

Rule 5: If the relative clause begins with a relative pronoun and some form of the auxiliary verb *to be*, followed by a verb phrase, you may omit the relative pronoun and auxiliary verb.

CORRECT: The energy *~~that~~ ~~be~~ consumed in overcoming friction* is not lost but is converted into heat.
INCORRECT: The man *~~who~~ ~~was~~ director of the program ten years ago* has since retired.

Exercise 5a. In the following sentences, reduce the relative clauses by omitting the relative pronoun and auxiliary:

1. Rutherford showed that the atom was largely empty space and hence had to contain something that was smaller than the atom.
2. "Water treatment" is a term which is traditionally applied to municipal treatment of water.

The fact that you may omit the relative pronoun and auxiliary verb does not mean that you always should. Good writers sometimes choose not to make such a reduction even though the opportunity presents itself. Why? It is mainly a question of whether or not you want to emphasize the word immediately following the auxiliary. If you do want to emphasize it, then retain the relative pronoun and auxiliary; if not, then don't.

Exercise 5b. In the following sentences, examine each relative clause headed by a relative pronoun and some form of the auxiliary verb *to be*. Is the next word an important one? If so, leave the relative clause as it is. If not, cross out the relative pronoun and auxiliary.

1. After mining and milling, uranium ore is concentrated into an oxide form which is known as "yellow cake."
2. As each electrical pulse reaches a synapse, chemicals which are called neurotransmitters come into play.
3. Decision simulations that are complex enough to be effective require a computer for making the computations involved.

Rule 6: If the relative clause begins with a relative pronoun and a main verb (i.e. non-auxiliary verb), you may omit the relative pronoun and change the verb to its *-ing* form.

CORRECT: *Persons ~~who~~ ~~do~~ doing research* find it very time consuming and sometimes impossible to make an adequate search of the literature *~~which~~ ~~bears~~ bearing on the subject they are investigating.*
INCORRECT: A seismograph is a device *~~that~~ ~~is~~ being used to measure the magnitude of vibrations that travel as seismic waves through and around the earth.*

This rule cannot be applied if the verb following the relative pronoun is an auxiliary verb, e.g. *is, was, are, can, may.*

Exercise 6. In the following sentences, reduce whatever relative clauses you can by omitting the relative pronoun and changing the verb to its *-ing* form. Make sure the verb is a main verb!

1. The model allows those who use it to compress real time, so that years of operation can be studied in a few minutes.
2. Some of the liquid refrigerant that emerges from the condenser can also be used for air-conditioning.
3. Today, marine engineers are responding to those ocean pollution problems that have now been widely acknowledged.

As in the case of Rule 5, good writers do not always apply Rule 6 whenever the opportunity presents itself. In fact, they avoid applying the rule more often than not. It depends mainly on how much emphasis or focus the writer wants to give to the relative clause: a full relative clause tends to command more attention than an *-ing* type of relative clause.

Rule 7: If the relative clause begins with a relative pronoun and another pronoun, you may omit the relative pronoun.

CORRECT: The exchange system *that we are most familiar with* is increasingly coming under scrutiny.

REFERENCES

Gass, Susan M. An Investigation of Syntactic Transfer in Adult Second Language Acquisition, unpub. Ph.D. dissertation, Indiana University, 1979.
Huddleston, Rodney D. The Sentence in Written English, Cambridge University Press, Cambridge, England, 1971.
Mukattash, L. "A Pilot Project in Common Grammatical Errors in Jordanian English," Interlanguage Studies Bull. 1978

PROFESSIONALISM WITHOUT LOSS OF INDIVIDUALISM

Judith Cheryl Johnson
Federal Highway Administration

Judith Cheryl Johnson
Federal Highway Administration
Department of Transportation
400 Seventh Street, SW.
Washington, D.C. 20590

Writer/Editor

B.A. in English, Fisk University, Nashville, Tennessee, 1977; Graduate Studies in Journalism, University of Missouri, Columbia, Missouri.

Technical communicators in the pursuit of professionalism need to remember their right to their individuality and not fall prey to group allegiance. Not to suggest that progress toward professionalism should be abandoned, but to ensure that desired professional objectives and goals are attained. Professionalism can inhibit creativity in language and hinder effective communication if its negative aspects are not carefully considered and examined. A technical communicator's understanding of language may enable him to produce a usable, effective message, but it is his individual creative perception and style that presents the message in a unique form.

Deadly as a hangman's noose, conformity steals the breath of life from an individual. It kills his creativity, leaving a robot guided by the collected will of others - contented to satisfy the majority. Technical communicators in the pursuit of professionalism should keep this image in mind for professionalism can foster this conformity that kills.

According to Random House College Dictionary, professionalism consists of a professional character, spirit or method of standing practice. The key phrase here is "standing practice." The American Heritage Dictionary defines a profession as an occupation or vocation requiring university training and advance study in a specialized field. A profession possesses characteristics such as professional associations, codes of ethics, work standards, specific credentials and regulation or licensing by government bodies. Again, the key phrases here have to do with specific practices, standards, or credentials.

These characteristics at first may appear desirable, but technical communicators should consider the negative as well as the positive aspects of the making of an occupation into a profession. The standing practices, rules, regulations, guides or codes of ethics that personify professionalism can lead to mass thinking, stifling innovation and rendering persons incapable of making individual judgments or decisions. Many times, members of professions begin to think as the group dictates without realizing they are losing their individuality. Others realize they are no longer acting as individuals, but fear to break the norm.

Communicators should be uninhibited, free thinkers seeking to reach their audiences through the best methods they can create. Technical communicators are no exception, and in today's world of growing technology, creative, imaginative technical communication is essential to meet the needs of a public overwhelmed by meaningless jargon and indecipherable terms.

John Campbell and G. L. Farrar write in their book, Effective Communication for the Technical Man:

> This is the age of rapid communication--instant television portrayal of events around the world via satellites, computers with increasingly effective software, rapid duplication facilities, and the rapid transmission of information. Each of us is beset with an ever increasing barrage of information to sort, think about, file away. One sometimes feels as if he is standing in the way of an "avalanche of paper."

If technical communicators are not

careful, their pursuit of professionalism can squelch the creativity needed to transform this "avalanche of paper" into clear, sensible language that informs, interests, and sometimes entertains their audiences.

Professionalism can provide a standard of education, ethics and specific skills that can benefit the future of technical communication, but standards that are too rigid can be detrimental to this future. Professionalism can produce communicators who dare not think beyond what they have been taught and limit the development of new, inventive communication methods. Yes, professional status can provide special services, educational programs, journals, employment information, higher salaries, and a special title - but do not bind an individual's neck in a noose, with standards that promote conformity. Allow room to breathe, without gasping for air.

Though the purpose of Campbell and Farrar's book is to give their readers effective communication skills, they understand the importance of allowing individuals breathing space. They stress that the ideas presented in their book ". . . are not 'recipes' to be blindly followed," and state, "In writing or speaking the personality and professional traits of the individual must somehow be projected. What is presented are guidelines for the development of an individual style. Style in communication, like style in other activity, is an art that is developed personally, never acquired second-hand."

In essence, rules do exist to guide technical communicators, but the rules can be adjusted, modified, and sometimes broken. It is the technical communicator's understanding of language that produces a usable, effective message. It is his individual creative perception and style that presents the message in a unique form.

A creative individual can be described as a person sensitive to problems, deficiencies, gaps in knowledge, missing elements or disharmonies, who identifies the difficulty, searches for the solution and communicates the results. Technical communicators must do this daily in their jobs. Therefore, creative individualism needs to be encouraged.

Stanwood Cobb, Importance of Creativity, believes, "Creativeness is as normal and as universal an endowment of the human race as is intelligence. It can find ways in which to express itself, or it can be inhibited by society." If not approached with caution, professionalism can be one inhibiting factor of society that destroys creative expression in individuals.

The effect of professionalism is similar to the problem of a child starting nursery school or kindergarten, who must learn school rules. The child is forced to conform to established patterns of thought and behavior, until the pressures of mass education and group activities block some of his creative urges.

Group activities and established patterns of thoughts can affect individuals in a profession. The individual conforms, follows the rules, and sometimes places the group's thoughts before his individual ideas. His creative urges diminish, and his work is a reflection of the profession's rules, ethics or regulations. In communications this is a dangerous situation, because of the power of language. Communicators, and this includes technical communicators, can control dissemination of information, influence public opinion or form public taste.

Conformity of thought, or rather mass thinking, can have a great social impact. "Packaged communications" designed to be consistent with one way of thought can decrease the flow of information in society, limiting public opinion. It also can bore its readers and listeners. One way of thinking and boredom can take away from the effectiveness of the message, which contradicts the purpose of communications.

Better communication is needed in the future, not language restricted by too many rules and regulations until the message is no longer communicated.

"As our world becomes ever more complex, there is a tendency for man to begin to feel himself alienated from it . . . Let's assume that there is a way of coping with complexity, however complex things and events become," writes Edward Matchett in The Making of Meaning in a Complex World. Matchett sees creativity as the answer to coping in a fast pace society growing faster.

Technical communicators are in a position to use this creativity--for creativity in language can aid a public bewildered by rapid change.

Technical communicators possess an essential role in the future of this society, and certain attributes of achieving professional status can aid the technical communicator's success in this role. Other attributes can cause failure.

Technical communicators, in the pursuit of professionalism, should consider the negative with the positive, or they may find themselves hanging from a noose.

REFERENCES

1. Blankenship, Ralph L., Colleagues in Organization: The Social Construction of Professional Work, John Wiley and Sons, New York, 1977.

2. Campbell, John M. and Farrar, G. L., Effective Communication for the Technical Man, Petroleum Publishing Company, Tulsa, Oklahoma, 1972.

3. Cobb, Stanwood, Importance of Creativity, The Scarecrow Press, Inc., Metuchen, New Jersey, 1967.

4. Dagher, Joseph, Technical Communication, Prentice Hall, Inc., Englewood Cliffs, New Jersey, 1978.

5. Dauw, Dean C. and Fredian, Alan J., Creativity and Innovation in Organizations, Kendall/Hunt Publishing Company, 1976.

6. Matchett, Edward, The Making of Meaning in a Complex World, Turnstone Books, 1975.

7. Merrill, John C., The Imperative of Freedom, Hastings House, New York, 1974.

8. Smith, Frank R., "In Pursuit of Professionalism," Technical Communication, 27:3,2 (1980).

NARRATIVES: A NEW METHOD FOR DETERMINING EFFECTIVENESS OF MANUALS

James Kalmbach and Jack Jobst
Michigan Technological University

JIM KALMBACH
JACK JOBST
HUMANITIES DEPT.
MICH. TECH. UNIV.
HOUGHTON, MI 49931
(906) 487-2066

ABSTRACT

Competently written manuals help insure that consumers are satisfied with their new equipment, but how can technical writers determine if their manuals communicate? One new technique involves narratives, or stories, obtained from those who use the manuals. In one study, a woman learning how to use a computer-text editor wrote a short narrative about her initial experience with the equipment and thereby revealed much more than could be obtained from a questionnaire. Narratives may well provide major assistance in working towards manuals that are both helpful and effective.

Not long ago, a young college student rushed into his professor's office, complaining bitterly about an instructional booklet designed to teach proper use of some text editing computer equipment which students had been assigned to use.

"I spent three hours typing something into that damn machine, and I got nothing for it," he said. The manual's no good."

This student's teacher had been promoting computer literacy and enjoyment of composing on a text editor, but if this student's reaction was typical, the teacher had obviously failed. In this particular situation, the equipment user was blaming the manual and the teacher for his dissatisfaction. In business and industry, however, the customer will blame the equipment and the manufacturer. If the problem isn't soon resolved, the consumer may cancel future purchases and tell other potential customers of this dissatisfaction. For these and other reasons, the effectiveness of manuals must be carefully evaluated.

This particular complaining student knew where to take his complaint because the teacher had written the manual. The instructor assigned his technical writing class to compose a simple letter of inquiry using one of the university-owned micro-computers. Manuals for this equipment were supplied with the product, of course, but they were much more detailed than most students wanted or needed to complete one short assignment. The instructor, therefore, had spent part of his summer's vacation composing a manual for his specialized audience: impatient students who wanted to assimilate instructional information quickly and perform the required task. The instructor called his booklet: "Using the Text Editor in Two Easy Lessons"

The instructor knew his manual was not flawless, but he was confident of reasonable success: he had, after all, implemented several suggested revisions offered by his colleagues. However, this one student who stood in his office didn't feel the manual was a success, for it had failed to write his material onto the "floppy disk" for storage. The instructor attempted to determine what went wrong, but the student was too upset to provide useful responses. The instructor then made a note to himself: "Check the instructions for writing to a disk--may be confusion here."

EVALUATING MANUALS

The teacher was receiving feedback on his manual's effectiveness, but he knew of better ways than having a "customer" come in and complain. All writers of manuals want to produce copy that is readily understood, but how can they determine if they are successful? The teacher could stand behind each student to watch for mistakes and answer questions, but this, of course, is not practical for large or even medium-sized groups of people. Nor does it guarantee learning. The students may complete the assignment with their instructor's help but not understand what they have done.

Next the instructor tried polling his students, asking for suggested improvements. Several individuals mentioned problem areas, such as ambiguous directions on how to move between the different editing modes. The instructor, however, had previously overheard several students mention unfortunate experiences with the machinery; but they said nothing in class.

There are several reasons for this poor response. For one thing, although the instructor assured the class that that no recriminations would result from their comments, several students nevertheless felt that they could be punished for criticizing something he had written. Other students failed to respond because they were intimidated by speaking in front of their classmates, while still others couldn't remember exactly what went wrong when they sat in front of the machine or could not put their problems into words.

The instructor then considered writing a questionnaire, but he was worried that the questions he might ask would not cover the problem areas. After all, he understood the manual; it was the students who were having problems. In addition, some students might also feel threatened by a questionnaire, even if answered anonymously, because they feel complete anonymity is impossible.

USING NARRATIVES

The instructor, however, wanted information and he couldn't obtain it. A colleague then told him of a new technique which could determine an individual's attitude without the person becoming unduly nervous, nor even realizing that he or she was performing a critical activity. It begins with a general, yet carefully composed question directed to an operator who has recently used the equipment. The colleague then offered to act as the third party and interview individual class members. Here is the first part of one such interview:

INTERVIEWER: Do you remember what happened the first time you used the text editor?

SALLY: Well, when I went in, no one was really around and I did find my own disk and figured out how to sign it out, figured out how to put it in. But then, the manual, I . . . I really couldn't follow what to expect from it when it said "Hit Execute" and "Hit . . . . um Insert" and things like that and the underlining part just kind of blew me away [laughs]. I didn't even try it out, um. . . .

Like her irate classmate, Sally had been required to write a letter of inquiry using a micro-computer. Her instructor's eleven page manual covered such basic knowledge as how to turn on the system, write, then file this material on a magnetically-recordable "floppy" disk.

In response to the interviewer's question, Sally told the story of her experience with the micro-computer. A narrative such as Sally's is an excellent source of information about the effectiveness of a user's manual.

A story, or narrative, is the telling of a series of events as those events are understood by the person speaking. A story is always a perception of events. Thus, to ask users about what happened the first time they used a computer is to elicit not the events themselves, but their <u>understanding</u> of the experience. In any experience, the multitude of details, sensations, and observations of events are effectively limitless. Narrators must select from this multitude only those events which they feel are relevant to the story, and exclude others.

While questionnaires often provide the technical writer with information about specific areas which might otherwise be mentioned, a narrative, generally, is a better source of information because it recalls a complete experience rather than just where problems occurred. Furthermore, narratives act as an inducement to memory and actually make remembering easier. In examining narratives about computers and about using manuals, we may infer that the sequence of events: (1)going in, (2) signing up, and (3) using a micro-computer, will be more or less the same for all users. The writer must then compare the expected activities with those activities described by the user. This comparison will reveal which steps the user was or was not able to accomplish.

ANALYSIS OF A NARRATIVE

Sally's narrative provides a good deal of specific information about what she could and could not do her first time on the computer. She was able to sign in, find her disk and insert the disk. She was not, however, able to follow instructions such as "hit execute," and underline. In another part of her narrative she tells us that she was able to type but could not record what she had typed. In order to type, Sally must be able to implement the "execute" and "insert" commands, which she was not originally able to understand (as she states in the beginning of her story). Regardless of flaws in the manual, Sally did eventually learn these commands.

In her story, Sally only reports the events which seem significant or important to her. It was significant to her that in the beginning she was able to sign in, find her disk, and insert this disk into the machine. It was an initial moment of triumph. She at least did something right. It was significant to her that she had trouble understand ing certain commands, but it was not significant, in her story, that she did eventually master the "insert" and "execute" commands. By combining those events which Sally reports, with the events which can be inferred, an accurate picture emerges of what actually happened.

Sally has explained what she learned from the manual, but she has also unconsciously revealed something equally as important: her attitude towards the equipment. This is particularly important in preparing the computer manuals, as the attitudes and expectations users bring with them affect their ability to understand and implement instructions. Sally portrays herself as a lone figure, almost helpless against the computer. She did sign in and find her disk, but from there the computer takes over. The instructor knew, from other sources, that an attendant was in the lab at the time Sally signed in and that other students from the class were working on other terminals. Yet Sally mentions none of these people in her story. The instructor also knew that Sally received some assistance from these students. Indeed, she probably learned the "execute" and "insert" commands from them, but that also is not part of her story. Instead, Sally piles failure on top of failure until she reaches a moment of complete desperation, followed by an interesting resolution to her conflict with the computer:

> . . .I spent about two hours on [the computer] the first time, just trying to figure it out. And I just felt like dropping the class, you know. This sudden fear that it is never going to work out. But since then I've used it a couple of times and, you know, I really, it's really easy now.

Here Sally reveals the point of her story, that she reached a high level of frustration: "And I just felt like dropping the class, you know. This sudden fear that it is never going to work out." Sally does not mention all the help she received because the fact that people had helped her would detract from this image of frustration. Similarly, if Sally had mentioned that she had learned some commands, her fear would not seem as legitimate. Someone listening to her story could simply ask: "So what? You got help; there were people there." In her narrative, Sally is a woman threatened by technology, and she carefully anticipates and refutes other possible interpretations of her story. This point is important for the instructor as it was Sally's attitude towards the machine as much as specific difficulties with the manual that lead to her troubles.

The ending, or resolution, to her story is especially interesting because Sally has overcome her resistance to the machine and is now comfortable with it: "But since then I've used it a couple of times and, you know, I really, it's really easy now." The resolution, however, is not surprising if we examine her narrative closely. Beneath Sally's cry of anguish, learning was clearly taking place, and with time and practice she was able to master the micro-computer.

The instructor soon realized that collecting narratives of this sort was valuable but had certain limitations. Because users sometimes suppress details which don't fit the point they are making, useful diagnostic information could be lost. He thus decided that a combination of narratives and questionnaires would be more effective.

He realized, however, that the narrative alone had revealed the essential cause not only of Sally's problem, but probably of the other students as well--those who grumbled in the hallway, and those who sat silent in the classroom. Their difficulty was not so much in the manual, but in their fear and apprehension of the computer.

This information was far more valuable than identifying the specific instructions which were not clear. More than clarifying the manual, the instructor had to do more to help students such as Sally and her reticent classmates overcome their initial fear. He also remembered the young man who had rushed into his office complaining so bitterly about the manual. He could see now that this student's frustration, like Sally's, stemmed not from the manual but from a more fundamental apprehension. If he had only asked the student to sit down and write out what had happened, perhaps he could have helped him.

"TECHNICAL CONTINUITY" AND "PROFESSIONAL ACCOUNTABILITY" – CONCEPTS THAT COULD BE STRESSED TO HELP UNIFY THE TECHNICAL COMMUNICATIONS COURSE

Dolores M. Landreman
Battelle Memorial Institute
Columbus Laboratories

Dolores M. Landreman
Battelle Memorial Institute
Columbus Laboratories
505 King Avenue
Columbus, Ohio 43201

(614) 424-5201
Proposal Specialist
STC Distinguished Technical Communication Award, 1979

The technical communications course can degenerate into little more than just another course in functional writing that includes some attention to technical communication forms and conventions. This might be true particularly if the instructor is an English teacher new to the technical communications field and accustomed to teaching ordinary writing courses. It might be helpful for such teachers to design their courses around some unifying theme such as one of the two suggested here: "technical continuity" and "professional accountability". This paper defines these terms and provides a number of suggestions for directed class discussions. Designing around such unifying themes could enhance the ability of the teacher to keep the course in focus. Class discussions based on such themes should stimulate serious thought about technical communication forms and networks and how they operate, and impress upon the students why accuracy and timelines are so important in technical communications.

INTRODUCTION

In every well-designed technical communication course -- whether presented in the high school or college or as an industrial training program -- important concepts such as "technical continuity" and "professional accountability" are undoubtedly included, most probably indirectly or by implication. However, these two concepts -- which are defined and explained in this paper -- are so significant that they should be dealt with explicitly. Teachers might do well to so design their technical communication courses that these concepts would be stressed throughout. Their relevance to the various forms, conventions, and networks of technical communication should be elucidated.

The design of technical communication courses to stress such themes would undoubtedly enhance the teacher's ability to maintain course control. The danger is always there of allowing the technical communication course to be deflected from its true educational role. It's such fun to discuss "horrible examples" of awkward and misleading sentence structure! The technical communication course can easily become not much more than just another functional or creative writing course that includes some attention to technical communication forms and conventions. Indeed, there might be a special temptation to allow such deflection on the part of teachers whose major professional background has been in the teaching of conventional writing courses. Directed class discussions on an appropriate unifying theme could stimulate serious thought about technical communication forms and networks and how they operate, and impress upon students why accuracy and timeliness are so important.

Although such matters as grammar, punctuation, and word usage may have to be dealt with incidentally with individual students, these matters are not proper subjects for emphasis in technical communication courses. Evidence of competence in these areas could well be a prerequisite for course enrollment. There is too much else to cover in the technical writing course! For example, I have noted some technical writing courses that fail to provide sufficient background concerning proposal preparation -- and proposals are at the very heart of the technical communication cycle.[1]

BACKGROUND FOR CLASS DISCUSSIONS OF THE TWO CONCEPTS

To help students grasp the significance of these two important concepts, it might be good to start the course with an exploration of the communication and recording implications of two basic facts about modern science and engineering. First, today scientific and engineering projects of any great importance are almost invariably group efforts. Accomplishments depend upon effective teamwork -- sometimes on an international scale and over many years. Second, modern projects can be very expensive. Even low-cost projects must usually be "sold", via a written or oral proposal, on a cost-effectiveness basis. The only "buyer" will be a decision maker with special interest in obtaining the contemplated results, adequate available funds, and willingness to bear the cost. This decision maker usually represents a Government agency or an industrial organization or segment thereof.

The decision maker must decide, for example, whether a certain line of inquiry or specific activities will go forward or not, whether a new idea will be explored, whether the results of previous work will be implemented. His decision concerning the future will be based in great part, on what has happened in the past. Thus, "technical continuity" extends backwards as well as forwards.

The proposed new project will be based, sometimes to a very great extent, upon the work of previous investigators. If that work has been carefully conducted and accurately reported, the base will be a solid one. If not, troubles -- and associated unanticipated expenses -- can constantly plague the new project. The new project, itself, will undoubtedly provide part of the base for subsequent related work. Thus, all the workers in the entire series of related efforts are accountable to each other for both high-quality work and high-quality reporting.

DETAILED CONSIDERATION OF TECHNICAL CONTINUITY

As implied above, the idea for a technical project seldom springs up in isolation, full-blown "from the head of Jove". And seldom does a project terminate absolutely with submission of a "final" report. With essentially every project there is a past, present, and future. Moreover, the special "present" represented by the span of time over which the project itself is conducted has many inherent past/present/future elements.

In practical terms, what does all this mean? Several simplified examples are discussed below, and some general suggestions for course attention or activities are provided.

Continuity With the Past

Ignorance of past technical accomplishments can lead to unnecessary costly and time-consuming repetition of work. Students should be encouraged to think of the beginning of each new project as a survey of the pertinent communications from the past. The depth and breadth of this initial step would depend upon a number of things. For example, are the results of a reliable and sufficiently recent pertinent survey available? Have specific information gaps been identified that might be closed by careful analysis of past work? Are there potentially serious hazards that might be avoided by utilizing past experience?

The look at communications from the past can be many faceted. Students will think immediately of reading books and journal articles. They might not think of checking such things as patents, advertisements, brochures, and catalogs. They might overlook possibilities for obtaining -- with a little persistence -- limited-access documents containing up-to-date data not yet disseminated through open channels. They might undervalue the importance of discussions with specialists in the field who probably possess or are aware of sources of important data not yet published -- the "newest of the new". In scheduling their technical projects, students might not allocate sufficient time for analysis of conflicting results or questionable conclusions of previous workers.

Sometimes, of course, exploration of the past is inhibited by, for example, proprietary industrial interests that block access to broad areas of information. Access can be blocked even though only part of the information sought is truly of a proprietary nature. Sometimes, too, unnecessarily stringent Government secrecy classifications prevent release of information that is not really sensitive.

Problems associated with unnecessary impediments to information flow gave rise to passage of the U.S. Freedom of Information Act. The Act is now being challenged and tested in practical situations. Students should know something about the major provisions of the Act and about the precedents being established in its implementation.

Communications from the past must be diligently sought and intelligently used. But -- what degree of diligence is appropriate or practical in specific instances? What is intelligent use, given time and budget constraints? Uncritical reliance on the results of past work can cause trouble if that work has not been carefully done and accurately documented. On the other hand, indiscriminant rechecking of all past work can cancel out potential savings of time and cost.

Continuity Within the Project

The plan for continuity of activities within a project is set forth initially in the project schedule contained in the proposal. It is also set forth, indirectly, in the proposed budget which usually shows the planned rate as well as amounts of expenditures.

The execution of the plan is described in progress and final reports; any deviations from the plan must be disclosed, as they occur. In a way, one might almost say that nothing happens "officially" in a project until it is recorded in some kind of document. In essence, the entire project is carried forward on a conveyor-belt substructure of documentation. The importance of this documentation is illustrated by the significance laboratory notebook records can assume in the case of patent applications.

The plan for project activities presented in the proposal might have to be flexible enough to allow for contingencies and even surprises. That is, the plan might have to take into account the potential for breaks in the continuity of activities. There might be a breakthrough discovery, for example, that forces a halt to await a critical decision. This decision might force a radical change in project direction. Or, there might be technical difficulties that delay pivotal decisions or bring activities to a temporary stop until these difficulties have been resolved.

A good way to deal with the potential for breaks in project continuity and the resultant needs for schedule (and probably budget) modifications is to divide the project -- at the proposal stage -- into phases. Phases might be based on arbitrarily selected time periods or on anticipated decision-making points. In the proposal, the Phase I schedule would then be described in detail, while schedules for subsequent phases would be described sketchily and in tentative terms.

Reports are the standard mechanisms for both requesting needed decisions and for putting them into the "official record" so that work can continue. The need for timely reporting is obvious. Written reports can often be mandatory for record purposes, even though an oral report might suffice for decision making.

Students should also be made aware of the evolutionary nature of progress reporting. Each progress report builds upon the one preceding it and forms the basis for the one following it. Progress reports should have a family resemblance. This should include format as well as organization and content.

Generally, progress reports have four distinct major sections: the Introduction, the Technical Discussion, the Conclusion, and the business status statement. The Introduction usually states the goals of the work done during the activity period just completed. It describes the problems identified at the beginning of the period and any that arose during the course of the work. The Introduction can be seen, in part, as a restatement and updating of the Conclusion section of the previous report. The Technical Discussion covers all significant activities and progress (or lack of it!) during the period. The Conclusion summarizes progress made, describes new problems that must be attacked during the next period, outlines the plan of attack, and states the specific goals of the anticipated activities. Thus, in essence, the Conclusion is the springboard for the Introduction of the next progress report.

The business status statement, which is concerned with budget and schedule, explains if necessary, and justifies if possible, any deviation from the business plan. It describes any contemplated adjustments to get back on schedule or within budget. It forms the springboard for the business status statement of the next period's report.

There are other aspects of continuity within a project that teachers might also want to discuss with students. For example, continuity of objectives is of basic importance. Sometimes in long or complicated projects there are temptations to get off target. Some exciting development opens up a fascinating new avenue for tangential activity. The basic objective of the project can get lost -- at least temporarily. Then, if monitoring or reporting is careless, or even neglected over a significant period of time, the project may be in serious trouble close to the end. Time and money might run out before the agreed-upon objectives can be achieved. The discipline of reporting helps ensure continuity of concentration on the real objectives.

Another aspect of project continuity relates to personnel assignment. Problems can arise, for example, if the Task Leader for a critical portion of the work is reassigned and a substitute is put into his place. If the first Task Leader fails to communicate adequately with his successor, it may be impossible for that successor to discharge his responsibilities. His ability to collaborate effectively with other Task Leaders may suffer drastically.

Students might be asked to think of good methods for ensuring project continuity. Such methods might include, for example, the holding of frequent staff meetings for timely exchange of information concerning progress and problems. Another might be the development of a project glossary to ensure that everyone is using the same terminology throughout the project. The underlying need for good communications should be stressed.

Continuity With Concurrent Related Activities

Concurrent related work can impact on a project. Government agencies that support technical projects are particularly sensitive in this regard. Government Project Monitors want to be sure that the Government is not paying for anything twice. They are keenly aware of the need for effective communication between related Government-sponsored projects. Sometimes proposals to the Government are required to provide a specific plan for effective "interfacing". Government proposal evaluators want assurance that the results obtained or the decisions reached in on-going related Government projects will mesh well with those of the project being proposed. In other words, they want overall continuity.

Students must understand that it is not sufficient to make a "good literature survey" just at the beginning of a project. It is necessary to maintain open communication lines throughout the project; the survey must be dynamic. It is important to be cognizant of new information being developed elsewhere, of new problems being encountered, of changes in the direction of related projects, of verification or rejection of significant data, etc.

Students might be asked to suggest a variety of measures for ensuring project continuity with concurrent related activities. A carefully planned, sustained program of professional reading would obviously be valuable. Membership in selected professional societies and attendance at their conferences could provide early access to useful new ideas. Participation in senimars and workshops and frequent enrollment in pertinent academic courses would also help. Correspondence with a dynamic network of others in the field might be better than any of the other methods for obtaining up-to-date information.

Maintaining cognizance/continuity means, in other words, planned, continuous, broad professional growth. Students might be questioned about their personal long-range plans for professional growth.

Continuity With the Future

Continuity with the future has two aspects: (1) continuity related to the immediate and the long-term future application of the specific project's results and (2) continuity with respect to the total related technical field (or fields). Both imply accountability to others.

With respect to the project itself, implementation of the results and feedback from that implementation are ordinarily of major concern. Students might be asked to consider, for example, the special significance of technical manuals for implementation of results. Manuals are designed for use in the future -- to ensure practical application new technology or new devices. They come in many varieties: installation manuals, check-out manuals, operation manuals, maintenance manuals, etc.

Persons responsible for manual preparation must be good at envisioning the future information needs of those who will be using them, perhaps over many years. Manual writers must be sensitive to requirements for consistency of instructions, alert for potential problems in interpreting illustrations, and aware of difficulties that might arise because of jargon obsolescence.

Although the final report on the project might be simple and unambiguous, satisfy completely the information needs of management-level people, and even be acclaimed as a landmark document, if the manuals intended for use by field personnel are not clear, the project can ultimately be judged a failure. Feedback from the field will be negative.

In the broad field of technical communications, manual writing is a distinct area of professional endeavor. Government agencies, in particular, support the production of countless manuals. Subject matter ranges from the monitoring of environmental pollutants to the maintenance of firearms. Many Government mandatory specifications have been promulgated to guide manual design. Thousands of manual writers are involved in Government-sponsored projects. Students might be asked to obtain and analyze examples of Government manuals or specifications for manual writing in their individual fields of career interest. They might be asked to estimate the useful life span of these documents.

Final reports often include a particularly important section obviously oriented toward the future: the Recommendations section. When preparing such sections, writers must be especially careful in assessing the future information needs of the decision makers addressed. They must also take into consideration their levels of technical backgrounds, their probable interests, their possible prejudices, and so forth. Not an easy analysis and forecasting challenge!

Most "good" Recommendations sections have common characteristics. Whenever possible they present complicated recommendations in several simple, independent statements. This helps to preclude future dilemmas arising from the need to make "maybe" decisions. Whenever feasible, statements are structured to permit discrete yes/no decisions. Skillfully written Recommendations sections point out alternatives and describe the future consequences of both positive and negative decisions.

Written reports can be future-oriented in another way. They are generally intended to be documents of record. Although a report might not stimulate action immediately, unanticipated future events might make it very timely many years later. The writer should consider the possible needs of future unknown readers. One question likely to come up, for example, could be the amount and nature of supporting detail that should be provided in appendixes.

On the very first day of a technical writing class, students might well be asked to consider the implications of timely report submissions. Project results cannot be used before they are made available to the intended users. Undue delay of a final project report can impede progress on related projects from their outset. The overdue report might be needed for decision making, for critical data synthesis or comparison, for component design configuration, or the like. If a required report is not available, costly schedule disruptions in the related project might result, and consequent problems might plague that project until its very end. The related project's results might be skewed or incomplete. A domino effect might be felt for many years and across many scientific or technical lines.

The importance that can be attached to the meeting of deadlines is illustrated by the attitude of the Government toward timely submission of proposals in response to its Requests for Proposals (RFPs). Government-appointed proposal evaluation teams simply will not (legally, they cannot) wait for late comers before they begin their evaluations. Proposals received after the cut-off time (specified in the RFP to the precise minute) are returned, unopened, to their senders.

Teachers would to well to establish deadlines for handing in class assignments and to stick with those deadlines despite loud student protests.

SOME ASPECTS OF PROFESSIONAL ACCOUNTABILITY

Unavoidably, the discussion above of the concept of technical continuity has involved indirect consideration of many aspects of professional accountability. The discussion that follows covers some other easily defined aspects.

Management Accountability

Students should be helped to realize that every technical project, no matter how small, involves management of and accountability for resources. Resource management requires planning, monitoring of progress throughout the project, and periodic and final reporting of resource use or disposition.

Some resources are concrete and expendable. These can include, for example, inexpensive chemicals or costly protective clothing. Other resources are intangible and, in a sense, indestructible. These are primarily the expertise of technical specialists. The latter resources, represented by the time spent by the experts in working on the project, must be allocated and monitored with care. This is because of the generally high cost of these resources and the competing demands on the available time of the experts.

In accounting for resources, the Project Manager ordinarily equates them with cost: salaries, purchase prices, rental expenses, etc. Effective cost control depends upon realistic budget planning, strict adherence to schedules, accurate recording, and good communications.

Scientific and Engineering Accountability

Students should also be helped to understand that in the future they may be concerned with many different types of scientific and engineering accountability. This will involve planning, monitoring, and recording, and it will require effective communications.

For some projects, such accountability can be of overriding importance. Typically, projects involving use of toxic or abused substances (e.g., heroin) or disposal of strategically important reaction products (e.g., nuclear waste materials) require expert planning, stringent accounting, and frequent reporting.

Students might be asked to relate communication skills and record keeping to accountability for such things as accurate instrument calibration, purity of samples, storage and transport of unstable materials, development of appropriate protocols and procedures, the writing of specifications, the use of statistics, blind and redundant testing, adequate equipment maintenance, tracing to standards such as National Bureau of Standards weights and measures, selection of controls and variables, introduction of contaminants, disposition of surpluses, subcontractor monitoring and control, and assignment of authority and responsibility. Obviously, communications with all involved members of the project team must be both accurage and timely.

Information Accountability

As has been stated, science and engineering accomplishments are the result of team efforts involving the past and present and having implications for the future. Therefore, accountability for communicating new and valuable information to a wide variety of collaborators -- some known and others never to be imagined by the workers on any specific project -- is an integral part of the overall picture.

Students might be asked to consider what the results would be of their personal failure to respect the information rights of their professional peers and colleagues -- both present and future. If there were a general lessening of the overall present sensitivity to the obligations of each professional to all other professionals, what would be the consequence? Is present sensitivity to these obligations sufficient to ensure long-range technical advancement for the good of mankind?

CONCLUSION

Teachers who weave throughout their plans and their class activities such broad concepts as technical continuity and professional accountability should find it easy to stimulate their students to exciting, worthwhile discussions of effective technical communications. Students should gain valuable insights into how technical communication forms and networks operate and why accurate, imaginative, and timely communications are so important. Peripheral discussions of grammar and punctuation should then have a realistic flavor and serve the purposes of the course rather than dominate it. Such a course should not be in danger at degenerating into just another conventional writing course.

REFERENCES

(1) Landreman, Dolores M., "The Case for Stressing Proposals in Technical Writing Courses", Proceedings, 25th International Technical Communication Conference, Dallas, Texas, May 10-13, 1978, p. 398.

IN SEARCH OF QUALITY COMMUNICATIONS

Gunther Marx
Johnson & Higgins

Gunther Marx
Johnson & Higgins
Asst. Vice President
95 Wall Street
New York, NY 10005
(212) 482-6057
B.A., M.A., Ph.D. cand.
Fellow & Past President, STC; Dir., Sigma Tau Chi

The continuing decline in quality of products and services -- a reflection of increasingly negative attitudes -- has reached alarming proportions. In this climate, slovenly communications have become the norm. Professional communicators must redouble their commitment to all the factors that constitute quality in communication. Only then will communicators be able to meet today's high expectations of business and industry. Educators charged with the training of professional communicators must broaden their scope of influence beyond traditional curricula to ensure that quality is achieved.

Discerning observers of today's economic and social scene view with increasing alarm an all-pervasive and inescapable phenomenon: the rapid and continuing decline in quality of products and services.

No one can be unaware of the poorly constructed, often unsafe office and apartment buildings, public and private housing, automobiles, home appliances, tools and machinery that have come to be accepted as the norm. No one is unaffected by those diminishing services that can be described only as inefficient at best, or insufficient at worst: services on which society as a whole depends -- such as public transportation, mail, telephone -- and those on which we rely as individuals -- ranging from the neighborhood drycleaner to the bank teller, from the cab driver to the checkout clerk at the local supermarket.

Negative Attitudes

Underlying the decline in quality of products and services, and perhaps even more profoundly disturbing, are the increasingly negative, thoroughly unprofessional attitudes we encounter daily, as revealed in expressions such as "It's not my job." "Your guess is as good as mine." "The truck broke down." "The computer made a mistake." If carried to the extreme, these, and a whole array of similar rejections of responsibility, threaten to destroy the very fabric of our society, to subvert the spirit of individual commitment and involvement that created this nation a little over two centuries ago and that carried it to greatness.

Along with many other products and services, communications are suffering from a rapid decline in levels of quality. No segment is spared from decay, whether in the media, in business communication, or in the literary or performing arts. Unfortunately, today, slipshod communications have become the rule, not the exception. They are so commonplace, so much taken for granted, as to be accepted as the norm. We are subjected on a daily basis to levels of writing and speaking that would have been unacceptable and held up to ridicule only a few decades ago.

In this climate, professional communicators have a special responsibility to create, maintain, and promote quality communications. It is a paramount and continuing challenge. But exactly what is quality in communication? How do you recognize it? How can you measure it? How do you create, maintain, and promote such quality? Clearly, we must learn to recognize and to deal with the malaise in thinking and in execution that readily label a communication as one that is lacking in quality.

Basics of Quality

Professional communicators simply are not doing their job if they are willing to ignore the very basics that constitute a quality communication: Does the communication do more than add to the merely obvious? Are the thoughts expressed logically? Are they stated coherently and in proper sequence? Are the statistical data clear, correct, and complete? Is the communication properly directed to the intended audience? Is the message clear and conducive to the expected reaction or response? Has full attention been paid to the meaning of words and phrases? Is the communication free from ungrammatical expressions, incorrect punctuation, misspelled words? Is it free from illegible, misleading, or superfluous illustrations? Are the graphics supportive of the words? Is the packaging appropriate?

All these considerations are the domain of the professional communicator, who must shoulder much of the responsibility for stemming the tide of slovenly communications.

A Glimmer of Light

Fortunately, the need for quality communications across the board is beginning to be more widely recognized. After several years of nearly zero growth in the American worker's output, business and government, as well as many concerned individuals, are becoming increasingly alarmed about the nation's productivity. At the same time, there is growing recognition that improved communication between management and employees has a direct and beneficial effect on productivity.

Business leaders throughout the country now recognize the need to communicate with their employees to explain what increased productivity can mean to the individual worker. These business leaders also realize that they must seek to encourage employees to offer their own suggestions and recommendations designed to improve productivity.

Moreover, leaders in business and industry recognize that they must communicate much more clearly the goals of the corporation, that they must communicate the concern of management for the needs of employees, that employers must communicate, for example, just what the company provides as a total compensation package for each employee. The list could be greatly extended.

The recurring theme of <u>communication</u> is no accident, of course. If the U.S. is to regain its preeminence, worker productivity must increase. Without effective communication, there is little chance for greater productivity in a democratic society such as ours. It becomes clear, then, that professional communicators are among those in the forefront of the movement to achieve a re-invigorated, recommitted America.

Expectations of Business and Industry

Armed with this strong charter or mission, what do industry and business require of professional communicators? Specific industries and businesses will have different requirements, of course, but certain attributes are common and are essential for effective communicators in the 1980s and beyond: Communicators must have the intellectual curiosity necessary to allow them to become an important and valued part of their organization. They must bring to bear an optimum combination of interpersonal and language skills. They must be able to articulate policies and produce a wide variety of communications with a maximum degree of personal flexibility and adaptability. They must be prepared to guide each communication project from initial concept, through all intermediate stages, to full and final realization.

Most important, industry and business need communicators who are truly dedicated to achieving and maintaining the highest standards of quality communication. Without exception, this will require a return and a rededication to basic values of excellence.

In the years to come, increasing attention is sure to be paid to manifestations of old virtues such as pride of workmanship, quality output, assumption of personal responsibility, exercise of proper discretion and judgment -- in short, a re-emphasis on total professionalism in all phases and types of communication. That this represents an enormous personal challenge to every professional communicator is obvious.

Education for Change

Those charged with preparing young men and women for positions as professional communicators must accept a special responsibility. Conventional approaches in our colleges and universities toward educating our future communicators may no longer be sufficient for today's and tomorrow's requirements. That is not to say, of course, that established courses and curricula don't have their place. Quite the contrary is true. The training of effective communicators now requires a conscious extension of the principles of quality communication into every classroom and every seminar.

Communication is too all-pervasive to allow it to be taught in formal classroom settings alone. Instead, the principles of quality communication -- whether written, verbal, graphic, whether print or electronic -- demand that attention be paid to them by every teacher in every classroom and in every subject. This integration can become an important beginning in reversing the degradation of our products and services, and in turning around the increasingly negative attitudes we encounter so frequently.

Advances in technology are far outstripping our ability to use technology effectively. The computer has become the status symbol of the 1980s. Professional communicators are no exception in coming under its spell. These machines are beginning to be used for certain copy editing, proofreading, and related functions. But obviously, computers can be only as effective -- or as defective -- as the programs that humans feed them. As a consequence, communicators needn't be concerned about being replaced by machines, for the "gigo" principle -- garbage in, garbage out -- will always apply. Professional communicators are among those who must always be mindful that no computer and no computer program, however sophisticated, can ever absolve the individual from assuming responsibility for the quality and value of the end product.

Striving for the Best

Occasionally, evidence emerges that points, if not to speedy solutions, then at least to a recognition of the need for quality. Even television -- perhaps the most persistent and most flagrant violator of all standards of quality communication -- on occasion has permitted at least a dim ray of hope to shine through a generally dismal picture. How else is one to explain the phenomenon -- though brief and commercially unsuccessful -- of law professor Kingsfield in "The Paper Chase"? John Houseman, creator of the role, describes the professor as "the enemy of shoddy work and sloppy thinking," a phrase that manages to encapsulate the whole battleground of professional communicators.

Pulitzer Prize-winning historian Barbara W. Tuchman, writing in The New York Times Magazine of November 2, 1980, comments persuasively and eloquently on "The Decline of Quality." After painting a deservedly gloomy picture touching on every facet of our culture, she nevertheless manages to find room for eventual optimism. "I cannot believe we shall founder under the rising tide of incompetence and trash," Mrs. Tuchman writes. "I have confidence in the opposite of egalitarianism: in the competence and excellence of the best among us. I meet this often enough... to believe that the urge for the best is an element of humankind as inherent as the heartbeat....As long as people exist," she concludes, "some will always strive for the best; some will attain it." Here, Mrs. Tuchman has articulated the very essence of the challenge facing professional communicators.

ENVIRONMENTAL IMPACT STATEMENTS AND SOME MODERN TRADITIONS OF COMMUNICATION

Carolyn R. Miller
North Carolina State University

Environmental Impact Statements (EISs) are a form of technical communication that has altered governmental decision-making processes and given explicit room in public policy to a new set of values. The historical antecedents for EISs lie in three distinct traditions: the public environmental movement, the limits and needs of the developing field of environmental science, and the requirements of administrative bureaucracy. Each of the three historical sources embodies a different set of values--with the result that early EISs were ambivalent and often successful.

In December 1969, Congress passed the National Environmental Policy Act, and President Nixon signed it into law on January 1, 1970. The act, usually referred to as NEPA, establishes national policy on the relationship between humankind and the "environment," establishes the Council on Environmental Quality, and specifies certain means by which the policy is to be carried out in the actions of the federal government. Probably the most widely known of the provisions is the one requiring the preparation of what have become known as Environmental Impact Statements on every "major federal action significantly affecting the quality of the human environment."

NEPA was formulated in a context of wide criticism of the fragmented way government had been handling natural resources and environmental affairs. This essentially formal problem of bureaucratic management became more important with the realization, in the late 1960s, that the problem was not merely a bureaucratic inconvenience but a substantial danger to the national health, economy, and heritage. The primary impetus for the legislation was the need to coordinate the piecemeal administrative treatment of the environment and to make that treatment more environmentally sound.

Environmental Impact Statements (EISs) were documents designed to change governmental decision making by informing decisions with information from the environmental sciences and environmentalist values that had hitherto had no explicit place in the decision process. EISs represent an important trend, in which scientific and technical information is used to inform social and political decisions. One member of an impact assessment team expressed it this way: "We are trying to supply scientific answers, translated into legal language, to what are social and political questions."

The earliest impact statements were small affairs, several pages long. But with the threats of litigation and the developing standards of adequacy in the courts, impact statements have grown--sometimes to be multi-volumed documents, ten volumes in the case of the Alaska pipeline, for example. One of the drafters of NEPA said about the beginnings of the EIS that,

> There are few clues in the legislative history concerning what NEPA's Congressional authors expected impact statements to look like. . . . It is certainly true, however, that [they] never contemplated anything so extravagant as the multiple volumed dissertations which now are commonly produced.

EISs have exhibited many failings and were, especially in the mid-1970s, subjected to much criticism. They were called "deadly, voluminous, and obscure"; they were accused by some of providing too much information and by others as providing too little, of being too conclusory and not conclusory enough; they have been accused of impeding administrative processes, progress, and clear thinking. In his 1977 Message on the Environment, President Carter said,

> To be more useful to decision-makers and the public, environmental impact statements must be concise, readable, and based upon competent professional analysis. They must reflect a concern with quality, not quantity. We do not want impact statements that are measured by the inch or weighed by the pound.

These difficulties with Environmental Impact Statements are due in part, of course, to the complexity of the problems they were called upon to analyze, and to that extent their length and detail are necessary. But the difficulties are also due, at least in part, to historical forces. EISs belong to at least three identifiable traditions: the public environmental movement, natural science, and administrative bureaucracy. Each has a set of values and a set of communication patterns and requirements that influenced the development of the EIS.

Public Environmentalism

The public environmental movement is deeply embedded in American history. It is the direct heir to the conservation movement of the late nineteenth century, which had two branches: wise-use utilitarianism (exemplified by Teddy Roosevelt and Gifford Pinchot) and preservationism (exemplified by John Muir and Thoreau). These two branches are extensions of the dialogue between the pioneers and the Romantics, those who put the needs of civilization first and those who saw transcendent value in wildness apart from man. A third, more contemporary branch of the movement is ecological naturalism, which sees the human race as completely a part of nature.

These several strands of environmentalism all surface in NEPA, to tie the law to its cultural antecedents and to suggest its confusion of purpose. The law's statement of policy, for example, includes language reflecting sentiments from the three environmentalist traditions: from utilitarianism, the law mandates the attainment of "the widest range of beneficial uses . . . without degradation, [or] risk to health or safety"; from preservationism, the law assures "for all Americans . . . esthetically . . . pleasuring surroundings"; and from ecological naturalism, the law promotes the general goal that "man and nature can exist in productive harmony." The public environmental movement establishes the general objective for EISs and a set of values for the documents to promote.

Natural Science

A second historical source for Environmental Impact Statements is natural science itself. NEPA, in fact, helped to create the scientific study of environmental problems. Most observers agree that environmental science is a complex, developing area of study, with both methodological difficulties and theoretical weaknesses. In 1971, the National Science Board reported to the President that:

> Environmental science, today, is unable to match the needs of society for definitive information, predictive capability, and the analysis of environmental systems as systems. Because existing data and current theoretical models are inadequate, environmental science remains unable in virtually all areas of application to offer more than qualitative interpretations or suggestions of environmental change that may occur in response to specific actions.

The research that has been produced in the service of environmental impact assessment has not allowed for the tentative nature of ecological information. Nor has it, according to some, observed good scientific methods in general. In general, impact assessment has produced reductionist, descriptive collections of data rather than integrated, predictive scientific studies. In addition, because the impact statement process has not allowed for the scientific requirement of peer review through publication, replication, and empirical testing of predictions, knowledge is neither organized nor cumulative, as it should be in a mature, productive science.

For the EIS, the tradition of natural science provides informational content and a method of inquiry that relies on objectivity and verification. But because of the tentative nature of environmental science and the nonscientific uses to which the EIS is put, both the information and the method become distorted in the document.

Administrative Bureaucracy

The administrative process is a creation of the rise of industrialism and of democracy. The complexity of the problems of industrialization and the democratic requirement for methods of control and procedure led to the creation of administrative agencies in the early twentieth century. Impact statements themselves were conceived as a means of control because the legislative drafters anticipated administrative indifference, or even outright hostility, to the policy statement in the law. In fact, the section of NEPA requiring impact statements was designed with economic decision documents as the model, and it was presumed that the Bureau of the Budget would supervise the environmental assessment process that leads to the document, just as it supervised all benefit-cost evaluations. Implementation of NEPA under President Nixon gave this supervisory role to the Council on Environmental Quality rather than to the Bureau of the Budget, but the precedent was set, nonetheless.

Systems theory tells us that administrative bureaucracies have effects on their human members. Bureaucracies create an "insider's viewpoint" because of their distinct separation from the environment; they promote contribution to the functioning of the bureaucracy as an end in itself; and they limit the way insiders perceive both the system as a whole and its environment, substituting a faith in specialized expertise for more general social knowledge. Thus, government agencies, which are responsible for preparing impact statements, are unable to comply with NEPA's requirements for impartial analysis. It is difficult, if not self-destructive, for an agency to challenge the means and ends of those whom they regulate or to question the mission that Congress assigned to the agency. The agency's self-maintenance objective outweighs the mandate of NEPA.

Administrative procedure relies on compartmentalized expertise, discouragement of innovation and change, and maintenance of the implicit values of the status quo. Documents produced by administrative agencies reflect these values and procedures. Bureaucracies use standardized and predictable methods for decision-making and seek to maintain stability by limiting the range of analysis.

Under this view, then, the decision by an administrator in the Department of the Interior, for instance, to build a flood control dam, rather than to seek nonengineering solutions for flooding or to work with the Department of Energy on siting and design to provide electric power as well as flood control is "rational." Concerns for wildlife preservation, socioeconomic disruption, the concentration of upstream pollution, and the like are minimal because the agency understands its function as that of building dams in the ways it has found dam building feasible in the past.

Tradition and Effectiveness

The three antecedent traditions that influenced the creation of Environmental Impact Statements are diverse and incompatible. The patterns of thinking in administrative bureaucracies create a set of values different from, and often at variance with, the values invoked by NEPA, values which are supposed to inform impact statements and administrative actions. American history bequeaths a fundamental ambivalence about nature to the environmental movement, which sees nature now as something to be used, and now as something to be worshipped. And in the absence of firm precedents regarding the status of scientific facts and their relationship to social policy, it fell to the courts of law to provide the final shape for impact statements.

The Environmental Impact Statement was created in response to political and administrative problems, not directly in response to communication problems. It was invented by an act of Congress--a product of technological change not of an evolving rhetorical tradition. When a document is legislated, rather than evolving, those who prepare and use it are often unaware of the traditions that ultimately determine its form and meaning. Such unawareness of the rhetorical sources of the EIS contributed to its initial ineffectiveness. The Environmental Impact Statement may be understood as a kind of rhetorical hybrid that required some time, much criticism and litigation, and, in 1978, a redesigning by the Council on Environmental Quality, to approach a form and style that were responsive to communication needs.

COMMUNICATION NEEDS IN HIGH TECHNOLOGY INDUSTRIES

Deborah Namm
Xerox/Diablo Systems, Inc.

Deborah Namm
Xerox/Diablo Systems Inc.
Management Development Mgr.
3526 Breakwater Court
Hayward, CA 94545
(415) 786-5188

Prentice-Hall publications,
Management Consultant
Fortune 500. Grad. degrees
communication.

Technological advancements have played an important role in the life styles of most people. This paper reviews research on business communication, highlights trends in the electronics industries of Northern California, and describes areas for improvement in the communication of those industries. Areas of communication needs include the need for: conflict resolution and interpersonal management skills, clear dissemination of company goals, greater sharing of information between research and production departments, communication of positive company image, and clear dissemination of technical information for nonspecialists.

INTRODUCTION

It is difficult to avoid recent technological advancements in the fields of electronics and computer science. Today most cars use semi-conductor devices for their ignition systems or dashboard controls; electronic components can be found in digital watches, refrigerators, telephones, elevators, radios, television sets, and other appliances. Semi-conductors and computers are directly or indirectly involved in the use of traffic lights, bank accounting systems, and cash registers in department and grocery stores (ABC 1979).

Despite problems in the U.S. economy in 1979 and 1980, the majority of electronics companies continue to prosper or maintain prior production levels (Entin 1981). The reason for continued growth in electronics related industries amid the recessionary and inflationary economy: electronic instruments are perceived as helping increase productivity during a time when it has steadily declined in the United States (Entin 1981). It is clear that high technology industries will continue to play an important role in the economy and life styles of Americans.

History of Growth in Electronics Industry

Electronics organizations made slow but steady progress in the 1950s; growth accelerated rapidly in the 1960s and 1970s with the development and refinement of the Silicon Chip.

In only a few years the Silicon Chip (which has remained smaller in size than an average adult fingernail), increased in ability to contain electronic components. A typical chip in the early 1970s contained two or three transistors; in 1979 a Silicon Chip housed 20,000-30,000 transistors (ABC 1979).

The growth of technology in electronics helped promote the burgeoning of companies specializing in electronic and computer products/services. The heart of the electronics industry settled in Santa Clara County in California. The Santa Clara Valley, more widely known as 'Silicon Valley,' was the site of three electronics firms in 1950; today it is the home for more than 500 electronics companies (Gruber 1981).

COMMUNICATION NEEDS IN HIGH TECHNOLOGY INDUSTRIES

This paper will examine trends in communication problems and needs of the high technology industries located in the 25 mile area known as 'Silicon Valley.' Content of the paper will include: a review of pertinent business communication research, highlights of trends in the electronics industry of Northern California, and a description of areas for communication improvement.

Research on Communication and Industry

Several studies have been conducted on the purpose, volume, and direction of managerial communications (Stinch and Combe 1974, Horne and Lipton 1965, Kelly 1964, Landseberger 1961, Burns 1959).

Research on the influence of communication on management decision making concluded that effective upward communication from the lower and middle management levels increased the quality of decisions (Lawrence and Lorsch 1969).

There have been several attempts to classify and identify important areas of research in business communication, including Bateman's list of eleven areas of study (Bateman 1977). Several articles have debated the subject of the business organization serving as a function of communication versus communication serving as a function of the business organization (Johnson 1977, Farace and MacDonald 1974, Porterfield 1980).

Despite the wealth of general research concerning communication and industry, there exists little verified information about the relationship between communication and organizational behavior in the technical organization (Miles 1975).

Much of the emphasis of existing research in technical communication has involved written communication (Gielselman 1980). For example, a recent review of articles published in TECHNICAL COMMUNICATIONS in the past five years, indicated an orientation toward "efficiency and effectiveness of written and graphic materials" (Gielselman 1980).

One study conducted in 1974 documented the importance of oral and interpersonal communications in the daily job activities of non-supervisory engineers as rated by their supervisors (Di Salvo 1980).

COMMUNICATIONS AND THE TECHNICAL-RESEARCH/MANUFACTURING CONFLICT

Studies centering on organizational behavior and communications in high technology organizations have revealed problems in communications between technical/research departments and manufacturing/production departments. Some of these problems include lack of understanding about activities in the other department, unclear pictures of how the other departments fit into the company's plans for production and profits, simplistic assessments of the value of other departments, and transmission of information to other departments in a form that is not readily understandable or useful. (See list of references)

Problems in communication between technical/ research and production segments of 'high tech' industries have been promoted by:

- A lack of recognition of the existence of different communication policies in each department
- Traditional conflict between production department's need to make existing products and the orientation of technical departments to move on to a new product
- Infrequent contact between technical/research and production departments.

Communication problems between research and development, and production departments exist in part because the type, frequency, degree of specificity, and frequency of feedback will vary according to the department/work unit. 'High tech' companies need to recognize differences in communication styles of departments and should implement communication policies that allow for the differences.

"There is a strong relationship between the technology of a work unit and the amount and type of information participants require to perform effectively. Mismatches between the information system and work unit technology account for a large percentage of information system difficulties" (Daft and MacIntosh 1978).

An example of the difference in the organizational and communications structure of manufacturing and research/technical departments includes a "research laboratory whose members deal with an uncertain environment that provides feedback about results only after long intervals" and "requires an organization design with low formality and infrequent but general measures of performance. On the other hand, a manufacturing plant in the same organization is faced with more rapid feedback and more certain information about its environment and requires more formality and more specific and frequent measures of performance" (Lorsch 1977).

High technology industries are concerned about the lack of timely, specific, useful, etc. communication between a research oriented and production oriented department largely because these firms depend on research departments to help give their companies a slight edge over competitors.

Most 'high tech' businesses in Santa Clara County spend as much as 6 to 10% of sales on research and development; totaling nearly a billion dollars yearly for research in the Silicon Valley companies. Pressure to come up with new technologies is greatest in the semi-conductor business, where companies need new developments on a yearly or biennial basis (Mitchell 1981).

The intense competition for technological development adds pressure to increase the traditional conflict between a company's need to make money by manufacturing products already in existence and the orientation/desire of the technical person to go on to the next product.

Lack of understanding about organizational goals and unclear explanations of company expectations concerning the relationships between manufacturing and the research/technical departments add to inter-departmental conflicts. Infrequency of contact and lack of information exchange can promote an 'us' versus 'them' mentality in organizations (Bulit 1979).

Communication problems between the technically oriented and production oriented departments of 'high tech' industries highlight the following areas of needed development:

- A need to reorganize company communication policies to address differences in the communication structure of research and production departments
- A need for greater use of conflict resolution and interpersonal management skills to deal with the traditional conflict of departmental concerns
- A need for clear dissemination of company goals and expectations (in a form that can be used as a guide for action)
- A need for greater sharing of information and concerns between research and develop-- ment, and other pertinent departments
- A need for a structural plan for liason between research and production departments.

Additional Factors Concerning Silicon Valley Firms

Although size and age of California based 'high tech' companies vary widely, some trends can be identified. These trends include:

- High turnover
- Demand for qualified technical specialists
- Rapidly developing technology with specialized set of terminology.

Demand for skilled and semi-skilled workers in the Silicon Valley has resulted in turnover rates averaging 20-50%. Pirating of competitors for technical specialists has become an accepted practice (confidential interviews with 'high tech' personnel executives in 1981, Entin 1981, ABC 1979).

A typical Sunday edition of the SAN JOSE MERCURY NEWS includes 50 pages of want ads offering enticements like free eye care, racquet ball and tennis courts on the work site, as well as cash bonuses for signing on as a new employee. With the demand clearly on the technical specialists side, companies need to communicate an image as a 'nice' or 'good' place to work to current and potential employees.

Rapid developments in technology have given rise to the use of acronyms, which can deter understanding for the nonspecialist. In addition, abbreviations are often used in place of complete names or titles. For example, you can easily find (or hear) the initials LSI used as a complete word, without any explanation or reference to 'large scale integrated circuits.'

The use of acronyms and abbreviations emphasizes the need to clarify and explain technological concepts in a more understandable way, especially since many customers and employees of 'high tech' organizations are nonspecialists in at least some areas of specialization.

Thus, the high turnover and demand for employees, rapid development of technology, and use of acronyms reveals additional areas for communication improvement:

- A need to communicate a positive company image for current and potential employees
- A need to communicate technological information so that it can be easily understood by the nonspecialist.

Information provided throughout this paper attempted to give the reader an overview of issues pertaining to communication needs in high technology industries. The development and expansion of research concerning communication in the technical industries will become increasingly important as technological advancements continue to impact the work environment.

REFERENCES

Bateman, David N., "What Business Communicationists Measure and How They Measure it," Guidelines for Research in Business Communication. Champaign, Illinois: American Business Communication Association, 1977.

"Bonanza in Silicon Valley," ABC, 1979.

Bulit, J.J., "Ways to Better Liason Between Corporate Research and Operations," Research Management, Vol. 22, 1979.

Burns, T., "The Directions of Activity and Communication in a Departmental Executive Group," Human Relations, Vol. 7, 1954.

Daft, R. and N. MacIntosh, "A New Approach to Design and Use of Management Information," California Management Review, Vol. 22, 1978.

Di Solvo, Vincent S., "A Summary of Current Research Identifying Communication Skills in Various Organizational Contexts," Communication Education, Vol. 29, 1980.

Entin, B., "Valley to Stay Silicon," San Jose Mercury News, February 2, 1981.

Farace, R. and D. MacDonald, "New Directions in the Story of Organizational Communication," Personnel Psychology, Vol. 27, 1974.

Gielselman, Robert D., "Research in Business Communication: The State of the Art," Journal of Business Communication, Vol. 17, 1980.

Gruber, S., "Prunes to Chips," San Jose Mercury News, February 1, 1981.

Horne, J.H. and T. Lupton, "The Work Activities of Middle Managers - An Explanatory Study," Journal of Management Studies, Vol. 2, 1965.

Johnson, Bonnie M., Communication: The Process of Organizing, Boston: Allyn and Bacon, 1977.

Kelly, J., "The Study of Executive Behavior by Active Sampling," Human Relations, Vol. 17, 1964.

Landseberger, H.A., "The Horizontal Dimension in Bureaucracy," Administrative Science Quarterly, Vol. 6, 1961.

Lawrence, P. and J. Lorsch, Organization and Environment, Homewood, Illinois: Irwin, 1969.

Lorsch, J., "Organization Design: A Situational Perspective," Organizational Dynamics, Autumn, 1977.

Miles, Raymond E., Theories of Management: Implications for Organizational Behavior and Development, New York: McGraw Hill, 1975.

Mitchell, James J., "R & D: Survival Key," San Jose Mercury News, February 2, 1981.

Porterfield, C.D., "Toward the Integration of Communication and Management," Journal of Business Communication, Vol. 17, 1980.

Stinch, Combe A., Creating Efficient Administration,New York: Academic Press, 1974.

ISSUES IN ESL AND THEIR RELEVANCE FOR TECHNICAL WRITING TEACHERS

Leslie A. Olsen
The University of Michigan

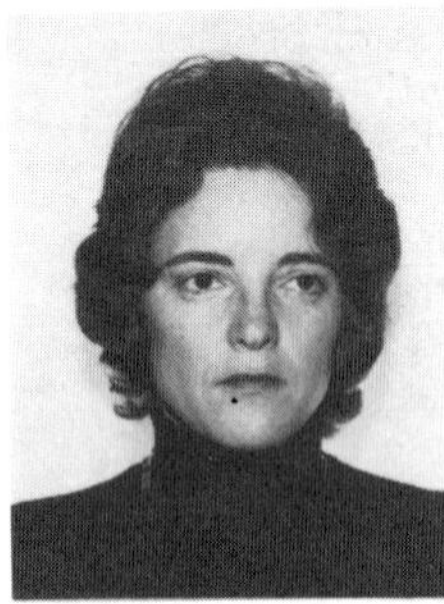

Leslie A. Olsen
Associate Professor
The University of Michigan
Department of Humanities
College of Engineering

Since technical writing teachers must be prepared for increased numbers of foreign technical students, this article surveys the issues and methodologies in English as a Second Language (ESL) instruction, the implications of ESL practices on technical writing classes, and the ESL resources currently available for foreign technical students. A bibliography of readings and resources in included.

INTRODUCTION

In the United States, technical writing teachers have traditionally developed courses for Americans, with little concern for the needs of foreigners. A quick survey of the available technical writing texts will confirm this. However, we are now faced with a large number of foreign students and professionals in our colleges -- an estimated 300,000, approximately 70% of whom are studying science and technology. Many require additional work in English as a Second Language (ESL) and, since they are primarily interested in technical English, turn to technical writing teachers for their advanced work. Unfortunately, when they arrive in our technical writing classes, we are faced with the challenge of providing appropriate instruction for them without much experience with their backgrounds, expectations, or previous ESL instruction; without training in the special areas of linguistics usually associated with second language instruction; and without any available reference materials to guide us.

Thus, the purpose of this paper is to provide some orientation for technical writing teachers on

1. the issues and concerns of ESL teachers,
2. those areas of ESL theory and methodology especially relevant to technical writing teachers trained in English, and
3. a survey of the best ESL resources available for foreign technical students.

CENTRAL ISSUES IN ESL INSTRUCTION

Theorists and practitioners in second language instruction have tried to understand the nature of language acquisition, the situations and processes interfering with it, and the methods which best encourage it. ESL instruction has developed from three different approaches to second language acquisition (Spolsky 1979): contrastive analysis developed by Fries and Lado; error analysis developed by Corder (1973) and Richards (1974), and interlanguage analysis initiated by Selinker (1972).

Contrastive analysis focuses on language description and compares features of a student's native language with the corresponding features in the new or target language. Contrastive analysis attempts to predict which features of the target language will be easy or difficult to learn (because of similarities or differences to the native language) and to explain the appearance of certain "mistakes" in the target language as residues of structures in the native language. Materials based on this approach stress constructions potentially confused by "interfering" structures from the native language.

Taking a different approach, error analysis focuses on language acquisition and learning; it attempts to discover systematic errors in language learning and thus to give evidence of the learning process. In so doing, it allows connections to be drawn between studies in first and second language acquisition and produces instructional materials based on studies in first language acquisition in children.

Finally, interlanguage analysis concentrates on the learner's knowledge of the target language as a whole, especially the various features of it which "fossilize" at intermediate stages of learning. The interlanguage is a system "distinct from both the native language and the target language" (Spolsky 1979), a system which is both rule-governed and approximate to the target language. Its fossilizations are assumed to be potentially much stabler than the transitional and ever-improving learning stages in contrastive and error analysis. Insights based on this approach appear in Interlanguage Bulletin.

All three of these approaches have concentrated on grammatical skills and methodologies for improving them, with relatively little attention devoted to higher level skills, that is rhetorical and pragmatic (or usage) skills. The main exceptions to this are theoretical stands by Widdowson (1978a), Wilkins (1976), and the advocates of the communicative and functional approaches to ESL instruction (Holden 1977 and Robinson 1980). These approaches view as the proper subject for study all of the elements operative in real acts of communication -- including illocutionary and non-verbal features, rhetorical conventions, and audience and situational constraints. Unfortunately, almost all of the instructional materials developed by these theorists falls far short of their goals, primarily treating the grammatical skills of prime interest to linguists (Huckin 1980).

Other questions under debate in ESL circles include the following, most of which are also debated by English teachers:

1. what are the appropriate skills to be taught -- reading (Ulijn 1980), conversing, writing, or studying (including lecture comprehension, note taking, and question asking)?
2. what type and level of English should be taught -- general English, English for Academic Purposes (EAP), English for Specific Purposes (ESP), or English of Science and Technology (EST)?
3. what are the best methods for reaching one's goals, eg., how authentic should communication situations be, should one use real or contrived texts? (Widdowson 1978a, Widdowson 1978b, Huckin 1980)
4. how exactly must one approximate the competence and performance of native speakers: does one need a rule of grammar or a rule of thumb (Berman 1979).
5. what are the special conventions of the type of English one is teaching, i.e., to what extent can and must one be able to specify the grammatical, rhetorical, and pragmatic (or usage) conventions and constraints? (Trimble et al, 1978, Selinker, et al, forthcoming)
6. what are the special needs and motivations of the student?
7. what are the most appropriate types of diagnostic and achievement tests? (Oller 1976)

IMPLICATIONS OF ESL INSTRUCTION FOR TECHNICAL WRITING TEACHERS

The prior ESL instruction foreign students have received creates expectations and needs of which technical writing teachers should be aware.

Expectations: Much ESL instruction, especially that overseas, is based on rote memory, testing, and lack of conceptual problem solving. Consequently, many students expect these features in their American courses. However, since a course such as technical writing often stresses problem solving or the application of general principles to novel situations, a teacher suddenly faced with foreign students might want to explain more carefully the course methodology and the reasons behind it; such a teacher might also insert some "extra" quizzes.

Needs: Given the uniformly strong emphasis on grammar and the limited acquaintance with Western audiences provided by most ESL instruction, many foreign students need first extra instruction on audience analysis as well as on rhetorical conventions, situational constraints, and non-verbal elements in technical communication. Second, irrespective of their scores on an English proficiency test, most foreign students in their first term or year need instruction in understanding American lectures. They often have difficulty in finding the main point of a lecture, or in discriminating among orienting material, main points, examples, asides, or quick jokes (Candlin 1976). Fortunately, the above are areas of relative strength for most technical writing teachers.

Third, foreign students and professionals almost always need work on the linguistically sophisticated uses of grammar, especially those grammatical functions based on rhetorical or pragmatic (usage) considerations. Unfortunately, these are areas of relative weakness for most teachers trained in English rather than in linguistics. (Indeed, given the sentence-level focus of much linguistics, these would be areas of relative weakness even for many ESL teachers.)

And fourth, foreign students enrolled in a predominantly American class frequently need class management practices different from those appropriate for native speakers of English. For instance, a newly arrived foreign student may well need extra time to complete an in-class test simply because that student takes longer to read, process, and produce material than does an equally "knowledgeable" American student. Further, a new foreign student may require somewhat different grading standards early in a course. Note: I am not arguing that one ought to give A's for E work. However, a student who has mastered all of the organizational material treated in class but made unexpected errors in grammar or style probably should be informed and perhaps penalized for making the errors, but also rewarded for mastering the material mastered. This is a very difficult grading situation, but some students who feel that they have no chance for a decent grade -- even if they work hard on their English -- will feel forced to plagiarize or somehow 'beat the system'.

SURVEY OF RESOURCES FOR ESL STUDENTS

There are three types of resources which might be of interest to technical writing teachers: a trained ESL teacher for consultation, materials on general ESL theory and pedagogy, and materials on the specific problems arising from even the relatively advanced students in a technical writing classroom.

The ESL consultant: The trained ESL teacher who would serve as a consultant is the most highly recommended resource, even if this person must be found at another institution. Such a person can save the technical writing teacher time and frustration and provide better instruction for the non-native student.

Materials on general ESL theory and pedagogy: Probably the best introductions to issues in ESL are the introductions to applied linguistics by Wardhaugh and Brown (1976) and Corder (1973) and the surveys of issues and resources in ESP (English for Specific Purposes) by Holden (1977) and Robinson (1980). ESP is a subfield of ESL; it develops programs to meet the special academic or professional language needs of its students and contains the sub-field English for Science and Technology (EST) in which we are particularly interested. Useful general journals in ESL/EST include Language Learning, the ESP Journal, The ESP Newsletter published at Oregon State University, Modern Language Journal, Modern English Teacher, Reading Abstracts, and Studies in Second Language Acquisition.

Materials on Specific ESL/EST Problems: Available materials include essay collections, bibliographies, journals, and conferences. There are two relevant collections of essays: Trimble, Trimble, and Drobnic (1978) and Selinker, Tarone, and Hanzeli (forthcoming). The most useful bibliography for non-linguists is in Robinson (1980) More specialized ones are by Trimble, Selinker, and Huckin (1974) and by Tomlin, Olsen, and Huckin (1980), the latter organized by main subject area (e.g., articles, modals, relative clauses). Journals in which useful and practical tips often occur include The English Language Teaching Journal, English Teaching Forum, and TESOL Quarterly. And finally, occasional conferences on ESL or EST are advertised in The EST Newsletter.

Diagnosing a student's problems is a difficult and tricky task since there are no good diagnostic tests available. The tests which are available are screening tests -- the TOEFL Test, the Michigan Test of English Language Proficiency, or the UCLA English as a Second Language Placement Examination -- and unfortunately, there is a tendency for these tests "or just about any battery of language tests to load on one major factor which can be identified only with global language proficiency" (Oller, 1976). In other words, unless technical writing teachers can find an ESL consultant, they will have to use their own observations for diagnosis and then route students to the materials suggested below.

Basic grammatical problems are probably more easily treated than any other problem because there are several ESL texts available; the most useful of these are probably Danielson and Hayden (1973) and Praninskas (1975). Sophisticated grammatical problems especially those which have a rhetorical or discourse base -- are much less frequently discussed but are treated by

Huckin and Olsen (forthcoming, 1981). They also treat special problems with rhetorical patterning -- overall organization paragraph organization, and connectives. It should be noted for our purposes that the scientific method is widely taught around the world and that most technical students should be familiar with western patterns of reporting, defining, and describing; some evidence for this is provided in Trimble, Trimble, and Drobnic (1978). However, in spite of this familiarity students will frequently be inexpert with some features which have a rhetorical base and thus require special help.

In addition to the areas described above, new foreign students often have serious problems with lecture comprehension and speaking in spite of "satisfactory" scores on an English placement test. Resource materials for this problem are available in a workbook by Sims and Peterson (1981) and in a cassette from them at the University of Colorado. Huckin and Olsen (forthcoming) also treat this topic and have video cassette practice materials available through the University of Michigan.

REFERENCES

1. Berman, R. A. Rule of grammar or rule of thumb? International Review of Applied Linguistics, 17(4): 279-302, November 1979.
2. Candlin, Christopher, and Dermott Murphy. Engineering Discourse and Listening Comprehension. University of Lancaster, Lancaster, England.
3. Corder, S. Pit. Introducing Applied Linguistics. Penguin Books, Middlesex, England, 1973.
4. Danielson, Dorothy, and Rebecca Hayden. Using English: Your Second Language. Prentice-Hall, Inc., 1973.
5. Holden, Susan, editor. English for Specific Purposes. In Modern English Teacher, 1(1), 1977.
6. Huckin, Thomas N. Review Essay: Teaching Language as Communication. Language Learning, 30(1): 209-227, 1980.
7. Huckin, Thomas N., and Leslie A. Olsen. English for Science and Technology: A Handbook for Non-Native Speakers. McGraw-Hill, Inc., New York,forthcoming.
8. Oller, John W. Language testing. In Wardhaugh and Brown (1976) below, pages 275-300.
9. Praninskas, Jean. Rapid Review of English Grammar. Prentice-Hall, Englewood Cliffs, N.J., 1975.
10. Richards, J.C., editor. Error Analysis: Perspectives on Second Language Acquisition. Longmans, London, 1974.
11. Robinson, Pauline. ESP: English for Specific Purposes. Pergamon Press, Oxford, 1980.
12. Sims, J., and P. Peterson. Better Listening Skills. Prentice-Hall, Inc., Englewood Cliffs, N.J., 1981.
13. Selinker, Larry. Interlanguage. International Review of Applied Linguistics, 10: 209-231, 1972.
14. Selinker, Larry, Louis Trimble, and Thomas Huckin. Annotated Bibliography of Research in English for Science and Technology. Center for Applied Linguistics, Washington, D.C., 1974.
15. Selinker, Larry, Elaine Tarone, and Victor Hanzeli, editors. English for Academic and Technical Purposes: Studies in Honor of Louis Trimble. Newberry House, Rowley, Massachusetts, in press.
16. Spolsky, B. Contrastive analysis, error analysis, interlanguage, and other useful fads. The Modern Language Journal, 250-279, 1979.
17. Tomlin, Russell, Leslie A. Olsen, and Thomas N. Huckin. English for Science and Technology: A Research Bibliography. The University of Michigan, Ann Arbor, Michigan.
18. Trimble, Louis, Mary Todd Trimble, and Karl Drobnic, editors. English for Specific Purposes: Science and Technology. Oregon State University, English Language Institute, Corvallis, Oregon, 1978.
19. Ulijn, Jan. Foreign language reading research: recent trends and future prospects. Journal of Research in Reading, 3(1): 17-37, 1980.
20. Wardhaugh, Ronald, and H. Douglas Brown, editors.A Survey of Applied Linguistics. The University of Michigan Press, Ann Arbor, Michigan, 1976.
21. Widdowson, H. G. Teaching Language as Communication. Oxford University Press, Oxford, 1978a.
22. Widdowson, H. G. The significance of simplification. Studies in Second Language Acquisition, 1(1), May, 1978b.
23. Wilkins, D.A.Notional Syllabuses: A Taxonomy and Its Relevance to Foreign Language Curriculum Development. Oxford University Press, London, 1976.

A LATE MEDIEVAL TECHNICAL DIRECTIVE: CHAUCER'S TREATISE ON THE ASTROLABE

George Ovitt, Jr.
Drexel University

George Ovitt, Jr.
Drexel University
Assistant Professor
Philadelphia, PA. 19104

Chaucer's Treatise on the Astrolabe, which is the oldest treatment of a technical subject in English, was written at a time when Western science was still in the process of developing an adequate language with which to deal with the complexities of the emerging discoveries of the previous two centuries. The problems that Chaucer encountered in describing the construction and function of the astrolabe may be seen to foreshadow the perennial problems of all technical communicators, and his solutions to these problems, though sometimes unsatisfactory, nonetheless provide the present-day technical communicator with a sense of his discipline's history.

BACKGROUND

Early medieval astronomy suffered from both a lack of tradition and a lack of mathematical rigor. The work of the Greek astronomers--including that of Ptolemy--was lost with the demise of the Roman Empire. Until the great age of translation and scientific revival in the twelfth century, astronomical knowledge was preserved primarily by the Latin encyclopedists, particularly Macrobius, Martianus Capella, and Isidore of Seville. Perhaps a single example taken from the influential Etymologies of Isidore (d. 636) will be sufficient to show the condition of astronomy during this time: "Mundus (the universe) is that which is made up of the heavens and earth and the sea and all heavenly bodies. And it is called mundus for the reason that it is always in motion (motus)." This simple and uninformative taxonomy, with its attempt to capture the essence of a quality or an object through the study of word origins (false in this case), may be seen as typical of early medieval astronomy.

During this scientific "dark age," observations of physical phenomena were considered to be both unnecessary and untrustworthy. The reality which underlay the material world was though to be discerned primarily in those rare moments when the truth of a text (e.g. the Bible, the Fathers) penetrated the veil of the senses. Indeed, observations and measurements could be positively subversive as in the case of the retrograde motion of the planets. Such an unexpected phenomenon led observers to posit theories whose sole merit was their ability to "save the appearances."

Another group of proto-astronomers, known to us collectively as the "computists," were concerned with the mathematics of astronomical calculations, but only to the extent that such calculations aided in the correct placement of the feast of Easter. This movement, though surely "one of the clearest indications of intellectual revival in the early twelfth century," was nonetheless more dependent on the interpretation of ancient "authorities" than on observation, and more concerned with the verification of a priori assumptions than with the discovery of physical laws. With both groups, the encyclopedists and the computists, concern with the structure of planetary motion was secondary to overriding theological imperatives, and knowledge of celestial mechanics remained sketchy.

It is in this context that the significance of the astrolabe must be considered. While references to this complex and beautiful instrument can be found in a text as old as Ptolemy's Planisphaerium, the Christian West did not have general access to either the theory or the instrument itself until the tenth century, and widespread familiarity with either did not come until the translation (from Arabic to Latin) of Messahala's Compositio et Operatio Astrolabii in the twelfth century. Yet, by the next century, the astrolabe was in use from India to Spain, England to North Africa, and its relative precision, combined with the fact that its use could be extended from the casting of horoscopes to the solutions of problems in applied geometry, insured that a long-interrupted tradition of astronomical observation would be permanently reinstated.

CHAUCER'S TREATISE: THE INTRODUCTION

Chaucer's Treatise is valuable for the student of scientific and technical communications for a number of reasons, not the least of which is the fact that it is the oldest work in English concerned with a complex scientific instrument. The Treatise is also useful in that it provides us with at least some indication of the kind of technical knowledge that a "non-specialist" of the fourteenth century--namely, Chaucer himself--might have mastered, and, rather more surprisingly, with an indication of the kind of technical information a ten-year-old child might have been taught.

Chaucer wrote his Treatise in 1391 to his son "Lyte Lowys," in order, as he himself tells us, to satisfy the boy's desire to "lerne scienes touching on nombres and proporicions." That the Treatise was a directive rather than a purely theoretical description is clear enough from the fact that Chaucer has given his son a "suffisant Astrolabie," and from the fact that he includes a series of "conclusions" or problems with the operating instructions so that Lowys can obtain immediate hands-on experience.

The Introduction to the Treatise details the content of the five parts that Chaucer had intended to write (he finished only two parts). A major portion of this introduction is taken up with an apology for the use of "naked wordes in Englissh" instead of Latin, the usual language of medieval scholarship. It seems clear that the use of the vernacular can be ascribed both to Lowys' tender age ("for Latyn ne canst thou yit but small") and to Chaucer's desire that other "lewd" men might "read or hear" his explanations. This rationale is highly significant, as is Chaucer's contention, also stated in the Introduction, that "true conclusions" can be articulated in a number of languages without having their value, or truth, in any way affected. The defense of the vernacular as a vechicle for the dissemination of scientia--in the broad sense of knowledge or belief--was not an altogether popular view in the Middle Ages. Indeed, at about the time that Chaucer was writing the Treatise, the followers of John Wyclif, in a movement that was to prove to be a dress rehearsal for the Reformation, were being castigated for making the "highest and subtlest" truths available to the unlettered or to those whose literacy was restricted to English.

Chaucer makes a further contribution to the dissemination of "hard sentence," and, without overstating the case, to the history of technical communication itself, when he admits that he has not only used the vernacular, but also has been willing to sacrifice an elegant style for a "superfluite of wordes" in order to insure that his audience understands what is written. This was truly an epochal admission in an age that often valued rhetorical flourishes far more than clarity of expression. In essence then, Chaucer attempted in his Treatise to solve the perennial problem of all technical communicators: he searched for a language and a mode of expression that would allow him to translate the "subtle and true" conclusions of a complex technical subject into a readily understandable form. One must now ask how successful he was in this attempt.

CHAUCER'S DESCRIPTION OF THE ASTROLABE

The astrolabe held by Lowys consisted of a circular disk, called the 'mother' into one side of which was inserted an alidade or pivotal sight. Chaucer describes the "moder" accurately as the "thickest plate, perced with a large hool," but he does not mention the alidade. It is worth noting here that Chaucer's description is written with "Lyte Lowys" pictured as holding and examining his own astrolabe ("Put the ryng of thyn Astrolabie upon the thombe of thi right hond"), and that, throughout this section, Chaucer assumes that the classification of the astrolabe's parts can be made more comprehensible through the use of figurative language--a usage unique to Chaucer and not a part of any of his sources.

On the astrolabe's other side were two thinner plates, the tympan and the rete. The tympan generally had inscribed on it a stereographic projection of the most important circles of heaven, while the rete, nicely described by Chaucer as "shapen in manere of a nett or of a webbe of a loppe," was an indicator that pointed out the locations of the brighter fixed stars. Very simply, the astrolabe that Lowys held could accomplish two kinds of operations. First, by using the alidade as a sight, he could measure the height of an object (e.g. a star) above the horizon and, by using the gradations from 0 to 90 degrees engraved around the rim of the mother, compute the angular distance between two objects. Second, by calculating the relative position of the sun on a particular day using astronomical tables, and by rotating the pointers of the rete to approximate the locations of the brighter fixed stars, a two-dimensional picture of the configuration of the heavens could be derived that was especially useful for the calculations of astrologers. That the astrolabe was also used for what we would consider to be serious science can be demonstrated both by the problems Chaucer appended to his Treatise and by the documented case of Walcher, who, on October 18, 1092, was able to determine the exact time of a lunar eclipse using his astrolabe. Indeed, the computation of the time of day was one of the most frequent and important 'hard' uses of the astrolabe until the invention of mechanical clocks late in the thirteenth century.

At this point one might justly ask if, his father's instructions in hand, Lowys would have been able to make the kinds of precise calculations that could be made with the astrolabe. While it is true that the step-by-step instructions of Part Two of the Treatise are clear enough, the description of the instrument in Part One is sometimes less than clear. For example, some of the parts are too briefly rendered: "These azymutz serven to knowe the costes of the firmament and to othre conclusions, as for to knowe the cenyth of the sonne and of devery sterre." Others are vague: "This zodiak, which that is part of the 8 speer, overkervith the equinoxial, and he overkervith him ageyn in even parties; and that oo half declineth southward; and that othir northward . . . " The azimuth of a star is, of course, the measure of the angular distance (measureed clockwise or from west to east) between the North Pole and the center of that star; Chaucer's description does not make it clear how the azimuth helps an observer to know the "region" (coste) of the celestial sphere, nor does it specify how the "cenyth" is to be determined. Likewise, there is some confusion in Chaucer's definition of the zodiac--that imaginary band sixteen degrees in width that was thought to span the heavens and to carry both the ecliptic (the sun's path) and the twelve zodiacal signs.

On the other hand, some of Chaucer's definitions are too prolix, especially when he allows his interest in astrology to overshadow his interest in astronomy: " . . . for 'zodia' in language of Grek sowneth 'bestes' in Latyn tunge . . . whan the planetes ben under thilke signes thei causen us by her influence operacions and effectes like to the operacions of bestes." Here, in a passage reminiscent of Isidore's 'scientific' etymologies and of the general medieval fondness for analogy, Chaucer cheerfully digresses for the sake of other kinds of scientia, other edifying lore. Later on in Part One, he writes more on the origins of the names of things, only this time the connection to the astrolabe is even further removed: "This signe of Cancer is clepid the tropik of somer, of tropos, that is to seien 'ageynward'." Indeed, what is most striking about the whole of Chaucer's description of the astrolabe is his clear preference for the language of description, for the terminology of the instrument, and his relative lack of interest in the kind of process description that would make the use of the astrolabe easier for a young, or old, student of astronomy. The problems of Part Two presuppose a familiarity with the configuration and interrelations of the astrolabe's parts that is only partially provided in Part One; yet it seems clear that Chaucer, like many other technical communicators, was relying on the actual manipulation of the object to do for the student's understanding what mere verbal description could not do. Again, one is tempted to see in the Treatise the adumbration of problems that technical communicators still face, and solutions that are still viable.

Historically, one could generalize further about Chaucer's Treatise on the Astrolabe by saying that it seems to straddle the increasingly impassable border dividing the earlier medieval preoccupation with the essences and ends of things for the fourteenth and fifteenth centuries' preoccupation with the empirical description of quantifiable material phenomena. That such a change was in the air even while Chaucer wrote his text for Lowys can be proven in a number of ways. One could consider the advances in technology of which the astrolabe is but a part; one could cite the logical controversies incited by the radical nominalists with their denial of the material existence of all universal categories; or one could consider the elegant experimental method of Theodoric of Freiberg (d. c. 1311), whose explanation of the mechanics of the rainbow effectively demythologized the most frequently symbolic of all natural phenomena. Thus it requires no act of historical daring to see the Treatise on the Astrolabe as a transitional work, nor does it take much daring to end this little treatise by drawing the obvious cautionary lessons. First, as this session will demonstrate, technical communications, like other disciplines, has a history. Of course, this is not, in itself, a surprising fact. But what might be more surprising is the length of this history and its variety. Second, and more specifically, there is the lesson of Chaucer's Treatise, a lesson only briefly sketched out in the course of this paper. Chaucer's key problem was to invent a language that would allow him to reduce the form and function of the astrolabe to a comprehensible series of factual statements. Because he had no predecessors, his difficulties were compounded. We may take a sort of rueful comfort in the intractability of the problems that face us--we still seek a language that will adequately render form and function--and we may also be gratified by both the successes and failures of those individuals, like Chaucer, who comprise the history of our discipline.

THE ROYAL SOCIETY, HENRY OLDENBURG, AND SOME ORIGINS OF THE MODERN TECHNICAL PAPER

James G. Paradis

Massachusetts Institute of Technology

James G. Paradis
M.I.T.
Assoc. Prof. of Technical Communication

The interaction of science and literary tradition in the mid-seventeenth century is an important source of the modern form of the scientific paper or periodical article. Some of the earliest standards in technical documentation can be traced back to those set forth in the Royal Society's Philosophical Transactions. The first periodical journal of England, this series, initiated in 1665 under the editorship of Henry Oldenburg, created a specialized audience and rhetorical models for which there was very little precedent. Three primary forces converged to establish what Robert Boyle, the patron and influential friend of Oldenburg, called the "experimental essay." First, the Baconian standards for technical language were formally recognized by fellows of the Royal Society and the plain style of language was promoted over the high oratorical style favored by the universities. Second, two forms of literature, the essay genre of the Renaissance and the letter model that had provided the traditional means of exchanging technical information among experimentalists, converged to provide a new short format for publishing experimental results. Finally, the institutional structure of the Royal Society itself furnished an administrative mechanism by means of which original work, first reported at meetings of the Royal Society, could then be printed in the formal literature and sold for profit and for the promotion of useful knowledge.

In this paper, I would like briefly to explore a number of scientific and literary developments in the mid-seventeenth century that transformed the traditional essay genre into a new model for reporting experimental findings. We can date, with reasonable precision, the origin of the modern technical journal article in the 1660's. It was a literary form that evolved in concert with the serial publication, beginning in 1665, of the Philosophical Transactions of the London Royal Society. This dual development of the first technical journal series and of a unique literary vehicle is a significant milestone in the history of technical communication. It shows us how, within the profuse literary experimentation of the Renaissance, the "experimental essay"--one of the most universal and prestigous vehicles of modern science and applied science--was consciously invented as a means of transmitting experimental knowledge. It also reveals the processes by which the literature of science and applied science decisively broke away from the mainstream literary traditions of the Renaissance.

THE ROYAL SOCIETY

The periodical journal, so popular in modern society, originated largely as a technical form. It was invented by natural philosophers (and their sympathizers) of the post-Renaissance scientific movement. The need for a new form of literature to communicate new experimental knowledge was created by the accelerating and increasingly collaborative development of a new body of mathematical, descriptive, and experimental knowledge. Much of this widespread fascination for organized "histories" or systems of knowledge was the result of Francis Bacon's call for the experimental renovation of traditional natural philosophy. In his seminal Novum Organum (1620), Bacon advocated an inductive methodology of moving from detailed observations to broad generalizations. Such a method, he argued, would enable natural philosophers collectively to expand "particular" knowledge in a variety of theoretical and practical subjects, corresponding to modern astronomy, geology, chemistry, physics, physiology, metallurgy, medicine, and the trades. [1]

By the mid-seventeenth century, Bacon's broad program for investigating nature and codifying useful knowledge had inspired informal study clubs in England and France. Such groups met periodically in London and at Oxford from roughly 1645 onwards. They studied agriculture, mathematical applications, pneumatics, astronomy, and anatomy, and planned and executed various group experimental demonstrations. The new clubs focused interest in experimental topics that were not treated within the classical Aristotelian curricula of the university natural sciences.

One of the most successful of the groups--and the prototype of the Royal Society--was the Oxford experimental science club, the inspiration of John Wilkins, Warden of Wadham College. [2] Formed roughly in 1648, this select group of virtuosi included Christopher Wren, architect-to-be of St. Paul's Cathedral in London; Robert Boyle, the most illustrious experimentalist of England; and, eventually, Henry Oldenburg, a German scholar, who was later to become the Secretary of the Royal Society. From 1648 to 1659, the Oxford experimental science conducted its meetings at Wadham College and at Boyle's private laboratory in the High Street of Oxford. The club then moved to London, where in 1660 it formed a regular assembly of members. It was chartered in 1662.

With its charter, granted by Charles II, the Oxford experimental science club became the "Royal Society for the advancement of the knowledge of natural things, and of all usefull arts, by experiments." This charter effectively consolidated the virtuosi of several experimentalist groups into a formal organization that was backed by the official, if not financial, support of the crown. The charter gave prominence and legal standing to a body of men, dedicated to the ambitious program for extending natural knowledge outlined by Bacon in his Novum Organum.

Henry Oldenburg and John Wilkins were elected secretaries to the Royal Society, Oldenburg being charged with recording the proceedings of meetings. These Society gatherings continued the discussions and experimental activities of the earlier private groups. In addition, Oldenburg undertook the management of the Society's correspondence with members in the English countryside and with a growing circle of European experimentalist virtuosi. Much of this correspondence was initiated by Oldenburg, who continually looked for European correspondents, who could keep him abreast of new developments in experimental science.

In his correspondence, splendidly preserved in the recent edition prepared by Rupert and Marie Boas Hall, Oldenburg set down the scope of the new Society. [3] Writing to Dutch experimentalist Peter van Dam (23 January 1663), Oldenburg described the large and ambitious goal of the Society:

> It is our business, in the first place, to scrutinize the whole of nature and to investigate its activity and powers by means of observations and experiments; and then in course of time to hammer out a more solid philosophy and more ample amenities of civilization. [3, II, 44]

To a French correspondent, Sorbiere, Oldenburg also declared that the aim of the new society was "to build a solid philosophy on the basis of observation and experiment, made with as great an exactness as is possible." The motto of the Royal Society, he noted, was Nullius in Verba ("on the word of no man"), which reflected the climate of doubt in the Society. Members were reluctant to accept as truth accounts and principles for which there was no clear and detailed evidence.

These combined emphases placed new and strenuous demands on the transmission of information, traditionally carried out through the exchange of letters. On the one hand, the aims of the Society called for detailed observation and experiment and for exactitude in the transmission of results. On the other hand, the motto stressed the need for independent confirmation of reported information. Together, these institutional requirements placed great emphasis on the careful wording of scientific accounts so that they either internally demonstrated their accuracy by describing reproducible experiments, or externally found substantiation in other, independently originated observations.

In dedicating themselves to the preservation and advancement of exact knowledge, the members of the Royal Society uniformily agreed on the necessity of adopting a "plain style" of language. Thomas Sprat in his History of the Royal Society (1667), written under the guidance of Secretaries Oldenburg and Wilkins, noted that members consciously avoided the ornamental language of high literary and philosophical discourse. As Bacon had, the membership viewed the received tradition of rhetoric and oratory as a corrupting force for the new experimental movement. This distrust extended to the language of the Aristotelian, astronomical, and alchemical traditions, which was thought to be vague and unreal.

The members of the Royal Society, Spratt reported, consciously avoided "this vicious abundance of phrase, this trick of metaphors, this volubility of tongue, which makes so great a noise in the world" [4]

> They have threfore been most rigorous in putting in execution, the only remedy that can be found for this extravagance; and that has been, a constant Resolution, to reject all amplifications, digressions, and swellings of style: to return back to the primitive purity, and shortness, when men deliver'd so many things, almost in an equal number of words. They have exacted from all their members, a close, naked, natural way of speaking; positive expressions; clearness; a native easiness: bringing all things as near the mathematical plainness as they can; and preferring the language of artisans, countrymen, and merchants, before that, of Wits and Scholars. [4, p. 113]

That the language of the practical arts and trades was featured as the ideal of experimental discourse pointed to the craft-oriented and realist priorities of the new experimentalist. Such confidence in practical application and simple experience constituted a fundamental departure from the learned tradition of the so-called school-men. This new institutionally-supported program to simplify language emptied conscious metaphor from technical discourse. It also laid the groundwork for the marriage of mathematics and technical language.

Thus, by 1667, some five years after the charter of the Royal Society, a program to focus on observable phenomena and to promote the experimental method had become a collective priority of a talented body of men. The same corporate zeal had been extended to the reform of scientific language, the ideal being that words should represent single, agreed-upon entities.

OLDENBURG AND THE Philosophical Transactions

The personal circumstances and talents of Henry Oldenburg, the Secretary of the Royal Society, led to a number of unusual innovations in the domain of technical publication. As acting Secretary (Wilkins, the First Secretary, was preoccupied with his own work), Oldenburg was obliged to implement the Royal Society's stringent requirements for the preparation and transmission of scientific knowledge.

Educated in Bremen, taking his Master's of Theology in 1639, Oldenburg followed a common Renaissance scholar's path and became an itinerant scholar-tutor, traveling to England in 1640. In 1656, after considerable travel in England and Europe, Oldenburg entered as a student at Oxford, where he continued an earlier commitment to tutor Robert Boyle's nephew, Richard Jones. Boyle himself had come to Oxford at the invitation of John Wilkins to participate in the research of the Oxford experimental science club. Also a prolific religious essayist, Boyle was a natural companion to the older, learned Oldenburg, whose German theological training and amateur interest in experimental science were supplemented by a thorough knowledge of French, German, English, Italian, and Latin. It was undoubtedly his intimacy with Boyle and his immediate family that drew Oldenburg to the meetings of the Oxford experimental science club.

By the early 1660's Oldenburg had also begun to work privately as Boyle's editor and publisher. He assisted him in the publication of both his Experiments and Considerations Touching Colour (1664) and his New Experiments and Observations Touching Cold (1665). Such a collaboration not only provided Oldenburg means of financial support, but it also made him editorial assistant to the greatest and most prolific living experimentalist of England and Europe. It created a collaborative relationship in which Oldenburg, a writer and translator of unusual versatility, learned how experimentalist literature was shaped and ultimately presented to an international, yet specialized, audience. This friendship and collaborative arrangement became one of the key defining forces behind development of the Royal Society as an international, institutional exchange for technical information.

As secretary to the Royal Society, Oldenburg's duties were primarily those of a recorder, correspondent, and general institutional organizer and promoter. He thus naturally became the Society's information specialist, its transmitter of technical knowledge. In a note, written in 1668, Oldenburg described his duties as Secretary:

> He attends constantly the Meetings both of the Society and the Councill; noteth the Observables, Said and done there; digesteth them in private; takes care to have them entered in the Journal- and Register-books; reads over and corrects all entrys; sollicites the performances of taskes recommended and undertaken; writes all letters abroad and answers the returns made to them, entertaining a correspondence with at least 50 persons; employes a great deal of time, and takes much pains in inquiring after and satisfying forrain demands about philisophicall matters, disperseth farr and near store of directions and inquiries for the society's purpose, and sees them well recommended etc. [5]

These responsibilities forced Oldenburg to develop a style and form of communication consistent with the stated priorities of the membership. The nature of his secretarial work was almost entirely communicative and it ranged over the entire body of technical knowledge. Oldenburg was required to explain scientific observers and experimenters to one another; be was also obliged to describe experiments carried out at Royal Society meetings. Hence, he was continually describing apparatus, procedures, and results in conformity with the stylistic ideals of the Society.

The correspondence, much of it carried out on his own initiative, was vital to Oldenburg's desire to make the Royal Society an international exchange for technical information. The personal letter, the pamphlet, and the treatise had long been the primary forms of technical communication. The letter, howeve was the most vigorous, immediate and flexible of thes forms. It constituted the primary means in the Renaissance by which scientific virtuosi exchanged views and sought commentary on their experimental findings. Oldenburg, himself, was an enthusiastic correspondent His letters allowed him to indulge his international interests, while he also pursued his technical priorities and his concern to promote the Society abroad. As Rupert Hall has noted, much of this correspondence formed the body of material from which Oldenburg was to shape his idea for a journal. [3, II, xx]

Typically, Oldenburg selected from his correspondence materials he thought would be of technical interest at meetings of the Society. He rarely read entire letters, but, rather, translated and then reduced them to an abstract, often of about one-tenth their original length. Relieved of the nicetie of their salutations and closings, these letters became brief articles on a variety of practical and theoretical matters. The topics on which Oldenburg corresponded ranged from German mining and Italian astronomy to American whaling and English agriculture In addition, he also corresponded on highly theoretical topics with the virtuosi, men such as Huygens, Auzout, Hevelius, not to mention Hooke, Boyle, and Newton. Most of Oldenburg's correspondents understood that the contents of letters were likely to be summarized and read at meetings of the Royal Society. In effect, then, the letter network of Oldenburg constituted a system of quasi-official claims to observational and experimental findings.

As the pressures of his correspondence mounted, Oldenburg was faced with the problem of translating letters and making a variety of copies that he could send to Society members and other investigators he believed could be sparked into responding. This burden as technical correspondent increased after the London Plague of 1664 drove Society members into the English countryside. With Boyle, his patron, who remained at Oxford until 1668, Oldenburg carried on a frequent correspondence in which he summarized recent letters from abroad and recounted experiments carried out at the Society's meetings. These long, newsy letters became prototypes of a technical journal.

In 1664, Oldenburg received news from his French correspondent, Adrien Auzout, that the French were planning a new "Journall." To Boyle, Oldenburg wrote (24 November 1664) that Auzout had asked him to provide information on anything of technical interest transacted at meetings of the Royal Society. The French, Oldenburg noted, were planning a "Journall of all what passeth in Europe in matter of knowledge both philosophicall and politicall: in order to which they will print." Although Oldenburg wished to participate, he also noted that he was unlikely to have the time to undertake such an exchange.

Auzout's letter referred to the soon-to-be-published Journal des Scavans, the first number of which appeared on January 5, 1665. This was the first of the journals that were to become so widespread in post-Renaissance Europe, England, and America. It was published by Denis de Sallo, a counselor to the French Parliament. De Sallo belonged to a small, informal circle of intellectuals much like the clubs of natural philosophers in London at the time the Royal Society was chartered. De Sallo's plan, according to Auzout, was to publish a variety of short accounts, obituaries, booklists, experiments, and disputes, all of which would range from matters of natural philosophy to current topics in politics and ecclesiastical affairs. Though its emphasis was to be on the exact sciences, the journal was also to treat the fine arts and literature, politics, and Church affairs. Its scope, then, was avowedly generalist. [6].

Although the need for a technical journal had long been growing, it was clearly Auzout's letter that moved Oldenburg and Boyle to consider publishing an experimentalist journal of their own. Oldenburg applied to the Council of the Royal Society, no doubt with the strong support of Boyle, for permission to publish an account of its transactions. He hoped thereby to earn a small income and to gain a wider audience for the growing body of practical and theoretical knowledge he had been gathering in his correspondence and Society journal-books. On March 6, 1665, some two months after the appearance of the first number of the Journal des Scavans, Oldenburg published the first number of his new journal. The full title was Philosophical Transactions: Giving Some Accompt of the Present Undertakings, Studies and Labours of the Ingenious in in Many Considerable Parts of the World. No development in the history of the Society would do more to advance its standards of discourse, its methodological ideals, and its commitment to the dissemination of useful knowledge. Almost immediately, the Philosophical Transactions became the premier forum in the world for the collective development of scientific thought. It gave birth to a new kind of technical audience, one that could participate in an ongoing process of serial publication. New submissions could either challenge or corroborate previously published results.

The first number of the Philosophical Transactions, of which 1200 copies were printed, contained some ten short pieces that were largely summaries of letters from Oldenburg's recent correspondence. The first short item summarized a letter from the Italian astronomer Guiseppe Campani, describing recent improvements Campani had made in a tool used to grind large optical glasses for telescopes. In addition, Campani confirmed Christian Huygens's observation, some six years earlier, of Saturn's rings. Campani also related a plan to determine whether Jupiter turned on its axis, by observing whether notable features on its surface moved. Campani's one-page "article" was followed by Oldenburg's one-paragraph report that Robert Hooke,

> with an excellent twelve foot telescope, observed, some days before, he than spoke of it (videl. on the ninth of May, 1664, about 9 of the clock at night) a small spot in the biggest of the 3 obscurer Belts of Jupiter, and that, observing it from time to time, he found, that within 2 hours after, the said spot had moved from East to West, about half the Diameter of Jupiter. [7]

Hooke's account confirmed Campani's hunch that Jupiter might revolve, and Hooke had included enough detail to fix the conditions and results of his observation.

The placement of these two initial pieces demonstrated the corroborative mechanisms Oldenburg, no doubt in consultation with Boyle, hoped to establish in the Philosophical Transactions. As editor, Oldenburg had thus inaugurated a publication policy in which he juxtaposed separate, detailed accounts of observations and experiments. Over an extended series of issues, this policy would produce a texture of information, some of it mutually supportive, some of it conflicting. The Transactions thereby would become an extended forum for the investigator, a literary vehicle that would itself conform to the cardinal principles of experimental investigation. Oldenburg had designed a publication that could live up to the Society's motto, "On the word of no man."

Of equal interest is the effort of Hooke to establish the date--and, hence, the precedence--of his observation. This dating of material was a priority Oldenburg established on the strong recommendation of Boyle. In November, 1664, he had written to Boyle, acknowledging the importance of dating correspondence: "The Society [is] very carefull of registering as well as the person and time of any new matter imparted to them, as the matter itself; whereby the honour of the invention will be inviolably preserved to all posterity." [3, II, 319] In the Transactions, the effect of requiring exact, detailed information and of dating it was to temporalize experimental knowledge. That is, it made the time element of experimental and observational accounts important in such a way as to assert the progressive nature of the scientific enterprise.

The two observations of Campani and Hooke in the first Number were followed by Oldenburg's long five-page summary of a "printed paper" or pamphlet by Auzout, the French Astronomer, in which Auzout partly fixed and then predicted the path of a comet on the basis of some four separate observations. By describing the comet's apparent trajectory, first relative to the earth's surface, then relative to familiar astronomical quadrants, Auzout hypothesized that, contrary to the popular opinion that the motion of comets was irregular, comets moved regularly and predictibly, according to laws. Auzout, Oldenburg announced, invited English astronomers to check his calculations and predictions, and either to "confirm the Hypothesis, or to undeceive him." Oldenburg included Auzout's detailed observations, "that the intelligent and curious in England may compare their observations therewith," for purposes of verification or, possibly, of suggesting a better hypothesis. This long, highly detailed account demonstrated the spirit of inquiry and objectivity that Oldenburg hoped to give to his new literary invention.

The remaining pieces of the first number are diverse, ranging from Oldenburg's announcement of Boyle's forthcoming Experiments and Observations Touching Cold, on which Oldenburg himself had worked, followed by an account of a "very odd mostrous calf," a note on a new lead-ore in Germany, a practical account from an American about a new whale-fishing enterprise in the Bermudas, a description of how Huygen's new pendulum watch could be applied to the navigational determination of sea longitudes, and, finally, an obituary account of a "Monsieur de Fermat" of Tolouse, with a list of his published mathematical works.

Although the Transactions might have had some limited appeal for a general audience, it was clearly, with its rigorous detail, aimed more at the investigator than the generalist; there were few niceties of style to provide a learned or polite context for the casual reader. The Transactions quickly accomplished several different objectives consistent with the institutional aims of the Royal Society. It projected the group spirit of the new scientific enterprise, a spirit that Bacon had considered axiomatic to the promotion of experimental knowledge. Relative to this intellectual collectivism, the Transactions also made the corroboration of facts a key element in determining the strength of any claim or argument. This tended to undercut the powerful traditions of oratory, rhetoric, and classical logic of the Renaissance. In addition, the Transactions exerted a considerable pressure to standardize the terms of scientific discourse, at the same time expanding the native technical vocabulary of the English and opening experimental and observational activities to tradesmen and other non-academic interests. This factor alone extended the spirit of experimentalism to a new audience and forged a new alliance between scientific investigation and the crafts. Finally, the serial publication of the Transactions, together with the policy of dating accounts as they were received, gave a distinctly developmental thrust to scientific knowledge, emphasizing its continuity and progressiveness. Oldenburg's new journal series became one of the earliest symbols of the visible progress of scientific knowledge.

REFERENCES

1. Bacon, Francis, Novum Organum, ed. F. H. Anderson, Indianapolis: Bobbs-Merrill, 1960, pp. 285 ff.
2. Purver, Margery, The Royal Society: Concept and Creation, Cambridge: M.I.T. Press, 1967, pp. 102-3.
3. Hall, Rupert, and Marie Boas Hall, eds. The Correspondence of Henry Oldenburg, 1-9, Madison: U. of Wisconsin Press, 1965-73.
4. Sprat, Thomas, History of the Royal Society, ed. J. I. Cope and H. W. Jones, 1667; rpt; St. Louis: Wash. U. Studies, 1958, pp. 113-14.
5. Hall, Marie Boas, "Henry Oldenburg and the Art of Scientific Communication," Br. Journal for the History of Science, 2 (1964-65), 290.
6. Ornstein, Martha, "The Role of Scientific Societies in the Seventeenth Century," Ph.D. Diss., Columbia U., 1913, pp. 235.
7. Oldenburg, Henry, ed. Philosophical Transactions, 1 (1665-66), rpt; New York: Johnson Reprint Corp., 1963, p. 3.

EVALUATING A SCIENTIFIC AND TECHNICAL INFORMATION PROGRAM: THE USER PERSPECTIVE

Thomas E. Pinelli,* Dr. Edward M. Cross,+ and Dr. Myron Glassman+
*NASA Langley Research Center; +Old Dominion University

Thomas E. Pinelli
NASA Langley Research Center
Hampton, VA 23665
(804) 827-2691

Assistant Chief
Scientific & Technical Information Programs Div.

Dr. Edward M. Cross
Old Dominion University
Norfolk, VA 23508
(804) 440-3575

Professor
School of Business

Dr. Myron Glassman
Old Dominion University
Norfolk, VA 23508
(804) 440-3561

Assistant Professor
School of Business

The NASA Langley Research Center has conducted an evaluation of its scientific and technical information (STI) program which included a survey of its engineers and scientists. The purpose of the evaluation was to assess the usefulness of the present program and identify areas and ways in which the program could be made more effective.

The survey questionnaire covered such topics as technical review of proposed publications, publication guidelines, research support services, STI products and services, and image/prestige of Langley STI. From the total user population of 1,036 engineers and scientists, 710 survey questionnaires were returned.

The survey produced valid empirical data establishing the adequacy of the NASA Langley STI program in meeting the needs of Langley engineers and scientists and providing a baseline for future evaluations. The survey can be readministered on a periodic basis and may serve as a model for similar evaluative efforts within R&D organizations.

INTRODUCTION

Founded in 1917, the NASA Langley Research Center (LaRC) is one of the leading national laboratories for research and development (R&D) in the sciences of aeronautics and space technology. For calendar year 1980, Langley's research population of 1,036 engineers and scientists produced 1,418 items of documented research. The research output of the Center is processed through the Langley Scientific and Technical Information Programs Division, which plans, organizes, and directs the activities and functions involved in the publication and dissemination of the Center's research output.

In February 1980, the Scientific and Technical Information Programs Division undertook the first comprehensive review and evaluation of the Center's STI program. The evaluation project utilized both survey research and systems analysis techniques, included all elements of the STI program, disclosed the strengths and weaknesses of the STI program, and identified ways in which the program could be modified to improve its overall efficiency and effectiveness. Phase I of the review and evaluation project employed survey research to assess the adequacy of the Langley STI program in meeting the information needs of Langley engineers and scientists.

RESEARCH METHODOLOGY AND PROCEDURE

The methodology for the survey combined the semantic differential technique, taken from communication research, with the concepts of classical and operant conditioning, taken from learning theory. Stage 1 of a two-stage survey procedure included personal interviews with 64 randomly selected Langley engineers and scientists. The interviews were taped and transcribed. The most frequently mentioned impressions were considered salient for the group, thus forming the basis for questionnaire development.

Stage 2 involved the collection of data through a survey questionnaire which contained 50 closed-ended questions and 3 open-ended questions. The questions elicited the respondents' knowledge of the NASA STI system and attitude toward the Langley STI program and employed a five-point attitude scale response. In addition, demographic material was solicited in the areas of publication history, years of work experience at Langley, and participation in the technical review process.

The survey questionnaire was sent to all engineers and scientists who had not participated in the interview or pre-test. A total of 710 surveys were returned, of which 647 were considered valid. From the valid questionnaires, a sample of 300 was randomly selected and analyzed.

FINDINGS

The study displayed all the earmarks of successful survey research. The response rate was high (76%), most of the responses were usable, and more than one-half of the respondents completed the open-ended questions. Favorable attitudes constituted the majority opinion for each survey topic covered in the questionnaire. Therefore, it was concluded that the Langley STI program was meeting the needs of Langley research personnel.

Because it is reasonable to believe that other R&D organizations have similar programs and characteristics, the information revealed by the survey may be of interest to other authors, editor/writers, managers of STI programs, or managers of R&D organizations. In addition to the findings, some of the questions used may be applicable to other R&D organizations. Eight summary questions are stated and their findings discussed.

1. Are the technical editing and organizational review processes perceived as effective?

The respondents indicated that the individuals who participated in these review processes were qualified and sensitive to deadlines. The general reaction was that better performance could be obtained by revising the processes in an effort to reduce the amount of time required to complete the review cycles.

2. Are publication guidelines available and clear and do they facilitate publication?

The respondents indicated that publication guidelines were available and clear. The general reaction was that in many cases the guidelines did not facilitate publication. Further, a strong majority indicated that an STI handbook containing guidelines for all publications and secretarial instructions was necessary and that an individual in each organization who thoroughly understood the publication process was necessary.

3. Are research personnel cognizant of the full spectrum of support services? Are the services adequate and timely?

Overall, Langley research personnel appeared to be satisfied with the performance of the research support services. However, as many as 30% of the respondents offered "no opinion" as to the importance of certain support services, indicating a possible lack of familiarity with these services.

4. How do Langley engineers and scientists perceive the quality and prestige of their research output?

An overwhelming majority of respondents indicated that the quality of material produced through the review and publication processes was high. Overall, the prestige of Langley STI was perceived as high, but somewhat mixed reactions were recorded for certain individual STI media. A majority of respondents indicated that the prestige of certain STI media was lower in their discipline. Since the overwhelming majority rated the quality of STI material high, the inference can be drawn that the minority respondents perceived their research output to be viewed with less prestige by engineers and scientists outside the Center.

5. Are research personnel familiar with and using selected NASA STI products and services?

The NASA STI system offers announcements, current awareness, and computerized data base services. Respondents strongly indicated that these products and services were important to their research. However, between 10-30 percent registered "unfamiliar with" for a specified product or service.

6. How important is publishing to promotion and how supportive are supervisors of publishing?

An overwhelming majority (80%) indicated that publishing their research results was important to their professional advancement. A strong majority (75%) considered their supervisors supportive of their efforts to publish through NASA formal series technical publications.

7. What is the publication "profile" of the Langley engineers and scientists?

While an overwhelming majority (88%) of researchers had published the results of their research, a slight majority (54%) had not published within the past 3 years. A majority of researchers (53%) utilized three media (NASA formal series technical publications, journal articles, and conference/meeting papers) for disseminating the results of their research.

Questions concerning specific publication media, attendance at conference/meetings, and participation in technical reviews specified "within 3 years." A strong majority (72%) had attended a conference/meeting (other than ones held at LaRC). A clear majority (63%) had published a conference/meeting paper and served on a technical editorial committee (67%).

8. How long has the average engineer/scientist been employed at Langley and at what supervisory level is he/she working?

A clear majority (63%) of researchers had been employed 16 years or more at LaRC. A strong majority (89%) were working as individual contributors rather than in a supervisory capacity.

EFFORTS TOWARD IMPROVEMENT

Since favorable attitudes consituted the majority opinion for each survey question, it was assumed that the Langley STI program is meeting the information needs of Langley engineers and scientists. Nevertheless, the findings revealed some areas of concern influencing the efficiency and effectiveness of the Langley STI program. The following are some of the program improvement efforts being undertaken by STIPD as a result of the survey.

- An analysis of the technical review process for Langley-authored formal series technical publications

- A review of Langley publication guidelines and the development of a comprehensive publications handbook

- The training of an STI coordinator in each division in the publication process

- The development of orientation lectures for research personnel explaining research support services

- A study to determine the acceptability of Langley-authored formal series technical publications as journal references

- The development of presentations explaining Langley STI products and services

In other areas, the appropriate authority and responsibility for improvement falls within the purview of the research directorates and the Center director. These areas include the following topics:

- Revision of the review practices used by the various research organizations

- Clarification of the role and function of Langley-authored formal series technical publications vis-a-vis journal articles and other reporting media

Finally, the overriding purpose of the Langley STI review and evaluation project was to determine if the dissemination of the Center's research output could be made more effective. Dissemination of the Center's research output in the broadest possible sense can be made more effective if the following actions are taken:

- Streamlining of processes involved in the review and publication of the Center's research output and sensitivity to production and dissemination deadlines by all personnel

- Elimination and/or correction of concerns associated with Langley-authored formal series technical publications

- Development of a secondary distribution plan through which copies of Langley-authored formal series technical publications are sent to engineers and scientists as part of the initial distribution

- Review of the Agency's STI dissemination policies and practices

FUTURE ACTION

As part of the review and evaluation project, a survey of recipients of NASA Langley STI will be undertaken. The survey questionnaire will be designed to determine the perceived image (value) of Langley's research output. Questions to be asked will cover such dimensions of quality as adequacy of data, effectiveness of report organization (format), and the quality of visual presentations. The use of and importance of Langley STI in terms of advancing the state-of-the-art will be assessed. The results of this external survey will be compared with those of the internal survey to determine if a positive correlation exists between the perceptions of Langley authors and the users of their published research.

CONCLUDING REMARKS

The Langley STI review and evaluation project represented the first comprehensive assessment of the Langley STI program. Missing from previous program evaluations were two elements: (1) a comprehensive evaluation plan which fully encompassed the program objectives and (2) the all important dimension of feedback from the user population.

The results of the user survey provided information to aid management in choosing the services and processes which are likely to produce a high degree of user satisfaction and the most efficient use of resources. The responses to the closed-ended questions provided concrete information about satisfaction with existing information and publication processes and services. Data concerning satisfaction with existing services established a basis for future evaluations and indicated which services required emphasis and de-emphasis.

The responses to the open-ended questions identified additional information products and services which should be provided by the STI program, portions of the program which needed improvement, and suggestions for improving the review and publication process. Taken together, the responses to the open-ended and closed-ended questions identified the changes likely to increase the effectiveness and efficiency of the total STI program.

The demographic information gathered from the survey provided a profile of Langley research personnel as authors and reviewers of Langley publications and additional information about the publication environment at the Center. The insight provided by the results of the survey was important because it enabled Center management to view the STI program in its totality and from the perspective of its users, Langley engineers and scientists.

The results of the survey of Langley engineers and scientists are documented in NASA Technical Memorandum 81893. This report, which contains the complete methodology and questionnaire used in the study, is available for sale from the National Technical Information Service (NTIS), Springfield, VA.

UNDERSTANDING JARGON

Daniel L. Plung
Exxon Nuclear Idaho Co., Inc.

Daniel L. Plung,
Exxon Nuclear Idaho Co., Inc.
Technical Editor
P.O. Box 2800
Idaho Falls, Idaho 83401

"Jargon" is a familiar term to us; yet our familiarity has not produced an understanding of what jargon is or how to eliminate it. This paper describes our misunderstandings about jargon, focuses on the five forms of pseudoscientific terminology that constitute the jargon in technical communications intended for scientists, and explains how understanding jargon will aid us in the performance of our professional duties.

I. INTRODUCTION

Jargon is a familiar term, especially to the professional technical communicator. This familiarity is only natural since technical/scientific literature is ostensibly considered by many to be the sole originator, benefactor, and promulgator of jargon. As a consequence, the sciences and -- through a form of guilt by association -- technical editors and writers are most frequently assailed as practitioners of the "black art" of jargoneering. These attacks come from many quarters--from the public who cannot decipher the languages of science, from authors whose works we must revise, and from members of our profession who oversimplify the problem. Yet, neither our familiarity with nor our vulnerability to the problem has prompted a complete, practical "understanding" of jargon.

The problem, understanding jargon, may at first seem oxymoronic; after all, how does one understand the unintelligible? There is, nevertheless, an understanding that can equip us to: 1) recognize and remove jargon; 2) defend well written technical materials from being falsely labeled as jargon; and 3) explain our actions in these instances.

Definitions

This understanding is predicated on three operational definitions:

1) Technical Communication: The presentation of scientific/technical information in its clearest, most concise form.

2) Technical Communicator: One responsible for producing (writing, editing, etc.) Technical Communications.

3) Jargon: Unintelligible or meaningless language which interferes with and thereby blocks the production of Technical Communications (as defined above).

II. THE PROBLEM

Keeping these definitions in mind, let us briefly examine the status of our current "misunderstanding" of jargon and the problem it represents for the technical communicator.

Numerous articles tell us that there is a simple solution to jargon: "eliminate it." Unfortunately, this solution incorrectly presupposes that we all know precisely what to eliminate and how to do it. Similarly, articles offering slightly more information on the subject suffer from a corresponding deficiency. Typically, they propose the use of fewer "long" words, shorter sentences, and more elaboration. For example: ". . a technical writer must become adept at writing good, simple sentences that develop into good, simple paragraphs. Simple words should almost always be used in lieu of complex words. Two simple sentences should replace one complex one."[1]

This tandem association that good is synonomous with simple, and simple with clear is very appealing. Until we remember our definitions. To place any value in this solution, we must incorrectly redefine a technical communicator as one whose sole responsibility is to serve as the expert's emissary to the layman. This, however, is not consonant with the broader scope of responsibilities assigned to many technical communicators. Rather, the technical communicator may be required to produce documents for a variety of audiences, audiences with different levels of technical expertise, experience, and training.

Although many possible audiences come to mind, for our purposes let us classify all readers into two broad categories: 1) layman, 2) expert. Accordingly, if we reevaluate the proposed solution we find it inadequate. While appropriate for eliminating jargon in materials prepared for the layman, the solution is inappropriate as a prescription for technical communications intended for the expert. When preparing materials for the expert, the precise (technical) terminology is required.

If we use simple language in communications among experts, we fail to satisfy the conditions for producing technical communications (definition 1); we fail to satisfy the responsibilities of our professional assignment (definition 2); and we fail to produce documents that are clear, precise, and concise. In other words, we produce jargon (definition 3).

III. THE SOLUTION

Granted, the attempts that have been made to understand jargon provide us with half our answer. The solutions correctly assess that: 1) for the layman jargon means technical terminology, and 2) to eliminate jargon in technical communications intended for the layman we must "simplify" our writing. These conclusions might be illustrated in the form of equations:

1) Technical Terminology + Layman = Jargon.

2) Simple Language + Layman = Technical Communication

We are still left to tackle the two issues that comprise the other half of the question: 1) What is jargon in communications intended for experts?, and 2) How can it be eliminated? The answers to these previously unanswered questions will be the focus of the rest of this paper.

Here again we should return to our definitions: Jargon is comprised of those writing traits that interfere with or block clarity, precision, and conciseness. In the case of the expert, technical terminology does not consitute jargon; rather, jargon is comprised of five forms of "pseudoscientific terminology."

1. Neologisms

The expansive nature of the sciences requires that, infrequently, new words or new meanings for existing words must be "coined" to provide a precise vocabulary for describing a new idea, item, or activity. However, the necessary new words -- those that contribute to the language's breadth and precision -- must be separated from the unnecessary words -- those that contribute nothing to the language. The former are not jargon; the latter are. The former reflect conscientious investigation of etymology and an understanding of the language's structure and construction; the latter reflect neither study of nor respect for our language.

The following passage from Popular Electronics will help illustrate this delineation:

> A hamfest, as its name implies, is a grand meeting of radio amateurs. . . Most hamfests are more than just swap meets, however. . . Discussions on such topics as Novice operating procedures, contest operations, DXing, antennas, radioteletype, slow-scan television, vhf operations, emergency communications, traffic-handling and amateur applications for microprocessors predominate.[2]

Among the relatively new words, several contribute to the language's growth, including: microprocessors and radioteletype. Two words that contribute nothing and are, therefore, jargon are: hamfest and swap meet. These two neologisms could have been replaced by appropriate, acceptable synonyms, thereby enhancing rather than compromising the clarity of this communication.

If we do not restrict the use and acceptance of neologisms, someday our dictionaries may be filled with entries such as the following:

> thruput (n), thru'-put, the midpoint between input and output, give or take a little, corresponding to various agreed upon objectives with regard to the proper utilization of input, in order to meet those objectives so that ultimate output is consistent with basic organizational policy, and the flow from input to output, if in accord with the above, falling as they do on either side of thruput, would if placed on a continuum intercept thruput moving from either end of the continuum toward the center, with little or no deviation from a previously plotted straight line.[3]

2. Popularized Technicalities

Popularized technicalities are the second type of pseudoscientific terminology. These words, which have a precise meaning within one particular field, through imprecise usage escape from within their prescribed definitional constraints; consequently, through continued abuse and misuse, they take on a less technical meaning and finally come to convey no more than a nebulous impression -- a ghost of their original meanings. This continued imprecise use, as H. W. Fowler suggests, may provide the user with some gratification but "at the cost of the harm done to the language by wearing down the points of words. . ." Their imprecise use is most costly in scientific communications: the distortion substitutes ambiguity for clarity and thereby drains meaning from our discourse.

"Viable," for instance, is perhaps today's most abused popularized technicality. The word was originally confined to a biological context: the capacity of a newly created organism, e.g. a child, to maintain its separate existence. However, in its popularized form, this technical term may mean workable, feasible, possible or have any number of vague meanings.

Fowler compiled a preliminary list of the more frequently abused popularized technicalities; the following selections from that list show how eclectic the borrowing and how successful the assimilation have been:

From Philosophy: optimism and pessimism.
From Religion: devil's advocate.
From Logic: dilemma, beg the question.
From Chess: checkmate, gambit, stalemate.
From Seamanship: dead reckoning, underway.
From Psychology: ego, psyche, phobia.[4]

3. Injudicious Language Skills

The English language, as we are often informed, is changing. Our previous discussion of neologisms is evidence of this truth. However, another form of pseudoscientific terminology also contributes to the language's growth and change: injudicious language skills. This problem is characterized by two activities: 1) altering a word's form, and 2) confusing words that sound or look alike.

The first of these activities is illustrated in the following sentence: "The material was still undecomposed after four hours." The word of interest here is "undecomposed." Think about that word for a moment. What exactly does it mean? Does it mean composed, partially composed, partially decomposed, or something entirely different? Clearly our difficulty does not result from an unfamiliar prefix or root; the difficulty lies in their injudicious combination.

These alterations of words to suit the author's purpose, irrespective of the appropriateness or the correctness of those modifications, is the jargon of this first subcategory. Several forms of modification are common:

a) Forming a noun by adding "ate" to a verb: shockate, pressate, washate.
b) Forming a verb by adding "ize" to a noun or adjective: innoculize, blenderize, rigidize.
c) Changing an intransitive to a transitive verb: react.
d) Forming new words by indiscriminantly adding prefixes: preincubate, rehydrolyze, undecomposed.[5]

The second subcategory results from the similarity in sound and spelling of many words in our language. These similarities often lead to making the mistaken assumption that words that are spelled or pronounced similarly are synonymous and, therefore, can be used interchangeably. These improprieties are numerous and almost all books on writing, technical or expository, include glossaries of word usage in which such combinations as "effect/affect" can be found.

Because these word lists are available in so many sources, I will not reproduce one here; rather, let us look at another component of this same subcategory: the confusion that results when the correct word is not considered adequate and a more imposing form of the same word is sought as its replacement; however, what is found is a word with an entirely different meaning. One example of this problem should suffice:

> "Reference: In regard to your recent inquisition from Machine Design."
> (Letter from manufacturer, 23 January 1980).
> Fortunately, the manufacturer had not been subjected to torture; a letter of "inquiry" had been sent and was being answered!

4. Scientific Shorthand

Scientific shorthand, like a stenographer's shorthand, is a method of substituting an abbreviation for a word or phrase. However, in scientific shorthand, it is not with the intention of simplifying, but of complicating and confusing. There are two forms of this activity: 1) partial statements, and 2) acronyms and initialisms.

The first subcategory, partial statements, consists of sentences as: "The next subject is organics." Here "organics" is used as shorthand, a partial statement of "organic materials." To demonstrate the prevalence of partial statements, let me quote a few examples from the medical fields:

a) A pelvic examination was done on the floor.
b) Have you done a urine on him?
c) The surgeon entered the vagina from the top.[6]

The second form of shorthand is more pernicious; it seems as if anytime a writer uses a phrase more than once in a report, he is given license to substitute an acronym or an initialism (a neologism) for it. It is true that some abbreviations of this sort simplify our reading; more-than-likely, however, there is more trouble added than simplicity gained.

Let me offer a single example, one that clearly indicates how far we have gone in our acceptance and use of initials as arbitrary substitutes for words. Each of us, through information on genetic research, has come to accept the use of the initials DNA to mean Desoxyribonucleic Acid. Yet, in a Nuclear Regulatory Commission handbook, two interpretations of those initials appear. The first is the generally accepted meaning; the second is more arcane, and, hopefully, would not be acceptable to most readers: Does Not Apply.[7]

Just consider the confusion this interpretation could create for the uninitiated!

5. Occupational Protectionism

I have purposely kept discussion of this type of pseudoscientific terminology for last for two reasons: 1) it is the most-often cited form of jargon; and 2) it is often incorrectly assumed to be the only form of jargon. However, having been introduced to the other types of jargon, we can now place this fifth form in its proper perspective.

Of all people, Archie Bunker of All in the Family has provided perhaps the most poignant remark about this form of jargon. While passionately arguing with his son-in-law, Mike, Archie Bunker exclaimed: "Why do you use all those big words to hide your ignorance!" This is the essence of occupational protectionism: the obscurity and mysticism of long words and complex sentence structure are employed to cloud and complicate the text. It is especially prevalent in the form of a dialect a discipline may affect, the substitution of lingo for language.

This form of jargon, often referred to as Gobbledygook, can be the most malicious because it may, at times, be motivated by the author's desire to obfuscate. Sometimes the blame is the author's alone; at other times, when the complex language and sentence structure are intrinsic to an entire profession, then all members of that discipline must share in the blame. Because, as I have noted, this type of jargon has received ample coverage, I will offer only a single example of how effectively an author can cloud his thoughts:

> "Given Freud's structural point of view, paper-writing can no longer be conceptualized without considering the intersystematic complications which threaten it from within. . . .The conclusion, then, is inescapable: the paper-writing function of the ego is not always the only ego function instrumental in paper writing. . . {S}cientific papers are sometimes written complementary part-object relationship between the paper-dictation function of the writer's ego and the dictation-taking function of a scribe or secretary."[6]

IV. CONCLUSION

We now have the answers to our two remaining questions. We know that in communications intended for experts, jargon is comprised of the five forms of pseudoscientific terminology. And we know that removing these five forms of ambiguity is the only method of eliminating jargon in these communications.

Here again, as we did with our earlier conclusions, we can effectively illustrate our new conclusions in the form of equations:

1) Neologisms
Popularized Technicalities
Injudicious Language Skills
Scientific Shorthand
Occupational Protectionism } + Expert = Jargon

2) Precise (technical) Terminology + Expert = Technical Communication

Knowing what constitutes jargon and how to eliminate it, we can now say we understand jargon. And, with this understanding of jargon, we are better able to fulfill our responsibilities as technical communicators. We are able to recognize and remove jargon; we are able to defend sound technical communications from charges that they are jargon; and, perhaps most important, we are able to explain what we do as technical communicators and why we do it.

References

1. Russ Bundy, "Preparing for a Technical Writing Career." Presented to the All Iowa Conference of Teachers of Business Communication, 7 November 1975.

2. Karl T. Thurber, Jr., "Hamfests are Fun." Popular Electronics, 17, no.6 (June 1980), 82, 84.

3. Gordon A. Zick, "To Throughput or Not to Thruput." Journal of Irreproducible Results, 24, no.1 (1978), 3-5.

4. H.W. Fowler, A Dictionary of Modern English Usage, 2nd ed. Revised by Sir Ernest Gowers. Oxford: Oxford U.P., 1965.

5. Council of Biology Editors Style Manual, 3rd ed. Washington, D.C.: American Institute of Biological Sciences, 1972.

6. Lois and Selma Debakey, "Medicant." Forum on Medicine (April 1978), 38-40, 42-43, 80-81, 83-86.

7. A Handbook of Acronyms and Initialisms (NUREG-0544). Office of Administration, U.S. Nuclear Regulatory Commission, March 1979.

SYNTAX, COMPREHENSION, AND BELIEVABILITY: IMPLICATIONS FOR TECHNICAL WRITERS

Dr. Charles H. Sides
Associate Director of Technical Communication
Clarkson College of Technology
Potsdam, New York

A pilot study designed to test the relationship of syntax to comprehension has indicated trends which suggest that some of what has been typically denoted as good writing in fact inhibits comprehension of text and that some styles of writing, though difficult to process in their surface features, enhance the retention of subject material. A series of cloze-recall tests were used to show the differences in the relative comprehensibility of the following styles: right-branching; left-branching; mid-branching; passive voice; inversions; and parallel structure. Questionnaires, also based on these styles, were circulated to college writing teachers in order to elicit responses which would indicate teacher biases for and against particular styles. Results of these indicated that college writing teachers tend to believe an author who writes bureaucratese even though they would not call such writing good writing. These results imply that teachers of writing, technical and otherwise, may have to reperceive their pedagogical stances with regard to what constitutes effective communication as opposed to stylistically pleasing prose.

As a part-time teacher of freshman composition during my first year in graduate school, I received a paper from a young engineering student which has profoundly influenced my professional career over the intervening years. The paper was written on the topic of retrofitting homes for solar energy production--a subject which holds some interest to me. The paper, regardless of the timeliness of its subject and my interest in it, was dull; not middle-of-the-road freshman mediocrity, but magnificent blandness. Since the hour at which I was reading was late, I--wishing to give the student the benefit of a doubt--went to bed, resolved to read the paper fresh the next morning. Nothing changed; the paper was incontrovertably soporific. Since we writing teachers have no corpus of knowledge to fall back on in situations such as these, I became inspired. In this moment of inspiration, I decided to count the number of sentences in the student's paper and then the number of words per sentence. I wanted to see if some pattern was recurring throughout the essay. The paper had thirty-eight sentences; of these, thirty-six had eighteen words apiece; one had nineteen words, and the other had seventeen words. I was incredulous, astounded--positive I had made an important linguistic discovery. Here, I thought to myself, was a young man who thought in eighteen-word segments. When I confronted him with his accomplishment and asked him if he indeed thought in eighteen-word segments, and when he responded rather nonchalantly that indeed he did think in eighteen-word segments, I was elated, certain in my naivete that I would be famous. He went on to tell me that when he was in the eighth grade he had had an English teacher who, along with the typical admonitions not to begin sentences with but, and, or because, casually remarked one day that the average American-English sentence contained eighteen words. Interpreting the teacher's words as a righteous path to perfection, my student had dedicated his writing life to producing eighteen-word sentences, and he was good at it. This event convinced me that to some extent form could be separated from content in writing for the purposes of study. What I want to address today is some of the ways this is being done and some of the implications it has for technical writing and writers.

The results of a recently completed pilot study suggest certain trends in the effects which syntax has on reader comprehension. A series of cloze-recall tests was used to discover these trends. For the tests, an original selection of writing was rewritten in six different structural styles: right-branching or cumulative; left-branching or periodic; mid-branching using centrally embedded modifiers; inverted; passive voice; and parallel structure. Test subjects, college student volunteers between the ages of eighteen and twenty-two, were allowed five minutes to read one of the stylistic versions of the original or, as a control, the original itself. Following the reading, the subjects were given five minutes to answer questions based on the style they had read. The results of the analyses of the tests suggest trends which, upon further, more detailed research, should prove useful to the educator and the communication researcher alike. As an example, the results for the most part replicated the conclusions reported by Fodor, Bever, Kintsch, Keenan, and others in earlier psycholinguistic research; in other words, passive constructions tended to be more difficult to comprehend than active constructions, negations more difficult than non-negated constructions. However, interesting and potentially important extensions of earlier research were also suggested: cumulative styles appeared more difficult to comprehend than teachers of writing have been led to believe, while centrally embedded styles, though laborious to process, seemed in some instances to promote reader comprehension. The possible reasons for and effects of these observations will be discussed later in this paper.

In addition to the tests, questionnaires were circulated to writing teachers at five separate colleges; these questionnaires were designed to complement the tests--though in a subjective and evaluative way. I intended to discover opinions and nuances which might lead to future research on comprehension and believability. Insofar as the questionnaires suggested that teachers of writing tended to believe formal style (which can be associated with bureaucratese) much more readily than informal style, they succeeded in identifying areas of much needed research.

TEST RESULTS

The following three bar graphs show, respectively, the percent of correct answers out of the total number of questions, the percent of correct answers out of the number of questions answered by the subjects, and the percent of questions answered out of the total number of questions.

Table 1.

Mean Values for Percent Correct of Total Number of Questions

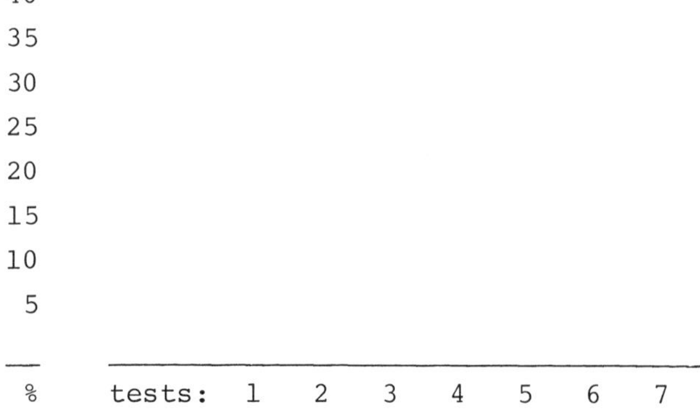

Table 2.

Mean Values for Percent Correct of Answered Questions

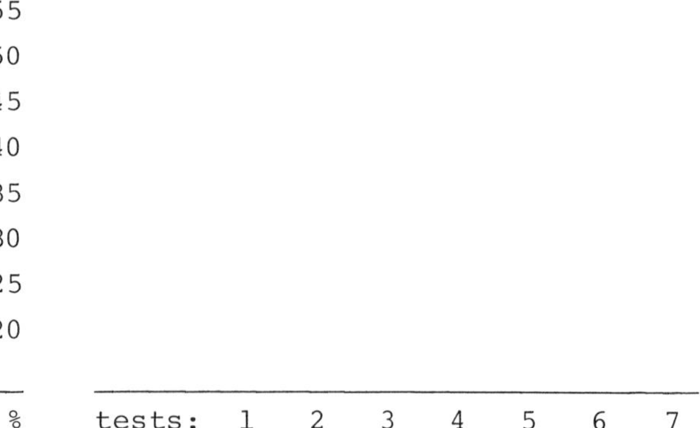

Table 3.

Mean Values for Percent Answered Out of Total Questions

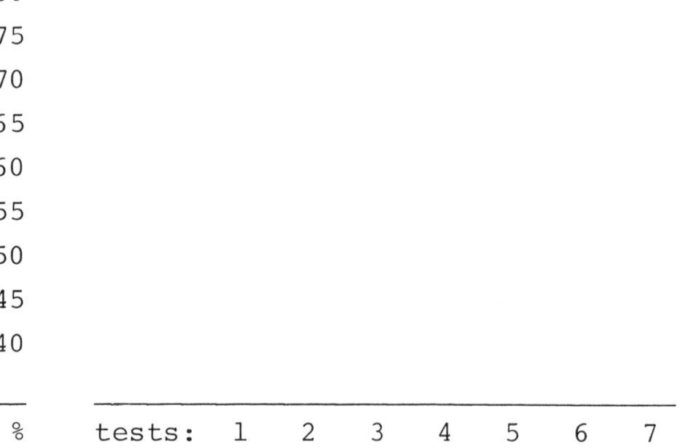

In each of the tables, test 1 refers to right-branching style; test 2 to left-branching; test 3 to mid-branching; test 4 to passive voice; test 5 to inversions; test 6 to the original or the control; and test 7 to parallel structure. Two interesting results can be seen from the graphs. First, the version developed in a mid-branching, or centrally embedded, style proved to be surprisingly easy to comprehend--almost as easy as the control; whereas, all of the rest of the artificially developed styles were much more difficult to comprehend. This should come as a surprise to those of us who have been told that cumulative style, based on right-branching modifiers, is not only a clear and effective way to construct prose but also the American idiom. Since the testing procedure used for this research was not designed to determine the reasons for the results noted, we can only speculate until further study can be conducted to provide those answers. One possible reason that cumulative style did not enhance the comprehension of the test subjects is the fact that it is the American idiom; in other words, readers are so accustomed to this particular stylistic patterns that their attention to what they are reading is dulled. But why should a centrally embedded style, a style which is notably laborious to process, appear to enhance comprehension? One possible reason is that the style is not what the reader expects and, hence, the reader decreases his rate of processing the language and consequently his attention to what he is reading is increased. The second interesting result to be noted from this research is in Table 3, for test 7, parallel structure. More readers answered more questions for test 7 than for any other test. Interestingly, however, this increase in processing speed was not accompanied by an increase in the level of comprehension. One explanation for this could be that when reading a style which has been developed through the overuse of parallel structure, readers become stylistically conditioned to the syntactic structure of the discourse. Once a reader of parallel structure style becomes aware of the structure, he may be lulled into a lower attention to content. In other words, the rhythm of this particular style dulls a reader's senses. Future study may show that when a reader becomes conditioned to parallel structure, he may even tend to ignore the material placed in the latter positions within the structure--much as readers tend to ignore material placed within parentheses. At any rate, the research presented in this paper suggests that syntax does affect comprehension in complex ways. Specifying these effects will benefit composition and technical writing pedagogies.

QUESTIONNAIRE RESULTS

As I mentioned earlier, I intended, through the use of a questionnaire distributed to college teachers of writing, to discover writing biases--emotional and at times illogical responses which would shed some light on the state of writing education in our colleges. The results exceeded my hopes. Perhaps the best method of communicating the types of reactions to the various styles I used in the tests is to give an example of a particular style and some representative reactions to it.

Right-branching.

> The search for intelligent life elsewhere in the universe is a timely and feasible undertaking with substantial potential secondary benefits which can be started now with only modest resources and can be expanded later to a much larger scale, if that turns out to be desirable and necessary.

Reactions.

> "Unfortunately, this is believable because one expects scientific writing to look and sound like this."
>
> "The writer seemed very objective because of his style."

The majority of the readers of this style felt that it was difficult, dull, and rambling; but they also admitted a tendency to believe what the writer had to say based on the style of the document.

Left-branching.

> Needing only modest resources which can be expanded later to a much larger scale if that turns out to be desirable and necessary, the search for intelligent life elsewhere in the universe is a timely and feasible undertaking with substantial potential secondary benefits that can be started now.

Reactions.

> "The writing sounds like the bureaucratese we expect in a scientific proposal. It avoids the suspicion of being trivial research by sounding like the typical, serious, and important research project."

One hundred percent of the respondents rated this style, which they classified as confusing and dull, as being moderately to strongly believable.

Mid-branching.

> The search for intelligent life elsewhere in the universe is, with substantial potential secondary benefits that can be started now with only modest resources and can be expanded later to a much larger scale if that turns out to be desirable and necessary, a timely and feasible undertaking.

Reactions.

> "The style gives the illusion of authority."
>
> "The aggravation necessary to follow the train of thought greatly overwhelms the amount of information conveyed.

Not surprisingly, the respondents found this style to be difficult to process; they also had a tendency not to believe the author, one respondent, going so far as to say that the writer seemed to be trying "to hide something."

Passive constructions.

> The objectives of the Science Workshops on Interstellar Communication are summarized by Morrison in his preface to the SETI report.

Reactions.

> "The style is very formal, but for this topic it is suitable."
>
> "The style is rather convincing."
>
> "The style sounds bureaucratic."

The respondents did not find a passive voice style as difficult to process as they did the three earlier styles; also there was some skepticism for believing the writer.

Inversions.

> It is thought that the search for intelligent life elsewhere in the universe is a timely and feasible undertaking with substantial potential secondary benefits that can be started now with only modest resources and can be expanded later to a much larger scale if that turns out to be necessary and desirable.

Reactions.

> "The style seemed proper."
>
> "The writer uses the objective style of science and technical writing, but he does not achieve credibility."
>
> "Though what the author has to say does not seem inane, the style he uses is a kind of scientific journalese: proper, measured, but without much vitality."

Inversions were disliked, but the style was not thought to be difficult to process. The respondents were much in the middle of the road as to whether they considered this style believable or not.

Parallel Structure.

> The search for intelligent life elsewhere in the universe is a timely and feasible undertaking with substantial potential secondary benefits. The search can be started now with only modest resources; it can be expanded later to a much larger scale if that turns out to be desirable and necessary.

Reactions.

> "As a mover of jargon, the writer has skill and conviction."
>
> "I like verve, grace, brevity, and forthrightness too much to read much of this sort of stuff--even for wages."
>
> "Most scientists use this heavy, abstract, sometimes circuitous style in scientific discussion, which is why this writer seems authoritative."
>
> "The phrasing seemed persuasive and sincere."
>
> "Clear statements about potentially difficult issues."

To say that the style developed through an exaggerated use of parallel structure was more well received than any other style would be damning with faint praise. The respondents considered this version to be moderately easy to read and moderately clear. They suggested a strong sense of believability in the author.

GENERAL CONCLUSIONS

The responses to the questionnaires point out an interesting paradox: the writing teachers tended to believe styles that they did not like and would not encourage their students to emulate as examples of good writing. With only a few exceptions, they did not point out the generally held conviction that bureaucratese can be and is being used to distort facts and withhold information. When we take these responses and compare them to the results of the tests, we see interesting and perplexing things. First, the tests suggested that a style which has been the vogue of writing education for some years may actually impede comprehension--for what reasons we cannot yet be sure. Second, many of the tested styles which appeared difficult to comprehend, and which the respondents concurred in the difficulty to process and understand them, the respondents nonetheless believed.

What implications do these results hold for us--teachers of technical communication. Their implications lie in their potential importance. If future research supports these findings, then we may have to reperceive our pedagogical stances on writing styles. Certain syntactic structures may prove to be advantageous vehicles for certain types of subject matter in particular writer/reader contexts, and the structures may not be what we have for years denoted as good writing.

THE TECHNICAL REPORT AS AN EFFECTIVE MEDIUM FOR INFORMATION TRANSMITTAL

Dr. Freda F. Stohrer* and Thomas E. Pinelli+
*Old Dominion University; +NASA Langley Research Center

Dr. Freda F. Stohrer
Old Dominion University
Norfolk, VA 23508
(804) 440-4016

Assistant Professor
Department of English

Thomas E. Pinelli
NASA Langley Research Center
Hampton, VA 23665
(804) 827-2691

Assistant Chief
Scientific & Technical Information Programs Div.

The NASA Langley Research Center is conducting an evaluation of its scientific and technical information (STI) program. This comprehensive evaluation includes a study of the NASA technical report as an effective medium for information transmittal. The evaluation of the technical report began with the compilation of empirical data through a survey of report producers and an assessment of report components, format, and organization through a study of current style manuals.

A mailing list obtained from the STC and NASA provided the research sample. A letter was sent to 611 organizations in industry; academia; government; and research associations. The letter requested examples of their technical reports produced for external distribution, their organization's style manual, and their publication guidelines. One hundred twenty-four replies were received within the established time limits; 99 were suitable for analysis and data extraction.

From the reports received, an exhaustive list of the structural components as well as the placement of the components within the report were determined. The survey produced empirical data and recognized publication standards which will be used to develop criteria to evaluate and possibly modify the NASA technical report.

INTRODUCTION

The National Aeronautics and Space Administration (NASA) uses a variety of media to fulfill its statutory obligation of making "...the widest practicable and appropriate dissemination..." of its research findings. Since the days of its predecessor, the National Advisory Committee for Aeronautics, NASA has relied upon the technical report for disseminating the significant findings of its research.

Founded in 1917, the NASA Langley Research Center (LaRC) is one of the leading national laboratories for research and development (R&D) in the sciences of aeronautics and space technology. For calendar year 1980, Langley's research population of 1,036 engineers and scientists produced 1,418 items of documented research which included the publication of 611 technical reports.

In February 1980, the Scientific and Technical Information Programs Division undertook the first comprehensive review and evaluation of the Center's STI program. As part of the project, a study of the technical report was undertaken to determine whether the NASA publication standards of style and organization made the technical report an effective medium for transmitting information.

NASA employs uniform publication standards designed to ensure the clarity, quality, and utility of its technical reports. These standards were designed to produce reports of maximum readability and ease of comprehension, written in a style that is both logical and familiar because of its wide acceptance in technical writing. However, an evaluation of NASA publication standards had never been conducted.

A survey of the literature disclosed that little, if any, documented research existed to support or suggest criteria for assessing the effectiveness of a technical report. Consequently, a survey to establish the present environment of the technical report and to produce empirical data against which NASA publication standards could be compared was deemed essential. This paper reports the preliminary findings of the study.

RESEARCH METHODOLOGY AND PROCEDURE

The survey sample was composed of individuals affiliated with organizations which were known to produce technical information. Addressees were compiled from two sources: the Society for Technical Communication (STC) membership and NASA's distribution list for technical reports.

A letter was sent to individuals representing 611 organizations in industry; academia; government; and research, trade, and professional associations. The individuals were asked to provide the following:

1. Copies of typical reports published by their organizations;

2. A copy of their style manual or the name of the commercially prepared manual (e.g., Chicago Manual of Style) if one is used;

3. A copy of the publications or graphics manual or standards covering such factors as design, layout, typography style, illustrative material, printing, binding; and

4. A form indicating the absence or presence of the requested information.

Approximately 200 pieces of literature were received from 124 organizations within the established time limits. Ninety-nine technical reports were suitable for analysis and data extraction. For the purpose of this study, the technical report was defined as the communication medium for the external dissemination of the results of investigative, analytical, or theoretical material of a detailed and complete nature. The data were analyzed according to established criteria. No statistical inferences were made from the findings.

FINDINGS

Of the 611 organizations contacted, 99 respondents sent material suitable for analysis and data extraction. The overall rate of return for the survey was 16.3%. The 99 responses were grouped according to organization. This grouping is shown in Table A.

Organizational Type	Requests	Responses	Percent Responding	Percent of Total Survey
Government	49	12	26.6	12.2
Industrial	426	54	12.6	54.5
Academic	76	11	14.4	11.1
Research, Etc.	60	22	36.6	22.2
TOTALS	611	99	16.3	100.0

TABLE A. SURVEY RESPONSE BY ORGANIZATION

The largest group in the survey population was the industrial organizations, followed by the research, trade, and professional organizations; government organizations, and academic organizations.

The data were then analyzed to produce an exhaustive list of report components. Ninety-nine report producers generated ninety-eight different structural components. Only five components were mentioned/used by 50% or more of the respondents. A breakdown of the five components, the percentage of use, and their location within the report is presented in Table B.

Component	%Use	Front	Body	Back
Cover	67.6	100.0	--	--
Title Page	80.0	96.2	2.5	1.2
Table of Contents	70.7	100.0	--	--
Introduction	57.5	17.5	82.4	--
Appendixes	59.5	--	1.6	98.3

TABLE B. COMPONENTS BY USE AND LOCATION

The location of the components is discussed in terms of the three areas of traditional book publishing: front matter, body, and back matter. The front matter consists of all material preceding the main text. The body of the report contains the investigative, analytical, or theoretical material. The back matter consists of reference material and other supplementary matter.

Of the five components listed in Table B, only the cover and table of contents were consistent in their location within the report. No agreement exists about placement of title pages, introduction, or appendixes while 50% or more of the respondents did not mention textual matter at all.

The diversity of the components was not the only difficulty encountered. The analysis revealed a general lack of agreement about terminology for parts of the technical report. Moreover, the components were located in almost every possible position throughout the report.

The components in the exhaustive listing were refined so that they could be compared more easily with the components covered by the NASA publications guide. Components which appeared to have the same function were combined. For example, "Lists of Drawings" were combined with "Lists of Figures." Any component mentioned by NASA was included. The number of components was reduced by eliminating any component used by fewer than five report producers.

The components derived from the exhaustive list were compared with the components and their recommended placement as specified in the NASA publication guide. The components and their placement as specified by NASA compared favorably to those contained in the refined list. The analysis did reveal variations in the number and placement of front matter components. Where body components were concerned, NASA placed the same elements in the body of the report as those contained in the refined list. A comparison of back matter components revealed certain variations, most notable in the placement of the glossary and index.

The respondents were asked to provide information relative to the use/non-use of style manuals and publication/production guides. Respondents were also asked to identify the use of commercially prepared style manuals. The responses were compiled and are presented in Table C. While the chart is phrased in use/non-use terms, mutually exclusive categories were not specified. Therefore, the percentages cannot be added to describe 100% of the sample.

	Chicago	AP	GPO	*Other Commercially Available Style Guide	Your Organization's Own Style Guide	No Style Guide Used	Publication/Production Guide	Do Not Use Publication/Production Guide
Government	16.6	16.6	83.3	33.3	41.6	25.0	58.3	41.6
Industrial	31.4	5.5	42.5	27.7	29.6	37.0	33.3	46.2
Academic	54.5	27.2	27.2	36.3	18.1	45.4	54.5	54.5
Research, Etc.	59.0	13.6	50.0	27.2	31.8	31.8	40.9	50.0
Total Survey Average	38.3	11.1	47.4	29.2	30.3	35.3	40.4	41.6

TABLE C. USE OF STYLE MANUALS AND PUBLICATION GUIDES BY ORGANIZATION

The majority of respondents relied upon a style manual to prepare technical reports; however, approximately 33% of the respondents used no style manual. The GPO manual was used by the majority (83.3%) of respondents from government operations. A majority of academia and research respondents, 54.5% and 59% respectively, used the Chicago Manual of Style in report preparation. Respondents were almost evenly divided in the use/non-use of a publication/production guide.

CONCLUDING REMARKS

The study represents an attempt to assess the NASA technical report as an effective medium for information transmittal. The evaluation included the compilation of empirical data through a survey of report producers and through a survey of report components, formats, and organization through a study of current style manuals.

A survey of the literature disclosed that little, if any, documented research existed to support or suggest criteria for assessing the effectiveness of a technical report. Consequently, a survey designed to determine the present environment of the technical report and to produce empirical data against which NASA publication standards for technical reports could be compared was deemed essential.

During the analysis of the findings of the study, wide variances in the technical report were discovered. Nearly one hundred components were identified, based on the bewildering array of terms. An attempt to reduce the number of components was made, based on the combining of similar components under one heading and the elimination of seldom used component terms. No standard sequence was discovered for the components, so only a general location for them could be made in terms of front, body, and back matter.

Future research will be devoted to assessing the adequacy of the reduced set of components, clearly defining the purpose of each component and developing evaluative criteria for each. The components of the NASA reports will be evaluated according to these criteria. Based on the tabulation of locations in common usage, alternate theoretical sequences of components will be developed and tested empirically. The NASA report will be tested against the sequence established by research.

GAUGING READER SATISFACTION: A CASE HISTORY

Frances J. Sullivan
Eastman Kodak Company

Frances J. Sullivan
Eastman Kodak Company
667 French Road
Rochester, NY 14618
(716) 724-3860

Editor

Secretary of the Society, 1975-77. Presently Chairman of Publications Committee. Coauthor of STC booklet *Academic Programs in Technical Communication.*

Faced with today's increasing costs of producing technical publications, writers need to know more than ever what readers think about their books, periodicals, and pamphlets. Precious budget money must go to publications that are on target.

Are readers satisfied with the publications that they have been receiving? What type of publications do they prefer? What would they be willing to pay for technical publications in the future?

This paper will describe the development and use of a questionnaire to gauge reader satisfaction with materials produced by an industrial publications department. It will also reveal the general results obtained through use of the questionnaire.

INTRODUCTION

Have you ever wondered, "What do readers really think about my publications?" Most technical communicators have had such thoughts. Not long ago, I was given the opportunity to find out what readers thought about the graphics publications produced by my company. My department was considering changes in the format and in the price of graphics publications, and wanted to know how readers might react to such changes. Of course, the best thing to do was to ask the readers. This article describes the way we went about doing that.

BACKGROUND RESEARCH

Before trying to design a reader survey, I spent a lot of time looking at sales and distribution records of the past year. This was to determine just how publications were reaching readers, and what publications were popular. It appeared that consumers obtained publications from the company's sales representatives, from dealers, and directly from the company. As a result of this research, we decided to do two surveys—one of the attitudes held by our consumer readers and one of the opinions of our sales representatives, as both readers and distributors of publications. There is not enough space here to go into detail about the survey of sales representatives. That survey dwelt on how and why salespeople were using publications, but also asked some of the same questions as the consumer survey about preferred format and price. This allowed us to compare what sales representatives thought consumers would want with what consumers actually said they wanted.

DESIGN OF CONSUMER STUDY

The next step was to decide how to design the study of the attitudes held by our consumer readership. Since we wanted to spend the least amount of budget money, I proposed a survey conducted by mail rather than in person. Also, I proposed that we use existing mailing lists in our possession, rather than generating or buying new ones. The mailing lists for three of our periodicals seemed ideal.

These mailing lists represented 142,000 consumers who should have some acquaintance with our publications. The lists also represented the three major segments of our market: the graphic arts, reprographics, and engineering fields. About half the consumers were on the graphic arts list, and one-fourth on each of the other two lists. To represent this population of consumers, a random sample of 5000 names was drawn which was weighted toward the graphic arts list—3000 names coming from that list and 1000 each from the other two lists. This weighting toward the graphic arts also reflected the fact that most of the publications we produce fall into that category.

For awhile, we considered three separate mailings of a questionnaire to the three separate groups. However, it seemed simpler to do only one mailing, using a questionnaire that could apply to all three consumer groups. On the questionnaire, an item would request respondents to identify the industry to which they belonged. Also, we decided not to provide an incentive to respond (such as a small amount of money or a promise of free product). Instead, a cover letter was prepared which simply asked for the consumer's cooperation so that we could better plan future publications. We did enclose a postage-paid return envelope to make responding as easy as possible.

QUESTIONNAIRE

Designing the actual questionnaire to be used was next on the agenda. The body of the questionnaire consisted of statements about Kodak graphics publications; for instance: "The topics in Kodak publications are covered in adequate detail." Respondents were asked to indicate how much they agreed with each statement by writing a number code next to the statement. The number codes consisted of the numbers 1 through 5, with 1 standing for "Strongly Agree" and 5 standing for "Strongly Disagree." Directions for filling out the questionnaire (with an explanation of the codes) were printed on the top of the first page.

Also printed on the top of the first page was an explanation of just what the term "publications" meant as it was used in the questionnaire. We wanted consumers to give their opinions on technical materials rather than advertising brochures. This explanation on the questionnaire reinforced a fuller definition of terms given in the cover letter.

Other parts of the questionnaire asked for some personal data from consumers; provided space for any comments about our publications; and gave our return address. The data section of the questionnaire was used to categorize respondents by age, education, industry, and source of publications. Also, respondents were asked in this section how long it had been since they had read a Kodak graphics publication. This would allow us to drop out those who didn't have recent knowledge of our publications—particularly if those respondents were a significant percentage of the whole.

The "Comments" section of the questionnaire was given a limited amount of space, to discourage lengthy replies which would be hard to handle at the tabulation stage of the study. In the "Comments" section, consumers were also given a brief list of topics on which they might want to comment (quality level, layout, illustrations, format). We included our return address on the questionnaire just in case the questionnaire got separated from the cover letter and return envelope in transit.

In devising the statements that made up most of the questionnaire, we used some actual statements made to our market researchers in interviews. This added some special validity to the questionnaire. The statements were all positive in structure, because experts in attitude measurement said that respondents often overlook negatives such as "no" and "not" in a sentence. The result of this inattention is, of course, a false answer. To get around the tendency of some respondents to agree with all the statements in a questionnaire, we worded some statements to elicit disagreement as a favorable response. (For example, we hoped respondents would disagree with the statement "I seldom refer to the Kodak publications that I have on file.")

HANDLING RETURNS

When the questionnaires were mailed, consumers were asked to return them by a date approximately two weeks away. This was to encourage consumers to fill out the simple questionnaire (one page, front and back) right away. Experts have found that giving people a lot of time in which to respond often results in fewer questionnaires returned.

Out of the 5000 questionnaires sent out, 1063 (21%) were returned—800 even before the deadline we had requested. Following the pattern of the questionnaires mailed, the returns were weighted toward the graphic arts field. A full 50 percent of the respondents were in the graphic arts field, with 23 percent being in the engineering field and 13 percent in the reprographics field. The remaining 14 percent of respondents were mainly in the fields of education and photography.

As each filled-out questionnaire was received, it was stamped with a number (such as 0001). This helped us keep track of the number of questionnaires already received and the whereabouts of an individual questionnaire. It was also helpful in the recording of comments.

Once a questionnaire had been stamped, it was briefly edited. This meant interpreting written answers and sometimes creating new codes. Batches of questionnaires were then sent to a computer service department for keypunching of the codes. The identifying number of each questionnaire was also keypunched. Both the keypunched cards and the original questionnaires were then returned to us.

At this point, I recorded the written comments by dictating them to a machine for later transcription. Comments were bunched under industry categories such as "graphic arts—commercial," and each comment was tagged with the questionnaire's identifying number. Recording comments in small batches—and categorizing them immediately—saved time and energy later in the reporting process.

DESIGNING COMPUTER PROGRAM

When all the punched cards representing the questionnaire data had been returned to us, they were taken to a person well versed in computer programming so that a special program could be designed for our particular research needs. First he designed a program that would simply tabulate the results, question by question. This gave us a basis for preliminary reports on the project.

After this first computer run, the programmer devised further programs that compared one question with another or compared different categories of respondents. These programs provided the information for further reports on the project. It is at this point that you can test any hypotheses you may have about the kind of attitudes you expect certain segments of your readership to hold. (For example, you may expect younger readers to be more favorable to publications that provide general information in their field than are older readers.)

REPORTING RESULTS

Basically, two kinds of reports were used in this project: a summary report for management and an individual report on each segment of our market.

QUESTIONNAIRE ON KODAK GRAPHICS PUBLICATIONS

By "publications" we mean technical materials such as data books, control aids, data releases on products, pamphlets on special techniques or applications, and instructional materials. Pamphlets and periodicals produced to advertise or promote products are not included.

After reading each of the following questions, please choose the appropriate response. Then (except for those items which require a written response) write the number corresponding to your choice in the column provided at the right of each question. Please be sure to mark only one response for each item.

For Questions 1-26, use the following codes to indicate how much you agree or disagree with each statement.

1 = Strongly Agree
2 = Agree
3 = Neither Agree nor Disagree
4 = Disagree
5 = Strongly Disagree

	WRITE YOUR ANSWERS HERE
1. The topics in Kodak publications are covered in adequate detail.	________(1)
2. I would rather buy a whole new book than keep up on data releases and technical pamphlets.	________(2)
3. Kodak publications are worth paying for.	________(3)

FORMAT FOR THE FIRST PAGE OF THE QUESTIONNAIRE. NUMBERING BOTH THE QUESTIONS AND THE RESPONSES MADE KEYPUNCHING OF THE RESULTS EASIER.

The summary report described the project and gave the general results, as well as some interpretation of the results. Instead of presenting the questions in numerical order, the report discussed similar questions together. Also, a few of the respondents' comments were included as being typical of the whole. Attached to the report was a copy of the questionnaire with exact results written under each question.

For the individual reports on each industry segment, a filled-out questionnaire presented statistical results and was accompanied by pages containing all the comments received from respondents in that segment. This was an easy and effective way of presenting complete results for each segment of the survey. The pages of comments were quickly put together by cutting and pasting the pages of comments originally dictated by category—then duplicating the pasted-up pages. These individual reports went to the appropriate publications editors preparing materials in each field. Copies were also made available to management, to advertising editors, and to customer-service people.

CONCLUSION

This investigation of the attitudes held by consumers toward Kodak publications turned out to be well worth doing. Both management and the publications editors who produce these materials appreciated learning the consumers' opinions. Of course, they were pleased to learn that our readers were quite satisfied with the quality of the technical information that they had been receiving from Eastman Kodak Company. They were also interested to learn, however, that the availability of technical publications could be improved—especially availability through dealers.

The survey gave my department the information it needed about consumers' preferences for type of publication, topics to be treated, and price ranges considered acceptable. We are now acting on that information, and feel that a similar study of reader satisfaction should be done every few years.

Perhaps you would be interested in doing a similar study. If so, the brief list of references which follows may be useful to you. Good luck!

REFERENCES

Berdie, Douglas R., and Anderson, John F. *Questionnaires: Design and Use.* Metuchen, New Jersey: Scarecrow Press, 1974.

Lemon, Nigel. *Attitudes and Their Measurement.* New York: John Wiley, 1973.

Oppenheim, A.N. *Questionnaire Design and Attitude Measurement.* New York: Basic Books, 1964.

Payne, S.L. *The Art of Asking Questions.* Princeton, New Jersey: Princeton University Press, 1951.

Rigby, Lynn V., and Gibbs, Verner E. "Measurement of Reader Satisfaction by Questionnaire." *Proceedings of the 19th International Technical Communications Conference* (1972): 17377.

Sonquist, John A., and Dunkelberg, William C. *Survey and Opinion Research.* Englewood, New Jersey: Prentice-Hall, 1977.

Warwick, Donald P., and Liniger, Charles A. *The Sample Survey: Theory and Practice.* New York: McGraw-Hill, 1975.

A Place for Humanities Majors in the Technical Writing Field

Lynne V. Swenson
Data General Corporation

Lynne V. Swenson
Data General Corporation
Technical Writer
4400 Computer Drive
Westboro, Mass. 01580
366-8911 X3028

Abstract

Software documentation departments should take advantage of talented college graduates with degrees in humanities subjects. Through extensive writing and the study of other writers, humanities students learn to write, to research, and to organize their ideas. A technical writer needs these skills, not just the technical facts, to produce comprehensive and readable manuals. While the ability to understand technical facts is necessary, the ability to write about them with clarity and style is most important for a technical writer.

Liberal arts graduates, especially those with humanities degrees, are a neglected source of potential technical writers. A writer with such an education may be technically uninformed, but still have the ability to understand technical concepts. The skills that a humanities education does develop are among the most important tools needed to write a comprehensive programmer's manual: the abilities to research, to organize ideas, and to report facts in a polished, consistent style.

To operate a computer, a user needs more than the technical facts. He needs these facts in an accessible form. They must be easy to find, well laid out, and above all, in readable English.

Humanities students are trained to communicate their ideas in readable English. Throughout their academic careers they must research, organize, and explain esoteric and detailed concepts. Their writing skills are finely honed through extensive writing and the study of other writers.

These writing skills are applicable to subjects other than English, philosophy or history. The subject matter of German philosophy may be far removed from the world of FORTRAN compilers, but it is no less complex. A writer who can explain philosophical theory can probably describe a programming language.

To research their papers, students must gather information from a variety of sources, both written and oral. These include books, articles, and conferences with faculty members. The writer must then select the information relevant to his subject and prepare it in one of several forms depending on his purpose and audience: the scholarly thesis, the short paper, the informal essay.

A technical writer must also gather information from a variety of sources, both written and oral. These include product specifications, other manuals, and conferences with program developers. The writer must then select the information relevant to his project and prepare it in one of several forms, depending on his purpose and audience: the system manager's manual, the beginner's instruction guide, and the product overview.

One technical writer who has a Ph.D in English and who taught writing at a college for ten years before changing careers said this about his current job: "Attacking a project is very much like taking or giving a course in an academic environment. As technical writers, we are reporters, teachers and researchers." His manager agrees: "The best preparation for technical writing is extensive writing and critiquing experience in all kinds of writing, including fiction and journalism."

The technical writer must, of course, understand what he is writing about. But he does not have to be a programming "guru". He must learn the concepts of computer programming, but this does not require years of study (programming is currently taught in some grammer schools.)

Someone who has never seen a listing before can soon produce his own, given a bit of instruction and guidance. A humanities graduate turned technical writer may never program at the level of a systems developer, but he can certainly learn what is necessary to describe the mechanics of language statements and runtime routines.

Furthermore, a writer's technical naivete can be an advantage. He can often anticipate his readers questions about the product because he is likely to ask these questions himself. A technical mind sometimes forgets what a reader might not know about a technical subject. An informed outsider's perspective is useful in manuals for both beginners and sophisticated programmers. Either user may be new to a given system.

People who are candidates for technical writing jobs and who have humanities backgrounds frequently fall into two categories. First are the newly graduated or about-to-graduate college seniors. Second are academics who are currently teaching, but are interested in a career change. Both types of candidates have potential as technical writers.

How does the manager of a technical writing group determine whether someone with little or no programming experience will succeed as a technical writer? Writing samples of all types and lengths are most important. From them, a hiring manager can assess the writer's ability to organize, report, and use different writing styles to suit different audiences. A manager should also ask if the candidate has taken math or science courses outside of his major. Success in these courses is a strong indication that he has the ability to understand technical subjects.

The interviewer may want to assign a short expository piece for the writer to bring to the interview. In it, the writer can explain how to use something mechanical. A telephone, a phonograph, or the standard transmission in a car are some appropriate subjects. Technical information on all of them can be obtained in most dictionaries and encyclopedias. This assignment will demonstrate the writer's ability to research and document technical subjects.

The interviewer should ask how the writer feels about computers. If he is curious and eager to learn, he will probably adapt very well to a technical environment. If he is apathetic or uninterested, his writing will reflect this attitude.

Students of the humanities are not the only college graduates who make good writers or even good workers. Some computer science graduates can articulate their ideas with clarity and style. Some English department graduates can't organize a paragraph. But because success in their major demands good writing, a successful humanities graduate is likely to be a good writer.

Computer companies should take advantage of talented humanities graduates. They are in abundant supply, but have traditionally been ignored by those in high technology. They bring many talants to software documentation departments that most computer science or engineering graduates simply cannot offer.

WHAT FOUR-YEAR COLLEGES DO FOR TECHNICAL COMMUNICATION

Thomas L. Warren
Oklahoma State University

Thomas L. Warren
English Dept. 205 Morrill
Oklahoma State University
Stillwater, OK 74078
(405)624-6142

Director, Technical
Writing Program

Academicians contribute to the technical communication profession in two ways: They prepare the students who will become technical communicators, and they produce articles, read papers, and give informal talks based on their research. Directors of technical communication programs monitor the needs of industry, business, and government through surveys and advisory groups. The Oklahoma State University program is typical of many in the country that prepare students and do research.

WHAT ACADEME DOES FOR THE PROFESSION OF TECHNICAL COMMUNICATIONS: FOUR-YEAR PROGRAMS

Prior to the 1950's, little formal education was available to anyone wanting to become a professional technical communicator. Colleges and universities had been offering a course in technical writing since World War II, but few degree programs existed. Tom Pearsall and Francis Sullivan[1] surveyed college programs for S. T. C. in 1976, and found 20 programs offering degrees in technical communication (6 offered graduate degrees, 10 offered baccalaureate degrees, and 4 offered associate degrees). A 1975 survey by the Council for Programs in Technical and Scientific Communication[2] identified 18 programs offering technical communication degrees (6 graduate, 9 baccalaureate, and 3 associate degrees). In a survey of science writing courses conducted by the Department of Chemistry, State University of New York at Binghamton[3], 34 programs were identified (10 graduate, 24 baccalaureate and 0 associate). Pearsall and Sullivan are currently updating their survey, and the authors of the science writing survey included a page whereby program directors can update their entries.

What these surveys suggest is that a number of schools are instituting degree programs in technical writing at all levels (associate, baccalaureate, and graduate) and that these programs are growing. The program at Oklahoma State University, begun formally in 1977, has increased from 12 students either majoring or minoring in technical communication in the 1979-1980 academic year to 29 in the 1980-1981 academic year. There has been a comparable increase in the growth of the service course enrollment (up 41% over a year ago.) Other schools are experiencing similar growth patterns as more and more students see the benefits of a career in technical communication. Such increases have direct benefits to industry in the large pool of qualified students from which industry can draw. Thus, one area where colleges and universities help the profession of technical communication is in training students. There are others that I will discuss later, but they are spin-offs from this primary benefit.

CURRICULUM IN FOUR-YEAR PROGRAMS

You will hear my colleagues discuss programs leading to associate and graduate degrees. My focus will be the program leading to a baccalaureate degree.

Typically, baccalaureate degree programs are divided into three very broad areas: general or university requirements, departmental requirements, and supporting electives. In a 124 semester hour program, such as at Oklahoma State, 56 hours (45%) are in general university courses, 24 hours (19%) in departmental courses, and 44 hours (35%) in supporting electives. Since each school develops a degree program in its own way, (and in different departments[4]), I will describe our program as typical.

We follow the division I describe above. Students may elect to follow the technical writing option as a major in English alone, using English as a second major, or as a minor with a major in another department and college. Since the three options became a reality, the number of students electing the option as either a major or a minor has increased 141%. The core of the program these students follow is the block of five technical writing courses listed below (using short titles):

Introduction to Technical Writing
Intermediate Technical Writing
Advanced Technical Writing
Internship in Technical Writing
Writing for Professional Publication

To these are added two courses in linguistics, including a course in grammar, and (under the present system) a course in literature. All eight of these courses are taken during the junior-senior year after the students have met the general university requirements. In addition the students will have 27 hours of electives from lower level courses and 16 hours of electives from upper level courses. These 43 hours of electives include one course in computer programming and the rest from areas such as graphics, business, engineering, technology, science, etc. The student may elect to become a "generalist" by taking a wide range of courses using the lower level electives to fulfill prerequisites for upper level electives, or may decide to "specialize" in a particular area. My advice to the students is to take as broad a base of courses as possible thereby bringing a multiple point of view to any writing project. In this way, I believe, we are providing graduates that fit in well with the needs of industry, business, and government.

I am not content to allow the program to be unresponsive to these needs and want to keep in touch with those who will be hiring my graduates. Other program directors are likewise concerned and have responded in various ways. Perhaps the two major ways of knowing what industry needs are the advisory committee and the survey.

John Brockman at Arizona State sent questionnaires to programs offering technical communication degrees asking about advisory committes. His results are published in the CPTSC Proceedings (1980). He found advisory groups working with programs to strengthen them. I am planning to form an advisory group at OSU drawn from the Oklahoma Chapter of STC and the Air Central Chapter in Wichita. Until then, I have relied on surveys. Recently, for example, the Curriculum Committee of the English Department has recommended revising the curriculum all English majors must follow (effective Fall, 1981). I believed that the curriculum followed by students majoring in English with a focus in traditional literary study, creative writing, film study, or linguistics was inadequate for students wanting to focus on technical communication. Consequently, I asked 20 publications managers and senior technical writers across the country and in Canada what they thought of the old curriculum, the new one taking effect in the Fall semester, and a new curriculum I devised. I have received 16 questionnaires of the 20 and find unanimous agreement that the Fall, 1981 curriculum for all English majors is inadequate for training technical writers. Some of my respondents even took the time and trouble to generate a fourth curriculum. (I plan to send the survey results to Technical Communications soon.)

In addition to answering the questionnaires, five of the respondents expressed delight that I would bother to ask the people who had to hire and work with my graduates what they thought of the curriculum and course content (I had also asked what they would teach or include in an advanced technical writing course.)

I plan to use the questionnaires when I present my new plan of study to the Curriculum Committee (by the time of the Conference, I will have already presented my proposal and had it acted on). I believe that the comments of those who hire or recommend hiring my students should be of major importance to the Committee. Happily, the survey contains comments about the broadness of my proposed curriculum, so that the student receives an education that cannot be characterized as vocational in the sense that trade schools produce narrow specialists. I will carefully consider the suggestions the respondents made for the advanced technical writing class--although it would take three semesters to teach everything they mentioned.

The point is that OSU's Technical Writing Program, as other programs, works cooperatively with industry, business, and government in setting up curricula and courses for students who will graduate and work as technical communicators. This area, however, is not the only area where academe helps the profession. While research is not as important an objective in an undergraduate program as it is in a graduate program, it still is done by faculty, who want to stay not just current but on the leading edge of research, and students, through papers they write and the problems they investigate in senior research problems courses. The faculty of four-year schools is normally under pressure from deans to do scholarly research, and they could choose to do it in some aspect of technical communication. The results of such research are communicated not only to colleagues through articles and papers, but also to students through lectures and projects. Faculty members publish their research in journals, such as Technical Communications, where STC members can have access to it. (Of volumes 26 and 27, discounting volume 27, No. 4 that was written by IBM authors, 14 of the 39 authors of articles were from the academic profession.) A wider forum is the ITCC where, in the 1980 proceedings, 87 academics gave papers representing 37.3% of the participants. (Academics gave 44 papers in stems other than Education and Research representing 26.2% of the papers. In Education and Research, they gave 43 papers or 66.2%. Nonacademics gave, consequently, a higher percentage of the papers in Education and Research than academics gave in other stems.) In addition academics communicate the results of their research in informal talks to STC chapters, formal presentations to organizations composed primarily of other academics (Conference on College Composition and Communication for example), and conferences of special interest groups (such as the Association of Earth Science Editors). In speaking to groups composed of people other than technical communicators and publishing in their journals, academics occupy a unique position: They can present views from both the academy and the technical communication profession.

Thus, academics in four-year colleges and universities contribute to the profession of technical communications in two major ways: First, they are involved in the degree-granting programs at the baccalaureate level that account for many members of the profession, basing their programs on a balance between the needs of industry, business, and government and the need to provide a sufficiently broad and liberal education. Second, academics produce articles, give papers, and make informal talks based on their research, thereby enlarging the research base of the profession and functioning as a bridge between their academic colleagues in other disciplines and their colleagues in the technical communication profession.

ENDNOTES

1. Thomas E. Pearsall and Frances J. Sullivan, Academic Programs in Technical Communication (Washington, D.C.: Society for Technical Communication, 1976).

2. Copies available from Professor Carolyn Miller, Treasurer, CPTSC, English Department, North Carolina State University, P. O. Box 5308, Raleigh, NC, 27650.

3. Sharon M. Friedman, et al. Directory of Science Communication Courses and Programs (Binghamton, NY: Department of Chemistry, State University of New York at Binghamton, 1978). The authors discovered 105 courses in science communication in 58 colleges and universities.

4. The table below summarizes where two of the surveys found technical writing programs housed.

	Journalism	Communication	English	Technology	Humanities	Tech. Illus.	Rhetoric	Tech. Writ.	Pharmacy	Physics	Science	Engineering	Natural Resour.	Chemistry
Pearsall/ Sullivan	3	3	1	3	6	1	1	1	0	0	0	0	0	0
Friedman, et al.	26	10	7	0	6	0	0	0	1	2	2	1	1	1

Cont'd.

	Geology	Medicine	Zoology	Unknown
Pearsall/ Sullivan	0	0	0	0
Friedman, et al.	1	1	1	2

USING CONTROLLED LANGUAGES FOR EFFECTIVE COMMUNICATION

E.N. (Ted) White
M and E White Consultants Ltd.

E.N. (Ted) White
M and E White Consultants Ltd.
Director
P.O. Box 257, West Byfleet
Surrey, England.

ABSTRACT

The paper is based on the work of the author over a number of years in developing controlled languages (in English and other European languages) to improve technical communication. A typical controlled language system consists of a limited vocabulary, composed of some standard words plus words selected for the particular application, some rules for producing clear, unambiguous information, and a training programme for the communicators. The results achieved in a variety of applications, for a number of important industrial companies, show improved writer-to-reader communication, removal of ambiguities, fast and accurate translations into other national languages, shortened training times for readers of training programmes in controlled English, and a new sense of pride and achievement in those communicators who are assisted by this technique to communicate more effectively.

INTRODUCTION

Controlled languages have been tested and proved beneficial in a wide range of applications to improve the effectiveness of technical communication. Typical applications include:-

(a) Simplifying instructions and descriptive information written by authors and engineers at post-graduate level for technicians or operators educated to a much lower standard.

(b) Presenting information prepared by experts in a manner easily comprehended by non-specialists.

(c) Producing standardised technical information for unskilled workers, or workers using a second or third language, so that technical words and phrases are firstly explained and then always used correctly.

(d) Preparing information for translation which can be quickly and easily converted by man or machine.

(e) Rewriting information which has been presented in a muddled, unclear manner and is creating difficulties even when used by persons of the same experience and qualifications as the writer.

READABILITY

The first application refers to examples where the writer of technical information may be educated to 12th grade and the reader is of 5th or 6th grade. This problem has been recognised by the United States Armed Forces and all future information prepared by suppliers of equipment will be subjected to a Flesch readability test before acceptance. The educational grade of the readers will be calculated before the test is applied and any information which is written at a high level and is likely to cause communication problems for the readers, will be rejected. These requirements are expressed in the military specification MIL-M-38784A (amendment 5).

The use of controlled language was tested by us recently during a three-day training course for highly-qualified technical authors engaged in writing technical manuals for computerised message systems. The authors prepared original texts of the kind normally provided for their customers; full of complex sentences, computer jargon and the idiomatics of the electronics workshop. The average Flesch count for this information was 12.

In a Controlled Language words have one clear meaning. LIGHT is a source of illumination. Not heavy is simply NOT HEAVY.

The same information was written in ILSAM, a simplified language based on Caterpillar Fundamental English, and the Flesch counts were 5 or 6. In just a few minutes, instructions written by 12th grade engineers for 12th grade readers were made suitable for 6th grade technicians.

The Flesch formula used for military purposes is:-

Grade level = 0.39 (a) + 11.8 (b) - 15.59

where (a) = average number of words per sentence and
(b) = average number of syllables per word

Military equipment contractors in the United States are unlikely to risk the high costs of rewriting military manuals and are certain to adopt Simplified English to ensure a first time pass on the Flesch test.

COMPREHENSION

The expert who is given the task of preparing information to describe the subject of his expertise for some other person's benefit is probably least likely to recognise that there may be problems of comprehension. Secure in his knowledge and assisted by the jargon and idiomatic expressions of his business he feels that his writings are such as to make an idiot into a skilled operator or maintainer. Unfortunately he so often neglects the first rule - to consider the reader. Full consideration of the reader (or of a variety of readers) can be lengthy and costly and can consume expert resources needed for other purposes. It is doubtful if the expert (excluding the professional technical author) will ever accept the requirement and the workload of considering a variety of readers.

The answer surely is to train experts in communication skills which can be generally applied to provide easily-comprehended information as normal practice. This can be done by introducing a controlled language and training the experts (and also professional technical authors) in its use.

Our experience is that the enforced grammatical rules and a controlled vocabulary are resisted at first but when the communication value of the work is demonstrated, the thrill of doing a good communication job takes over. A class of electronic experts on an ILSAM training course recently found the first day boring but at the end of the second day admitted they had discovered new and satisfying ways to communicate. Later on in the programme they were happy to commit themselves to using the system as standard practice. Publications experts in another company, after some months of using Simplified English said they just felt satisfied by doing a better communication job. In our experience, the experts can do a consistently better job in ensuring comprehension by the reader, by adopting controlled vocabularies and the associated rules for clear writing.

CONTROLLED TERMINOLOGY

In earlier times the English acquired the reputation of being a direct, clear-speaking race of people who always called a spade a spade. Unfortunately in these days of high technology they have invented a number of different terms and words for a spade, and they delight in using them all. This is very sad for the immigrants, of whom some 300,000 working in factories are estimated to have very limited vocabularies. Some may have difficulty in reading safety notices and some in following written procedures. We saw a recent example where a class of workers were trying to follow a procedure in which one switch was referred to in successive paragraphs as a switch, a push-button and a control. The one-word, one-meaning rule in a simplified language and the adoption of standardised nomenclature appear as constraints only to the unprofessional writer. The professional should view them as essential steps in the art of effective communication.

One of the rules introduced by the training programme for a Controlled Language system says "keep sentence structures simple. Do not introduce second ideas, particularly if not explained.

NOT – The control unit, which is duplicated, has low input resistance.

BUT – The control unit has low input resistance. There are two units with automatic selection of the ....

COMMUNICATION VALUE

The problem referred to above, where workers of different nationalities have to work together and communicate, is experienced by many countries, to some of whom it is a new experience. Further interface problems are experienced by countries exporting goods or commodities to an ever wider range of international markets. Countries are exporting today the high technology goods they formerly imported in only small quantities. New technologies are introduced into developing countries in weeks instead of years, and the guidelines for effective product support for technological exports from these countries represents a very different problem with a much longer time scale.

In our view the real answer is for importers and exporters to recognise product support as a potential problem and to plan for it. A typical plan might include:-

(1) Production of product support data to support the logistics strategy, using a national language which can accommodate the technology.

(2) At all times use a simplified language system with a controlled vocabulary, grammatical rules and controlled, simplified terminology.

(3) Use machine translations, with human editing, for those areas where the source language is not acceptable.

In this way, the communication value of the information can be high even when using new technologies and crossing language barriers.

TRANSLATIONS

There is much talk of translation by computer and some computer programmes exist for complete parsing of technical text, and subsequent translation into a number of languages.

However, experiments in the USA have indicated beyond all doubt that the translation output from the computer is infinitely more reliable and more acceptable when the input is prepared in controlled language. The process is also much faster and less costly. We are often told that the problem with computer translations from simplified languages is that human editing of the computer output is necessary. Our view is that human editing of the output is always necessary, whatever the input, and we see that situation existing for the foreseeable future.

Many computer users seem to have missed one important fact. That no advantage results from writing a computer programme of the type mentioned above when a simple word-searching and word-changing programme will also produce a translated text for editing purposes. A costly parsing programme is unlikely to eliminate the human editor, so it makes sense to accept the cost of editing and to use a simple word-changing programme which can be written in BASIC or other simple computer languages. The only other requirement for such a system is that the input information must of course be prepared in the ILSAM language, so that the words used are already contained in the computer files for the source language and for the target language.

As for the human translator, he or she spends some 30% of the time in searching for word equivalents, especially when technical terms, jargon and idiomatics are encountered. This waste of time can be prevented or greatly reduced if the input is in controlled language. Moreover, the input can be expected to be clear, unambiguous and easily understood.

CLARITY

Even when persons of equivalent qualifications and experience have to communicate, there can be problems. One example occurred in the corporate legal department of a large multi-national

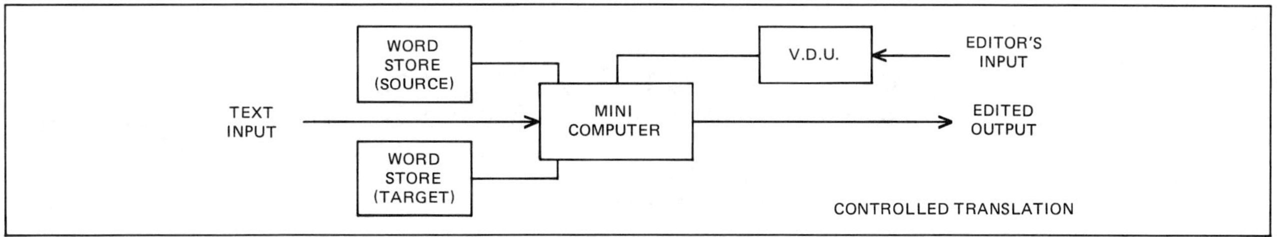

company where we had to rewrite and recast a legal document to achieve clarity. The document had been prepared by the best legal brains but it proved impossible to understand even for the few highly-qualified readers. The problem was solved not only by controlling vocabulary and terminology, but by structuring the paragraphs, sub-paragraphs and lists of procedures so that subjects were no longer mixed and confused. In such examples, the original writer is usually astonished that one experienced in simplified languages can apply the techniques so successfully to his work. The simple truth is that communication problems usually respond to similar basic rules; it is just the applications that vary.

CONCLUSION

The case for simplified language is proved by a wide variety of applications in a number of companies, in a variety of technical disciplines and in a number of national languages. The changing national and world conditions affecting human skills, training, language barriers, product support, technology transfer and educational programmes indicate a growth in demand for simplicity and clarity in technical communication. The communicator can become a COMMUNICATOR by accepting the controlled language in consideration for the target - the reader.

REFERENCES

(1) Maintenance and Servicing Manuals

"Maintenance Planning, Control and Documentation"by E.N. White, F.I.S.T.C., S.M.S.T.C.

Published by Gower Press Ltd., England.

(2) Effective Technical Writing

"Good Style" by John Kirkman, F.I.S.T.C.

Published by Pitman, England.

(3) Technical Manuals Specification

"Technical Manuals - A Guide to Users' Requirements"

Published by Institution of Plant Engineers, London.

(4) ILSAM Language System

"An International Language for Product Support"

"A Simplified Training Language"

Free publications from:

M and E White Consultants Ltd.,
P.O. Box 257, West Byfleet,
Surrey, England.

MAINTAINING PURITY OF LANGUAGE IN TECHNICAL COMMUNICATIONS

Barbara B. Whitehouse
The Foxboro Company

Barbara B. Whitehouse
Copy Editor
The Foxboro Company
Dept. 387N
Foxboro, MA 02035

(617) 543-8750

Of all the animals on earth, man is unique in using spoken and written language for communication. Our language, American English, differs from British English in spelling, vocabulary, pronunciation, and usage, but shares the same basic rules of grammar. Like any living language, American English changes constantly, the changes now accelerated by the recent influences of mass media like radio and television. As technical writers and editors, we should be discriminating in accepting changes which do not follow the rules or in using new words with which our readers may not be familiar. We should set high standards in using our language, the better to deliver our message to our audience so that it will not be misunderstood.

WHAT IS LANGUAGE?

When I looked up "language" for this paper, I discovered that not one of my dictionaries defines the word in exactly the same way as any of the others. The best definition, in my opinion, reads: "The expression and communication of emotions or ideas between human beings by means of speech and hearing, the sounds spoken or heard being systematized and confirmed by usage among a given people over a period of time."(1) My own definition is a little more simplistic: Language consists of the words we use and the rules which govern their use. The results depend upon what words are selected and how they are combined.

ANCIENT HISTORY

The earliest humanoid may have communicated with grunts like those made by a chimpanzee, but as time passed, evolving man(2) learned to make distinctive sounds for separate purposes. Over the centuries he learned to expand his language by creating words to express his emotions and to indicate past, present, and future. The ability to communicate with words is unique to man, who alone has both the brain power and the particular vocal equipment to make language possible. The crow can warn the flock of danger, but only man can describe the peril.

As civilization advanced, the need for a written language became manifest. Early written languages were pictographic, with a separate picture or symbol for each word. Later, the alphabet - a letter or combination of letters representing each different sound - made it possible to combine the letters in the same order as the sounds they represent to make a symbolic "picture" of each word. Eventually somebody decided that the words in use should be listed in alphabetic order and be given one "proper" spelling. In England this occurred in about 1600, and the first dictionaries were directed toward foreigners and those Englishmen of the lower class whose business brought them into contact with "gentry" (who, of course, knew all the words and thus had no need of the book).

Meanwhile, others studied how people arranged the words in speech and came up with books of rules to govern all future utterances and writings, and we had our first grammars. All too often, as users of the language, we forget that the language was there before the rules were written. The rules of grammar do not usually differ that much from one expert to another, and we almost become slavish in following the rules just because there is so much agreement among grammarians.

Dictionaries, on the other hand, do not advocate conformity to hard and fast rules. Modern lexicographers record what is happening in the world of words, without making judgements. The latest dictionary merely tells us which current usage is most frequently favored in speech and literature. We cannot go to our dictionary to find whether our language is pure.

Probably the only "pure" language in present use is scientific Latin, simply because it is a specialist language designed to serve a limited purpose and is not subject to the day-to-day changes which affect a living language. This is not the Latin of Ovid or Caesar, which must have been as prone to change as any other language in daily use. Called "Modern Latin," it has been used in scientific literature since 1500 A.D. If, in the year 2500, a scientist locates algae or lichens on Mars, they will be described just as Linnaeus described the flora of the 1700's. A millenium will not have changed Modern Latin.

AMERICAN ENGLISH vs BRITISH ENGLISH

Not so with English and, more specifically, with American English. In many respects British English and American English are almost separate languages. When our forefathers came from England to the new world, they brought their language with them, but it could not have been long before it began to change. The first American-edited dictionary(3) altered some of the spellings, and over the years the "u" disappeared from most of the English words ending in "our" - colour, flavour, honour, etc. - and the "i" dropped out of aluminium, and there are numerous examples of differences in spelling. American English has been more enriched by the addition of words from other languages, first from the various native Indian languages and later from the languages of other nationalities as they came to our shores. Such words as moccasin, toboggan, picayune, corral, poncho, kindergarten, hoodlum, kibitzer, bedspread, snoop, spook, jazz, and tote are all adopted from other languages, and don't forget pizza, chow mein, chowder, pretzels, coleslaw, squash, and tacos!

The more obvious differences between American English and British English show up in the names given to items which did not exist in 1620. Here both languages have usually gone separate ways. For instance, our radio is the Englishman's wireless, our elevator is his lift, and our truck is his lorry. The automobile has a hood, trunk, and defroster and runs on gasoline in America. It has a bonnet, boot, and demister and runs on petrol in England. American tourists traveling to London almost require an American-British dictionary. When you stop to think of it, it is remarkable, that the languages spoken in America and Britain are still so similar despite 361 years and 3000 miles of ocean.

THE LIVING LANGUAGE

With the exception of Modern Latin, languages are always changing. New technologies add new vocabularies almost overnight. Old words acquire new meanings. Nouns evolve into verbs. Other words are so frequently misused that they reluctantly gain acceptance. And just as new words are added, old ones are used less and less and gradually become obsolete. Later they, too, may get a new lease on life by being revived or given a new meaning.

Despite these evolutionary changes, the basic structure of both American English and British English remains virtually the same. The fundamental rules governing grammar, syntax, and punctuation are practically identical in both languages. This is because grammar, while still evolving, does so much more slowly than does vocabulary. Forty years ago, splitting an infinitive was strictly forbidden in both speech and writing. Now it is becoming accepted even in scholarly prose. But while this change has been slowly taking place, thousands of new words have been added to our vocabulary, starting off perhaps as jargon or slang, but becoming accepted in the general language. Can you imagine trying to function in the United states in the 1980's without solid-state, video, supersonic, penicillin - or boner, unisex, or rock?

Thus does a living language evolve, and the American language is definitely alive! It is a dynamic language, dynamic in the sense that it is "marked by continuous . . . productive activity or change."(4) New words are added almost daily as a result of our continuing introduction of new products, new methods, new procedures. The electronics industry, space exploration, and computer technology have added new words and new meanings for old words, and the same can be said for medicine and many other fields.

The dictionaries cannot keep up with all these new words. Webster's New Collegiate Dictionary does not yet include "microprocessor," "microcomputer," or the new meaning of "chip," yet these are now everyday terms in the computer industry and even to a large segment of the general public. Most new words of this type begin as jargon, of course, and our second speaker will have more to say on this subject.

Up until the second quarter of the 20th century, our language evolved at a leisurely pace because communications between different sections of our vast country were physically so difficult. The advent of radio and television changed all that. Even more than compulsory education, radio and television have served to coalesce our language. Regional idioms and accents tend to disappear as we are exposed daily to a language pronounced deliberately without any individuality. Most broadcasters strive to lose any accent they may have had and to become virtually accentless. To coin a word, "univoice."

This commonality is both good and bad. It dilutes the colorful idiomatic speech which made it so much fun to converse with "the natives" when traveling. In New England, where I come from, it is getting harder and harder to find young people who speak with the regional twang. Listening to radio and watching television has made them all speak alike. But this is probably a relief to foreigners visiting in the United States. In the past they must often have wondered if we spoke English at all in some areas.

There is one facet of life in America today, however, which I deplore. Advertising agencies evidently believe that for their sales pitch to be effective, it must be couched in poor English. Personally, I avoid any products whose advertising insults my intelligence, but no single individual who refuses to buy a product because of bad grammar in its advertising is going to reverse this trend. In the words of Rex Stout, "Changes made by the genius and wit of the people are inevitable and often desirable and useful. Those imposed by ignorant clowns such as advertising copywriters and broadcasters are abominable and should be condemned by all lovers of language."(5) Regrettably, however, such machinations are part of the world we live in, and we must learn to live with them. Charlton Laird(6) points out that repeated use of "like" for "as" in advertising (remember "...tastes good like a cigarette should"?) has already made this use of "as" almost obsolete.

With the very foundations of our language being undermined by the tides of inept teaching, indifference, carelessness, and deliberate misuse, we who love our language have a responsibility <u>not</u> to contribute to its degeneration.

"PURE" ENGLISH?

With this brief review of the history of language fresh in our minds, we can now attempt to define "purity" as applied to American English. Since the applicable definition of "pure" reads: "Free from what vitiates, weakens, or pollutes; containing nothing that does not belong"(7), we know that our language, by its very dynamism, can never be "pure." We have seen that our rules of grammar, while strong, are not rigid but can be bent. Consider also the following:

* Punctuation guidelines change, too - remember when we used brackets outside of parentheses when we had one parenthetical expression within another? Now double parentheses are acceptable.(8)

* Usage is almost as changeable as vocabulary. For a fascinating book on this subject, I refer you to the "Harper Dictionary of Contemporary Usage."(9)

* Dictionaries do not agree on a single correct spelling for an estimated 1500 words(10), some listing one variant first, some another. Nor can they agree on hyphenation; for example: pro·cess, proc·ess, mea·sure, meas·ure, stand·ard, stan·dard. These are such useful, everyday words that I find the disagreement in how they should be hyphenated quite incomprehensible.

EFFECT ON TECHNICAL COMMUNICATIONS

Evolutionary changes in our language inevitably affect technical communications. We cannot summarily adopt all new words as soon as they appear. Nor can we always do without <u>all</u> of them. We must be discriminating in adapting old words for new purposes or in using nouns as verbs. We have an obligation to our reader to keep our writing clear of snags that jerk his attention away from our message - that interrupt the smooth transfer of information from writer to audience.

The audience for technical documents ranges between PhD's and high-school dropouts but is usually somewhere in the middle. Certain technical papers will always be documents so abstruse as to have meaning to a very limited number of specialists. Others are "how-to" guides written to instruct users to operate uncomplicated machines. In between are many informational, educational, or instructional documents of various degrees of complexity written for this middle audience. Our writing is usually explanatory, with nothing reminiscent of spellbinding oratory about it. In fact, it quite often falls into the bored-to-death category. Only the writer who is composing for a mass audience may swerve from the straight and narrow path onto some of the more exciting byways offered by picturesque adjectives, scintillating adverbs, and provocative verbs!

The main tools available to the technical writer, besides graphics, are words and the rules that govern their use. Let us be cautious when considering the bending of any of our basic rules of grammar. While we admit that the language came before the rules, in the English language, at least, those basic rules have served us well for almost four centuries, and we should not be in too great a hurry to abandon them. They have structured our language so satisfactorily that we in the 20th century can still enjoy the works of early 17th century writers - once we have learned the outdated vocabulary.

We can assume that each of our readers has been introduced to the rules governing the parts of speech, verb tenses, syntax, punctuation marks, and spelling. This will be true whether English is his native language or whether it is an acquired tongue. Therefore, by adhering to the rules as closely as possible, our writing should contain elements of English usage familiar to all readers, The flow of the sentences, the selection of vocabulary, the punctuation - all should contribute to clearer writing, especially when the subject matter, by its very nature, may confuse the reader. As Marguerite Krupp might say, "Eschew obfuscation!"

Any document containing necessary jargon should have a glossary containing clear explanations of all new terms. This is only common courtesy to the reader who is unfamiliar with the subject, and it should be mandatory for any document which must undergo translation. More and more technical documents written in English will someday appear in other languages. The inclusion of jargon, colloquialisms, or words with several meanings may mean a final document which tells a different story from the original.

Consider the poor translator. If, as frequently happens, English is his second language, he probably learned it "by the book." Accordingly, our use of that language should be simple and straightforward. Otherwise, our message may well be garbled or misleading when it appears in a second language. While we cannot possibly anticipate every pitfall which may occur during translation, we can avoid as many as possible by writing plain English. Use a standard vocabulary and simple, grammatical sentences. Any effort to avoid repetition by using several synonyms for the same object may result in confusion and possible misunderstanding to non-English readers. We must keep it simple, both in vocabulary and in structure, if our message is to come across in other languages as clearly as in our own. This may well have a beneficial effect on the English version, too. It will probably be crisper and more understandable - even if not as "literary" as we might like.

CONCLUSION

Our basic job is to communicate. If we use obscure words which our readers must look up in the dictionary, if our sentences are so involved that they must be read more than once before they can be understood, if we fail to conform to the basic rules of grammar so that readers become more conscious of sentence structure than of the intended message, then we are not doing our job. Instead, we should follow the rules as far as the constraints of our subject matter and our medium will allow. If we must use jargon, we must do so with discrimination. We must select words with the exact meaning needed for our purpose. We must follow the rules of grammar, spelling, punctuation, and usage, because by doing so we are less likely to place obstacles between us and our readers.

I leave you with the words of Isaac Asimov: "The English language is the finest tool for communication ever invented. Since it is used indiscriminately by hundreds of millions, it is no wonder that it is badly misused so often. All the more reason for those who can to hold the standard high."(11)

Notes:

1. The Reader's Digest Great Encyclopedic Dictionary (including Funk & Wagnalls "Standard College Dictionary"), Reader's Digest Association, Pleasantville, NY, 1966.
2. With apologies to the women in my audience, I have used masculine nouns and pronouns throughout this paper for expedience only!
3. Compiled by Noah Webster and published in 1806.
4. Webster's New Collegiate Dictionary, G. & C. Merriam Company, Springfield, MA, 1976.
5. Harper Dictionary of Contemporary Usage, Ed. by William and Mary Morris, Harper and Row, New York, 1975, p. xxiv.
6. "Language and the Dictionary," Webster's New World Dictionary of the American Language Simon and Schuster, New York, NY, 1980, Page xviii.
7. Webster's Collegiate.
8. Style Manual, U. S. Government Printing Office, Washington, DC, 1973, p. 142, 8.105.
9. See Note 5.
10. Harper, p. 640.
11. Idem., p.xv.

G
Graphics and Production

Stem Managers

Robert S. Burns
Senior Proposal Writer
Air Products and Chemicals, Inc.
Allentown, Pennsylvania 18105

Senior Member, STC
Member, IEEE

Joseph E. Smith
Manager of Proposals
Salvucci Engineering
Pittsburgh, Pennsylvania 15222

Member, STC

GRAPHICS AND PRODUCTION

Visual information, whether it appears on the printed page or on a projection screen, is as vital to the communication process as the text. We should, therefore, not only have an active interest in graphics and the production process, but we should also understand it, because only together do art and print create a unified message. To accomplish these goals, the Graphics and Production Stem offers topics that will cover essential areas of the technical communication field.

For example, the Graphics and Production keynote address will analyze a successful television program and apply its philosophy to technical communication. Other sessions will explore the relationships which must exist between artist and text. We will also examine visuals and the spoken word, as well as methods for creating visual interest in what might otherwise be mundane.

State-of-the-art technologies involving video in industry and education and computer-generated graphic art will be introduced, discussed, and demonstrated. In contrast, poster session theory and practice will be the topic of a panel which will analyze this age-old graphic art form.

Production, an area which is neither strictly text nor art, will be explored in sessions for use in creating an end product, the proposal; justifying money, materials, and equipment; and measuring employee effectiveness. Vendor, client, and agency relationships will be discussed by a panel from the viewpoint of the vendor, a design studio. Another panel will explore the business of audiovisual contracting from both the client's and the contractor's perspectives.

A tutorial will focus on three-dimensional isometric drawings and their application to technical manuals. In addition, art-text coordination will be examined in a special international session.

Another tutorial will explore the nonfiction film pertaining to its values as a propaganda tool. In addition, the camera will be the subject of a special slide-talk on comparative photography.

We will "Meet the Winners" when we view winning international audiovisual entries consisting of filmstrips and side-tapes. The Graphics and Production Stem will also conduct a two-evening Film Festival.

We believe the Graphics and Production Stem will get to the point when it explores the crucial business of creating the unified message both in process and as an end product.

HISTORY, SCIENCE, AND PEOPLE: SHAPING TODAY'S TECHNICAL MESSAGE WITH YESTERDAY'S ART

Mary Fran Buehler and Andrea Stein
Jet Propulsion Laboratory
California Institute of Technology

Mary Fran Buehler
Jet Propulsion Laboratory
Supervisor, JPL Publications Group
4800 Oak Grove Drive
Pasadena, California 91109

(213) 354-2295

Chairman, STC Goals Committee
Associate Fellow, STC

Andrea Stein
Jet Propulsion Laboratory
Technical Writer
JPL Publications Group
4800 Oak Grove Drive
Pasadena, California 91109

(213) 354-2619

Member, STC

People-related artwork can be used to humanize the technical message, especially for the general public. At the Jet Propulsion Laboratory, a deliberate effort is made to incorporate artwork that shows people in relation to technology, whenever this is appropriate. Historical artwork can be used to show both the relation of people to technology and the place of that technology in history. Two examples are given: Project Galileo, a mission to explore the planet Jupiter and its large moons in the 1980s; and the International Halley Watch, a program to coordinate international observations of Comet Halley, due to return in 1986.

INTRODUCTION

Science and technology can seem forbiddingly impersonal, especially to nonscientists. Hardware and mathematical equations and computer printouts can seem to have an existence of their own. The truth is, of course, that there would be no science and no technology without people. Somewhere in the dim past, people struggled with theory and concepts, or tinkered with Rube Goldberg-like contraptions, that made possible the scientific advances and the gleaming mechanical marvels we enjoy today.

One of our responsibilities in technical communication is to help people understand the importance of science and technology in their lives so that they can participate—through understanding—in today's scientific advances. This paper describes some efforts to promote this understanding by using people-related art (in this case, historical artwork) to provide human interest in scientific publications that are designed for lay audiences. The authors believe that this interest, in turn, allows the reader to identify with and understand the scientific message more easily, since the experience of humanness is universal to humans, but the knowledge of science is not.

Obviously, this approach will be effective only if it is used selectively in appropriate situations. Artwork, like any other device, will be perceived as artificial unless it is an integral part of the message. The two examples described in this paper lent themselves very well to the use of historical art.

The publications discussed here were produced by the Jet Propulsion Laboratory (JPL), which is operated by the California Institute of Technology for the National Aeronautics and Space Administration. The Laboratory is primarily responsible for the exploration of the solar system, along with associated research and development, as well as Earth-bound activities in energy, transportation, medical technology, and related fields.

One of the chief factors in producing successful publications at JPL is the teamwork that combines the talents of managers, engineers, scientists, graphic designers, artists, writers, and editors into a seamless, unified message that transcends the individual contributions of its creators. Therefore, when the authors say "we" in this paper, all of these people and their special talents are implied and acknowledged.

PROJECT GALILEO

Project Galileo, which is expected to explore the planet Jupiter and its four largest moons during the 1980s, is a follow-on to JPL's successful Voyager encounters with Jupiter in 1979. Project Galileo is a dual mission consisting of a Probe and an Orbiter. The Probe will plunge into Jupiter's atmosphere, making scientific measurements until it is silenced by the planet's increasing pressure. The Orbiter will circle Jupiter, observing and photographing the planet as well as the major moons as the spacecraft makes repeated flybys of the moons. It will relay both its data and the Probe's data back to Earth. The Probe itself is the responsibility of the Ames Research Center of the National Aeronautics and Space Administration.

Project Galileo is named for the man whose observations it will carry forward—Galileo Galilei, who, in 1610, first observed the four large moons of Jupiter (Io, Europa, Ganymede, and Callisto) that will be the Orbiter's satellite targets. Therefore, when we at JPL planned a brochure on Project Galileo, it seemed natural to carry forward Galileo's image in our publications.

Here we found a rich source of early scientific communication—both in text and in graphics: reproductions of Galileo's notebook, for example (Figure 1). Galileo recorded his observations of the little moons he saw around Jupiter—in different positions on different nights— and gradually concluded that the moons were orbiting the planet. This led to the larger conclusion that the planets were orbiting the Sun, thus supporting the Copernican model of the solar system.

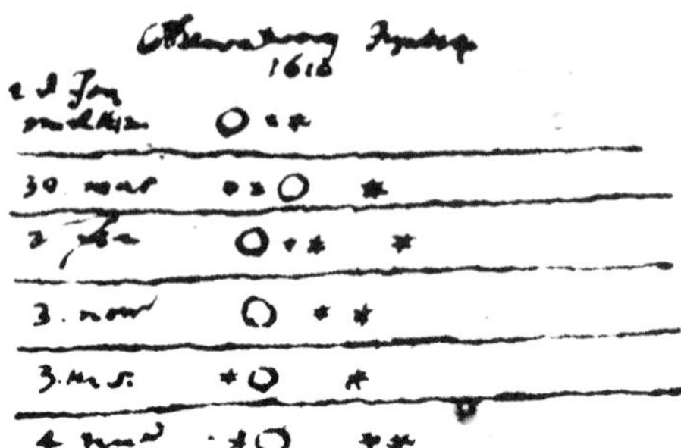

Figure 1. Notebook Entries of Galileo in January 1610, Showing His Drawings of Jupiter and Several of Its Satellites.

Galileo's drawings of the moons were also included in his scientific report of the discovery, in the publication *The Starry Messenger,* published in 1610. On the inside cover of our first Galileo brochure, we featured a portion of this text, with its early illustrations of Io, Europa, Ganymede, and Callisto, the "Galilean" satellites (Figure 2). We also pointed out that the text of *The Starry Messenger* was written in the common language, Italian, rather than in learned Latin: Galileo was making his scientific discoveries available to the people, not just to scholars.

Because this brochure combined the efforts of a graphic designer and an artist, we were able to incorporate portraits of Galileo and drawings of his inventions and theories into illustrations (Figure 3). This feeling of Galileo's time and its art was achieved throughout the brochure; even the present-day illustrations of the spacecraft were done in an ancient calligraphic style.

Further graphic couplings between Galileo's era and our own came about as the result of brainstorming sessions. We searched for analogs between the technology of Galileo's day and the late 20th century: the sailing ship and the space ship, the cannon trajectory and the spacecraft trajectory, the pendulum clock and the computer readout, the weight dropping from the tower of Pisa, in the famous story, and the Probe dropping into the atmosphere of Jupiter (Figure 4). All of these elements, and more, were tucked away in the artwork of the brochure.

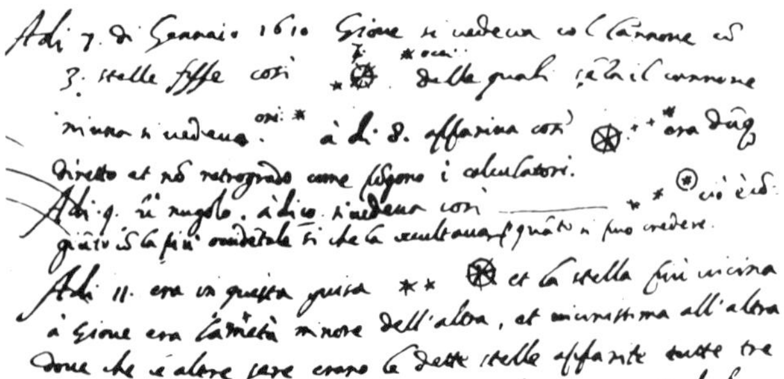

Figure 2. A Portion of Galileo's Notes, Reporting His Observations of Jupiter and Its Satellites.

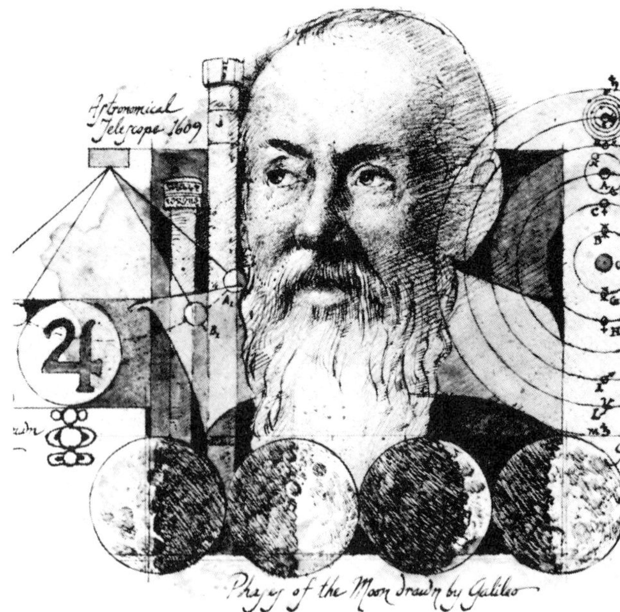

Figure 3. A Brochure Illustration Combining Galileo's Portrait, His Telescope, His Drawings of Saturn and the Phases of the Moon, and His Version of the Copernican System.

Figure 4. Brochure Illustrations Show Analogies Between Galileo's Time and Our Own: Among Them, the Story of the Weights Dropped From the Tower of Pisa (Left) and the Galileo Spacecraft Probe Dropping Into the Atmosphere of Jupiter.

The brochure was received enthusiastically by project personnel, the scientific community, and the general public, so far as we have been able to learn. It has won several awards in art and publications competitions.

THE INTERNATIONAL HALLEY WATCH

The International Halley Watch, a program coordinated by the Jet Propulsion Laboratory, will collect, analyze, and report the findings of professional and amateur astronomers around the world who will be observing the return of Comet Halley in 1986. Data from spacecraft missions to study the comet will also be included.

Obviously, from the publications viewpoint, there are similarities and differences between Project Galileo and the

Halley Watch. The two are similar, in that they have rich historical backgrounds and a rich supply of artwork. They are different, in that our Galileo heritage, on the one hand, came from one period and one scientist, with a finely focused theme: Project Galileo's return to Jupiter to solve some of the mysteries first tackled by Galileo, with only an early telescope, in 1610. The Halley Watch, on the other hand, will record a comet that has been visiting the neighborhood of Earth every 76 years throughout history, inspiring curiosity, superstition, forebodings of doom, and depictions of the comet and its viewers by artists working in many centuries, many countries, and many mediums.

Thus there are renderings of Comet Halley, and the people who watched it, in many forms—woodcuts, broadsides, paintings, cartoons, and—perhaps most famous—the Bayeux tapestry, which records Comet Halley's appearance in 1066, when it was popularly supposed to be a warning of doom to King Harold of England, who, true to the forecast, fell in the Battle of Hastings in the same year.

The involvement of people in the Comet Halley story thus becomes not only a panoply of watchers, but a dedicated succession of artists—graphic technical communicators in the Middle Ages, for example, who were faithfully recording the greatest astronomical occurrence of their time.

The origin of the Bayeux tapestry, which is believed to date from the 11th century, is uncertain. Some legends say that it was embroidered (it is an embroidery, not a true tapestry) by Queen Mathilda, wife of William the Conqueror (winner of the Battle of Hastings), and her ladies in waiting. In any case, the tapestry, almost 231 feet long, graphically depicts the Battle of Hastings and has been called "the longest cartoon strip in the world." It is now housed in a museum in the city of Bayeux, in Normandy.

One panel of the tapestry records the appearance of Comet Halley, and the astonishment and wonder of the people who viewed this ill omen. A cropped section of this tapestry panel, redone as a line drawing, appears in one of the JPL leaflets on the Halley Watch (Figure 5). The Latin inscription tells us that "they wonder at the star." In the same leaflet, a cropped section of a cartoon announcing Comet Halley's return in 1905 was used for the cover art (Figure 6).

A woodblock illustration of Comet Halley in 684 A.D. was published in the *Nuremberg Chronicles* in 1493 (Figure 7). This graceful, star-like comet illustration was used in a JPL report of the Science Working Group for the International Halley Watch.

Figure 5. A Line Drawing Adapted from the 11th Century Bayeux Tapestry, Showing Comet Halley in 1066.

Figure 6. A Cartoon Depicting Comet Viewers in 1905.

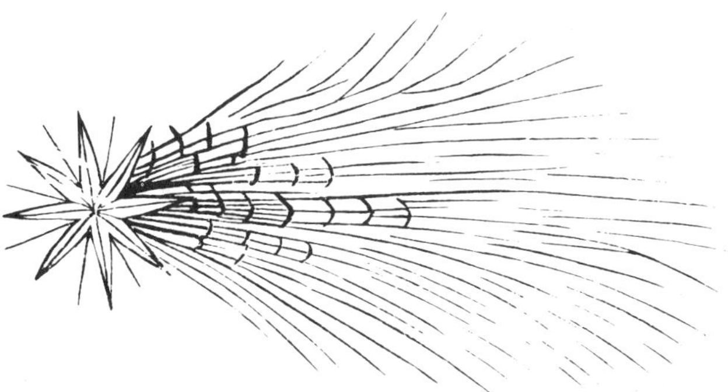

Figure 7. A Woodblock Illustration Printed in 1493.

The cover of the same report featured the panel of the Bayeux tapestry that shows the appearance of Comet Halley (Figure 8). But this embroidery does not date from the 11th century—it was created only recently by Mrs. Delphine Delsemme, an astronomer and the wife of Professor Armand

Figure 8. Part of a Modern, Reembroidered Version of a Panel in the Bayeux Tapestry, Created To Commemorate Comet Halley's Visit in 1986.

H. Delsemme of Belgium (now of the University of Toledo, in Ohio). Professor Delsemme is a distinguished member of the Halley Watch Science Working Group. In a true labor of love, Mrs. Delsemme visited the tapestry at Bayeux, matched yarn colors and stitch patterns with the original, and spent 400 hours embroidering the new Comet Halley panel.

These are only a few examples of the rich historical supply of Comet Halley artwork. Since we will be producing publications on Comet Halley for most of the 1980s (including results of the observations from the comet's visit in 1986), we are grateful that the supply is so large, and we hope to find and use more examples. Fortunately for us, comets have captured people's imaginations throughout history—just as they do today.

USING HISTORICAL ARTWORK

In our experience—and we are still learning—the appropriate use of historical artwork gives the technical message interesting new dimensions, which we hope will attract, stimulate, and involve our readers. Some pointers are given below on why, when, and how to use this kind of artwork.

Why?

Advantages of this artwork include the following:

1. It is available, already created. (Permission must be obtained to use it, of course, as with any other type of material, if there is a copyright in effect.)
2. It puts science in the stream of human experience—i.e., history— and, we believe, makes science more humanly familiar and comfortable.
3. It does not "date," since it is dated already. (For example, the clothing and hairstyles from the Middle Ages help to date the period.)
4. It can be informative and reinforce the scientific message. (For example, Galileo's notebooks show not only what he saw, but how he recorded his observations.)
5. It can be used to authenticate the treatment of a historical period. (For example, we have used the typeface on the title page of Galileo's *The Starry Messenger* as a model in our choice of typefaces for publications and displays for Project Galileo.)

When?

It may not always be possible to use this kind of artwork in an integral way. Artwork—like anything else—should not be used if it is inappropriate.

Some suggestions:

1. This kind of art seems to be more appropriate in publications designed for the general public, or a specialized public, than for strictly scientific audiences, although even for scientists it would seem to add depth and interest if used wisely (and probably sparingly).
2. Whoever has the responsibility for choosing the name of a project should consider the connotative possibilities of the name. The early, temporary name for Project Galileo, for example, was "Jupiter-Orbiter-Probe," which certainly identified the mission's purpose, but lacked the beautifully evocative qualities of "Project Galileo." But—like the artwork—the name must seem right and should reinforce the message. It should not be added just for effect.

How?

Again, some suggestions:

1. To blend the text and art into a unified message, team members should work closely, brainstorm, strike sparks off each other. One person's stray idea may set off an inspiration in another. The distinct talents of writers, editors, and artists can cross-pollinate to great effect.
2. Some research may be necessary. We ask our clients (project personnel, scientists); they may have fine collections on their special areas. Our comet scientists at JPL have been an extraordinary source of information and artwork.

   Another good starting place: a reliable encyclopedia, such as the *Encyclopedia Britannica*, which gives good leads. Public city libraries, especially the children's departments, can be very helpful; children's books may have good, simple, interpretative graphic treatments.
3. The team should always think of the art and text as parts that work together: "How does this picture help to tell our story? Can it be sharpened by cropping? Where will it do the most good in telling the story?"
4. It helps to think of a piece of scientific hardware as more than just hardware. Think: "This hardware would not exist if people had not created it. How does it relate to people? Who discovered the principles that made it possible? What did the early models look like? Who tinkered them into shape?"

SUMMING UP

In our view, science does not exist—*cannot exist*—in isolation from people. If nothing else, people today pay taxes that support much of our scientific work. We should help them to see how science affects them in ways that make their lives richer.

And—for us as technical communicators—there is an extra bonus. We find ourselves in touch with other technical communicators who were working hundreds of years ago—like the artists who illustrated the comets of the Middle Ages—and we see that our own specialty, like the scientific accomplishments it records, has roots deep in history and in the efforts of people in all ages to make sense of the world in which they live, and to tell other people about it.

ACKNOWLEDGMENT

The research described in this paper was carried out by the Jet Propulsion Laboratory, California Institute of Technology, under NASA Contract No. NAS7-100.

EMERGING ROLE OF DOCUMENT COORDINATORS

Rochelle R. Calinson

Environmental Science and Engineering, Inc.

Rochelle R. Calinson
Marketing Coordination Manager
Environmental Science and Engineering, Inc.
904 372 3318

In 1976, ESE employed 2 document coordinators of a total staff of 150. In 1981, coordinators total 24 of 400 employers. Document coordinators fulfilled a company need which resulted in rapid expansion of the group. Examination of ESE's history provides: (1) definition of document coordinator, (2) need for coordinators, (3) operational and potential roles for coordinators, and (4) characteristics of good document coordinators.

INTRODUCTION

Document coordination is an expanding field of technical communications that encompasses editing and writing as well as many other skills needed for total document production. Many A/E firms and other technical and nontechnical organizations are using Document Coordinators to solve the production problems that can be associated with marketing and project work.

Environmental Science and Engineering, Inc. (ESE) is a firm that has successfully instituted a document coordination program. Headquartered in Gainesville, Florida, ESE provides environmental services to public and private clients.

In 1976 ESE instituted a document coordination program that has grown rapidly in the past 5 years. The program is successful in fulfilling a basic company need for efficient, cost-effective document production, and has expanded to include many additional tasks.

In this paper, experience with ESE's program is used to: (1) define the role of a document coordinator; (2) identify the need for coordinators; (3) outline operational and potential job roles for coordinators; and (4) indicate characteristics of good coordinators.

SIMPLIFIED DEFINITION OF A DOCUMENT COORDINATOR

A Document Coordinator (D.C.) assumes primary responsibility for document preparation and ensures that schedules are met. At ESE, a coordinator is assigned for each document. Coordinators act as the main liaison between technical personnel (engineers and scientists) and document production personnel (word processing specialists, graphic artists, and technical editors) from planning stages to final printing. This role can expand to encompass responsibility for communications with clients and regulatory agencies. The D.C. becomes the focal point for instructions, revisions, decision-making, and quality control related to the document. Information about format, content, and revisions is communicated through the D.C. and the D.C. resolves any problems.

Document size can range from a 3-page letter to a 3,000-page Environmental Impact Statement (EIS). Types of documents include: short reports, technical papers, brochures, proposals, expert testimony, speeches, in-house manuals, and detailed reports. Time schedules for production can vary from 10 minutes to 3 years. In addition to document production, D.C.'s coordinate presentations to clients, sponsorship of conventions, company meetings with client, meetings with regulatory personnel, and many other marketing or project-oriented functions.

Coordination is a simplified term to cover the complexity of what the job means at ESE. In addition to normal tasks, the D.C. organizes emergency production; obtains input to meet schedules; deals with problems; and does whatever is necessary to complete a job. Miscellaneous tasks can include final xeroxing, checking final copies, and making transportation arrangements for document delivery. A D.C. also may write major sections and make presentations to clients. The job can be stressful and thankless, or infinitely rewarding. Weld Coxe has called a coordinator "an aggressive mother," and, no matter how unflattering it may be, the term is fairly accurate.

NEED FOR COORDINATORS

ESE began using coordinators to: (1) serve as key assistants to project and proposal managers, (2) relieve engineers and scientists from organizational and production tasks that they were ill-suited or overqualified to perform, and (3) improve efficiency and cost-effectiveness of producing documents. Informal company policy dictates that responsibility for client satisfaction should be shared between technical and support staff personnel.

All of these goals have been fulfilled. High-level personnel can concentrate on managerial and technical responsibilities and delegate organizational and production tasks. In a short time period, coordinators can be trained to expedite document production, recognize potential problems, and ask more experienced coordinators for guidance.

Well-trained D.C.'s can accomplish tasks more effectively and efficiently than someone who performs them only occasionally. One indication of the success of ESE's program is that project and proposal managers use the services of document coordinators regularly and frequently request that a D.C. be assigned to special tasks such as convention planning and regulatory liaison. Technical personnel who have joined ESE from other firms cite the document production system as a definite improvement over other firms' methods for producing documents.

JOB ROLES FOR DOCUMENT COORDINATORS

In addition to the major roles previously mentioned, D.C.'s can be assigned a wide variety of tasks in an organization.

In project management and project-related tasks that the company is under contract to perform, coordinators can:

*Assume responsibility for document production budgets;
*Monitor and maintain schedules;
*Meet with clients to plan format and make decisions on document specifications;
*Attend meetings with regulatory personnel to discuss requirements and revisions;
*Hold technical writing seminars for company personnel; and
*Maintain project and cost files;
*Assist with records search and bibliography preparation;
*In-house preparation of manuals (QA/QC, SOPs, Field Ops, etc.);
*Arrange and monitor printing of documents; and
*Attend project team planning meetings.

Coordinators who have marketing-oriented roles can:

*Write and update resumes;
*Write and update corporate experience listings;
*Monitor schedules;
*Establish and maintain mailing lists;
*Read listings of available jobs and send for "requests for proposals" (RFP's);
*Route RFP's to appropriate personnel who decide whether or not to "Go" on a proposal;
*Maintain marketing files including client files and proposals;
*Attend strategy-planning sessions for proposal preparation;
*Help plan presentations, speeches, and corporate participation in conventions; and
*Write and update brochures, slide shows, and other marketing materials.

For both project-oriented and marketing-oriented areas, there are many other tasks which D.C.'s can perform. At ESE, the only limiting factor of the job sometimes is the physical capacity of the D.C. Coordinators continually are challenged with new assignments. The job is seldom boring.

CHARACTERISTICS OF GOOD DOCUMENT COORDINATORS

Coordinators continually work under pressure and must meet short schedules while always maintaining good working relationships with the total company. Selecting the right individuals for D.C. positions is essential to ensure program success.

Besides document coordination experience and a solid English/grammar background, certain other characteristics are important.

Sense of humor and flexibility--The ability to "roll with the punches" is important because everything that can go wrong will. The copying machine will break during a critical deadline, the manager will rewrite his prose five times, and people will discover errors in the document after it's sent to the client. A coordinator must be able to handle these situations and keep the rest of the project team working smoothly.

Organizational skills--Sometimes the only person on the project who is organized is the D.C. The coordinator must be able to devise systems to accommodate each particular document, maintain orderly files, remember innumerable details, delegate tasks, and simultaneously complete many different assignments.

Ability to deal effectively with people--A coordinator constantly urges personnel to complete work and keeps people informed about problems and errors. There are infinite opportunities to irritate other personnel. Sometimes bad situations are unavoidable, but a tactful coordinator minimizes these problems and gets documents completed efficiently and on time. D.C.'s also must work closely with clients and potential clients.

Personal commitment--Since coordinators have the primary responsibility for the production of a document, they become essential to the production process and must stay until the document is completed. Deadlines and schedules may conflict with personal plans; coordinators often spend long hours and 70-hour weeks preparing documents on schedule.

THREE-DIMENSIONAL INTERACTIVE GRAPHICS SYSTEMS: REVOLUTIONIZING TECHNICAL ILLUSTRATIONS PRODUCTION IN THE 1980's

John A. Cuozzo Jr.
Computervision Corporation

John A. Cuozzo Jr.
Computervision Corporation
Product Documentation Support Specialist
201 Burlington Road
Bedford, MA 01730
(617) 275-1800

Award of Excellence (1977-78)
Award of Merit (1979-80)
Boston Chapter STC Publications Competition

Three-dimensional (3D) interactive graphics systems can revolutionize the production of technical illustrations and publications. Originally developed as computer-aided design (CAD) and computer-aided manufacturing (CAM) tools, the 3D CAD system is fast gaining recognition for the powerful assistance it can provide technical illustrating and publications operations. The 3D CAD system can address many of the needs and problems inherent to the process of generating and maintaining line art, and of merging this art with the text of a publication.

This paper examines the advantages of using a 3D modeling system. It outlines the potential engineering and manufacturing applications that can be supported by a 3D CAD/CAM system. These applications represent the source of input matter for the technical illustrator. The focus then shifts to specific ways the system helps the technical illustrator convert that input into a high-quality illustration. A typical workflow for CAD generation of technical illustration leads to a suggested approach to the merging of CAD graphics with automated text files to produce the final document.

INTRODUCTION

Just as text processing systems have advanced the text handling methods of publications departments in the 1970's, three-dimensional (3D) interactive graphics systems will be revolutionizing the handling of technical illustrations in the 1980's.

Interactive graphics systems, or computer-aided design/computer-aided manufacturing (CAD/CAM) systems, were originally developed to aid engineers, designers and detailers in creating standard engineering drawings of mechanical objects and technical artwork for applications, such as printed circuit board design. Three-dimensional CAD systems provide exceptional power and flexibility for creating and revising line drawings and art. (NOTE: The term CAD will often be used in lieu of the term CAD/CAM).

Today, the benefits of using interactive graphics to process technical illustrations is becoming more widely recognized and accepted. This acceptance is largely due to (1) the exceptional productivity increases that computer-aided design offers technical illustrators and other users, and (2) the integrating force that CAD/CAM exerts within the corporate organization.

Technical illustrators can think of a CAD system as a replacement for the traditional drawing board, pen and paper. In lieu of these tools, using the Computervision system as an example, the CAD user works with a pen-like electronic digitizing stylus on an electronic tablet that records the location and movement of the stylus. The results appear on a high-resolution television-like viewing screen, the Cathode Ray Tube (CRT). A 3D CAD system quite literally adds another dimension to this picture. The illustrator can, in effect, use the power of the 3D system to "walk around the side of the drawing board", inserting geometry in true three-dimensional space. The resulting artwork is kept in an electronic file cabinet, called a database.

THE PROBLEM-SOLVER

What are some of the technical publications problems that 3D CAD systems can help to solve? At the global level is the creation of line art describing a 3D object. Along the way, one might find that the illustrator is the last link in the corporate chain, destined to receive the requirements for an illustration needed "yesterday". Or, the illustrator may be given a preliminary specification, which no doubt changes only when the original illustration is near completion.

Perhaps the presentation of the final illustration has to meet rigid standards, or, it may be subject to the desires of a customer. And what could be more natural than the need to modify an existing illustration, or a copy, to reflect changes to the object it describes. Then there are those illustrations that almost, but not quite, match what is needed.

Of course, the art of illustrating often becomes the labor of what seems like the endless drawing of identical parts within one or more pictures. The list goes on and on. Even the physical storage, retrieval and security of final artwork is a major consideration. Finally, the completed art might be merged with the text in a publication. Historically, this has meant long hours of cutting and pasting. By passing completed artwork to phototypesetters in a form they can accept for automatic merging with text, a CAD system can eliminate this time-consuming, often monotonous task.

The point is that a 3D CAD system is the one tool available today that has the power, accuracy and flexibility to overcome many of the major stumbling blocks inherent to the traditional technical illustrating environment. By solving these problems, it frees the illustrator to be more creative *and* more productive.

This paper will review the advantages of a three-dimensional modeling system over a two-dimensional drawing system; demonstrate how a 3D CAD/CAM system can help throughout the user organization; and discuss specifically how the 3D CAD system can help the technical illustrator. Stepping through typical phases in the computer-aided technical illustrations workflow, the paper will also highlight additional capabilities of the CAD system and lead to a suggested approach to merging the resulting graphics with automated text files to produce a completed document.

ADVANTAGES OF 3D CAD SYSTEMS

Why discuss the differences between two-dimensional (2D) and 3D CAD systems? After all, the output will be displayed or printed on a 2D medium, be it the CRT screen, film or paper. Although some interactive graphics systems process data only in a two-dimensional XY-plane, they do not have the power or offer the flexibility of a three-dimensional system. What are some of the advantages of a 3D system?

A 3D system allows the illustrator to achieve a true representation of a 3D object or part by creating an actual line art model of the part in three dimensions. By manipulating that model in 3D space, the illustrator can access and present a view of the part from any angle or perspective. To change the viewing angle or perspective, the illustrator does *not* have to redraw the object, as in a 2D system. Since the true geometry has been captured, the illustrator simply tells the system what rotation and/or perspective to define, and the system generates the new picture. Taking this one step further, Computervision users can interactively control the dynamic movement and scaling of the displayed illustration. Using camera-like dynamic zooming, they can change the scale with which the model is viewed. The model's location on the screen or the particular vantage point from which it is viewed can be interactively modified by dynamic horizontal and vertical scrolling and by dynamic three-dimensional rotation. The illustrator receives immediate visual feedback of the results of each adjustment of the view, ensuring the most desirable presentation can be produced in minimal time.

A 3D system substantially assists the illustrator in the creation of sectional cutaways and assembly explosions, either along an axis or for use with the sectional views. The 3D system also simplifies the task of drawing odd-shaped parts.

Because they can process 2D drawings, and then go beyond into 3D processing, three-dimensional CAD systems offer greater flexibility than their two-dimensional counterparts. The 3D systems permit creation of a database for a 3D model of an object—a requirement for many applications. They also enable the illustrator to efficiently work with true 3D pictorial illustrations. While providing the capabilities to do both 2D and 3D illustrating, the 3D database serves as a primary source of input to the illustrator, and as a starting point for the artwork.

3D CAD/CAM: HELPING THROUGHOUT THE ORGANIZATION

The 3D CAD/CAM system provides a single database that truly integrates engineering and manufacturing groups and their supporting corporate organizations. As an example, the CAD/CAM system can assist people with many functions.

Starting with an engineering design concept,

- the designer can create, analyze and modify product designs
- the drafter can produce corresponding layout and assembly drawings
- the detailer can dimension and annotate the part drawings
- the technical illustrator can create artwork for publications, sales proposals and the like.

Optionally,

- the designer can generate finite element models, calculate the mass properties of the design and perform other types of engineering analysis
- the manufacturing engineer can create numerical control (NC) tapes for NC machining operations, create flat pattern layouts for models of sheet metal parts, and produce manufacturing assembly documentation and bills of material
- the accounting organization can receive extracted parts lists (bills of material) and system utilization (project) accounting information

Applications Areas Supported

Using CAD/CAM systems, productivity can be increased across a wide range of design applications areas. In addition, the technical publications organization can more effectively and efficiently fulfill the documentation needs of those applications areas it supports, such as: Mechanical Design and Drafting; Finite Element Modeling; Manufacturing Engineering & Numerical Control (NC); Printed Circuits & Electrical Schematics; Wiring Diagrams; Integrated Circuits; Mapping; Piping/Plant Design; and Facilities Planning.

HOW 3D CAD/CAM SYSTEMS HELP THE TECHNICAL ILLUSTRATOR

Three-dimensional CAD/CAM systems address a number of the technical illustrator's concerns and needs. These are outlined below, and will be demonstrated by the capabilities of the Computervision system. They include:

- creating and updating illustrations
- protecting original artwork and providing secure storage
- ease of locating and retrieving existing illustrations
- observing documentation standards and special customer requirements
- interacting with other functional areas of the organization
- reducing tedium caused by repetitive tasks
- increasing productivity

Creating and Updating Illustrations

The guidelines for creating technical publications are often dictated by customer requirements, such as the "New-Look Handbooks" for the U.S. Military, as well as by corporate publications standards. The New-Look manuals make extensive use of illustrations to show equipment from several different angles. A three-dimensional database is invaluable for creating the multiple perspectives required by these military specifications.

On a CAD system, the illustrator can create a complete illustration, either from scratch or by selecting and modifying an existing engineering design model, detail drawing or a previously-created illustration. Accurate descriptive artwork can easily be presented in isometric or perspective views. To eliminate duplicated effort, any portion of the line art can be readily copied elsewhere on the drawing. Only half of symmetrical geometry need be drawn. The remaining half can simply be "mirrored" to complete the geometry. Standard or commonly-used parts can be kept in a library for on-demand insertion into any drawing.

Updating illustrations is greatly simplified by the fast graphics editing capabilities and the consistently high output quality of a CAD system. Virtually any editing function can be performed on the graphics: change existing line art; combine parts; modify perspectives; delete from a drawing; assemble/explode an assembly of parts; change line patterns (e.g., solid to dashed); edit annotations and other text; and define additional views.

The illustrator can also control the display of the illustration to facilitate easier viewing and editing. One powerful display control tool is the use of layers. Layers can be thought of as transparent overlays which can be selectively used to separate text or annotations, graphic elements, formats and dimensions. Viewing and editing can be controlled by selecting specific layers for display on the user's terminal screen.

Protecting Original Artwork and Providing Secure Storage

The artwork editing process is performed on a temporary copy of the original artwork. The illustrator can choose to replace the original with the edited version, or to create a new editied version and keep the original intact. Access to the artwork that is stored in the system's database can be restricted by the use of passwords to ensure that only authorized individuals can use or modify the artwork. Physical security is provided by storing copies of the original drawings on magnetic tape at a separate storage location.

Ease of Locating and Retrieving Existing Illustrations

Each illustration is stored in the system under a unique name so the illustration can be easily located when needed. Different levels of file management sophistication for the storage and retrieval of the graphics data are available. The most fundamental method is to store the data as a record or part with a unique name in a hierarchical (tree) structure.

An illustrator can also select an existing drawing as the starting point for a new illustration and modify it, rather than create a totally new illustration.

Observing Documentation Standards and Special Customer Requirements

Standard parts, symbols and illustration formats can be prestored in the system to facilitate standardized documentation. The illustrator simply requests the required standard data, whenever needed. This feature significantly shortens artwork creation time.

Because the system operates on a true three-dimensional database, the illustrator can quickly and easily alter any view or perspective to meet special customer requirements without having to redraw the original.

If desired, the artwork can be plotted in one or multiple colors on a plotting device to produce a camera-ready copy. The system can also produce output in a format suitable as input to a computer output microfilm (COM) device.

Interacting with Other Functional Areas of the Organization

With all information stored on a single database, it can be easily transferred from one department to another. The single database gives a clear foundation for interdepartmental discussions of the work to be done. It allows one department to capitalize on the efforts of another, reducing or eliminating interdepartmental redundancies.

Reducing Tedium Caused by Repetitive Tasks

When multiple parts to be illustrated are similar in shape but vary in dimension, the illustrator can define the geometry of a "family of parts" that share the geometric description, and use parameters to describe the characteristics that can vary. The geometry is defined only once. Then, each time a part is to be drawn from this family, the illustrator simply specifies the desired parameter values in response to questions asked by the system. The CAD system then generates the specified part.

Increasing Productivity

Overall productivity gains in the first 6 to 12 months of system use should realistically range from 2:1 to 4:1. For individual tasks, the ratio can even range as high as 20:1.

TYPICAL WORKFLOW FOR CAD GENERATION OF TECHNICAL ILLUSTRATIONS

Normally, an engineer uses the CAD system to design and analyze a conceptual product model. A detailer uses the model as the basis for creating detail drawings to guide the product's manufacturing cycle. The illustrator can choose to use either the model or the detailed drawings as a starting point for the illustration. Typical phases in the technical illustration CAD workflow include:

1. Recording data on the system (i.e., graphics creation)
2. Retrieving existing graphics
3. Editing graphics (and text)
4. Storing the results
5. Merging the illustrations with text

Recording Data on the System

There are many ways to create publications artwork on a CAD system. To directly create the line art, the illustrator can use a digitizing stylus or specify explicit geometric coordinate information. As a second approach, a sketch of the desired illustration can be traced using a device called a digitizer. The digitizer translates the sketch into graphic data coordinate information and the system generates the corresponding line art. A third method is to retrieve from storage and edit an existing engineering model, detail drawing or technical illustration.

Retrieving the Graphics

When retrieving the graphics, the CAD system simply searches the file catalog for the specified name and displays the graphics on the CRT.

Editing Graphics

Once retrieved, the graphics can quickly be regenerated, with or without changes. The power, speed and ease with which a CAD system edits, and the consistently high output quality provide some of the greatest justifications to use a CAD system to create and maintain technical illustrations.

Storing the Results

To create a new copy of the artwork in the database, the illustrator can simply specify a new, unique name. To replace existing artwork, the illustrator specifies the original filing name.

Merging Text and Graphics

Historically, text and illustrations have been manually prepared and merged into a composed final document. Current technology makes practical, in fact advisable for most medium or large publications operations, the use of specialized computers to aid in text and graphics generation. In the 1980's, the trend will be to merge automated text and CAD graphics with minimal human intervention. This represents a convergence of three technologies: interactive graphics processing (CAD/CAM), interactive text processing and photocomposition. Overlapping capabilities of vendors in each of these areas presents interesting evolutionary possibilities for publications processing in this decade.

For the immediate future, a sound approach to further automating technical publications would be to:

1. Create and edit the graphics on a three-dimensional CAD system
2. Create and edit the text on an interactive text processor or a photocomposition system
3. Merge the resulting text and graphics on the photocomposition system

CONCLUSION

Three-dimensional interactive graphics systems offer a powerful tool to publications users throughout industry and government. For the technical illustrator, computer-aided design offers many benefits:

- ability to use existing line drawings as a starting point for illustrations
- faster creation and editing of drawings to keep current with engineering changes
- an infinite selection of drawing views
- consistent high-quality, accurate output
- minimum effort to produce repetitive art
- fast, easy retrieval and storage of drawings
- potential for automated merger of text and graphics

Other benefits may include:

- easier recruiting of qualified technical illustrators
- an enhanced creative environment
- smoother interdepartmental exchanges of information

Overall, the combination of interactive graphics, text processing, and phototypesetting has the potential to dramatically increase the productivity of the technical publications user and decrease the time and expense required to generate a completed document. The results: better documentation that is available sooner at lower cost.

Perhaps the National Science Foundation Center for Productivity has summed it up best: "CAD/CAM has more potential to radically increase productivity than any development since electricity." And, 3D CAD/CAM systems *are* increasing productivity, combining proven results with continuing innovation at the leading edge of this technology. With the right tool to do the job, people throughout the organization can become more productive, and the organization can be a more profitable, exciting and challenging environment in which to work.

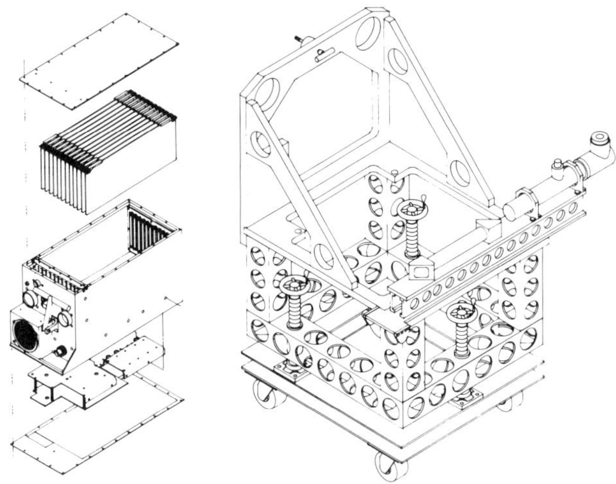

Illustrations Created on a Computervision System
Courtesy of Texas Instruments

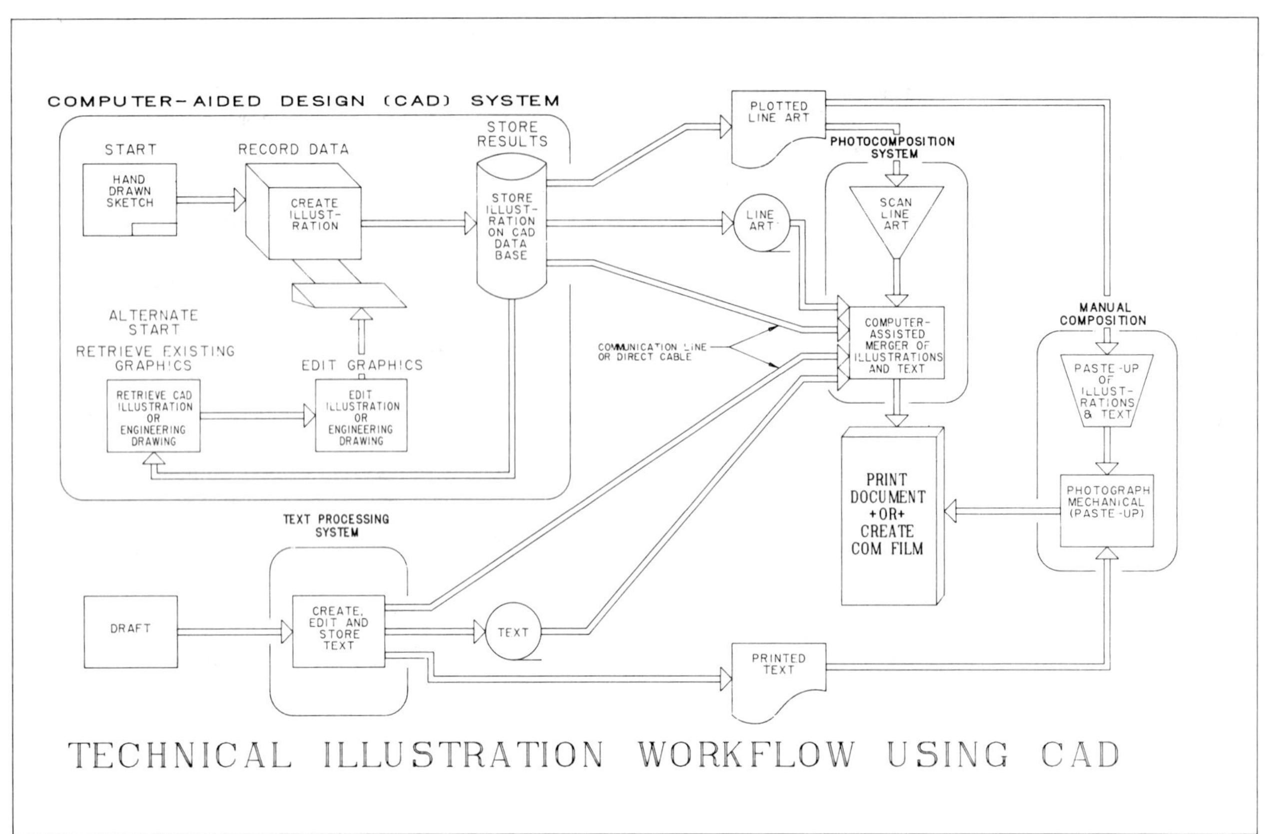

COMPUTER-AIDED DESIGN (CAD) SYSTEM
START
HAND DRAWN SKETCH
RECORD DATA
CREATE ILLUST-RATION
STORE RESULTS
STORE ILLUST-RATION ON CAD DATA BASE
ALTERNATE START
RETRIEVE EXISTING GRAPHICS
RETRIEVE CAD ILLUSTRATION OR ENGINEERING DRAWING
EDIT GRAPHICS
EDIT ILLUSTRATION OR ENGINEERING DRAWING
PLOTTED LINE ART
PHOTOCOMPOSITION SYSTEM
LINE ART
SCAN LINE ART
COMMUNICATION LINE OR DIRECT CABLE
COMPUTER-ASSISTED MERGER OF ILLUSTRATIONS AND TEXT
PRINT DOCUMENT +OR+ CREATE COM FILM
MANUAL COMPOSITION
PASTE-UP OF ILLUST-RATIONS & TEXT
PHOTOGRAPH MECHANICAL (PASTE-UP)
TEXT PROCESSING SYSTEM
DRAFT
CREATE, EDIT AND STORE TEXT
TEXT
PRINTED TEXT
TECHNICAL ILLUSTRATION WORKFLOW USING CAD

AVOID DUPLICATION OF EFFORT IN PROPOSAL WRITING

Celeste F. Davis
Environmental Science and Engineering, Inc.

Celeste F. Davis
Environmental Science
& Engineering, Inc.
P. O. Box ESE
Gainesville, FL 32602

904-372-3318

technical writer/
document coordinator

Much duplication of effort in writing proposals can be avoided by maintaining basic "shells." Shells contain previously written material common to most proposals such as: capabilities and facilities of different areas of the firm, corporate experience, project management descriptions, quality assurance procedures, and resumes. With the use of shell material, only the introduction summary, and scope of work must be written specifically for a given proposal. Since only a few sections are original, more time and effort is spent on getting to the point.

INTRODUCTION

Much duplication of effort in proposal and report writing can be avoided by maintaining basic "shells" of previously written material common to most proposals or reports. With the use of shell material, only the introduction, summary, and scope of work must be written specifically for a proposal. Since only a few sections of the proposal are original, more time and effort can be spent on getting to the point.

CREATION OF SHELLS

Shells are created by collecting general information from proposals or reports that have already been written. Shells are kept in two forms: Lanier text-editing disks and 3-ring binders. Shell material is duplicated from proposal disks to shell disks. The copy is typed out from the disk and placed in a 3-ring binder. Technical personnel also give input for any missing areas of expertise. Shell information is circulated to appropriate staff members at least annually for updating.

The 3-ring binder form enables proposal writers to locate pertinent information without using word processing equipment. Codes are put on each page giving the name of the disk and the name of the page the information can be found. For example, MKEXP11.2/SRCTST.1 is the heading code for information on disk MKEXP11.2. MKEXP11 is the name of the subject (Air Experience). The "2" designates the second disk in that subject. SRCTST is the name of the type of air experience (Source Testing). The "1" designates the page number of that particular section. The 3-ring binders have tabbed dividers with the disk and page names written on the dividers so that if information is missing from the shell it can still be retyped from the disk.

When writing proposals, selected shell information is duplicated from the shell disk onto the proposal disk, allowing the information to be adapted for a particular proposal by deleting, inserting, and/or rearranging information without changing the information on the shell disk. The following flow chart illustrates the process of creating shells and using this information in writing proposals.

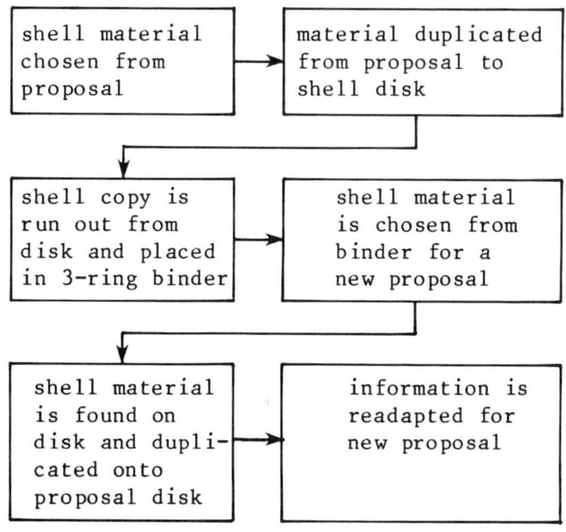

TYPES OF SHELLS

The Marketing Document Coordination Department of Environmental Science and Engineering, Inc. (ESE) maintains the following shells for use in producing marketing materials:

1. Resumes,
2. Experience Summaries,
3. Technical Summaries,
4. General Marketing,
5. 255 Resumes, and
6. 255 Experience Summaries.

These can be grouped into three main categories: resumes, experience, and general marketing shells.

Resume Shells

The Resume Shell contains both a 1-page and a longer version resume of each member of ESE's staff. When appropriate, slanted versions of resumes emphasizing particular disciplines also are maintained.

Short paragraph resumes are filed in another shell for use in the project management and personnel sections of proposals. A short summary of the person's educational background is given as well as specific project management and related technical experience. These can be adapted easily to specific proposals by emphasizing relevant project experience.

The 255 Resume Shell contains resumes in the format required for the Government Standard Form 255. The Government Standard Form 255 lists a firm's qualifications and experience on a standardized form divided into 10 sections. Resumes in this shell are typed only on the left half of the page. This enables any two resumes to be typed on a page by using word processing equipment. Both resumes are temporarily recalled and memorized on the same page and then typed on the Government Standard Form 255.

Experience Shells

The Experience Shell contains short project descriptions of important corporate experience, filed by discipline or project area. The project name, reference name and address, project location, and a paragraph summary of the project are given.

Technical Summaries give 1-page descriptions of all projects. The information listed includes client name, address, telephone number, date of project, contract amount, reference name, contract number, company project number, and name of the project manager. Summaries are filed by client name and have alphabetical tabs to separate the material. As new projects are begun, a technical summary is written by the project manager, put in the shell, and expanded when needed by the technical writer. This shell provides a very useful reference tool for the entire company. Information can quickly be obtained on all projects that have been done for a particular client. These summaries are useful when filling out prequalification statements because of the available contract information.

The 255 Experience Shell contains all of the project experience summaries used in both Sections 8 and 9 of the Government Standard Form 255. Proposal writers look through this shell to obtain relevant experience that already has been typed in correct format. The Lanier operator then duplicates the desired experience listings from the 255 Experience disks to the same page and has it typed on the correct form. This shell is divided into different corporate discipline and project areas.

General Marketing Shell

This shell contains all other information that is commonly needed when writing marketing materials. Some of the corporate materials included are: long and short summaries of facilities and capabilities, a description of the quality assurance program, personnel summaries of all ESE technical staff (a table containing name, number of years experience, degree, and areas of specialization), description of ESE's project management structure, confidentiality statements, business costing forms, general financial and organization information, and a listing of staff professional registrations and certifications. The facilities and capabilities sections can be adapted to emphasize particular disciplines by duplicating a long write-up of a particular area into the short summary.

SHELL ORGANIZATION

Shell materials are placed on separate word processing disks for organized insertion and deletion of information. Large shells are divided by appropriate categories. Resume shell disks are arranged alphabetically and contain no more than three last name initial letters per disk, allowing names to be located easily and additional resumes to be added in an organized manner. Experience shell disks are divided by specific discipline and project areas.

In order to keep shells intact, it is important that information is duplicated from the disks before it is changed for particular proposals. Only the technical writers have access to 3-ring binder shells. They give photocopies to proposal writers and replace originals in the shell. Access to shell disks also is limited to technical writers and Lanier operators.

If word processors are not available, the 3-ring binder can be used to "cut and paste" information together. All information in the shell should be kept in the same format to reduce reformatting time.

ADVANTAGES AND DISADVANTAGES OF SHELLS

Current information is more accessible since it is concentrated in one place. Since all shell material is contained on only a few disks, it is easier to duplicate all desired information onto proposal disks to create new proposals. Companies with regional offices that have compatible text-editing equipment can transmit shell materials to their regional offices. Shells are helpful for new proposal writers because they contain much current information in a compact form.

Some of the disadvantages of shells include the added time spent on general overhead work but shells reduce writing time for other marketing materials. It is difficult to keep shells intact and up to date but much duplication of effort is avoided when the material is available in one place.

SUMMARY

By maintaining shells, much time and effort can be saved in writing proposals. Shells also provide the most up-to-date information on corporate facilities and capabilities, and corporate and personnel experience.

EVERYTHING YOU WANTED TO KNOW ABOUT GENERATED GRAPHICS, AND LESS

R.M. Field
Westinghouse Electric Corporation

Ron Field
Manager, Communication Programs
P.O. Box 355
Pittsburgh, PA 15230
(412) 373-4727

STC First Vice President 1980-81
STC President 1981-1982
STC Associate Fellow
Who's Who in the East
Who's Who in Finance and Industry
Technical Delegate, Pa. Governor's Conference on Libraries and Information Services

The trail from the "gee, that's nice" stage to the first slide produced on a generated graphics system is long and replete with obstacles. All the lectures we've ever heard on planning really pay off in this area; indeed there cannot be too much planning. Each stage of the program, from initial investigations to operation, must be carefully planned. By far the easiest of all the stages is the initial investigation; probably the most difficult is the procedure development and conversion from the old system to the new. The paper discusses the various stages of the process in terms of the author's actual experience, and provides some helpful hints.

For these purposes, generated graphics is defined as follows: the preparation of artwork - for business, marketing, or technical slides, overheads, movies, or video - generated by an artist using an interactive computer system. This definition excludes those systems described at previous ITCCs, such as CAD/CAM for engineering drawings and various terminal/mainframe combinations for displaying financial and other data. We are concerned only with those systems capable of generating technical artwork as we know it in our daily communicator's work.

INTRODUCTION

That point in time when your capabilities are almost saturated by your customers' demands has arrived. No more slides can be produced with your available resources. How do you solve the problem? Automation is one answer. But with so many equipment options available, how do you make a choice, and after the choice, what then? As you can guess, a lot of planning is involved in the process. Indeed, the path to the first slide produced on a generated graphics system is replete with obstacles, obstacles that can be eliminated only by careful planning. The various stages of the planning process (Table 1) and some helpful hints are provided below.

TABLE 1
GENERATED GRAPHIC PLANNING STAGES

1. Initial investigation
2. Selection criteria development
3. Selection/justification
4. Site preparation
5. Personnel training
6. Work procedure development
7. Initial transition period
8. Trans-transition period (transition evaluation period)
9. Permanent operations period

INITIAL INVESTIGATIONS

Once you've decided on automation to solve your work loading problem, what next? Explore the subject at a professionl society meeting? Discuss your needs with colleagues? Go to an equipment exposition? Call a vendor? All of these approaches are valid and should be followed at one time or another during your investigation. I suggest, however, that the first step is to write down what you want to accomplish with this new toy. This should be a brief statement of your objective both for the system and for you (Figure 1). When you have prepared your initial statement, you are ready to organize and to proceed with your investigation.

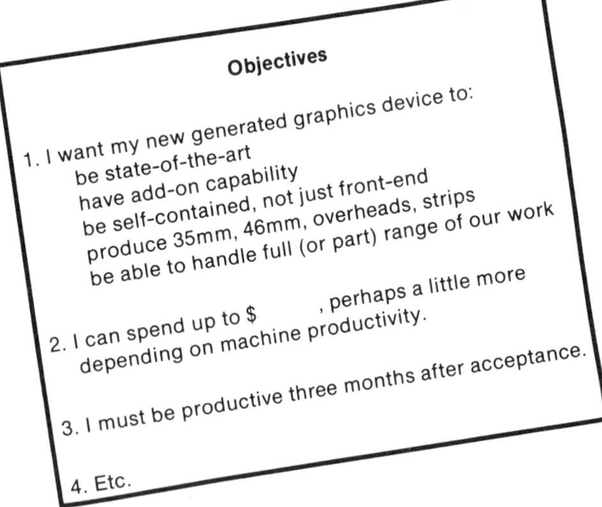

Figure 1. Initial Statement of Objective

The investigation should be organized into segments to reduce backtracking and duplication of effort. The segments should then be arranged to permit the maximum amount of data gathering with the minimum number of trips. I suggest as the first step talking with people in your locale (perhaps at an STC meeting), and then with some vendors. A visit to an exposition should come next, followed by a look at some operating installations in organizations similar to your own. This modus operandi permits a progressive increase in your knowledge and should enable you to make more meaningful evaluations as you progress through the various planning stages.

The important goals of this stage are (1) to gain a knowledge of generated-graphic equipment systems, (2) to see as wide a spectrum as practical of the available systems, (3) to permit reevalution of your objective, and (4) to build a base for developing your selection criteria.

SELECTION CRITERIA DEVELOPMENT

One of difficult parts of the entire process is the development of criteria for equipment selection. The development process requires you to possess (1) an intimate knowledge of every facet of your operation, (2) an understanding of the strengths and weaknesses of your people, (3) accurate department production records including an accurate breakdown of the mix of work, (4) an accurate estimate of future work volume and mix, and (5) a reasonably firm expectation of what is to be produced on the system in terms of volume and type of work.

Again, I suggest you write down your anticipated goals, but in a form suggested in Figure 2. This is a base from which you will refine your goals as your investigations continue. It is also the position from which you will make the necessary compromises in your final selection process. We at Westinghouse NES spent almost four years in our investigation and selection process and we refined our criteria several times to meet our changing requirements.

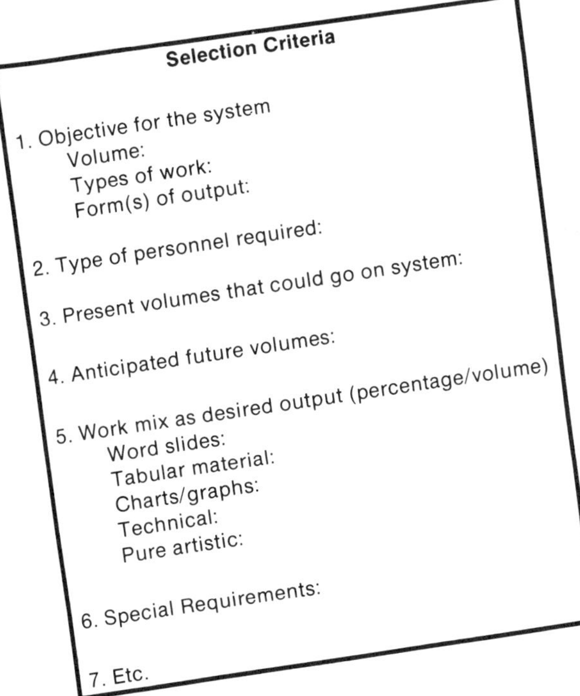
Selection Criteria

1. Objective for the system
   Volume:
   Types of work:
   Form(s) of output:

2. Type of personnel required:

3. Present volumes that could go on system:

4. Anticipated future volumes:

5. Work mix as desired output (percentage/volume)
   Word slides:
   Tabular material:
   Charts/graphs:
   Technical:
   Pure artistic:

6. Special Requirements:

7. Etc.

Figure 2. Selection of Criteria Worksheet

SELECTION AND JUSTIFICATION

After your have investigated and tried out as many systems as practical, you should quickly eliminate those that are patently not capable of meeting your requirements. In fact, you should be looking to eliminate equipment from future consideration during your entire investigatory period, if for no other reason than to leave more time for detailed evaluation of the final contenders.

Until now, system cost really hasn't been a consideration. Once you have identified one, two, or three potential systems, the purchase/lease/rental cost and operating costs of each become important. Because each company has its own system of financial justification, only one general statement can be made: be absolutely certain you know all the costs of acquisition, installation, training, and operation. Ask for specific breakdowns in writing, so there will be no recriminations later. The hidden costs in each of the systems must be evaluated as part of the overall ROI evaluation to assure yourself and executive management that a thorough evaluation was performed.

Once your technical and financial evalutions are complete, you have to make whatever compromises necessary to obtain the most system for the amount of funds made available by your organization. After the choice is made, and the purchase order is issued, the fun can begin.

SITE PREPARATION

Now that the evaluation process is complete, the justifications have been approved and the purchase order has been issued, you probably think the hard work is done. Not so. Site selection and preparation, personnel selectioin and training, and procedure development are among the most difficult obstacles to clear.

Site selection and preparation are unusually important for a number of reasons (Table 2). Power must be adequate, voltage constant and alarmed. There should be a sprinkler system for equipment protection. It is desirable to have all cabling under the floors or in cable trays away from the aisles and other walkways. Many systems operate better in "computer environment," where air temperature and humidity are closely controlled. Adjustable lighting is desirable to eliminate glare on the CRT face(s) and to reduce operator eye fatigue. Many other factors should be considered for your local conditions. At all times keep in mind the truism: Excited electrons are not always happy electrons. Only happy electrons are productive electrons.

TABLE 2
SITE PREPARATION CHECKLIST

1. Adequate power and constant voltage
2. Dedicated power lines
3. Special air-conditioning/humidity control
4. Raised flooring for cable runs
5. Sprinkler and fire alarm system
6. Soundproofing for office-adjacent walls
7. Adjustable lighting fixtures
8. Immediately adjacent work and storage facilities
9. Adjacent storage facilities
10. Low traffic area
11. Visitor control

Site preparation, above all, is not the place to save money!

PERSONNEL TRAINING

Personnel selection and training is vital to the success of any automation program — more so, even, than the equipment itself in most cases. For generated graphics systems, a good operator should be (1) an experienced artist, (2) curious about new approaches to art preparation, (3) moderately familiar with a typewriter keyboard, (4) of innovative personality, and (5) interested in working with the system. The last qualification is the most important.

Once personnel have been selected, some difficult choices remain in selecting the best method of training. Should the training take place at the vendor's facility or at yours? Should the training be provided by the vendor (at a price) or by a specialty training firm? You owe it to the success of the program to thoroughly investigate all of your options before making a decision. All other factors being equal, I suggest that you go with vendor training at his site. Being away from their everyday environment will help your artist operators to get maximum benefit from the training process.

Remember to train a few more people than needed to operate the system, to assure full-time coverage.

WORK PROCEDURE DEVELOPMENT

Developing your new departmental working procedures will probably be the most difficult phase of the entire program. This is true if for no other reason than that this may be the first time you and your department will be working in an automated environment. No matter how much advice you receive from experienced people and from your system vendor, nobody but you and your people can develop workable procedures.

The best suggestions I can offer are the following:

1. Carefully review your present procedures.
2. Carefully analyze the operations required to produce slides on the automated equipment.
3. Determine the similarities between your present system and the new one.
4. Determine whether your existing format is satisfactory for system output, or a new format is needed.
5. Draft new work procedures to achieve the desired output.
6. Review existing log and work request forms.
7. Modify log and work request forms as necessary to accommodate the new system.
8. Discuss the system with all department personnel, even those who will not be using it.
9. Run a parallel test operation on two or three small, actual jobs that are being done using your existing procedures.
10. Compare the results of the test run.
11. Analyze with all the operators what went right and what went wrong.
12. Make any obvious adjustments.
13. Repeat steps 9 through 12 until you think you have it right (you won't!)

This may or may not work for you, but it does give you some feel for the complexity of developing the new procedures. As discussed in the following paragraphs, refinement will be a way of life.

INITIAL TRANSITION PERIOD

The initial transition period is essentially that period of time during which the newly drafted work procedures are implemented, evaluated, and modified. This process is carried out in parallel with your existing system to assure no lost work. Two important goals absolutely must be achieved during this period:

1. No regular work can be delayed.
2. The shift from old to new must be orderly and cautious in scope.

If too much work is attempted on the new system too quickly, major problems and lost schedules are likely to result. Start with one small project at a time and build up the load as your operators and you build confidence.

It is a good idea to do these selected projects with both the new equipment and your existing methods, in parallel. Obviously, this will cost more. But the added cost is a good investment if it prevents any of the many problems that are likely to occur during this period.

Good communication between the operators and management is vital during this period to ensure that all involved are aware of how effective the interim procedures are. All problems — no matter how minor — should be written down and discussed thoroughly to determine what changes may be required.

Remember that caution is a virtue at this point in the overall program.

TRANS-TRANSITION PERIOD

This period overlaps the last quarter of the initial transition period and the beginning weeks of the permanent operations period. It is basically a time of evaluation and fine tuning. The equipment, personnel, and work procedures should be carefully evaluated to make certain that every facet of the program is working as intended. The more care you take with this evaluation, the greater is the likelihood of success.

Equipment should be checked out in accordance with its specifications and operating procedures at least twice, the second time about 10 days after the first. You should be present with the vendor's representative at each test.

Each of the operators should be given the opportunity to work on the equipment for a set amount of time each day. This will build proficiency and provide an opportunity to observe who may need some additional training. Again, the system is only as good as the people are proficient.

Evaluation, reevaluation, and more evaluation of the initial transition procedures in actual practice will pay large dividends in a smooth running, efficient system when you finally go operational.

PERMANENT OPERATIONS PERIOD

All the hard work that you and your staff put into setting up your generated graphics program have brought it to fruition — a complicated system of equipment, people and procedures. It is now operational for permanent use.

But do not expect this to be an ever-running system requiring little or no attention. Because it was important enough for you to entrust with your organization's work, the system deserves periodic evaluation, first to see that it is indeed achieving those goals you defined long ago, and second, to look for any possible refinements that could make it work even better.

SUMMARY

Preparing to place into operation a generated graphics system can be straight-forward and rewarding if careful planning is done before and during each of the stages. Set goals should be established at the earliest possible point to ensure that any progress achieved is in the right direction. Above all, you must have patience and be cautious in progressing along the way to the first generated-graphic visual.

MAKE YOUR SLIDES WORK FOR YOU

Walter Golman
Center for Naval Analyses
An Affiliate of the University of Rochester

Walter Golman
Center for Naval Analyses
Chief Editor
2000 N. Beauregard Street
Alexandria, VA 22311
(703) 998-3886
Conduct seminars and prepare articles about technical writing and uses of visual aids.

It's easy to ruin a talk by preparing poor slides or using good ones badly. It's just as easy – and a lot more rewarding – to reinforce a talk through the right use of proper slides. How to devise slides that work, how to use them to best effect, when to use slides and when not to – all this is discussed and illustrated. The point of view is that of the user, not the artist.

Who needs slides? After all, every organization represented here turns out large quantities of printed reports – by the half-ton, if not the ton. Aren't these printed reports enough?

The answer, of course, is that the people we want to reach – especially those at higher levels – get much of their information in the form of talks. Or, to use our slightly pompous expressions, "oral presentations" or "briefings." When you give a talk, it is sometimes useful to show slides. So, we're talking about slides.

Our ways of dealing with slides are about the same, whether they are elegant 35-mm. slides in color, overhead-projector slides in black and white, or the cruder type that you and I prepare in a hurry with a grease pencil on a piece of acetate.

Let's begin with a simple question: *When* should you use a slide? Or, conversely, when shouldn't you? The answer is that every slide must carry its own weight. If it supports your talk, use it. If not, don't. Dump it, burn it, give it to the cat to play with. Do whatever you like, but don't inflict it on your audience.

IF A SLIDE ADVANCES YOUR TALK,
USE IT.

IF NOT, DON'T.

Glance at a blank wall. Do you find it offensive? Not at all. No matter what some people may tell you, there does not have to be something on the screen all the time.

In fact . . .

A BLANK SCREEN
IS NOT A SIGN
OF
MORAL TURPITUDE

SLIDES

- TEXT
- TABLE
- ART
  - GRAPH
  - MAP OR CHART
  - DIAGRAM
  - PICTURE: PHOTO, DRAWING

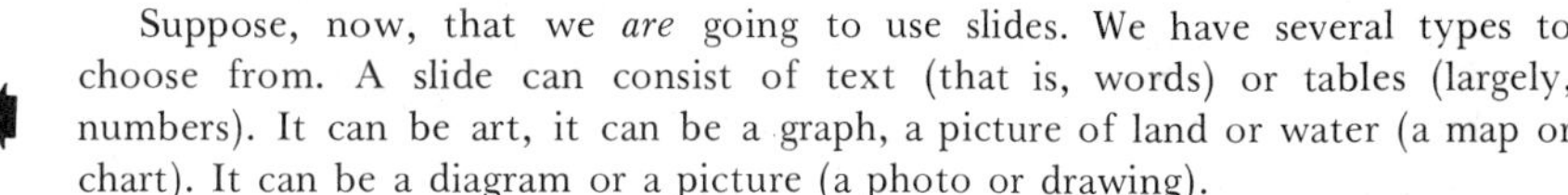

Suppose, now, that we *are* going to use slides. We have several types to choose from. A slide can consist of text (that is, words) or tables (largely, numbers). It can be art, it can be a graph, a picture of land or water (a map or chart). It can be a diagram or a picture (a photo or drawing).

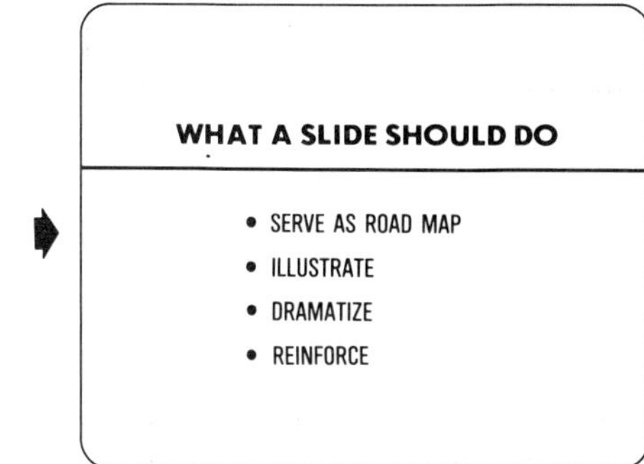

What can a slide *do* for you? Four main things: First, it can serve as a kind of *road map* – telling your audience where you are, where your talk is going, and how you intend to get there. You want to keep your audience with you all the time. A slide can *illustrate*. A picture can be a lot clearer than words alone. A slide can *dramatize*. It can do this in several ways. A well chosen cartoon may meet your need. Or, on occasion, some attention-getting language – not necessarily of the Pow!-Bam!-Zowie! variety – will lend special emphasis. Suppose, for example, you have analyzed a policy – or tactic, or gadget – to find out whether it can work. You conclude it can not. On a slide, you can state the question simply. Then, project on the screen a single, large word: "NO." Of course, you will follow with an explanation, but your main finding will have been stated dramatically. It will be remembered long after everyone has forgotten the means by which you reached it. The last – and perhaps most important – use of all is to *reinforce* your message. What the audience is *hearing*, it is *seeing* at the same time. When you appeal to two senses instead of one, you have a better chance of making your point.

MATCH SLIDE TO SCRIPT

- CONTENT
- TIMING

For a slide to do its job, it must be coordinated with the text. It is obvious enough that you do not show a bathtub while talking about anchors or encyclopedias. But there's a second point that's not always so obvious: You want to make sure the slide goes *on* when that subject comes up in your talk. When you're done with the material, *take off the slide*. As we saw earlier, a blank screen is not an ugly screen.

[An aside here to whoever is flipping slides: When it's time for the screen to be blank, insert a blank slide or block the light with a piece of cardboard. Do not turn the projector on and off. There's no reason to distract the audience with the sound of the switch and – what's worse – the fan.]

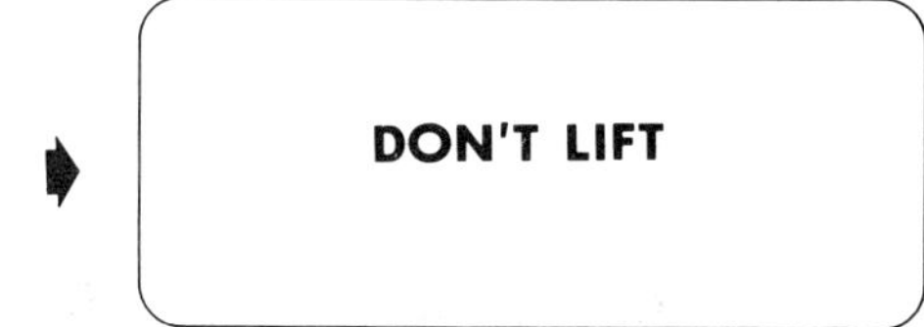

Another point: Someone who needs slides in a hurry – particularly someone who is not experienced – may yank something out of a book and have it photographed directly. There's a word for the result: "terrible." Don't lift. Slides must be produced as slides. The reason is the difference between the *reader* and the *listener*. The reader has time. He can take a complicated graph, turn the book around, check all the curves, no matter how many there are. For better understanding, he can turn to another part of the book. He can set his own pace. The listener, by contrast, can not. This means that you have got to keep the audience informed all the way. You have got to keep your message clear enough for them to understand immediately. Let's look at a couple of bad examples.

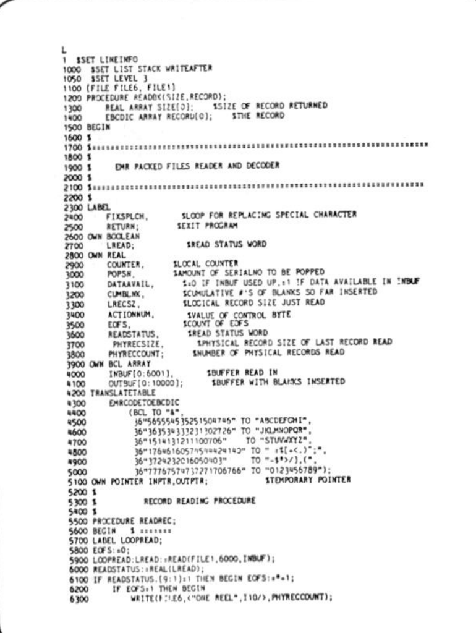

Here's a thing of beauty, a printout that was valuable beyond measure — *in print.* On the screen — a large nothing.

DON'T CROWD

(NO MORE THAN A DOZEN LINES PER SLIDE)

That horrible example illustrates another point about slides: Don't clutter them. Don't try to jam everything into one slide. A reasonably good rule of thumb is to limit yourself to about a dozen lines of type on a slide. Spread your message among several slides if you have to. But don't crowd any one of them.

Something else that may interfere with getting your message across is language that is not as clear as it should be. The point is to stick to straight, human language. Poor wording is bad enough in a printed publication. It is intolerable when it is imposed on a helpless audience, which can not take the time to puzzle out the meaning and lacks the nerve to leave the room and head for the bar.

TALK STRAIGHT ENGLISH

IN THAT THE AMATA* HAS INITIATED AND IMPLEMENTED THE APPROPRIATION OF CIVILIAN RESIDENTIAL FACILITIES BY MEANS OF THE ASSIGNMENT OF EXTENSIVE THOUGH FINITE FORMATIONS OF ARMED AND UNIFORMED SERVICE PERSONNEL, ASSIGNED AND CASUAL, COMBAT-ORIENTED AND SERVICE-ORIENTATED, ACTIVE-DUTYWISE AND LIMITED-SERVICE WISE, FOR QUARTERS AND OTHER SERVICES THEREUNTO APPERTAINING, WITHIN BUT NOT RESTRICTED TO THE PRESENT THEATER OF ACTIVITY;

Aforementioned monarchical and tyrannical authority.

FOR QUARTERING LARGE BODIES OF ARMED TROOPS AMONG US;

To illustrate the point, I've taken a single thought out of a well known document and — to put it baldly — mangled the language. Though this is admittedly a fake, you may detect a faint resemblance to slides that you and I have sometimes been exposed to. Thomas Jefferson wrote the original words more elegantly — in the Declaration of Independence. Here they are, on the bottom line.

But it is not only in noble documents of great historic importance that clarity is important. Here, for instance, is a sign on the observation platform atop the Louisiana State Office Building. I'm sure you'll agree that it is sharp and clear and makes its point beyond all doubt.

ANYONE WHO THROWS AN OBJECT FROM THE TOWER TO THE GROUND WILL BE ARRESTED, PUT IN JAIL, AND FINED.

BE TERSE

The sign demonstrates one more point of importance — the need for brevity in slides. In a text, you use as much language as you need to make your point. In a slide, confine yourself to a sort of shorthand. Be concise, even to the point of terseness. The fuller explanation is given by the speaker.

Something else you don't want to do is irritate your audience. You are probably careful not to use the pointer in some way that distracts. You don't swing it around or keep tapping it on the lectern. When you point to something on the screen, you rest it on the screen so that it won't skitter around. Generally, speakers don't misuse pointers.

Yet, many speakers, including good ones, do something that can be just as irritating. They irritate with a word. You and I know that what we are looking at is a *slide*. It looks like a slide, it projects like a slide – it *is* a slide. Still, all of us have heard speakers who keep referring to slides as slides: "On this slide we prove," "the right column of the slide," "the top of the slide," "the bottom of the slide," "the center of the slide," "see the pretty slide," "see the slide run." You can avoid this irritation – and very simply. Since everyone knows it's a slide, just disregard the fact. Say: "in the right column we have this," "the graph shows that," "take a look at the value at the bottom." You can even use a four-letter word: *here*. "Here we show . . ." whatever it is we are showing. We've removed the irritation.

We've said enough about bad slides. Let's look at a few proper ones. We'll take a real event and make up a few examples of slides that the situation would require.

Put yourself back about four centuries. The year is 1588, early spring. The place is England. England is rife with rumors of invasion and war. There are good reasons for these rumors. Philip II, the King of Spain, has been boasting all over Europe that he intends to invade England. He has been assembling supplies, massing troops at Channel ports, and – most important of all – putting together a mighty fleet. His propagandists have named it the Invincible Armada.

So, Charles Howard, the Lord Admiral of England, turns to his favorite think tank – here we are, in this room – and says, "I want you to look into this. Come back with some findings, and tell me what you think I should do."

This is a meeting of the advisory committee. We are giving our report. The chairman of the committee, is of course, Sir Francis Drake.

We begin with a road-map kind of slide. It tells where we are and where we are going. There are two bad ways to present such a slide, and we'll consider them. Then, I'll suggest a way that may be better.

OUTLINE OF THIS PRESENTATION

- BACKGROUND
- ASSUMPTIONS
- SPANISH FORCES
- ENGLISH FORCES
- MOE, OUTLINE OF ANALYSIS
- FINDINGS
- RECOMMENDATIONS

One bad way is to say, in effect: "Here is a list. While you clods are picking your way through it, I'm going out for a beer." Now, that is not only discourteous, it also does you harm, because you are relinquishing control of your talk. And you want to keep as much control of that talk as you can.

Another bad way goes to the other extreme. We've all seen it. The speaker clears his throat, turns his back, and reads us every syllable: "O-u-t-l-i-n-e o-f t-h-i-s p-r-e-s-e-n-t-a-t-i-o-n, b-a-c-k-g-r-o-u-n-d, a-s-s-u-m-p-t-i-o-n-s . . ." By the time he's read us two lines, we want to strangle him. Doesn't he know we can read silently a lot faster than he can read out loud? Besides, who cares to be told that being smart enough to be in this room does not imply enough intelligence to read?

What *do* you do? You want to walk your listeners through the slide, and yet you don't want to insult them. Here's a reasonably casual, informal way:

"Sir Francis, gentlemen . . ." [To forestall complaints, let me point out that though Elizabeth was on the throne, it is certain that in those benighted times there would be no women in the room. Sorry about that.] "Gentlemen, this presentation will begin with some background material and assumptions. Then, I'll list the two forces – the enemy's and ours. I'll touch on our measures of effectiveness and say something about our analysis. Then I'll tell you what we found and what we recommend."

We've done what we wanted to. We've walked the listeners through the slide without insulting their intelligence.

QUESTIONS

1 DO THE SPANIARDS PLAN TO INVADE ENGLAND?
2 CAN WE BEAT THEM ON LAND?
- WHAT ARE THE CONSEQUENCES OF FIGHTING THEM ON LAND?

3 CAN WE BEAT THEM AT SEA?
- WHAT ARE THE CONSEQUENCES OF FIGHTING THEM AT SEA?

4 WHAT TACTICS ARE BEST FOR OUR FLEET?

"Sir Francis, we've divided the problem into four main questions: Do the Spaniards really intend to invade? If they do and if they land on English soil, can we beat them, and what are the implications? If we fight them at sea, can we win? What will happen? Finally, if we do choose to fight at sea, what tactics should we use?"

We move on a bit and say something about the enemy forces.

THE ARMADA: MAJOR SHIPS, BY TYPE

TYPE	NUMBER
GALLEONS	64
ZABRAES AND PATACHES	35
GALLEASSES	4
GALLEYS	4
CARAVELS (OARED)	20
"GREAT FLEMISH HULKS"	23
TOTAL	150

We walk through our list: "The Spaniards have 150 major ships: 64 galleons; of zabraes and pataches, 35; galleasses, 4; galleys, 4; oared caravels, 20; and 23 of a type of ship that our intelligence people merely describe as 'great Flemish hulks.' They can not tell us more. These, in summary, are the warships of the Armada."

[When a list is on the screen, it's a good idea for the speaker or — if the projector is of the overhead type — the projectionist to point out each item as it comes up for discussion. The projectionist can use a pencil, but a narrow piece of clear, colored plastic can highlight each item without blocking words or numbers.]

I told you earlier that a slide that does not advance your talk should be dumped, and that is generally true. But, this time, you've decided that the data from which you've drawn the slide may prove of interest to the audience, though not as a part of your speech. So, you produce a detailed slide as a backup, and you carry it with you, out of sight but ready for use. You'll do something like this from time to time. Generally, you take it home again, unseen by any audience.

But, in this one case — because I'm writing this script — something pleasant happens. Sir Francis looks at your list of ship types, and says, "This is very interesting, but it would be useful for us also to know who the Spanish commanders are. We understand something about their individual fighting style, how they might react in a given situation. Do you happen to know who the commanders are?"

THE ARMADA:
MAJOR SHIPS, BY COMMAND

SOURCE	COMMANDER	SHIP TYPE	NUMBER
PORTUGAL	MEDINA SIDONIA	GALLEONS	10
		ZABRAES	2
BISCAY	JOHN MARTINES DE RICALDE	GALLEONS	10
		PATACHES	4
GUIPUSCO	MICHAEL DE OQUENDO	GALLEONS	10
		PATACHES	4
ITALY AND THE LEVANT ISLANDS	MARTINE DE VERTENDONA	GALLEONS	10
CASTILE	DIEGO FLORES DE VALDEZ	GALLEONS	14
		PATACHES	2
ANDALUZIA	PETRO DE VALDEZ	GALLEONS	10
		PATACHES	1
FLANDERS	JOHN LOPEZ DE MEDINA	"GREAT FLEMISH HULKS"	23
-----	HUGO DEMONCADE	GALLEASSES	4
PORTUGAL	DIEGO DE MANDRANA	GALLEYS	4
-----	ANTHONIE DE MENDOZA	PATACHES AND ZABRAES	22
-----	----	CARAVELS (OARED)	20
	TOTAL		150

At this moment, it's important for you not to look smug, not to smile — on the outside. Ever so casually, you bring out your slide. Sir Francis finds what he wants. He and his colleagues look it over carefully, discussing the details among themselves. Are you worried about this interruption of your talk? Don't be. This is on *his* time, *not yours.* The customer has asked for some specific information, and you have given him what he needs. Later, when you get home, you may want to celebrate by pouring yourself an extra shot of buttermilk. Such happy moments are rare.

You move on to your findings.

"Sir Francis, our findings are divided into two parts — the Spaniards' advantages and ours. First of all, they have more major ships than we do — 150 of them, including 64 galleons. England has more ships — possibly 191 — but many of them are smaller. You can see this disparity in size in our second item. They have twice as much tonnage as we do.

"Their ships are practically invulnerable to boarding. This is a significant fact, Sir Francis; we'll go into it later.

"They have more naval guns aboard. I stress the word "naval" because they are also carrying artillery pieces, presumably for use during the invasion. We, on the other hand, have more long-range guns than they do, and this is very important.

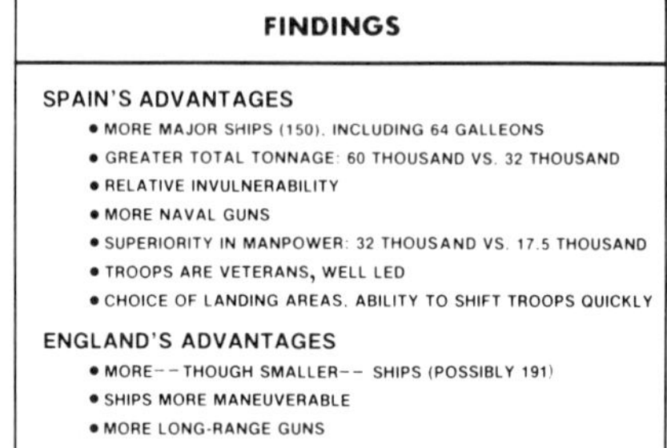

"In manpower, they outnumber us, almost two to one. Their troops are seasoned veterans, the best on the Continent. Their commander is Alexander Farnese, the Duke of Parma, regarded as the finest commander of infantry in Europe. And I must add this, Sir Francis: Our troops, as you know, are green, and they are commanded by the Earl of Leicester, who is lacking in experience. [At this point, Sir Francis has a few words to say: *'Lacking in experience,* hell! He's a blithering idiot.'] Quite so, Sir Francis.

"The Spaniards have great mobility. Their ships can pick their landing areas. They can shift quickly from one landing area to another, far more quickly than we can march troops to oppose them. This capacity gives them an enormous advantage. Next, England's advantages: Though our ships are smaller, we have more of them. This means, among other things, that our ships — or the Dutch flyboats — can blockade Parma's amphibious ships, keep them in port, and still have enough left to fight the enemy's warships.

"Our ships are more maneuverable. This is extremely important. We also have more long-range guns, and we'll show you that that, too, is highly significant.

"You may have noticed, Sir Francis, that we are definite in our knowledge about the Spanish forces. We say they have 150 ships. Yet, about our own forces, we say we have 'perhaps' 191. With your experience, sir, you know that we often have better intelligence about the enemy than we do about ourselves."

"Earlier, I stated the four questions we would try to answer:
We believe the Spaniards do plan to invade England.
Can we beat them on land? We believe so, but the cost would be very high.
Can we beat them at sea? Yes — we are reasonably sure of that — and the losses would be tolerable.
What tactics are best for our fleet? We shall recommend some."

QUESTIONS AND ANSWERS

1 DO THE SPANIARDS PLAN TO INVADE ENGLAND? [YES]
2 CAN WE BEAT THEM ON LAND? [PROBABLY]
- WHAT ARE THE CONSEQUENCES OF FIGHTING THEM ON LAND?

3 CAN WE BEAT THEM AT SEA? [YES]
- WHAT ARE THE CONSEQUENCES OF FIGHTING THEM AT SEA?

4 WHAT TACTICS ARE BEST FOR OUR FLEET?

TACTICS RECOMMENDED

JAB--DON'T SLUG
- USE FIRESHIPS
- FIRE AT LONG RANGE
- HIT AND RUN
- DO NOT CLOSE
- DO NOT TRY TO BOARD

"The most important thing, Sir Francis — and we'll borrow words from another time — is to jab. Don't slug it out at close quarters. As for the tight formation of the Armada — break it up with fireships. Make good use of your long-range guns. Keep firing at long range; don't get too close to the enemy ships. Take full advantage of your greater maneuverability. Do not close. Above all, do not try to board."

With the help of this fine think tank, Queen Elizabeth's navy defeats the Armada. The history of Europe — and the world — changes enormously. Evidently, when you call in a smart think tank, you can expect that sort of result.

Now, to summarize.

- First of all, make sure that any slide you use carries its own weight. If it doesn't, don't use it.
- Don't worry about having the screen blank some of the time.
- Make sure that your slide and text match, that the subjects are the same, and that the slide is in view only when it should be.
- Don't yank material out of a printed publication and think you've got yourself a slide; you don't. Prepare your slide specifically as a slide.
- Use a pointer.
- Don't jam too much onto a single slide. A reasonable limit is about a dozen lines of type per slide.
- Make sure your language is clear. Don't confuse your listeners.
- Be brief, even to the point of being terse.
- Don't irritate your audience with distracting overuse of "slide, slide, slide."

Finally, a question: Why don't we show the architect of the victory over the Armada? After all, Francis Drake was a great mariner, one of the towering military figures of all time, a man who left his lasting imprint on the world. Why not display his picture?

You know the answer. Adding his portrait would not add strength to this talk. And . . . *if a slide does not advance your talk, don't use it.* Sorry, Sir Francis.

POSTER SESSIONS: ONE-TO-ONE TECHNICAL COMMUNICATION

George W. Griffith
Department of English
The University of Tennessee at Knoxville

George W. Griffith, Ph.D.

Dr. Griffith works at the Oak Ridge National Laboratory where he has been a senior technical writer and editor for seven years. In addition to working at ORNL, he teaches technical writing, technical editing, and popular science writing at the University of Tennessee, Knoxville. Before coming to ORNL he taught technical writing for 10 years at the university level.

An active senior member of the East Tennessee Chapter of STC, Dr. Griffith founded (and chaired for the first two years) the Practical Conference on Communication (PCOC), held annually in Gatlinburg, Tennessee.

The poster session, a relatively new and effective form of one-to-one technical communication, has its own requirements of organization and presentation which must be met but which are usually unknown to potential users of this medium. Professional communicators need to understand and master this medium which combines technical information with technical art and technical writing and editing.

Poster sessions are a relatively new way to present technical oral reports–a way that has become a significant form of technical communication. Few technical speakers are trained, experienced, or competent as speakers. In any case the traditional approach to technical oral reporting (excluding the Perry Method) tends to be inadequate even with the best of speakers. Although the poster presentation has its own difficulties, it is an effective way to communicate technical information on a one-to-one basis.

Traditional oral technical reporting presents many problems. The method tends to be a one-way lecture with questioning from the audience almost impossible. Many speakers are untrained, inexperienced, and incompetent; their idea of presenting a paper is to simply read it aloud. And nothing is so boring as listening to a badly read paper, especially if it were badly written in the first place. Usually an audience comes together to hear only a few papers out of all that are being given. Individuals often have to sit through many unwanted papers to hear only those few in which they are actually interested. This problem is compounded by simultaneous sessions with a Hobson's choice of attending only one of several interesting papers being given at the same time.

The poster session format avoids the problems common to the more traditional oral technical reporting. This format allows an attendee to concentrate on papers of interest. Also important, the poster session provids an exchange of information on a one-to-one basis. This arrangement permits true two-way discussion in great detail to any extent and in any direction desired.

A poster session has certain difficulties because this medium is new to most participants; as a result, most authors don't know how to prepare effective posters. Some naively tack up typewritten pages. To avoid poor presentations, careful specific guidelines must be supplied to all who are scheduled to give a poster presentation. These guidelines include a reminder that the poster exhibit be manned the entire time (usually 2-4 h) by the author or coauthor.

An effective poster session requires careful planning. The number of participants is strictly limited by the time and space available. Usually a poster display requires a minimum of four feet of space on an aisle for several hours. More elaborate arrangements require even more space. In any case, individual presentation areas must be arranged for maximum privacy and effectiveness.

Posters can be arranged in different ways; the simplest is a row of panels (usually 4 x 8 ft) arranged along an aisle like pictures hanging on a wall. But often there is no wall, so the posters must be self-supporting which usually means a zig-zag arrangement of panels on table tops (Fig. 1). This way each person has one side of a two-panel unit. Someone else uses the other side of the panels.

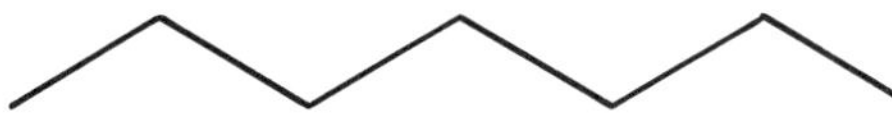

Fig. 1. Self-supporting arrangement of two-panel poster exhibit.

A more elaborate geometry is used with the three-panel unit which is also arranged to be self-supporting (Fig. 2). Again both sides can be used. Other arrangements are possible but these three are the most common.

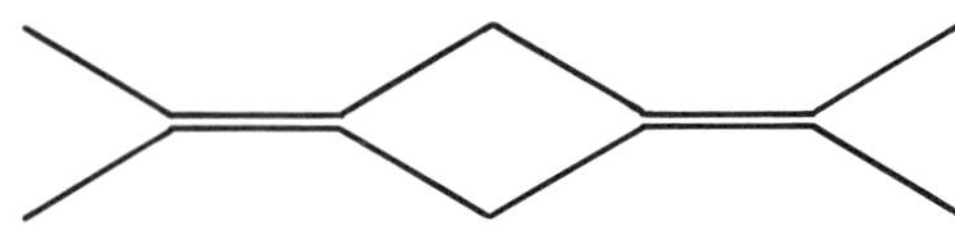

Fig. 2. Self-supporting arrangement of three-panel poster exhibit.

In addition to exhibition time and space, the host organization provides certain basic supplies such as thumbtacks, scotch tape, marking pens, rubber cement, note pads, heavy grade paper, scissors, ashtrays, water pitcher, glasses, and an identification card with the title of the presentation and the name of the author(s). Most of this is unnecessary if the author has brought finished posters to the session instead of unattached poster components.

Guidelines should be given to instruct the author on setting up his exhibit, presenting it, and finally on taking it down. The author should bring his poster (or poster components) to his assigned location at least one hour before opening time. This allows time for mounting the components and also reassures the person who is in chage of the session that you will be there.

The importance of an informal atmosphere cannot be overstressed. Interested persons casually stroll through the entire exhibit of posters, often with a cup of coffee or tea in hand. Like Saturday shoppers they walk slowly, mostly looking. But at each exhibit there is usually someone in the crowd pausing to look closer, perhaps to read the text or to look more closely at the graphics. At this point the author has an audience.

To start a discussion the author always has ready a short statement summarizing the main points of the display. It is often good psychology to end this summary with some question designed to better engage the audience in the discussion that follows (or should follow).

The discussion should be kept conversational. The author answers all questions fully but without lecturing. At the same time the questioner must be allowed silence in which to study the poster exhibit and given an opportunity to respond to the information he is getting.

Sometime during the discussion a handout sheet may be given. This handout summarizes the presentation and lists the name and professional address of the author. This handout makes it unnecessary to take notes-a desirable consideration because notes taken under such circumstances tend to be cryptic and often illegible.

A poster exhibit brings together technical research, technical art, and technical writing. Quality research deserves quality presentation which means, in part, quality technical art and writing. Often authors presenting in a poster session fail to provide that quality of art and writing which can only be given by professionals in these fields. Few technical people are competent to do their own posters without outside professional help.

The person who wants an effective poster exhibit must view the technical artist and technical writer/editor as helpers, not adversaries. The expertise of both artist and writer should be accepted in this spirit and acknowledged. Both artist and writer need information from the technical author-one to plan the individual graphics and the overall layout, the other to write titles, text, and captions. Both need adequate time to do their job. Always, close cooperation among the author, the artist, and the writer is essential.

Professionally prepared posters are not cheap. The writer's time is not the major cost; it is the graphic artist who plays the most expensive role, and so it is he who must make the overall cost estimate. Sometimes a single 4 x 8-ft poster may cost $500 or more.

Poster design is governed by principles of organization and presentation. Organization is the overall combination of components (layout) which include titles, text, and graphics. A dynamic rather than a static organization is achieved by grouping components, connecting them (often by careful use of color or texture), and focusing the reader's attention on cetain of them. Components are best pre-assembled in stick-on modules. Overcrowding must be avoided.

Presentation requires perfect work-no mistakes, no sloppy hand lettering, no last minute make-do. Everything must be clearly legible to the reader at the distance of several feet over which posters are normally viewed. Text should be kept to a minimum with careful editing to eliminate all jargon and to make the titles, text, and captions correct and concise. Avoid humor; if the subject is not to be taken seriously, then it should not be presented.

Without exception, poster titles, text, and captions must meet the standards of correct technical writing. These standards require correctness, consistency, conciseness, clarity, continuity, construction, and content (the Seven C's of Technical Writing).

Captions require one special comment: never insult the reader by telling him the obvious in a caption. Never say "This is a photograph (or photomicrograph, graph, or table) of. . . ;" any idiot can look at a photograph (or photomicrograph, graph, or table) and identify it as such. Instead say something meaningful about the *subject* of the photograph (or whatever).

Never use tables in a poster; they are simply too hard to read. Instead convert needed data into a graph because graphs can be comprehended at a glance. Trends and fluctuations are obvious in a graph but not in a table.

The poster session is a popular and effective form of one-to-one technical communication. As individuals become familiar with this medium and more expert in using it, its popularity and effectiveness will become even greater. Professional technical communicators need to master this medium for their own use and to help the technical community do better what is now does rather poorly—communicate!

THE BUSINESS OF CREATIVITY

J.B. Hancock
JBH Productions, Inc.

J. B. Hancock

JBH Productions, Inc.
2810 Adams Mill Road, NW
Washington, D.C. 20009
(202) 667--4911

President
Producer/Director

Audio-visual production is often thought of as a glamorous endeavor. In fact, the business side of audio-visual production comprises seventy-five percent of the total effort. This paper proposes some good business management skills involved in successful audio-visual production from the perspective of both the client and the producer.

Most of us, as technical communicators, have been involved in an audio-visual production either as the client or the producer. Some of us have even dabbled in more than one of the audio-visual formats: the simple photo-montage of plant employees for an in-house exhibit, the slide/tape presentation that dazzled the masses at the annual meeting and resulted in your promotion, or the sophisticated 16 millimeter film that brought your client's customers off their seats screaming "I'll buy, I'll buy!" Looking back at our combined production experiences, both client and producer are more apt to remember the kind words, the pat on the back, and the occasional award. We all too quickly forget the other side of the coin. The hours you spent wondering, "Why doesn't the producer understand what I am saying?" The time I spent working and reworking my proposed budget to figure out, "How much money am I going to lose on this $10,000 version of 'Gone With the Wind'?" The myriad explanations to management rationalizing why, "The show for the annual meeting on February 1st will be ready for delivery on March 15th!" Me, throwing up my hands at 3 A.M. and screaming, "Everybody wants to be a producer!" Ah yes, those wonderful, exhilarating moments known collectively as the production of audio-visuals.

Although some individuals have given cursory attention to existing information on the aesthetics of audio-visual production, most involved in the process pay little or no attention to the business side of production. In most instances the results are disastrous, since audio-visual production and services are the only experience in the world totally governed by O'Toole's Commentary on Murphy's Law: i.e., "Murphy was an optimist."(1)

Preparing the Request

The audio-visual process begins even before the initial bidding procedure. The audio-visual contract is merely the end result of a long chain of events. Generally it is the client who initiates the contracting process by issuing a bid or request for proposal (RFP). In essence the client is saying, "I need help and here is what I need." Our problems thus begin.

Problems because many clients are not sure what they need. Sometimes they know what is needed, but refuse to divulge the information, asking instead for the creative consultant to suggest an approach or format that may, or more likely may not, fill the bill. These approaches are unfair to both the client's needs and the producer's capabilities. But the situation can easily be remedied.

After years of responding to myriad bids and RFP's, our firm now makes it a policy to seek out those clients who are willing to take the time to be specific about their needs and objectives. We have arrived at a short list of standards by which we evaluate the requests we receive. If the bid or RFP does not meet most of the following, we generally do not respond.

- Is the bid of RFP clearly written? Is the language understandable?
- Are there hidden clauses?
- Does the client know what he/she wants? Are objectives clearly stated? Are they realistic?
- Is the audience defined? Is it realistic?
- Is the format indicated? Is the client open to suggestions?
- If format is given, are specifications given? Does the client indicate a working knowledge of audio-visuals?
- Is a level of effort indicated? Is it realistic?

Some clients, having had audio-visual production experience, will indicate a dollar amount, or level of effort. Most do not. Clients should keep in mind that buying audio-visual services is very different from purchasing a truck or even a service like typesetting. You are buying the creative talents of one or more individuals, and, unless the client has a written script or detailed storyboard, the creative person(s) is being asked to guess on a number of variables in a nonentity. It is similar to asking a real estate agent, "How much does a house cost?" After he or she stops snickering, you will probably get down to some basic questions and arrive at a accurate cost estimate of a house that fits your needs.

In terms of audio-visuals, indicating needs, objectives and level of effort means the producer can, in turn, use his or her talents to provide creative ideas which both meet the indicated parameters and result in the best product for the specified amount of money. Remember, the more initial information the client can provide the producer, the easier it will be for the producer to estimate the effort, time and cost involved. In the long run, both the producer and the client will profit.

Should the Producer Respond?

Once the bid or RFP has been received by the production company, the staff must ask itself some probing questions about the company's capabilities. The most obvious is, can we perform the services requested? Are we comfortable with the format, and can we offer the client a quality product? If not, can we recommend another firm that can supply this service? When making recommendations, however, be especially careful to choose only firms that have similar or higher standards. After all, because of your recommendation, their work will reflect on you!

If it is decided that the requested services can be provided, the production company should take a close look at the required credentials. Perhaps the client wants a training film on chicken plucking. But the specifications call for expertise in the area of chicken plucking. The production company may have produced on that topic, and, although it wasn't a film, they feel confident that after one look at their chicken plucking filmstrip the client will be sold. Or, if the company has not produced on that topic, they may investigate the possibility of a "joint venture" with the National Association of Chicken Pluckers. If, however, the producer has no credentials or experience in that area, and feels that the requirement is justified, the producer probably should not pursue that project.

Assuming that the producer determines that the credential or reference requirements can be met, the projected time frame should be examined. Is it realistic? If not, is the client open to suggestions on a more logical production schedule?

Now for the big question: can the client afford us? Or will we end up losing money? I am especially suspicious of a bid or RFP which states, "the contracting office will provide the narrator" (or photographer, or music, etc.). I am delighted that your company's president has a nice voice, and that one of the administrative assistants looks great in snapshots, and that the in-house newsletter has a stock photo library of slides from the plant bowling league. But the client who nickles and dimes professional and creative talents will be treated accordingly by the producer. If you can't afford to do it right, don't do it at all!

The last question the production company should ask is: what paper work is involved in preparing the proposal? Some jobs are very small, but the potential client still wants a fifty pound proposal. Other clients, never having purchased creative talent before, require the bidding firms to use the budget form included in the bid or RFP. This is usually the same budget form the company or agency uses to order widgets. Other clients require copies of past productions, with no indication that the materials will be returned. The client should remember that the preparation of a proposal is both time consuming and dollar consuming. Our estimate is that a well prepared proposal costs from $200 to $800 in staff time, office supplies, telephone charges, copying, packaging, and miscellaneous charges. If the potential client is asking for more than this level of effort, especially on a simple project with low profit margin, the producer should think twice about responding to the request.

There is one additional topic in which the producer must be interested. Is the bid "wired"? For the uninitiated, this means that the client has already promised the work to another producer. But, because of company or agency policy, other production firms have to be contacted and allowed to submit a proposal, usually to no end. How can a producer tell if the bid is wired? Quite frankly, sometimes you can't! But there are some telltale hints.

Look for the issue date and corresponding due date. A complex proposal issued on the 3rd, posted on the 4th, received at your firm on the 6th, and due back on the 7th, is probably wired. Likewise, look for restrictions governing the location of the production company. If the specifications call for a firm located between Main Street and South Street in the same town as the issuing agency or company, you can bet the contract is wired.

Look at the technical specifications. If there is detailed information about required staff, special production techniques, and/or specialized in-house equipment that is not state-of-the-art, you can assume the contract is wired. Find out if this audio-visual production is the second or third task in a continuing contract, or the result of a completed study. Chances are the original contractor will get the job. But don't be put off. Approach the original contractor, and inquire about subcontracting possibilities.

An outrageous cost estimate for production is another excellent indication that the contract has been wired. Only very large concerns can afford a half a million dollars for a ten minute film. Even then, they rarely spend extravagently. If the producer receives a request for a proposal, costs it out according to the specifications, and arrives at a figure totally out of line with what the agency or company can be expected to spend, the producer should drop it like a hot potato! Chances are someone has already convinced the client that this complex production can be accomplished for a nominal price. Undoubtedly one or all of the following will take place. The client will be unhappy with the final product and will forever curse all producers. The producer will lose money. Huge cost overruns will be incurred. In both the long run and the short run, no one will be happy. A general rule is: if the producer has any gut feeling about the bid being wired, drop it. It is simply not worth the trouble.

Preparing the Proposal

If the producer has determined that the bid is legitimate and well worth the production company's investment, then it is time to get down to work. It is important for producers to keep in mind that this is your chance to let the potential client know that you are an excellent business administrator as well as a veritable genius when it comes to creativity. This is the producer's opportunity to shine.

A successful proposal is concise and to the point. It addresses the specific needs of those individuals who are in a position to give approval to the proposal. The winning proposal answers, as completely as possible, the client's questions: namely, what do I need to solve my problem, how will it be done, and how much will it cost? These three questions translate into the three major components of a good proposal: the technical proposal, the business proposal, and the budget.

The first component, the technical proposal, should be a clear, complete and concise statement of how the producer plans to meet the client's objectives. The importance of this section cannot be overstated. The client is not asking the producer to parrot back the statement of work included in the bid or RFP. Instead, the client is asking the bidding firms to use their creativity and business expertise to suggest a solution to a communications problem. Most importantly, the technical proposal will ultimately serve as a guide throughout production, reminding both client and producer of the agreed upon objectives and approach.

Oversimplification, misinterpretation, lack of understanding, and vague generalities are only a few of the significant reasons given for proposal rejection. It is important, therefore, that the creative bidder give ample time to examining the problem thoroughly, and then selecting an appropriate approach. The technical proposal must exhibit a clear understanding of the task, its objectives, and the intended audience. If this understanding does not exist, an Academy Award winning crew, working twenty hour days at no cost will not save the project! If there is no evidence of a basic understanding of the problem on the part of the producer, then the client need read no further.

The second part of the proposal, the business component, should give the client total confidence in the producer's ability to manage the administrative and financial aspects of the project. Especially helpful to the client are sections which detail the capabilities of the production company, the experience of key personnel involved in the production, a production schedule, a schedule of payment, and a management plan if the project is complex or involves a series of productions.

While most producer/directors cringe at the thought of preparing a business proposal, the true role of a producer is as business manager of a creative production. Individuals calling themselves a "producer/director" should be fully capable and willing to perform the business, as well as the creative, duties implied in the title. And the client has every right to expect a complete and detailed business proposal. After all, audio-visual production is an important investment.

The cost estimate, or budget, is the final section of a complete proposal. The objective, of course, is to arrive at an accurate estimate of the total production cost. This requires a great deal of time and energy on the part of the producer, but attempting any production without a detailed budget can be devastating.

In preparing a budget, expecially one where no script exists, Sodd's Second Law must apply: i.e., "Sooner or later, the worst possible set of circumstances is bound to occur!"(1) The producer must include not only the actual filming, but must factor in time and money consuming variables such as pre-production meetings, creative think-tank time, weather, post-production, perhaps distribution, and maybe even plain old bad luck. Regardless of the number of variables, the producer should be able to provide the client with an up-to-date estimate. A wise producer, however, will include a gentle reminder that prices do not stay constant. Increases in the price of oil and silver, and general inflation as well as client delays may affect the cost estimate. This is particularly true when there is a large time lapse between proposal submission and final contracting.

There are some general rules for proposal preparation and evaluation.

Producer Do's

- Be honest. If the client suggests production schedules, formats, or other variables that are not realistic, suggest alternatives. But be gentle. You don't want to leave the impression that you think the client is stupid or the suggestions ridiculous.
- Remember the "boiler plate". This is especially true of government proposals. Forgetting to check the box on compliance with the clean air regulations could cost you the contract.
- Be creative. This applies to all sections of the proposal. The client needs input and is asking for your help. Don't be too stingy up front, it may leave the impression that creativity doesn't exist at your firm.
- Be concise and neat. A well written proposal does not require inordinate length or elaborate packaging. If your proposal is clean, neatly typed, and understandable, it will stand out.

Client Don'ts

- Don't be fooled. Packaging does not make the proposal. The time taken to design a flashy cover is wasted if there is no substance.
- Don't be conned. Certain contractors have been known to claim "big name" references and/or "award winning" key personnel. Check references

and ask to see samples of work.

- Don't be put off by a request for progress payments. Generally this implies that the production company is a small business and needs an on-going cash flow to survive. Keep in mind that a small business is generally more appreciative of each client, and will work harder and longer to meet the client's needs.
- Don't be gullible. Once you detail your needs and objectives, a producer should be able to give you an accurate cost estimate. Also, don't fall prey to the "lowest bidder" syndrome. If the bid is ridiculously low, it will eventually be revised upward, usually when it is too late.

Overall, a thorough proposal will provide a basis for the working relationship between the client and the producer. If done right, both the client and the producer will benefit.

Contract Types

Although there are variations on the theme, there are two basic types of audio-visual contracts: fixed price and cost plus fixed fee. Our firm generally suggests that clients stay away from the latter, since it is a nebulous agreement that can open wide the door to cost overruns. Sometimes, however, a cost plus fixed fee contract is unavoidable. For instance, the nature of the subject may be such that the producer will be responsible for determining the client's needs through extensive research. Or the production may require shooting locations or special talent which cannot be defined until a script has been finalized. In those instances where a complete budget cannot be prepared for legitimate reasons, the client may wish to contract on a cost plus fixed fee basis.

We usually prefer to have our clients agree to the inclusion of a contingency fee in the approved fixed price contract. We suggest a fee of between ten and twenty percent of the total estimated cost to cover the unexpected. If everything works according to the original plan, we come in under budget, and the client is delighted. If the unexpected happens, we come in on budget, and the client wasn't pressed into coming up with eleventh hour funding.

There are distinct advantages and disadvantages to both types of contracts. The fixed fee contract is based on a maximum price. The more quickly and efficiently the production company works, the more money it makes. Needless to say, if there is a prolonged delay in production or a cost overrun not covered by the contingency, the production company stands to lose money. Sometimes a lot of money.

The cost plus fixed fee contract is based on an estimation of direct labor costs, direct material costs, an overhead percentage, and a set fixed fee. The production company is, up front, saying, "This job is so complex and/or sketchy that all we can do is give you a guesstimate of the cost." The company spends up to the estimated amount, and if production is not complete, the client must find additional funding or stop production. Although producers may be delighted with this type of contract since they are relieved of the responsibility for cost overruns and inflation factors, the client stands a good chance of being very disappointed. In the long run this can be detrimental to both the client and the producer.

Obviously the worst kind of contract is no contract. It is important, therefore, that a written contract be agreed upon, and that, regardless of type, it fit the needs of the production and protects all the involved parties.

Contract Control

Which brings us to contract control. Two of the world's largest commodities are producers who have trouble getting paid, and clients who didn't get what they expected. If you are having these types of problems, you should be working on improving your contracts. Although oral agreements are nice in theory, to quote Samuel Goldwyn, "A verbal agreement isn't worth the paper it's written on."(2) A well written contract, on the other hand, will insure a good business relationship from beginning to end, and will provide its own controls.

Let's assume that the producer and the client have arrived at a clear, complete and concise contract. All the principle parties have signed the contract. Neither party has allowed a secretary or office assistant to sign and then initial for the boss. Provisions for approvals and progress payments have been spelled out. Perhaps the producer has included a clause which allows the client to take a discount for prompt payment. There is also a clause which clearly states that the producer is entitled to stop work if payment is not received. Interest on overdue accounts is detailed. The contract is in place and work has begun. But there are additional controls that both parties can exercise to protect themselves.

First, keep records. This is especially true of client approvals. Date everything, preferably by machine. Write letters confirming telephone conversations, especially those where the client has given verbal approval or where either party raises questions. Take a witness along to meetings, especially sensitive ones. Then write a memo to everyone involved recapping what was discussed and decided. Although these controls mean additional paperwork, these records will keep everyone in production and out of court.

In the same vein, both parties should know the people with whom they are dealing. If only for self-interest, both parties must be in close touch with each other. Get a sense of how things are going. Get together and work out problems. A solid proposal, an agreed upon contract, and a close working relationship should give both producer and client a basis for discussion and/or negotiation before the problems become too big to handle.

Client Management

Speaking of control, let's talk about clients. From an audio-visual producer's point of view, the client who knows audio-visual production and how to work with a producer is a rarity. Most clients fall into one of two catagories: those who know little about audio-visual production but pretend to be experts; and those who know nothing and refuse to learn. Weiler's Law applies to these two types: i.e., "Nothing is impossible for the man who doesn't have to do it himself."(1) In order to help these clients get through the production without suffering a stroke or firing the production company, the producer must provide both information and guidance. The great awakening comes when both realize that any production is, in fact, a series of compromises between what each would like and what must be.

The pre-production phase is critical to the total development of the production. Details are planned, re-planned, torn apart, and re-planned. Clients usually just want to get going! A deadline looms in the not too distant future, and the client is anxious to see some results. But both the producer and the client must ignore the calendar long enough to insure that proper attention is being given to the various pre-production tasks: research, scripting, talent search, lighting set-ups, etc. The producer, familiar with the process, must help the client understand the importance of these tasks to the total project.

Once filming actually begins, the client may feel downright excluded. Everyone works better without someone glaring over their shoulder, and producers are no exception. But the client should not be left out of this phase, if for no other reason than because it is the most fun. Even though some clients may claim to be too busy or simply not interested in the production phase, most expect to be involved. And rightfully so!

The wise producer will include the client in the actual production -- as a technical advisor, the executive producer, or even a goffer. First of all, clients will enjoy themselves immensely. After all, this is totally different from their normal routine. But even more importantly, the client who has worked a fourteen hour day with the production company, and who has struggled through a two hour lighting set-up for a ten second shot, an uncooperative child actor who refuses to deliver the correct emphasis on a line, and/or a one hour cleaning session on a piece of factory equipment for one still photograph, will be much more appreciative of the producer and the production team as a whole. An uninformed, uninvolved client, on the other hand, becomes an obstacle.

During post-production, again, involvement makes for a happy client. If the client is not consulted during this phase, the producer should not be surprised if the final product is rejected. Producers should seek both opinion and written approval. If the producer and the client have worked together continuously, the final product will reflect that cooperation.

The Business of Creativity

What we have examined is the business side of audio-visual production. And, although we would all like to spend our time involved in the creative aspects of the process, both the successful producer and the satisfied customer know that creative time may account for only twenty-five percent of the overall effort.

In closing I would suggest that clients remember that successful producers are both business consultants and creative consultants. If you hire them, use them. They are there to help you, take their advice. They don't try to do your job, don't try to do theirs.

Producers, remember that your clients are human beings with real needs and valid restrictions. They are not ogres, nor are they the Bank of England. Treat them carefully. Listen to them. And give them your best. They are, after all, your bread and butter.

To both I would say, establish rapport. Learn from each other. And have fun. If you combine creativity with common sense and good business management, you will.

(1) Bloch, Arthur, Murphy's Law and other reasons why things go gnorw! (Los Angeles: Price/Stern/Sloan Publishers, Inc., 1978).

(2) Johnston, Alva, The Great Goldwyn (New York: Arno Press, 1937).

THE SHORT NON-FICTION FILM: PROPAGANDA THAT PREACHES/TEACHES

Joseph B. Kirkish
Michigan Technological University

Propaganda is not a dirty word. When properly used, it can provide religious comfort, stir political beliefs, encourage social improvement, or stimulate thinking on any number of issues. Propaganda films can do the same. Building on Serge Eisenstein's theories, modern non-fiction filmmakers have refined and pruned the documentary film, so that in just a few minutes a message can be driven home. Non-fiction films need not be restricted to the classroom or the television set; they can be used in any situation in which enlightenment on a relatively foreign topic is needed to encourage discussion.

It's no news that communication skills are on the decline. It's also no news that people are reading less and looking more. It's also common knowledge that people prefer entertaining newscasts to hard facts, lighter press features and comics to hard news, and "junk food" gratification to thought-provoking movies and literature.

Perhaps all this is somehow connected. A general statement might be made that a gradual reduction in disciplinary thinking very easily results in misleading (non-specific) and downright erroneous communication, and that statement should be of concern to the academic and business world alike.

But slang, grammatical errors, poor spelling, and the inability to construct a clear sentence are only the obvious problems, merely the tip of the iceberg, a tangible warning of disintegration in expression. There is more. The frightening suspicion that creeps to the surface is that the thoughts being expressed aren't worth much to begin with.

A society that emphasized emotions, immediate gratification, and the present really doesn't have much of importance to say. That is the unpleasant idea that rears its ugly head.

While a study of good literature, history, sociology, economics, politics, the arts, etc., could prepare the mind for profitable thinking, there is another way - a kind of fighting fire with fire method - which can in a fast food age stimulate the potentially effective yet unenlightened person into useful discourse. It is a form of propaganda, and it works.

Now, contrary to some opinions, propaganda is not a dirty word. It is, simply put, a one-sided approach to any issue and can be either "good" or "bad" depending on its use. A religious message is propaganda. So is a political speech. And so are the commercials seen on television. All of them, good or bad, have one thing in common: regardless what their motivation, they either preach or teach.

The non-fiction film - or "documentary," if you will - can be a powerful communicating tool that can either teach or preach - or even combine the two in a single brief but effective blow. If one picture is worth a thousand words, think of what a five-minute film could do at 24 pictures per second, especially when it has been meticulously thought out, carefully edited, with the right sound track - and in motion.

It was the great filmmaker of the Twenties, Serge Eisenstein, who perfected the feature length propaganda film in Russia, but it took World War Two to bring out the best in its potentials through men like John Grierson and Basil Wright when they adapted Eisenstein's propaganda techniques to suit England during those troubled times. Perhaps more educational than polemic, their films simultaneously built national pride while they promoted the home war effort. Then came the more universal academic and commercial applications; in the United States in particular, classrooms and television sets were "enriched" with propaganda films that could, with refined montage techniques, produce the same psychological and educational results in far less time than the earlier films.

Universities and advertising agencies alike saw the potential and utilized it. The University of Michigan's audio-visual center produced a series of four-minute "trigger" films that impressively exposed the generally ignored problems of old age far better than any lengthy prose or discourse could. Advertising agencies like Foote, Cohen and Belding in Chicago touted their capabilities with a complex audio-visual presentation that effectively bombarded the senses of potential clients in less than a quarter of an hour.

Good short non-fiction films are not hard to find; they are available from countless academic or commercial distributors and are relatively inexpensive to rent. They can also be home made with some practice and on a relatively small budget. They are more portable and durable than most other forms of mass communication; and in a visually oriented society they are, for better or worse, far more effective than any other medium in stimulating thought and discussion.

Consider, for example, the trigger film made by the University of Michigan that follows an elderly lady on what appears to be a modest day on the town. Dressed in her most elegant but old-fashioned clothing, she shops in a supermarket for the few staples she needs. She barely notices - or pretends not to notice - the contrast with others indifferently stuffing their baskets. At the checkout counter, her smile freezes on her face and the light in her eyes vanishes as we hear the cashier call impersonally across the floor, "Hey, Charley, how do I handle these food stamps?"

In four short minutes, the viewers have been witness to the humiliation and embarrassment suffered on a human being due to ignorance or sheer lack of concern. Such a film not only propagandizes (and does it well), but with immediate emotional and intellectual impact it triggers thoughts and ideas, and it stimulates a discussion on a number of topics from those dealing directly with the need for awareness of and consideration for others to feelings about members of society we usually choose to ignore.

Consider, too, a non-verbal, animated film called MORE. Crying babies are placated with toys, country roads are clogged with cars, housing tracts usurp open areas, eating becomes gluttonous. We demand more, and as more is turned out, pollution becomes greater and the earth begins to crumble. In less than three minutes, the film has dealt with conspicuous consumption; the message is loud and clear, again a final signal for thinking and discussion.

An interesting experiment could be made with any group of participants. A film like MORE is shown, after which the group is asked to use ten minutes in jotting down thoughts - rambling or organized - prompted by the film. Then comments are requested from the group. The flow, if it is normal, will be both spontaneous and active; it will be thought-provoking, too, and likely enlightening. It might be restricted to the topic of over-population, but, should time permit, it also could move easily into thoughts prompted by some particularly impressive sequence - thoughts that may be tangential or even unrelated to over-population - but thoughts all the same.

What happens is quite understandable. The emotions and the intellect of the viewer are stimulated by what is carefully compressed in the film to react at first interiorly, then transitionally on paper, and finally, with momentum still growing, exteriorly in open discussion. The trigger effect of seeing the film, then writing immediate reactions or observations, and then expressing them in communion with others is an almost effortless progression.

The film, in other words, not only brings a calculated bit of knowledge to the viewers, but it encourages them to quickly assess it and then respond to it. And they do.

There is nothing quite so effective and effortless as a film of this sort. And the topics for presentation are endless. The effect works on groups of all sizes and ages, and in any areas of expertise. Even the making of such films could be a valuable educational experience; it could be done by a single individual or made into a group project.

While this paper has limited itself to films of less then five minutes in duration, it might be said in conclusion that while the shorter films do seem to have the stronger impact, others running up to half an hour and pertaining to more complex issues are also available. But the short film, when well made, is not only less expensive to rent or purchase, and quicker to view, but can also with the right creative touch drive home a message with uncommon force and accuracy. And it can do it with clarity.

GRAPHICITIS: CAUSE AND CURE

Allen L. Kundratic
The Johns Hopkins University Applied Physics Laboratory

Allen L. Kundratic
The Johns Hopkins University
Applied Physics Laboratory
Design Artist/Illust.
Johns Hopkins Road
Laurel, Maryland 20810
(301) 792-7800 Balt.
(301) 953-7100 Wash.
President IGI Balt./Wash.

There is a recurring problem throughout history associated with getting artwork, known as "Problems with the Artist". Van Gogh, Michelangelo, even the 'guy' who decorated cave walls was an agitation to someone. So what? We all have work related problems in every profession. Well, I would like to give you tips on avoiding problems when requesting art and illustration services, some insights picked up during the 25 years I have been out-standing in the field of graphics; a guide for helping you get it when you need it.

Engineers, scientists, project managers, authors and advertisers are required to deal with a band of notorious vagabonds when in need of graphics for papers, talks and ads. With a better understanding of what an artist is, you will be able to avoid some common foozling. I will explain these slanted proverbs; "A happy artist is worth a thousand pictures", "The quickest way to an artists heart is through his ego", "You can lead an artist to the board but you can't make him draw it your way!" and (foozling).

INTRODUCTION

My purpose for presenting this paper was to escape a little work at the Laboratory. Gad-zooks!, was I ever wrong! The job of writing, illustrating and presenting this "High Class" talk was a whole-soul, gut wretching experience. Unaccustomed to public writing as most artists are, caused me no-end of panic and anguish. I wore out two dictionaries, several pads of paper and a perfectly good typist before I had anything written down. The problem of authorship was compounded by the usual difficulties with the illustrator, even though they were the same person. My written thoughts often conflicted with the illustrating ideas, and therein lies the tale. Leave us look into this fascinating relationship of the Originator and the Artist. Come with me through the looking glass and watch as the characters develop. The Originator - anyone requesting services from an Art Department for the purpose of enchancement of written statements and concepts. The Artist - anyone providing illustrations, diagrams, tables or charts to make an otherwise dull document worth reading.

THE RELATIONSHIP

The Artist and the Originator, a natural conflict of professional egotism. Each looking for fulfillment of expression, the plaudits of an audience, the creating of the memorable talk. The Originator stringing his thoughts into words, the Artist attempting to replace those words with illustrations. The following drawings satisfy the Artist in me, while the captions seem inadequate to the writing half. I could say more than:

The Artist as seen by
The Originator

The Originator as seen
by the Artist

My dealings with both sides over the twenty-five years I have been an illustrator/artist have revealed a pattern of resentment, animosity even; my views on this may not help, but they couldn't hurt.

PERSONALITIES IN THE RELATIONSHIPS

The Artist relating to an originator

- He's about as brilliant as a no-watt bulb.
- Not unlike a boatman with only one oar in the water.
- He's not playing with a full deck.
- I'd say he's lost his marbles, but that model didn't come with any.
- He's like wet paint without a sign.

The Originator describing an Artist:

- He's not wrapped too tight.
- The string on his Yo-Yo's two feet too long.
- He's got five out of four screws loose.
- He's definitely two bricks short of a full load.

The view now is that something in their make-up works like sandpaper on silverware. Maybe we can isolate a typical example. The originator asks for an organizational chart, this is what the Artist comes up with:

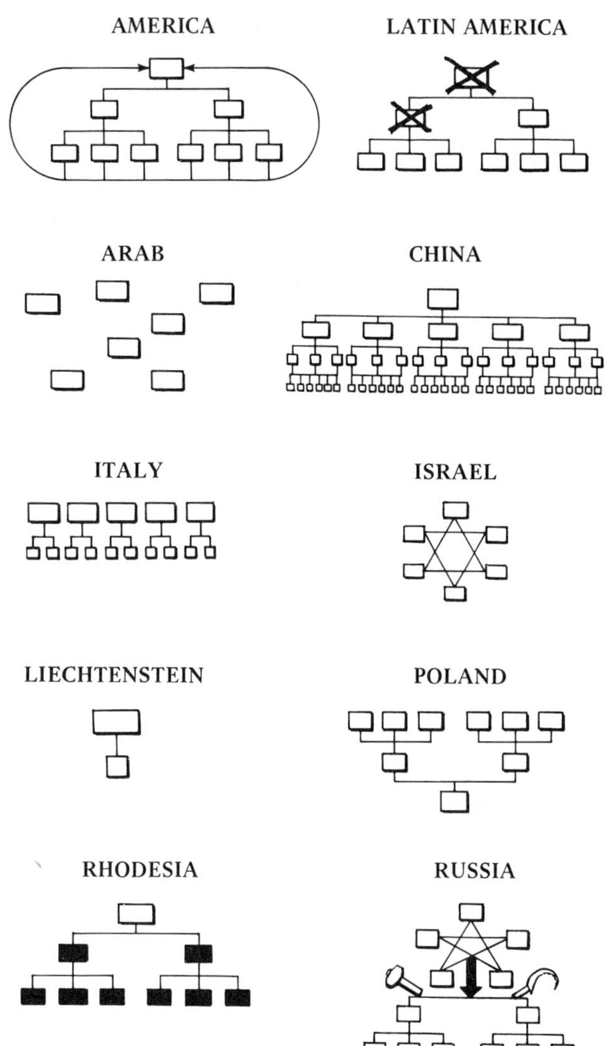

Next time be specific!! the Artist.

Now we are getting down to some of the basic problems. The Artist may feel frustrated because the Originator decides all the ideas to be explored and calls the shots. The Originator feels threatened that the Artist may be more talented than he, thus stealing the show he feels is all his.

By eliminating some of the negatives in the relationship, we can begin to lay down rules of co-existence.

No! Artists are not temperamental, childish and immature, as long as you don't touch their crayolas.

No! Originators are not know-it-all's who think the sun shines only on them because of their smarts.

No! Artists are not mushrooms to be kept in the dark and fed a diet of manure.

No! Originators are not Gods that commute from Mt. Olympus.

THE MOTIVE OF THE RELATIONSHIP

Like so many fragments of a once beautiful statue, our work ethic is crumbling around us and in us. The ability and desire to produce quality work is replaced with a "time is money," "good enough for what it's for," "profits for profit sake," syndrome. Working in this environment, you are suspect when you demand good work from someone. 'What is the matter with Jones?' 'He wants a good job, who does he think he is?' 'We don't even do a good job for the President!' My advice here is to get the art people involved early in the project, ask for suggestions, give them the feeling of being part of the project. When a presenter puts up a slide that is poorly executed, with dirt and lint marring its content, your project is weakened by association. This is not just in communications it is rampart in all areas of America. If your new car starts falling apart, who cares? If the buttons fall off your new suit, who cares? If the first time you get your new shoes wet, the soles go flop-a-dee-flop, who cares? We all should and we must change our profit hungry system and get back to what made America great, Craftsmanship. Start in your own area of expertise and motivate others to do a good job everytime, not just when it is demanded. Set the example by being very good at what you're doing and others will follow.

SUMMATION OF THE RELATIONSHIP

All Originators do not have problems with the purveyors of art services. This could be because they have followed some or all of the following:

- Be lovable and happy around the art group.
- Treat them as professionals ignoring idiosyncrasies.
- Provide legible, detailed and common sense instructions.
- Give praise when deserved and complain when required.
- Use good logical approaches to all problems as they develop and solve them quickly.
- Do not wear out your welcome by checking up to often. Show confidence in their ability to deliver.

Most of my comments are about in-house art groups, as vendor or contracted art services are delivered on a "pay as you get" basis. In-house people are more able to say 'go fly a kite' or on the other hand give you personalized service. They are there to serve you and are being paid by your company. Use their talents effectively. The more you use them the better they become and thus are more profitable to have around. Most of all, use their years of experience to your advantage.

It is not uncommon to have an art group with a cumulative years of experience of 80 to 100 years sitting around tying tails on flies or making paper airplanes. So much of our communicative material needs the talents available to make them more effective. The advertising people are making use of art, comedy, photos, and 'good looking women and men' to sell a product. You, technical people, must learn to use all of the talent you have to draw on and be innovative in their deployment. Don't be afraid to be better.

THE REWARDS OF THE RELATIONSHIP

You, Originators, are prime movers of your projects and need the support of the Artist to produce an excellent end product. You must coordinate, motivate and sometimes agitate to reach the mutual goal. As the presenter, you collect all the comments both good and bad, and you are first to know the results. The Artist, however, is back at the office working on another project, wondering how your project went over. Did the art provided, make a difference? Did it do the job? For pity-sakes tell him!

Tell Him Now

If with pleasure you are viewing
Any work a man is doing,
If you like him or you love him,
Tell him now.

Don't withhold your appreciation
Till the Parson makes oration,
As he lies with snowy lillies o'er his brow.

For no matter how you shout it
He won't really care about it:
He won't know how many teardrops you have shed.

If you think some praise is due him
Now's the time to slip it to him
For he cannot read his tombstone when he's dead.

Steve Ward

Have the courtesy to allow the ones who worked for you in on the results and status of the project. A simple phone call, a note or in person, run down, can be helpful. Create a feeling of mutual respect and a stimulating rapport with the Artist on one project and look forward to enthusiasm on future collaborations. A good Artist and intelligent Originator go hand in hand, but not where everyone can see them.

Presenting Publications Production Data:

DON'T BUILD A BATTLESHIP TO CROSS A RIVER

James W. McCafferty
Rocketdyne Division, Rockwell International

James W. McCafferty
Manager, Reprographics

Rocketdyne Division, Rockwell International
6633 Canoga Avenue
Canoga Park, CA 91304

(213) 884-2218

Publications: "Your Style Guide: A Publications Pole Star," 26th ITCC, STC, 1979; "Plant Vigorous Seeds," 16th ITCC, STC, 1969; "Creative Editing," Handbook of Technical Writing, John Wiley & Sons, 1971.

In this modern age of statistics, the gathering, recording, analysis, and presentation of data can become such major tasks that the production of the product *to which all of these data relate can occupy an also-ran position in the allocation of priorities. A concept of publication production statusing is discussed that attempts to keep production data in a proper perspective and still provide management with the necessary facts for them to make intelligent decisions.*

INTRODUCTION

I have always regarded myself as a creative writer—CREATIVE writer. The great American novel still resides somewhere in the recesses of my now somewhat atrophied brain. I accepted a position in industry only as an interim measure until I could afford to spend full time at my real profession, CREATIVE writing.

That was a good thirty years ago. Over those thirty years I have progressed—or drifted, depending on how you look at it—from the technical WRITING aspects of technical communications into all those other areas that make up today's complex and sophisticated technical publishing field. But still, outside my office I mentally hang a shingle that says "CREATIVE writer."

Now, a writer has certain personality characteristics associated with him that are time-honored, and to a certain extent valid. He is introspective, sensitive, a dreamer of his own dreams, and as a consequence somewhat vague as he relates to the real world around him. His checkbook never balances. This last certainly applies to me, and I have long since come to the self-justifying conclusion that that's the way it is because I'm a WRITER. My mind is busy with other things.

Unfortunately, in industry you are not judged by what your dreams might be, or even how poetically you can describe a sunset. Industrial management wants to know what, in terms of dollars and cents, they're getting from their employees. "What have you done for me lately?" seems to be the essence of their attitude. Narrow, but practical. And antithetical to the nature of the writer.

A letter recently came to a large Eastern university, addressed to its Department of Literature. "Dear Sirs," it said, "Please send me as soon as possible your latest literature on how to make sauerkraut." It shook the ivy-covered dons of the Department of Literature. My boss does the same kind of thing to me—constantly. And I have found that a casual "huh?" does not satisfy him. It has thus become necessary for me, against my every instinct, to become somewhat knowledgeable on production capacities, costs per unit produced, and a bunch of other stuff that bores the hell out of me. But it really turns him on.

Count Helmuth Karl Bernhard von Moltke, Prussian Field Marshall (1800-1891), once said, "I divide my officers into four classes: the clever, the stupid, the industrious, and the lazy. Every officer possesses at least two of these qualities....The man who is clever and lazy is fitted for the highest command; he has the temperament and requisite nerve to deal with all situations."

In my case, I have the *lazy* qualification to an outstanding degree. To implement the *clever* has always been my weakness. By the way, von Moltke went on to say, "But whoever is stupid and industrious is a danger, and should be taken out and shot." There's a guy who knew how to deal with people.

Returning to my point—I'm sure there was one—the problem was to define and maintain a sufficient data bank of production statistics to satisfy my boss that things were progressing in a satisfactory and cost-effective manner, without expending a great deal of effort and manpower to do it. When I first became involved in this problem, or rather when I first recognized that there was a problem that I had to deal with, I was responsible only for the Publications function—writer/editors and production typists. However, the Publications function is central to the whole publishing cycle for printed documents at Rocketdyne—we call it prime—and therefore had to know all of the costs associated with the publishing cycle.

PUBLICATIONS COSTS

Monitoring publishing costs for the many programs—we have close to 100 with contractual report requirements—

was quite difficult and complex. Various computer tabulations provided hour and dollar expenditures by departments and groups against specific contracts, but it was impossible to separate the various functional costs (from areas such as Publications, Illustrations, Photographic, and Printing) without considerable effort and time, and not nearly quick enough to satisfy program managers who believe their documents were costing too much.

Since time charges on employee timecards are the source of the computer-tabulated information, the accounting system was refined and the timecards more fully utilized to provide explicit and meaningful expenditure data on a *unit* level.

First of all, a permanent two-digit number was assigned to each specific function of Reprographics. This enabled us to monitor total contract costs by each functional unit; but there remained a dark, problem area. In some instances there were units which were overrunning budgets established by requirements set forth in contractual work statements. It was not possible for these budgets to accommodate unforeseen publication requirements. Program managers, not realizing the extent of unscheduled or larger-than-planned publication efforts, would assume that Publications costs were too high. To justify overruns, we had to conduct lengthy audits of job sheets and time charge records to find costs on individual publication efforts for a specific contract. It was then possible to show program management that our publication costs were not out of kilter and often, because of other factors, were lower than standard. In many instances, it was found that overruns were caused by a significant amount of additional but requested effort. To arrive at this conclusion, however, required many hours of statistical scrambling by all the units concerned.

The need, then, was for a system that would provide weekly (with the capability of daily, in some instances) expenditure tabulations by department, group, functional unit, and *specific jobs* against all contract or proposal charge numbers. Because of its "prime" function with regard to formal documentation, Publications assumed the responsibility for monitoring all charges.

To accomplish this objective, an unused 3-digit column on the employee timecard and the last three numbers of a Publications job number assigned sequentially to each specific job are used to key the effort for *all* contributing units. For example, if the Publications job number 5658 is assigned to a contractual report, then the last three digits (658) of this number become a part of the total charge number used by every employee working on this effort (Fig. 1). Each week, the actual hours and dollars expended against the contract with the number 658 identifying the specific task are accumulated and shown on the tab run (Fig. 2). Because of this one additional entry on the tab, the entire expenditure in hours and dollars by all units involved in that particular job can be extracted with minimal effort. It is now possible to derive complete Reprographic costs by any time increment against any contract, either by total expenditure or as the respective task is being processed.

A significant aspect of this procedure is the fact that the responsible writer/editor becomes aware of the costs for all phases of his reports. Quality, schedule, and coordination of report/proposal production have long been the writer/editor's responsibility. Now, readily available cost figures provide a more detailed control of his function.

Because of his involvement with the mechanics of editing and coordinating the reports, proposals, or other publication documents, it is impractical for the writer/editor to spend time retrieving cost data. Instead, this responsibility is assigned to one individual, a publications coordinator who receives the tab runs, extracts the appropriate information, and maintains cost records showing hours and dollars expended for each job by function (editing/coordination, illustrating, and printing). In addition the coordinator is in a position to monitor the costs for any task on a daily basis, if necessary, in situations where budgets are very restricted.

Now that we have the total cost for the job, it is a simple task to determine cost-per-page by dividing the total cost by the pages in the job. To us, a page is a page. We could differentiate types of pages—standard, per style and format guide; to

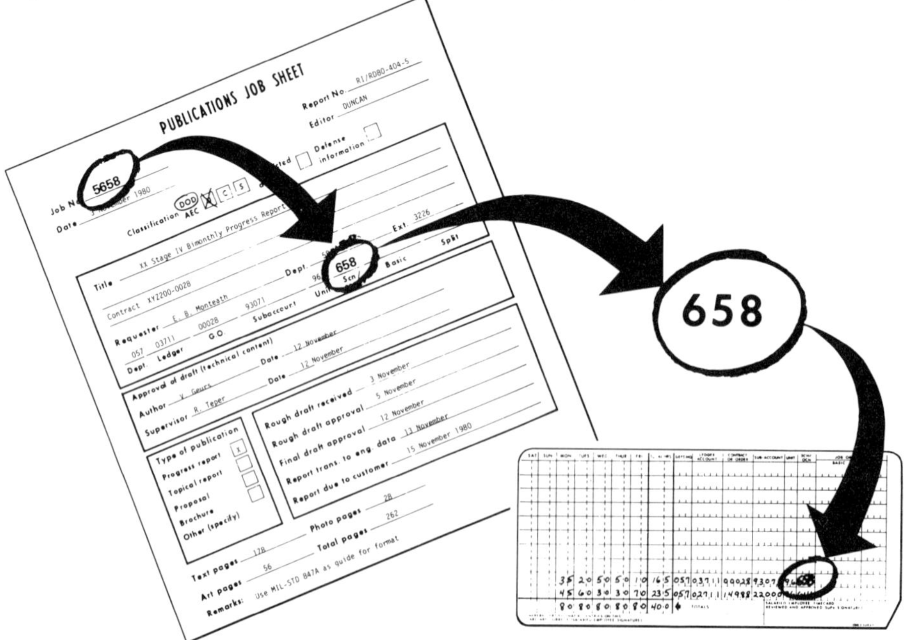

Figure 1. Job Number to Charge Number

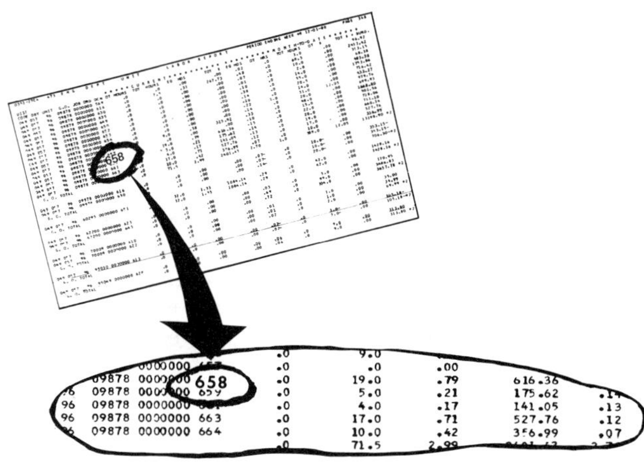

Figure 2. Tab Results

special format, such as for page-limited proposals; or photocomposed with double column copy, in-text figures, etc. But as a rule, we don't. Taken across a period of time, these refinements don't mean much when we're looking at *trend.*

However, the statistics can be cut a number of ways. We can evaluate the performance of the writer/editor based on cost-per-page. If the costs for one writer/editor are consistently higher than the rest, it is worth finding out why. Is the program area to which he's assigned more complex, resulting in more complex reports? Are there more drawings used in his family of reports, resulting in higher per-page costs? Or is he dogging it, and needs to be jogged up a bit? The numbers are *indicators*, not absolute criteria.

Over the years that we've kept these data, they have been rather consistent. A page at Rocketdyne is about one hour's worth of effort; a typical basic line drawing (a simple flow diagram, power curve, or the like) averages about an hour and a half. Using these criteria we can come up with a pretty fair estimate—*if* the estimated page count is at all accurate. (Often we're estimating the cost of the briefing or report before it's conceived, and a 100-page estimate can double if the engineer/authors wax verbose.)

There are some basic operating concepts that should perhaps be explained here. First of all, Reprographics supports all of Rocketdyne, all programs, centrally and with one basic, standard format. There are variations, of course, but they are treated as exceptions that have plus-minus impact on the average time/cost data. The estimating standard is a typewritten report, space-and-one-half, with a 15% factor of illustrations to text. That is, the average 100-page report will have 15 drawings in it. The editing, production typing, and illustrations staff all report to the same manager, and thus work closely together to accomplish the mission.

When we get into a major documentation effort—a major proposal, for example, with many volumes, page limitations, type size specified in the RFP, and other special requirements, the estimate has to take those elements into account. Luckily we now have historical data to rely on. In the early '70s we produced a very large proposal for the Space Shuttle Main Engine, and decided to use photocomposition, double-column, justified, with other refinements in terms of binding, packaging, and so forth. Flying blind, we estimated that job to the best of our ability, and the estimate was wrong. Luckily, we won the proposal and I was forgiven. In late 1977, we produced another major proposal for the Missile X Stage IV. The technology supporting photocomposition had advanced and we now had a basic capability for text processing in-house, but the cost data from the Space Shuttle proposal was nevertheless invaluable in estimating for that proposal. We were more nearly right on that job, which we won based on "the excellence of the proposal submitted."

ILLUSTRATIONS COSTS

Volume production in Illustrations is largely in support of reports and proposals (line drawings, curves, diagrams, schematics, and the like), and briefing charts, which average about 1500 units per month. With volume like this, statistical evaluation is meaningful; and again, a chart is a chart, and is, coincidentally, also a report or proposal line drawing. Such a standard "unit" costs us an hour and a half, and that number has held good for some years—practically since we got rid of the Leroy lettering system, in fact. When you get into an isometric drawing, a rendering, a sales brochure or display, there's no such standard. It's up to the manager to monitor the production, and keep aware, by the seat of his pants, of the production trends and who's producing. I know that "seat-of-the-pants" accounting methods are going to bother the time-and-motion study folks, and I've had my time in the soup with the auditors over this, but there are just some things that can't be quantified on a ledger page.

REPRODUCTION COSTS

On the other hand, in the areas of printing and duplicating we can run statistics enough to satisfy the most enthusiastic auditor. The presses do the unit counting for you, the operator's time can be accumulated in any conventional way, and you've got a meaningful measure of production. If we want to compare apples and oranges—that is, how many 8-1/2 x 11-inch units off the multilith presses versus how many off of a blueline machine, we can divide the blueline square footage produced by 8-1/2 x 11 and call them units.

Our monthly statistics for copy production are shown in Fig. 3, and this is what I show my boss—and he shows the division president. The division president is very concerned about the proliferation of data, which he feels has a tendency to be self-generating and, to an extent, counterproductive. So he looks at the bottom line, copies per employee. The population at Rocketdyne is growing to accommodate a growing business, so it's logical that the volume of paper should also grow, but only in proportion to the headcount. We have stayed in the middle to high 20's for some time, and that's fine. But what happened in October of our fiscal year 1981, when per capita copy count went to 33? I had to explain that real fast. Note that the curves are in terms of daily average to eliminate the pips caused by 5-week months.

PHOTOGRAPHIC COSTS

Another area in the Publications/Reprographics field that lends itself admirably to statistical analysis is the Photo Lab. Their cost-per-unit—whatever the unit might be—is made up of the manhours or labor involved, the cost of the material, and any vendor costs applied. This total cost, divided by the

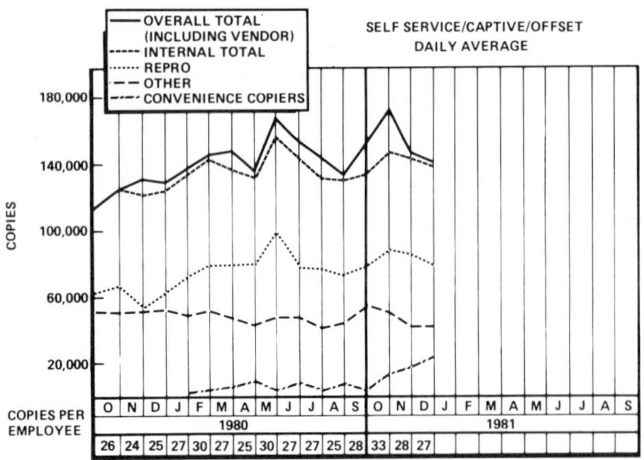

Figure 3. Reprographics Copy Production

units produced, gives some interesting statistics on individual types of units, which can be very effective in convincing your customers to use the most cost-effective end product. For example, Fig. 4 shows us that an 8 x 10 color view-graph costs $5.75, but a 35-mm slide costs $1.19. Those bare numbers need some interpretation for the requester, but still have basic validity. With the cost of silver and other photographic supplies inflating at their current rate, these items need constant attention and concern. Because of the price of film, we recently converted from 4 x 5 cut-film cameras to 120-mm and 35-mm, sacrificing some quality to achieve lower costs.

CONCLUSION

We keep a lot of records—raw data. They are retained pretty much as they are generated, by job, by function. Certain basic data (finished pages produced per month, daily averages of copies produced per month) are reported to management in a standard monthly summary. But the value is in the bank of data that can be cut as required for specific studies requested. A few months back we were asked what percentage of our overhead budget was devoted to the various departments of the division— by department. Since we are not in a charge-back mode for overhead costs, we had no reason to maintain this sort of statistic; but the job records show requesting department. We were able to generate a pie chart that gave the necessary information. It is cheaper to spend a couple of hours of analysis to respond to such specific requests once in a while than to spend a couple of hours every week or every month trying to anticipate what the next request might be.

In compiling and presenting publications production data, I think there's one rule, standard to all good presentation techniques, that should be paramount: *understand your audience*. What do they need and want to know in order to properly make decisions about your organization? And in many cases that audience breaks down into two basic areas: your boss and you. You will, of course, require greater detail than he does on which to base your decisions, and he'll appreciate the care and clarity with which you synthesize and present the facts to him. Give him the simplest, most straightforward analysis you can, but let him know that you have the data to back it up. Don't try to snow him with volume, or he'll think you can't separate the wheat from the chaff. Keep it simple, appropriate, and to-the-point. You'll find it beneficial to your boss, your professional status, and most significantly to your own understanding of what you're doing.

ACKNOWLEDGEMENT

The author wishes to thank Mr. H. B. Warren, Publications Project Coordinator, Rocketdyne, for his assistance in preparing this paper.

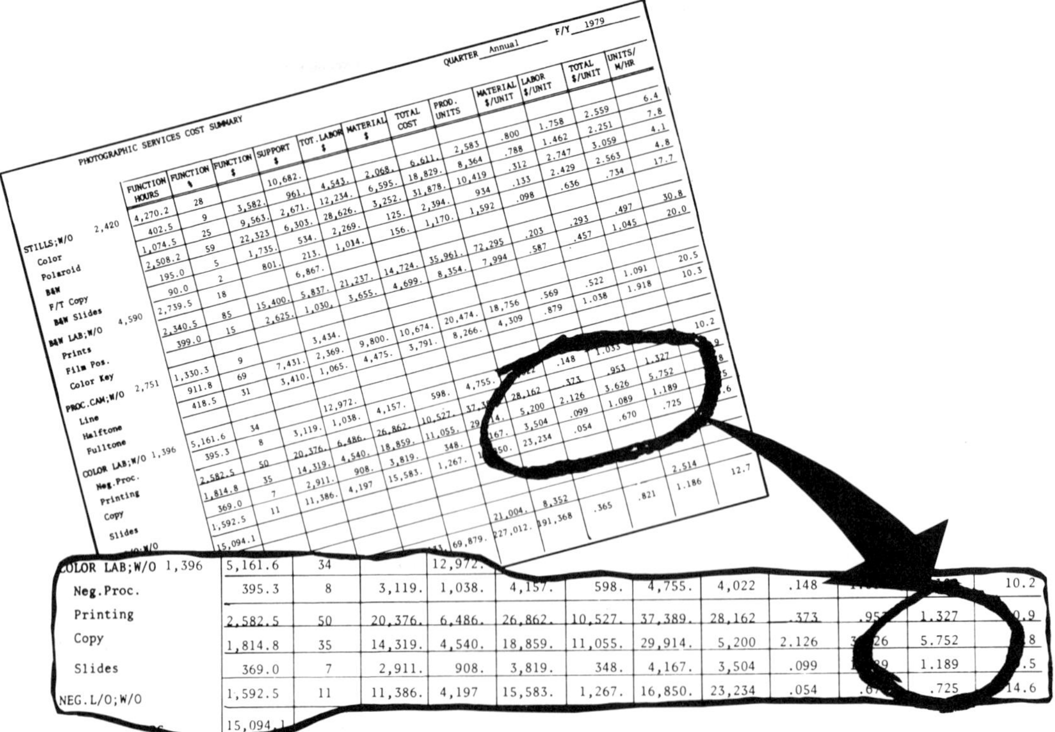

PHOTOGRAPHIC SERVICES COST SUMMARY

QUARTER Annual F/Y 1979

	FUNCTION HOURS	FUNCTION %	FUNCTION $	SUPPORT $	TOT. LABOR $	MATERIAL $	TOTAL COST	PROD. UNITS	MATERIAL $/UNIT	LABOR $/UNIT	TOTAL $/UNIT	UNITS/ M/HR
STILLS;W/O 2,420	4,270.2	28		10,682.								
Color	402.5	9	3,582.	961.	4,543.	2,068.	6,611.	2,583	.800	1.758	2.559	6.4
Polaroid	1,074.5	25	9,563.	2,671.	12,234.	6,595.	18,829.	8,364	.788	1.462	2.251	7.8
B&W	2,508.2	59	22,323	6,303.	28,626.	3,252.	31,878.	10,419	.312	2.747	3.059	4.1
F/T Copy	195.0	5	1,735.	534.	2,269.	125.	2,394.	934	.133	2.429	2.563	4.8
B&W Slides	90.0	2	801.	213.	1,014.	156.	1,170.	1,592	.098	.636	.734	17.7
B&W LAB;W/O 4,590	2,739.5	18		6,867.								
Prints	2,340.5	85	15,400.	5,837.	21,237.	14,724.	35,961.	72,295	.203	.293	.497	30.8
Film Pos.	399.0	15	2,625.	1,030.	3,655.	4,699.	8,354.	7,994	.587	.457	1.045	20.0
Color Key												
PROC.CAM;W/O 2,751	1,330.3	9		3,434.								
Line	911.8	69	7,431.	2,369.	9,800.	10,674.	20,474.	18,756	.569	.522	1.091	20.5
Halftone	418.5	31	3,410.	1,065.	4,475.	3,791.	8,266.	4,309	.879	1.038	1.918	10.3
Fulltone												
COLOR LAB;W/O 1,396	5,161.6	34		12,972.								
Neg.Proc.	395.3	8	3,119.	1,038.	4,157.	598.	4,755.	4,022	.148	[illegible]	[illegible]	10.2
Printing	2,582.5	50	20,376.	6,486.	26,862.	10,527.	37,389.	28,162	.373	.953	1.327	[illegible]
Copy	1,814.8	35	14,319.	4,540.	18,859.	11,055.	29,914.	5,200	2.126	3.626	5.752	[illegible]
Slides	369.0	7	2,911.	908.	3,819.	348.	4,167.	3,504	.099	1.089	1.189	[illegible]
NEG.L/O;W/O	1,592.5	11	11,386.	4,197	15,583.	1,267.	16,850.	23,234	.054	.670	.725	14.6
[illegible]	15,094.1											
[illegible]						[illegible]	21,004.	8,352	.365	.821	2.514	12.7
[illegible]						69,879.	227,012.	191,368			1.186	

Figure 4. Photographic Costs, Labor & Material ÷Units

A BEGINNER'S GUIDE TO PORTABLE-VIDEO-EQUIPMENT AND ITS USES

MICHAEL M. MIRABITO
BOWLING GREEN STATE UNIVERSITY

Michael Mirabito
Bowling Green State Univ.
Instructor
c/o 413 South Hall
Bowling Green, Ohio
419-372-2138

The use of video equipment by private companies and industries has increased over the last ten years. If your organization is also considering using such equipment, you should be familiar with some basic terms and concepts before this equipment is purchased. These terms include: what portable-video-equipment is, how it works, and how your organization can save money by selecting specific types of equipment. With this basic vocabulary you will be better prepared to examine the different brands and types of video equipment currently for sale.

Introduction

> Private television expenditures by business, government, and non-profit organizations for communications and training--including salaries, overhead, program production and equipment...totalled $279.2 million at the end of 1976 and is rising at an average annual rate of over 43 percent. Nearly two-thirds of this total represent expenditures by business and industry, where they are expanding at a rate of over 50 percent a year.[1]

When Judith and Donald Brush wrote their book in 1977, they predicted the future use and growth of video equipment in the industrial/corporate complex. The use of such equipment to create programs ranging from company news shows to visual demonstrations on "how to operate" new pieces of equipment became a powerful tool in the communicator's hands. It provided the communicator with a medium of sight and sound to convey a specific message.

Many books appeared on the market at that time explaining how video equipment worked and how it was used in producing programs. Most of these books, however, assumed the reader had prior knowledge of working with this equipment. This was not the case in many instances.

Today, with new technologies ushering in its innovations in the video field, this problem is more acute than it was a few years ago. New video equipment is introduced on the market, and the reader who lacks any basic knowledge of this equipment is lost in a maze of terms. This essay, therefore, is an introduction for these readers to video equipment. It provides simple explanations and descriptions of the primary pieces of Portable-Video-Equipment used in industrial settings.

Portable-Video-Equipment

The term portable-video-equipment usually refers to the Video Camera, the Video Tape Recorder (*VTR*), and the Microphone.

In contrast to most cameras which use film to record a picture, a video camera uses an *electronic pickup tube*. In film cameras, light enters through the camera's lens and focuses on the film's emulsion, thus recording the picture. Light entering through a video camera's lens, however, focuses on the front surface of the pickup tube. The light (the pictures the camera is shooting) is changed by the tube into electrical signals. These signals then travel on a cable to a Video Tape Recorder (VTR) where they are recorded on Video Tape.

As a video camera changes light into electrical signals, the Microphone changes sound. These electrical signals travel to the VTR on a separate cable and are recorded on the video tape. The sounds or audio signals (which may include people talking), are recorded SIMULTANEOUSLY with the video signals on the same video tape.

Video tape itself is the material on which the VTR records the pictures and sounds. Like audio tape used for tape recorders, video tape is a plastic based material coated with metallic-oxide particles. It is to video what film is to cinematography.

Once the pictures and sounds are recorded, the video tape is *played-back*. The recorded tape is played through the VTR, and as it passes through the machine, an electrical signal is created which travels on a cable to a television set. The set changes this signal back into the original pictures and sounds.

Summary of Steps

1. The video camera changes light (pictures) and the microphone changes sounds into electrical signals.
2. Both of these signals are recorded on the same video tape by the VTR.
3. The recorded video tape is played-back on the VTR creating an electrical signal.
4. A television set changes the signal into the original pictures and sounds.

Field/Studio Video Camera

The video camera is the crux of the portable-video-system. The price range and available features vary from camera to camera, and one should be bought based upon your organization's needs. This includes the possible future expansion of your video department where the camera you buy should be able to meet these future needs. In this writer's opinion, the Field/Studio video camera is such a camera.

It is part of a modular system where different features are added to the basic camera body. In its *field configuration*, the camera operates on a portable battery supply, is carried by one person, and may weigh less than fifteen pounds. The camera is used for On-Location-Shoots in this form. This is when your organization dedicates a new office building, for example, and you Shoot (record the pictures and sounds) a video tape of the dedication ceremony. The field camera and its companion VTR enable you to shoot a program wherever the subject, such as the dedication program, may be.

If your organization builds a Television Studio, the camera is modified for this new location by replacing the field modules with those designed for studio use. These may include a different Lens and Viewfinder, and by connecting the camera to a Switcher and power source in the studio (See Annotated Terms).

To sum up, the field/studio camera performs TWO jobs and can be used in two different types of productions. It is a good investment, for as your video operations grow, the camera can change and grow along with it.

Annotated Terms and Descriptions

As was promised in the introduction, the following is a annotated guide and introduction to additional video terms and descriptions.

Editing: The video programs you shoot can be edited. This means that different scenes can be joined together on one tape. The on-location scene of the dedication ceremony, for example, can be edited (electronically joined) to a scene taped in the studio. Both scenes are edited together to create ONE program. In contrast to film editing, the video tape is not cut to join pieces of tape (scenes) together. Rather, one scene is electronically transferred from one tape and is recorded on another tape.

Lens: The lens focuses and projects light (pictures) on the video camera's pickup tube. The lens normally used on a video camera is a *Zoom* or *Variable Focal Length* lens. It provides wide angle through telephoto shots.

When the lens is at the *wide angle* mode (setting), it provides a wide field of view. If you shoot the dedication ceremony, for example, at the wide angle mode, the whole building will be visible on television when you play-back and view the tape. In the *telephoto* mode, on the other hand, the whole building will not be visible. Only the building's door might appear on the screen, but it will fill the whole screen. Looking through the camera at a telephoto setting is similar to looking through a binocular, for it magnifies the object.

Lights: Video cameras for in-studio and on-location-shoots require a specific number of footcandles (a measurement of the intensity of light) to operate. Special lights (lamps) provide this light or illumination.

For on-location-shoots, if additional illumination is needed, portable lights are used and can be attached to light stands. In most studios, however, the lights (designed for studio use) are hung on a Lighting Grid. This is a series of metal pipes bolted to the studio's ceiling.

Studio: In its simplest form, the studio is a soundproofed room where a program is shot and recorded. The soundproofing prevents any outside noise from entering the room and consequently being recorded on the video tape during the shoot. The studio is a controlled environment. As your video department expands, you can also add more features (such as the lighting grid) to your studio to serve your needs.

To save money, an existing office can be modified into a simple studio, and if your organization already owns portable lights used for on-location-shoots, these lights can be adapted for this location.

Switcher: A switcher is an electronic device which enables you to use two or more video cameras (which are connected to the switcher) to record a program during an in-studio or on-location-shoot. If two cameras are used, for example, either image the camera is shooting is recorded on the video tape.

Consequently, when the button controlling the first camera (Camera One) is pressed on the switcher, camera one's picture is recorded. This can also be considered "editing on the spot." You choose *which picture* you want recorded in the *sequence* of your choice.

Viewfinder: In contrast to a film camera where you look through the lens via an optical system to see the picture, with a video camera, you look through a small black and white monitor (television) mounted on the camera. The viewfinder electronically enables you to see the picture the video camera is shooting.

Video Equipment and Format Sizes

Different video tape sizes are used with portable-video-equipment, and the most common sizes or *Formats* are 3/4 and 1/2-inch tapes. The term format also lends its name in identifying the format size of video equipment. A VTR using 3/4-inch tape, for example, is part of the 3/4-inch format.

The 3/4 and 1/2-inch video tapes are packaged in cassettes. The cassette is a plastic housing which serves the same function as a cassette used in a tape recorder: it holds and protects the tape. The 3/4-inch cassettes are called *U-matic*, while 1/2-inch tapes are either VHS or Beta (used in Sony manufactured equipment).

It is important to be familiar with these formats as they can serve your organization in a variety of ways. Your organization can save money by using a specific format of tape and equipment for a specific purpose. Each format also has its own particular strengths and degree of cost efficiency.

Presently, the 3/4-inch format is the standard size for portable-video-equipment. The advantages of this format are:

* 3/4-inch video equipment is flexible in its uses. An example of this is the field/studio camera where it can be used in two different production locations.
* As the industry standard, many organizations already own 3/4-inch equipment. If you eventually plan to market your tapes, there is an existing outlet.
* 3/4-inch video equipment provides technically superior video tapes. This means the recorded pictures and sounds will faithfully reproduce the event that was shot.

The sophistication of 1/2-inch equipment is not as great as the 3/4-inch equipment at the present time. There are not, for example, 1/2-inch field/studio video cameras that can be modified for use in both locations. Yet, this format has one *definite* advantage: it can save your organization money if your video tapes are dubbed and distributed.

A dub is a copy of your program on another video tape. As a "Xerox" is a copy of an original document, a dub is a copy of the original video tape. One-half inch tape is less expensive than 3/4-inch tape, so these dubs will also be less expensive to make. Another cost saving factor is that a VTR designed to use 1/2-inch tape is also less expensive than its 3/4-inch counterpart. Thus, if your organization has twenty branch offices and each branch will have a VTR to play the distributed tapes, the total cost for all the VTRs will be lower than if you bought the more expensive 3/4-inch machines.

In deciding which format your organization adopts, this writer suggests that you use both sizes. The 3/4-inch video equipment can be used to tape your programs to produce technically superior tapes in addition to the flexibility this format provides in terms of equipment uses. If your organization then decides to distribute its programs, the original 3/4-inch tape can be dubbed on a less expensive 1/2-inch tape, which in turn will be played on the less expensive 1/2-inch VTR. Consequently, you benefit from 3/4-inch's flexibility and technical quality and from the 1/2-inch's cost efficient operations.

Appendix

As the scope of this paper does not permit an in-depth examination of video equipment and all of their functions, the following books provide this information:

Barwick, John and Kranz, Stewart. *Profiles in Video*. New York: Knowledge Industry Pub. Inc., 1975.

Bermingham, Alan; Talbot-Smith, Michael; Symons, John; and Angold-Stephens, Ken. *The Small TV Studio*. New York: Hastings House, 1972.

Brush, Judith and Brush, Douglas. *Private Television Communications*. Boston: Herman Publishing, Inc., 1977.

Burrows, Thomas and Wood, Donald. *Television Production*. Dubuque, Iowa: Wm. C. Brown Co., 1978.

Millerson, Gerald. *The Technique of Television Production*. New York: Hastings House, 1972.

Zettl, Herbert. *Television Production Handbook*. Belmont, Ca.: Wadsworth Publishing Co. Inc., 1976.

Footnote

[1]Judith Brush, and Douglas Brush, *Private Television Communications*, (Boston: Herman Publishing, Inc., 1977), p. 19.

DOCUMENT COORDINATION: A NEW ROLE FOR THE TECHNICAL WRITER

Rebecca G. Montevideo
Environmental Science and Engineering, Inc.

Rebecca G. Montevideo
Environmental Science
and Engineering, Inc.
P.O. Box ESE
Gainesville, Florida

904/372-3318

Technical Writer/
Document Coordinator

A new role of the technical writer is that of document coordinator. This includes involvement in a document from the time it begins until it is delivered to the client. The document coordinator can be defined as a "Jack of all Trades" because of the many tasks involved in coordination of a document. Once a technical writer/document coordinator is familiar with the company's capabilities and limitations, he/she is able to guide the inexperienced technical personnel, in systematic, smooth, and consistent document preparation, saving both production time and costs. This paper illustrates duties and responsibilities of the technical writer/document coordinator in producing marketing proposals and the advantages of this concept.

INTRODUCTION

Streamlining proposal preparation is particularly important today because of the rising costs of document production. One way to cut these costs is to employ a staff familiar with every aspect of the document production process. This will save time in producing documents.

A new role of the technical writer is that of document coordinator. This includes involvement in a document from the time it begins until it is delivered to the client. This paper illustrates duties and responsibilities of the technical writer/document coordinator in producing marketing proposals.

RESPONSIBILITIES INVOLVED IN DOCUMENT COORDINATION

The first responsibility of the document coordinator is to gain a good working relationship with the proposal manager. To do this, the coordinator should have an initiation meeting with the manager to discuss the strategy to be used to produce a quality document in a timely, cost-effective manner.

Initiation Meeting

All staff members who will be technically involved in the proposed document should be present at the initiation meeting. The document coordinator should have suggestions or questions prepared prior to the meeting so that everyone leaves the meeting with a clear understanding of their responsibilities in producing the document.

After the strategy has been planned, the proposed document should be outlined. The document coordinator can assist in this task by providing examples of previously used, successful outlines. The outline for a proposal should include the following: Introduction, Scope of Work, Schedule, Personnel or Project Team, Work Experience, Company Facilities and Capabilities, and Business Section. The outline for a proposal may vary, depending on client needs specified in the Request for Proposal (RFP).

After the outline has been established, sections of the document should be assigned to appropriate personnel for preparation.

Projection and Coordination with Other Document Production Groups

Since a number of groups are involved in producing a document, all document production steps should go through the document coordinator to ensure proper scheduling. Groups involved in the physical production of a document include: word processing, graphics, accounting, quality assurance/quality control (QA/QC), and copy center. The document coordinator should inform these groups, as soon as possible, if any problems are anticipated, such as a tight schedule or unusual format.

Use of Shells in Document Preparation (Proposals)

Shells (notebooks containing boiler plate material) such as descriptions of company facilities, technical capabilities, and related experience should be maintained to include accurate, up-to-date information. Appropriate sections can be used directly from these shells and save many hours of production time.

Resumes for both staff members and consultants also can be maintained in shells and "pulled" for use as needed. Usually, resumes are tailored to meet the specific needs of each proposal.

It is the responsibility of the document coordinator to ensure that all marketing shells are kept up-to-date.

Systems Manual

Companies should have a systems manual outlining the standard format used for reports and proposals, as well as standards production procedures. Elements of the format should include: line spacing, headings, type style, and table and figure format. The document coordinator uses this manual to ensure consistency in style and answer any questions in format that may arise later in the production of the document.

Proofreading and Editing

After the document coordinator has extracted relevant information from the marketing shells and the proposal manager has written the introduction and scope of work, the document is edited and put in proper format using standard editing symbols. The document coordinator submits them for typing or word processing. After typing, the document coordinator proofreads and edits, as needed.

Use of Word Processing Equipment

The use of word processing equipment can save a tremendous amount of time in document production replacing the old "cut and paste" method. Pages of text can be stored on memory disks and later be recalled, updated, and rerun to incorporate changes with a minimum of effort. This allows the document coordinator to receive quick turnaround and also allows more time to address other obligations in producing the document. The document coordinator should be trained to operate the word processing equipment in the event of an emergency.

Graphics

Graphics are an important element of the document. It the document coordinators' responsibility to schedule production of graphics and ensure accuracy of the finished product. If a special cover is required for the document, scheduling is important to allow time for drawing and printing. The document coordinator should act as a liaison between graphics personnel and the report/proposal manager to ensure a consistent graphics format. Other graphics that may be used in documents include maps, flow charts, critical path schedules, organization charts, etc.

Review

In most cases, the proposal manager's supervisor will review the document for technical accuracy. It is the responsibility of the document coordinator to review the document for editorial accuracy and format consistency. Some companies maintain a QA/QC system to ensure both technical and editorial accuracy. If a QA/QC system is used for review, the document probably will be reviewed again by the proposal/report manager to answer any questions or comments from his supervisor or QA/QC. It is the responsibility of the document coordinator to know which stage of review the document is in at all times.

When the proposal manager has addressed all questions or problems, the document is returned to the document coordinator for final editing. After the document has been edited and changes have been incorporated by word processing, it is proofread again by the document coordinator and returned to the proposal manager for final review. When the proposal manager's review is complete, the review process begins over and the proposal manager's supervisor reviews the document for final approval, as well as QA/QC. The document is then returned to the document coordinator for final processing.

Checking Responsiveness to Client's Needs

It is the responsibility of the document coordinator to ensure that the document meets the client's requirements. In the case of a proposal, the document coordinator reads the RFP carefully, checking the document to ensure that all requests from the client are addressed.

Flexibility

The document coordinator can assist in smooth production of a document by using shells, systems manual, and word processing equipment. It is important for the document coordinator to follow a routine system when producing a report or proposal, yet remain flexible enough to allow for deviations in the routine if it does not allow for fulfillment of the client's needs. Also, some proposal managers may have their own style, requiring flexibility by the document coordinator.

Transportation of the Completed Document

It is the responsibility of the document coordinator to decide how the document will get to the client. If the document is ready to be transported during normal working hours, the document coordinator may assign this task to someone else. In the case that it is necessary for the document to leave the company after business hours, the document coordinator makes the delivery. In any case, the ultimate decision of transportation is the coordinator's responsibility. Transportation methods include: bus, mail, UPS, airlines and courier service, or hand delivered by the document coordinator or a designated company representative.

It is the responsibility of the document coordinator to ensure the arrival of the document to the desired location on or before the specified deadline.

Post-Proposal Tasks

The responsibility of the document coordinator does not end when the document has been submitted to the client. The document coordinator ensures that all appropriate in-house personnel receive copies of the document, including the company library for central filing. The document coordinator keeps records of all work completed and the history of the document, including: (1) when the document left the company, (2) when the document was received by the client, and (3) the name of the person receiving the document on behalf of the client. The document coordinator should keep all rough drafts of the document that could be useful at a later date, i.e., cost sheets, and major corrections by the proposal manager, his supervisor, or QA/QC.

If new information, such as experience or capabilities of the company were written for this proposal, a copy should be placed in the appropriate shell for use in other similar proposals.

SUMMARY

The document coordinator can be defined as a "Jack of All Trades" because of the many tasks involved in coordination of a document. An important advantage of this concept of document production is the one-stop information source for style and content. Another advantage is in the training of inexperienced technical staff in proposal preparation by an experienced document coordinator. Once a technical writer/document coordinator is familiar with the company's capabilities and limitations, he/she is able to guide the inexperienced technical personnel, in systematic, smooth, and consistent document preparation, saving both production time and costs.

"SEX" IN TECHNICAL ART

—An Approach to Content Development

Robert H. Nielsen

THE BOEING COMPANY

Robert H. Nielsen
The Boeing Company
Graphics Coordinator

417 Prospect St.
Seattle WA 98109
(206) 394-3862

Art Chairperson
Puget Sound Chapter STC
1981 Graphic Arts Competition

Abstract

Attractive and arresting artwork can contribute to making a technical document or presentation communicate better. Techniques for visualizing more distinctive art are discussed, as are some considerations for choosing effective styles. Suggestions for improving standard graphic devices are offered. The appropriateness of art to a particular method of presentation is addressed.

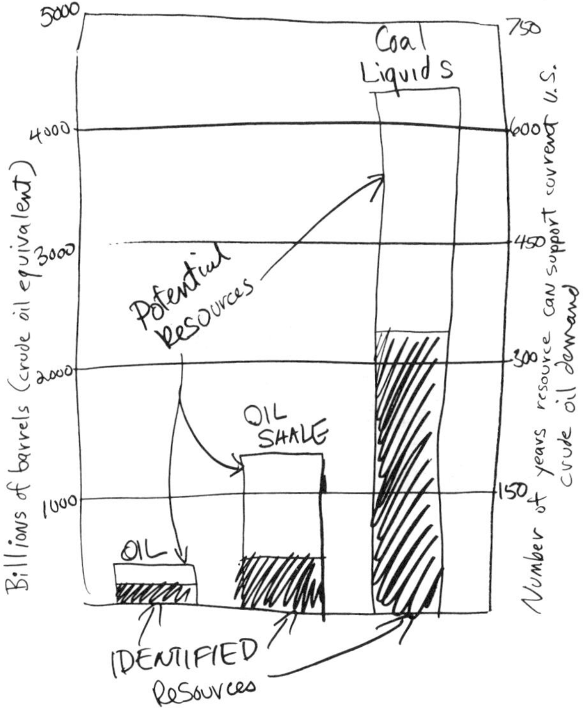

The Problem

That homely little bar chart with the cryptic note scrawled above it isn't an indelicate proposition; it's a plea for help in improving a drab piece of technical art. A technical presentation or document can't do a good job of communicating unless it captures the attention of the audience. Attractive and intriguing visuals can command that attention; and graphics specialists, writers, and editors are the people who can contribute most to making artwork attractive, lively, and informative.

The audience for your technical presentation is bombarded from every direction by information. Messages come from television, newspapers, magazines, trade journals, textbooks, reports, and your business rivals. Even when a subject is in the field of a reader's specialization, even when he is *paid* to read it, he is more apt to respond favorably to, and absorb, information that is clear and visually stimulating. The audience is already conditioned by media; your technical presentation must somehow distinguish itself from all the other data streaming through our society.

How To Fix It

Accepting the challenge to "sex up" a piece of technical art can be rewarding and fun. Look critically at each rough input, read it carefully and try to think of ways to streamline and pictorialize it. Don't be afraid of using a decorative approach to some of the art, and even try being outrageous. If your approach is too outrageous, someone is sure to vote it down, but in the process you may loosen that person up just enough to have him accept something more innovative than he might have at first.

Techniques for making technical art distinctive are limited only by imagination. Some are purely graphic, such as bending the bars of a chart into new shapes or substituting a picture for boxed words in a flow diagram. Other techniques are more sophisticated, such as developing an illustration to replace a wordy table or using a piece of art work and an explanatory caption to replace a number of paragraphs of narrative. Captions, by the way, are extremely useful tools, sadly underused in most technical art. Fine examples of the caption device abound in *Scientific American, National Geographic* and *Popular Science*, where they are extensively used to tell an important part of a story. A narrative caption under an interesting photograph or illustration is an excellent

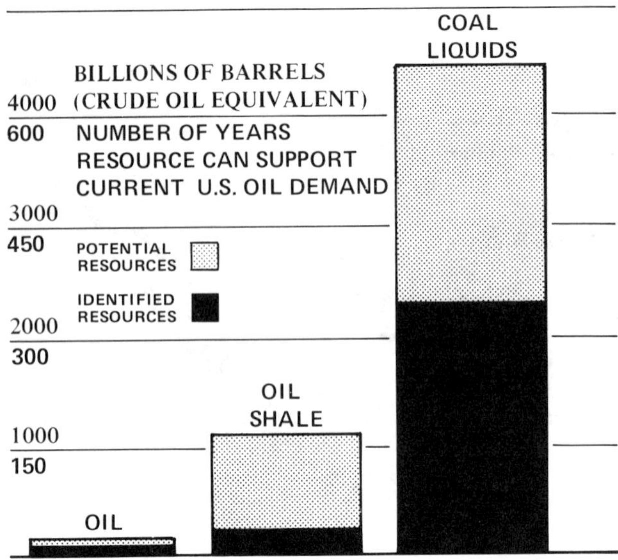

Same Old Bar Chart, Permutation 1

Same Old Bar Chart, Permutation 3

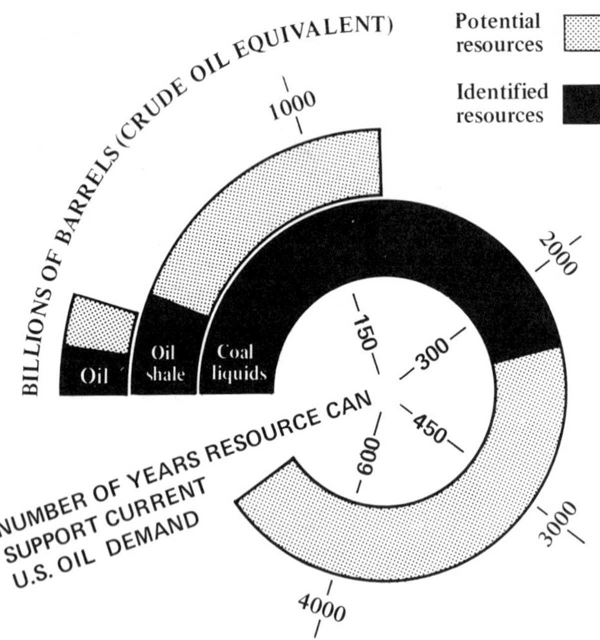

Same Old Bar Chart, Permutation 2

Things to do with bar charts. *All of these charts are the homely little bar chart from the first page in different costumes. The first is quite ordinary with the lines drawn straight and a key added to tell one part of the bar from the other. It makes something of a demand on the reader in that he has to figure out that one particular typeface relates to a particular set of numbers on the ordinate scale, and those numbers always appear above the grid lines. The second example has the same old bars, but this time they have been wrapped around a center point and have become curves. The chart is about as easy to read as the first, but it is certainly an unusual shape which the viewer might find entertaining. The third example is still the same little bar chart with the bars made into cylinders and the scales distorted in perspective. It's not nearly as accurate in giving data as the other two, but it certainly points out the fact that there seems to be a tremendous supply of fuel to be derived from coal liquids. Is accuracy of data or plotting really important in this kind of a chart? Suppose the data are off by 50 years in either direction, will it matter 600 years in the future? The magnitude of coal-derived fuel is the real message. A strong graphic statement can convey this kind of information much more effectively than can words.*

way to draw the reader into the body of the presentation.

Making technical art more pictorial usually makes it more interesting because viewer attention is more easily attracted by a picture than by words, no matter how splendid the writing nor how artfully it is presented. (Do you know anyone who can resist looking at the cartoons first in a new copy of *The New Yorker*?) Words, too, are an important tool in technical art. A picture with words on it may very well be the *definition* of technical art.

The trick is to use as few words as possible, making each one a vital element of the illustration and the message it is communicating. The reason for this approach is that words are not, by themselves, illustrations—nor can they be. Words are comprehended in a serial or linear way. One after the other, they gradually build an impression in the reader's or listener's mind. A picture, however, is understood as a gestalt: it is projected directly and as a complete entity into the viewer's consciousness. There's no debate about whether something is big or little in a picture or any puzzle about its shape. The fundamental difference in the way words and pictures are perceived gives each a particular strength in communicating certain types of information. The synergy that takes place when words and pictures are combined can produce extremely effective communicating devices.

A table or flow diagram is a particular way of using words that is not the same as narrative, but is peculiarly unlike pictorial art as well. Words in boxes become discrete parcels of information that relate to other parcels of information because of their graphic arrangement or interconnection. The relationships created by these juxtapositions make implications about the data that would be difficult to communicate in a narrative. Words in boxes are read in a linear fashion, then compared as a gestalt. As useful as this type of graphic quantification of information is, however, the visual product is essentially dull. A table looks like every other table, and an organization chart by any other name is still an organization chart.

Some old and pedestrian techniques can add a surprising amount of sparkle to technical art. Visual presentations, such as slide shows, are particularly fertile ground for improvement. They can be enhanced with little cost or complication. If presenters insist on using word charts (one suspects that word charts are no more than speaker's notes shared with the audience), the words can be superimposed over an appropriate background: pictures of facilities, pictures of products (we in the airplane business are particularly fortunate to have lots of pictures with an abundance of blue sky in which to write messages) or textures (grass, wheat fields, carpet, cloth, almost anything without an overly busy pattern). "Found" images can be incorporated into slide shows—especially useful are old prints or reproductions of artwork on which the copyright has expired or never existed.

A 35mm camera and a bit of imagination can produce effective and unusual visuals. Charts can be done on large boards that are then set up in interesting places—perhaps somewhere in the company's facilities (laboratories, production areas, computer rooms) or in purely decorative surroundings—nailed to a tree, tacked on a sailboat, surrounded by flowers. If a background can be found that is particularly appropriate to the presentation, terrific.

Is It Better or Just Different?

A really successful illustration is one that meets the tests of attractiveness and the ability to communicate clearly, and these tests must be administered with rigor and lack of prejudice. Sometimes technical people who are absolutely superb in their own disciplines fall in love with an ineffective graphic approach to express their ideas. Their subjective bias becomes so strong they will simply not entertain the idea that their approach cannot communicate a perfectly worthwhile message. Every piece of technical art should be presented in such a way that a person of reasonable intelligence cannot misunderstand the essence of the message. Perhaps all the exquisite subtleties need not be explained, but the basics must be clear—two quantities are being compared, or we start at A and arrive at Z, or a new shape is different from an old shape.

Dumb Question

Often the first step toward more communicative pictorial technical art is to search for a visual equivalent to an idea. *What degree of risk is associated with this new technology? How soon could the new facility be at full production?* Writing those statements on a viewgraph with a big question mark in the center *is* more pictorial than the statements alone, but it's also trite, obvious, and sophomoric. And no matter how brilliant a solution the engineering department evolves, the light bulb is a lousy metaphor for a terrific idea.

Bad Idea

A good graphic communicator can be of enormous assistance to a befogged technical person because he can look at the elements of a piece as abstract quantities that interact. A graphic specialist often can see and impose a visual order or pattern that was hidden in the technical person's rough sketch. If a piece of art that crosses your desk makes you groan, chances are it won't look much better if it gets the standard layout and pasteup treatment with straight lines and neat type. *Talking* with the technical person who came up with the basic concept will frequently reveal good ways to simplify and enhance both his illustration and his message. Near miracles of communication have resulted from asking: "What do you really *mean*?"

Table

	AIRCRAFT LOCATION	GATE OR RAMP	TAXIWAY	RUNWAY	IN-FLIGHT
	SYSTEM CHECKS	PREFLIGHT, ELECTRONIC	PREFLIGHT, MECHANICAL	(LANDING) DATA TRANSFER	CONTINUOUS
CONDITION	NORMAL	NO FAILURE DEPART	DEPART	DEPART	NORMAL
	PREFLIGHT FAILURE	FAILURE DETECTED DEPART	DEPART	DEPART	NORMAL
		FAILURE DETECTED DEPART	DEPART	DEPART	RESTRICTED
	PREFLIGHT FAILURE NO-GO	FAILURE DETECTED ROUTE TO HANGAR OR GATE FOR REPAIR REPAIR AT GATE	FAILURE DETECTED ROUTE TO HANGAR OR GATE FOR REPAIR		
	IN-FLIGHT FAILURE	NO FAILURE DEPART	DEPART	DEPART	FAILURE DETECTED MODIFY FLIGHT PLAN

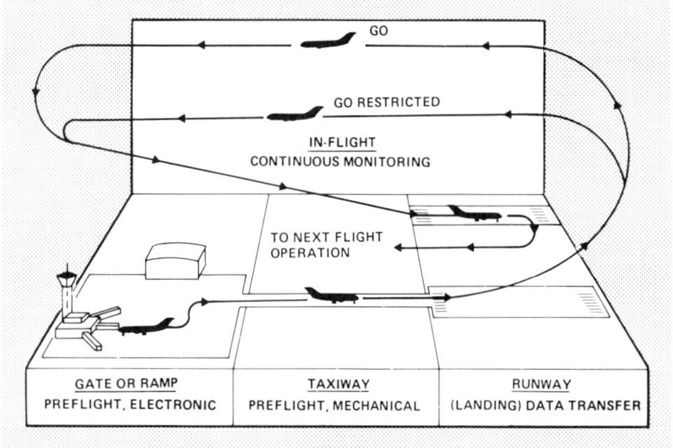

Figure

The table is a fairly standard way of presenting data, but the data in this case are more interesting than one may think. The illustration was developed to depict the Preflight Failure condition described in the table. The accompanying discussion (which is necessary to both the picture and the table) explained how airplanes could still operate even if a particular piece of equipment was malfunctioning. Three other illustrations were developed for other parts of the table, and all were used in a slide show. The picture does considerably more to illustrate what the speaker is talking about than does the table.

The most obvious visual equivalent for an idea usually is the absolute worst choice for an illustration because it's probably already a cliche. (The *road* to higher profits; the *mountain* of difficulties; the *train* of events.) And some equivalents just don't have much visual impact. Pictures or drawings of crude oil, coal, and oil shale are likely to be little more than dark, amorphous shapes; errors in computer software are virtually invisible; and illustrating an odorless, colorless, tasteless gas (such as hydrogen) is an exercise in futility. An *effect* or *property* of hydrogen gas, however, could be nicely illustrated with a drawing of a balloon or a zeppelin, especially if the low molecular weight of hydrogen is an issue. Illustrating *liquid* hydrogen is an even greater challenge. It might be possible to take the airship (zeppelin equals hydrogen) and combine it with a faucet (faucet equals liquid). Does the result *really* say liquid hydrogen in a pictorial way, or have we created a rebus that muddles things with obscure symbology?

Liquid Hydrogen or Obfuscation?

Show Business

Really effective technical art should be tailor-made to suit its application. For slide shows and visual presentations, few or no words should appear on the screen. There will be, after all, a live human being speaking during the presentation. Let the *person* provide the words and explanations, and let uncluttered visuals illustrate his points clearly. Never insult the audience by projecting words on a screen and reading the words as well.

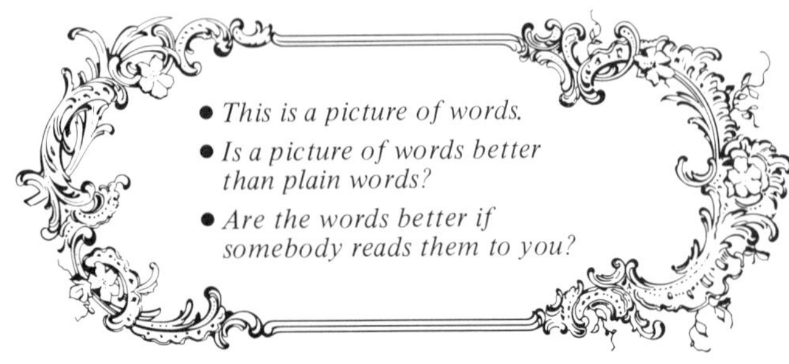

Words as Art

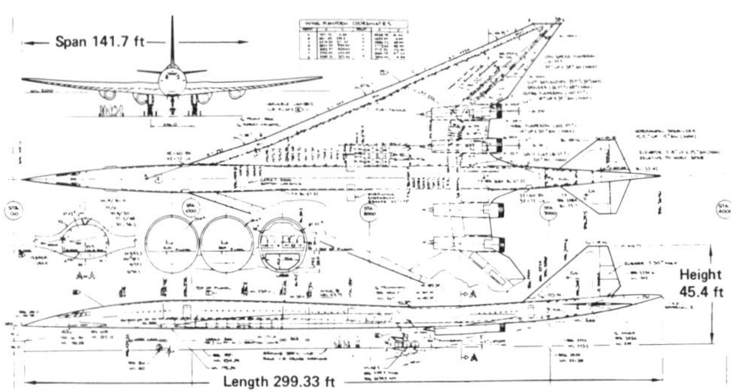

Somewhere in that rat's nest an airplane is hiding. And the visual clutter is not accidental. The technical specialist who used this illustration in a visual presentation said the intention was to show that the design had been carried on to a level of considerable detail; an admirable and worthwhile idea to communicate. But does the picture itself really convey that lots of work had been done, or is it just hard to see?

The concept of considerable design work having been done is a matter of credibility, a quality difficult to address in graphic terms, at least in a drawing like this. To do so, perhaps an overall view could be shown, and then an enlarged detail such as a particularly innovative flap system could be displayed on subsequent slides with the speaker pointing out the level of design undertaken for this model.

Technical people often request that illustrations for a visual presentation be able to stand alone, to contain sufficient written information so that the artwork is completely self-explanatory. The rationale is that copies of the artwork are distributed as "souvenirs." That logic suggests that the presenter could avoid the inconvenience and expense of travel, and the discomfort of standing up to talk before a group of people. He could simply send a tray of slides off to the symposium. The slides are sure to be about as interesting as he would be in person. Or, the presenter might even write a technical paper, have it printed, and then have the pages photographed. The photographs could then be displayed by an automatic slide projector.

Nonsense, of course. The point is that art for a visual presentation must be uncluttered and arresting and must *support* the speaker. Art for a document that is to be read and considered must have sufficient data built into it so that it really can stand alone. Artwork just cannot be effective when used interchangeably in two different media.

If part of a presentation simply can't be illustrated, consider this innovation—projecting a neutral picture (a generic system, product, or facility). How about projecting no picture at all, (an opaque slide or the projector shut off) and letting the attention of the audience fall on the speaker? Too theatrical? Hardly. A good presentation should be good theater as well.

Perhaps the most important consideration for improving technical art is to look at it from the point of view of the audience. Is it done with style and with a level of professionalism befitting the subject? We in the technical communications field have much to learn from our colleagues in magazine publishing, electronic media, and film. The word technical is, after all, only an adjective; it is the noun communication that should concern us most.

AN INTRODUCTION TO PRESENTING PUBLICATIONS PRODUCTIONS DATA

Lou Perica
SRI International

Lou Perica
SRI International
Manager, IMEG Publications
333 Ravenswood Avenue
Menlo Park, CA 94025
(415) 859-2933

Numerous published papers and ITCC presentations

This paper presents the reasons why publications production data should be compiled and reported and suggests ways for doing it. Data can be collected in all publication areas: graphics, composition, reproduction, and writing and editing. What should be counted and standards for accumulating the data are discussed. For instance, what constitutes a page of editing or composition? Should illustrations be segregated by complexity or type? What is meaningful data in the area of reproduction and binding? Accurately collected data presented in a useable form can be used to justify more staff, new equipment, or a well-deserved raise or promotion. In some cases the production data are used to establish production rates which are combined with salary information to provide estimating guidelines. The data can also be used to identify the "peaks and valleys" of production as an aid to scheduling work. Suggested forms for the collection of data are included.

INTRODUCTION

As a youth (My, that must have been a long time ago!), I remember passing a store that had a gallon jar full of beans in the front window with a sign that offered a prize to the person who came closest to guessing the number of beans in the jar. The memory of that scene returns when I hear the term "bean counter." A bean counter is not satisfied that the jar is full of beans. The bean counter has a passion to know the exact number of beans in the jar.

The subject of publications production data in some ways fits the bean-counter mentality. It is clear when the jar is full. When the jar is partially full, how many beans are needed to fill the jar? Understanding the answer to this simple question requires a detailed explanation. Similar questions pertaining to production data are explored in this paper. Knowing the number of beans required to fill the jar satisfies curiosity. Knowing publications production data serves more serious purposes.

WHY COLLECT PRODUCTION DATA?

The reasons for collecting publications production data include:

- Having a measure of effectiveness
- Knowing maximum capacity of the system
- Justifying more staff or new equipment
- Establishing production rates
- Providing cost estimating guidelines
- Substantiating a request for a raise or promotion.

Though this paper presumes that collection of data is a management function, the astute employee will collect his or her own data to provide a measure of effectiveness and to justify a raise or promotion.

WHAT IS A PRODUCTION UNIT?

Using the analogy of the jar full of beans, the jar represents total publications production capacity, and the beans represent production units. Now there are beans and there are beans. Lima beans will occupy more space than kidney beans. Clearly, a page is not a page is not a page. A final, single-spaced typewritten page will have more text than a double-spaced draft page. Generally, a publication will enter the production system as a certain number of draft pages. What is a draft page?

Normally a draft page is double-spaced text within a prescribed image area. The image area we use is approximately 6 inches by 8-1/2 inches. The larger the image area, the more text that can be accommodated. Type size, of course, is also a factor, just as size was a factor with the lima beans and the kidney beans. With smaller type, more characters can be crammed on a page. (Another way to define a page in the context of word processing is to establish a standard based on number of keystrokes.) Exhibit 1 illustrates a page-counting system that makes allowances for tabular material and partial pages. Exhibit 2 gives conversion factors to simplify data collection.

EXHIBIT 1 GUIDELINES FOR ESTIMATING DRAFT PAGES

- A standard page of text is 25 lines.
- Any pages with 6 lines or less will be accumulated until 25 lines have been reached.
- Illustrations are counted as 1 page.
- Four photographs equal 1 page.
- Each three columns of a table equals 1 page.

EXHIBIT 2 HANDY DANDY GUIDE FOR PAGE CONVERSIONS

Number of		Number of
Double-spaced pages × 1/2	=	single-spaced pages
Double-spaced pages × 3/4	=	1-1/2-spaced pages
1-1/2-spaced pages × 4/3	=	double-spaced pages
1-1/2-spaced pages × 2/3	=	single-spaced pages
Single-spaced pages × 3/2	=	1-1/2-spaced pages
Single-spaced pages × 2	=	double-spaced pages

The standard established above can be applied to editing, composition, proof reading, and production coordination (assuming these are production areas in which it is worthwhile to collect data), and each area may be considered a separate jar of beans.

Text and tabular material fit the bean counter approach, but what about the area of graphics? The complexity of illustrations varies greatly. A simple graph takes much less time to produce than an orthographic rendering. One method of data collection is to segregate simple line drawings that fit within the established standard page image. Oversized illustrations are counted as multiples of the standard page. Specialized graphics are separated by category, depending on the complexity of the drawing or the creativity required.

In the printing and binding area, the standard page is generally in terms of 8-1/2 by 11-inch pages. Foldouts and other oversized sheets are either segregated (if there is sufficient volume) or converted to equivalent standard pages. In addition to the number of sheets, the number of copies may be important for the bindery. The type of binding can also be segregated if significantly more effort is required for different types of binding, e.g., perfect bind and staple and tape.

The subareas to be surveyed will depend on the sophistication of the printing operation. Color printing requires different standards for production measurement. If diazo production is part of the printing operation, square feet of material produced may be the best gauge of activity.

After standards (the number of jars and kinds of beans) have been established in the areas of interest, how can the data be collected?

COLLECTING THE DATA

The source of editing data is usually the records kept by individual editors and production coordinators. Composition, graphics, and reproduction data normally are obtained from production logs kept in each area. The log serves the primary purpose of keeping track of work. Exhibit 3 is a sample log sheet used both by individual editors and by functional areas. We require that the log sheet be submitted each week to ensure that the data are current and accurate. Exhibit 4 is a sample form used to solicit the data on a periodic basis. The data can also be collected on a centralized basis by extracting data from a work sheet such as the one shown in Exhibit 5. The organizational level of data segregation is decided by management on the basis of need and ease of collection.

The frequency of data collection can range from daily to yearly. Our accounting system is organized into 13 equal 4-week periods. Because labor charges are reported by our management information system each period, we have chosen to collect our data on a period basis. Note that data collected on a monthly basis has to be adjusted for the number of days in the month to be consistent for comparison purposes.

REPORTING THE DATA

The frequency with which data are reported will vary with need. In fact, management may choose to collect the data but to report it only upon request. We collate and report publications production data on a period basis (every 4 weeks). More frequent reporting (daily or weekly) may or may not be meaningful if there are wide variations in production that tend to smooth out over a longer period.

A simple tabular presentation is acceptable for the period reports. Exhibit 6 also incorporates a comparison with the period average for

EXHIBIT 3 PUBLICATIONS SERVICES PRODUCTION LOG

PROJECT NUMBER			AUTHOR OR DESCRIPTION	ORG. A	ORG. B	OTHER	DRAFT PAGES	IN MO/DAY	DUE MO/DAY	OUT MO/DAY	EDITOR, TYPIST, OR REMARKS

EXHIBIT 4 DATA COLLECTION FORM

(Name) (Period)

	Org. A	Org. B	Other	Total
Jobs carried over (period start)				
Pages carried over (period start)				
Jobs received (this period)				
Pages received (this period)				
Jobs completed (this period)				
Pages completed (this period)				
Jobs in process (period end)				
Pages in process (period end)				

EXHIBIT 5 PUBLICATIONS WORK ORDER

Prod. No. | Charge No. | Sub. | W.O. | Author/Requestor | Loc. | Ext. | Date of Order

Org. | Job Title | Out of Press

COORDINATION

COORDINATOR ______ EX. ______ DATE IN ______ DATE COMP ______

DRAFT PAGES ______ NO. COPIES ______ COVER/TITLE PAGE ______ TYPE BINDING ______

☐ PRESS

☐ COPY CENTER

EDITING

EDITOR ______ EX. ______ DATE IN ______ DATE DUE ______ ______ A.M. ______ P.M.

DRAFT PAGES ______ TABLES ______ FIGS ______ DATE COMP ______

COMPOSITION

TYPIST ______ EX. ______ DATE IN ______ DATE DUE ______ ______ A.M. ______ P.M.

AUTHOR REVIEW ______ CORR. RECD. ______ DATE COMP ______

FORMAT ☐ One Side ☐ Single Space PAGINATION: ☐ Consecutive FIGURES: ☐ Interleaved TABLES: ☐ Ruled

☐ Two Sides ☐ Space & 1/2 ☐ By Section ☐ Inserted ☐ Unruled

☐ Std. Page ☐ Dbl. Space ☐ At End

GRAPHICS

ILLUSTRATOR ______ EX. ______ DATE IN ______ DATE DUE ______ ______ A.M. ______ P.M.

CHECK PRINT DUE ______ CORR. RECD. ______ DATE COMP ______

Illustrations ______ No New Figs. ______ Orig. Art ______ PMT ______ SIZE: SIZING SHEET

Simple ______ No Reusable Figs. ______ Vugraphs ______ Plates ______ ☐ Standard ☐ Yes

Average ______ Captions Only ______ Slides ______ Glossies ______ ☐ Special ☐ No

Complex ______ File Numbers ______

INSTRUCTIONS AND JOB HISTORY

DATE TO PRESS ______

JOB COMPLETE ______

the previous year and the average for the current year. To emphasize trends, a graph presentation on a quarterly basis is useful. Exhibit 7 is a graph presentation that emphasizes the peaks and valleys during a year's production. These data can be used to avoid losing key personnel during a temporary downturn in business, such as that experienced in the 1980 recession.

SUMMARY

In summary, determine the need for publications production data collection, establish standards, provide the forms to collect the data, collect the data on a periodic basis that is meaningful, and report it in an appropriate manner. To avoid becoming a "has been" in publications, become a person who "has beans" and knows the jars to keep them in.

EXHIBIT 6 PERIOD REPORT IN TABULAR FORM

Period 10

PRODUCTION COMPARISONS

	19XX Average	Period 1-9 Average	Period 10
Coordinated new pages completed	7,686	4,833	4,758
Edited pages completed	4,141	3,770	3,042
Typed pages completed	1,539	1,394	1,325
Illustrations completed	189	179	170
Biographies completely processed	40	56	26
Total time charges, hr	3,482	2,899	2,931
Leaves (sick, vacation, etc.), hr	564	411	367
Production realization*	85.2	78.1	69.1

*Percent of available time not charged to Org 000 burden.

COORDINATION

	Org. A	Org. B	Other	Total
Jobs carried over (period start)	8	11	0	19
Pages carried over (period start)	1,366	1,433	0	2,799
Jobs received (this period)	68	74	2	144
Pages received (this period)	2,051	2,356	156	4,563
Jobs completed (this period)	70	58	2	130
Pages completed (this period)	2,862	1,978	156	4,996
Reprinted pages completed (this period)	184	54	0	238
New pages completed (this period)	2,678	1,924	156	4,758
Jobs in process (period end)	6	27	0	33
Pages in process (period end)	555	1,811	0	2,366

EXHIBIT 7 PERIOD DATA FOR COORDINATION PRODUCTION IN GRAPH FORM

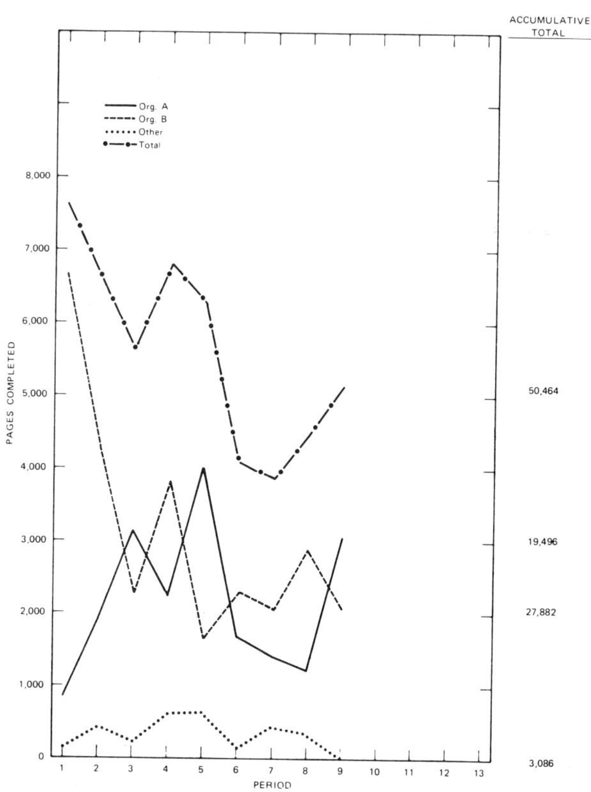

An artist's view of the team concept to presentations

Don Ryder
Lawrence Livermore National Laboratory

Don Ryder
Lawrence Livermore National Laboratory
P.O. Box 808
Livermore, CA 94550
(415) 422-5856

Senior Technical Illustrator

Our laboratory has recently instituted a team approach to editing and producing the presentations of our speakers. The team consists of an editor, an artist, and the author, who jointly edit the presentation visuals. We find that the team concept works exceptionally well and that the resulting visuals are of high quality. Acceptance of the team-concept is shown by the increased requests for our help by new and previous authors.

I am the artist-member of a three-person team that prepares some of the Lawrence Livermore National Laboratory's technical, scientific, and recruiting visuals. (Visuals include viewgraph, slides, and flip charts for presentations.) The other two members are the editor and the author. Our presentation format is based on a methodology that emphasizes a single-thought, illustrated visual to communicate clear, retainable information.

Storyboarding is the beginning

At our first meeting, the author storyboards his draft visuals on a conference table (Fig. 1). (Storyboarding is the spreading out of the individual drafts in sequence, similar to that done in animation and movie making.) To organize and simplify an author's thoughts into flowing, informative visuals, the editor edits the author's message into a series of statements that serve as thesis titles for each visual. The titles are simple, complete sentences—written as they might be spoken—and containing a single, supportable thought, best remembered when written in 11 words or less.

On-the-spot creativity

My main job is to create "on the spot" the art that substantiates and supports the thesis title. First, I help visualize the author's thoughts. Then I generate the rough sketches that most simply illustrate and support the thesis title. The more complex the subject, the greater the need to have simple art. Because the author will be addressing each portion of the visual in his talk, it is often possible to oversimplify the art and have it still be effective. Three to four art elements are ideal, especially if the purpose is to present easily retainable information. Each element should be complete, descriptive, and individually understandable. Color and a one- or two-word callout may aid clarification, while adding interest and variety.

Fig. 1. Storyboarding the author's draft visuals is the first step in editing his visuals.

A comparison demonstrates this point. We all may have seen, or have been guilty of using, the type of visuals shown in Fig. 2. These visuals contain category-type titles, which are incomplete sentences that usher in a listing or a confusing schematic or block diagram. What is important or significant in the visuals is not readily apparent. And the significance of the visuals is left to the audience to determine. Figure 3 shows the visuals in Fig. 2 reworked to our format.

Whether simple or complex, good art can go a long way toward carrying a marginal speaker or saving a mediocre presentation, but it will always complement and polish a well-given presentation. Well-edited, professional-looking visuals can be dramatic, controversial, beautiful, and humorous; they can capture and hold attention, and therefore, increase retention. And well-designed visuals provide ready-made notes for the speaker, which add to his confidence in his delivery.

Brainstorming the visuals

The drawing out of thoughts, ideas, and information from the author and putting them into words with appropriate supporting art is referred to by us as brainstorming sessions. The end product from

COMMONLY USED FORMAT

- NOT TOO INTERESTING
- USUALLY HARD TO READ FROM MIDDLE OR BACK OF ROOM
- HARD TO INGEST & RETAIN ALL THIS INFO
- IMPOSSIBLE WITH SEVERAL "LISTING" TYPE VU-GRAPHS
- AUDIENCE IS FORCED TO READ WORDS & LISTEN TO SPEAKER SIMULTANEOUSLY

Organization chart

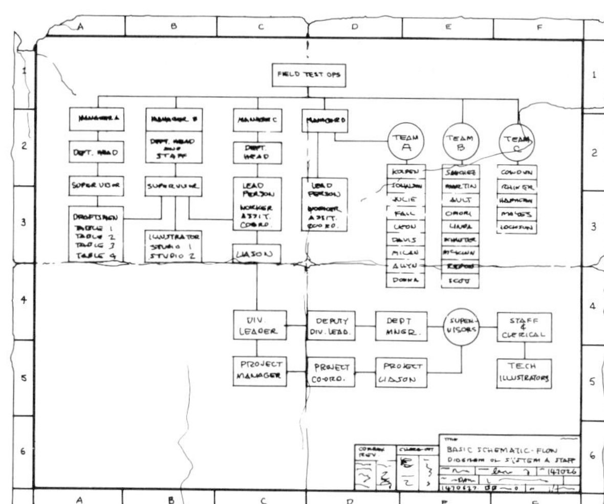

Fig. 2. Examples of often typical visuals, which contain category-type titles followed by a listing or a too-detailed drawing.

these brainstorming sessions is an orderly series of draft visuals that portray the author's message.

Two separate brainstorming methods are used by editors. From both methods, a draft is created featuring a quickly but well-edited thesis title and a fast but nicely drawn supporting sketch. Both methods have advantages and disadvantages.

In one method, the editor will, with the author and artist, endeavor to make a "clean" draft with a well-developed thesis title and a complete supporting sketch. The draft is then ready to go directly to finished art, where the title will be set and the art will be completed. Except for a proofing cycle, the art is ready to be made into a visual. Each member of the team participates in preparing this clean draft. The clean-draft method requires more brainstorming time, and the rate of production is two or three clean drafts per hour.

By comparison, the second method involves preparing a "rough" draft. It requires the same work to develop the thesis title and the ideas or rough notes about the supporting art. However, instead of all three members working together to finish a clean draft, only ideas or roughs of the art are formulated at this time. Then the editor and author move on to other visuals, while the artist either goes aside during the session to prepare clean sketches from his rough notes, or he may chose to remain an active member of the team to help create thesis titles and other art ideas. The final clean-draft art would have to be prepared at a later time. Using this method, we complete from four to eight visuals per hour. However,

This format emphasises simple titles with supporting graphics

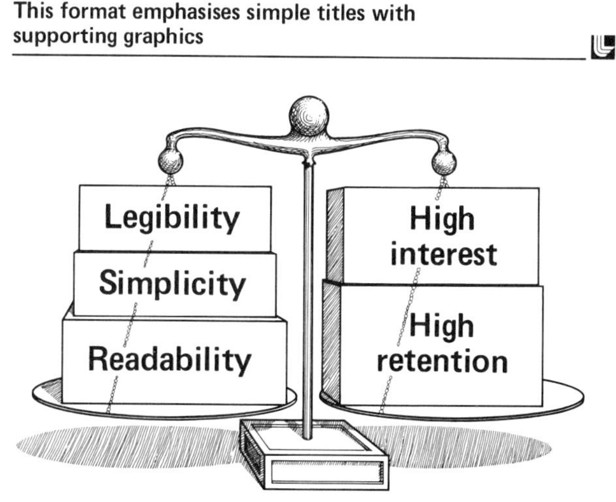

Field manager and his project and diagnostic leaders manage teams A and B

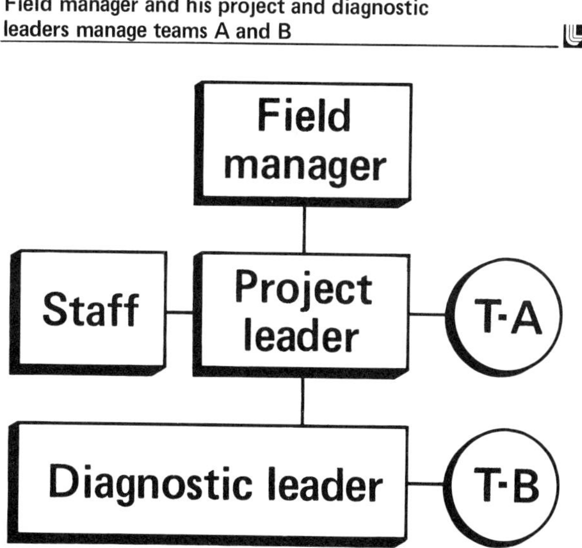

Fig. 3. Examples of the visuals in Fig. 1 reworked to contain single-thought thesis titles with supporting graphics.

additional art time must be put in later to arrive at the clean-draft stage.

There are mixed feelings about the two methods, and the editors are evenly divided. Authors like the second method because they can spend less time in brainstorming sessions, but they uniformly wish to proof the clean draft sketches before final art. This creates an extra proofing and correction cycle.

This artist feels that the first method is preferred for the reason that the additional proofing and correction cycle is eliminated. Also, it requires that the artist participate in all facets of the brainstorming session, which is both interesting and challenging.

Dry-run time

After the final brainstorming session, the clean drafts are again storyboarded. However, this time the author's supervisor and colleagues also attend. The author delivers his presentation as he normally would. The audience checks for proper flow, continuity, ease of assimilation, inappropriate jargon, and any ambiquities. Also, his platform skills are evaluated. Questions during the presentation are resolved, and the resolutions are incorporated into the talk, if applicable.

Skills needed by the artist

"Must-have" skills for the team artist include (1) the ability to visualize abstract concepts and quickly complete line or multicolored, three-dimensional sketches of people and objects; (2) knowledge of graphics production, including costs, procedures, and tunaround times; and (3) the ability to work and get along with others. "Should-have" skills include basic word communication and platform skills. "Nice-to-have" skills include some exposure to the technical subject matter.

Cost of visuals

There are many variables that dictate the tempo and rate of production of each brainstorming session. Although the degree of abstractness oı technical content of the material is important, it is usually how well-prepared the author is and his attitude that makes the real difference. Ideally the author arrives with very rough individual drafts of each of his visuals, some supporting sketches, and lots of good ideas. Also, it helps if he is sensitive to editorial constraints and production problems, and remains completely flexible.

A study of a year's budget and manpower shows that the breakdown of our costs for a complete presentation package (25-30 visuals and some platform skills evaluation) is as follows: editorial time 14%, artist time and art production 79%, and photographic laboratory 7%. The hours per visual according to the levels of presentation quality—fair, good, and top quality, respectively—are as follows: Editor—0.4, 0.8, and 1.7; artist and production—2.25, 4.5, and 9.5; photographic laboratory—0.2, 0.4, and 0.85.

Acceptance of the team approach

As in any large technical information group, we provide many communication services to a variety of different in-house customers. This latest service—the team appoach to presentations—has been well-received. Those who used this service have either returned or have recommended it to their colleagues.

AUDIOVISUAL CONTRACTING—THE CLIENT'S PERSPECTIVE

S. Martin Shelton
Naval Weapons Center

S. MARTIN SHELTON
Manager, Film Projects Branch
Technical Information Department
Naval Weapons Center
China Lake, CA 93555

With the increased emphasis on contracting for audiovisual products and services, it is imperative that the client and the contractor have a mutual understanding of the legal aspects of a contract and the types of contracts used to procure audiovisuals. From the client's perspective, this paper explores the legal ramifications of a contract, discusses the key elements of procurement planning in the contracting cycle, and examines the major types of contracts used.

INTRODUCTION

To communicate more effectively in today's hyperkinetic information environment, organizations—corporate, governmental, and private—are using audiovisuals more often and in more diverse and extensive ways. Audiovisuals, recognized as practical and powerful communication tools, have become an integral part of organizational life.

Objectives of audiovisuals span the broad spectrum of all aspects of communications. They range from technical communication to teaching (perceptual motor skill development, for instance,) to attitude and behavior modification ("work safely," "buy bonds," etc.) (1).

To meet the need for timely and effective audiovisuals, many organizations have developed their own in-house capability. Others fulfill their requirements from vendors, either through open purchases or contracts. The process of purchasing audiovisual products and services can be as simple as buying a roll of film or as complex as contracting for a complete audiovisual production.

To set the perspective for this paper, my background in audiovisual contracting is limited to governmental experience. Although governmental procedures may differ somewhat from those in the private sector, the principles are essentially the same.

At best only a cursory treatment of this broad topic can be developed in this paper to highlight some of the points common to all contracting and to alert managers to contracting complexities, pitfalls, and rewards and the need for expert counsel from appropriate procurement and legal departments in any contracting operation.

PROCUREMENT PROCESS

It is axiomatic that a successful contract can result only from good planning. And that effective procurement planning must begin far in advance of the need for the goods or services produced by a contract. To be evaluated critically are the client's needs, the technical requirements, and the fiscal aspects of the proposed procurement. Vital to a successful contract is agreement between client and contractor on the work to be accomplished; results expected in terms of quantity, quality, and schedule; responsibilities; contract monitoring; contingencies; and payment schedule. A successful contract is profitable to the client and the contractor in terms of professional satisfaction and growth, and fiscal advantage. Lastly, a good contract is administered easily.

With few exceptions, the procurement process of contracting comprises the following basic functions:

- Definition of requirements
- Procurement planning
- Preparation of the solicitation document
- Selection of contract type
- Announcement of procurement
- Evaluation of proposals
- Contract award
- Contract administration and performance monitoring

These functions are not to be considered as definitive and having finite limits, but as overlapping areas along the contracting continuum.

DEFINITION OF REQUIREMENTS

One of the most difficult and essential tasks a manager has in the contracting process is to analyze the communication goals in order to define the audiovisual requirements exactly and develop the procurement specifications. This task can be accomplished by precisely answering such fundamental questions as: What exactly is needed? Why? When is it needed? How is the work to be accomplished? Where?

Obviously, not all the pertinent questions can be stated here. Those posed are but some of the most critical. Others would need to be developed and answered to meet specific contract requirements. Inherent in this analysis is, therefore, a precise identification of the services or products (or combination) needed, the goals to be achieved, and an indication of how they are to be accomplished—the requirements statement.

Audiovisual Production as Example

Definitions. If the requirement is for an audiovisual production, one of the first steps is to define the communication goal in terms of the target audience, including minimum rate of achievement, or level of performance, required of the audience in the near- and long-term.

After defining the communication goals and the target audience, decisions must be made regarding medium selection, the distribution plan, and the contracting scheme.

Medium Selection. What type of audiovisual (or combination of audiovisual and other media) will best achieve the goals—motion pictures, filmstrips, or sound-slide shows, for instance? Background on this topic is discussed in the author's treatise "Do You Need a Film?" (2).

Distribution Plan. What is the distribution plan? Some of the factors to consider are audience composition, size, and location; viewing environment; equipment needs; projectionist and discussion leader, if required; laboratory coordination; release print logistics, distribution, and maintenance; and feedback (evaluation) (3).

Contracting Scheme. What is the contracting scheme? Important questions to answer here are which of the major audiovisual phases (scripting, production, and distribution) should be contracted for (singly or in combination) and which, if any, should be done in-house? These questions are old, much debated and have no pat answers. There are as many "solutions" as there are situations and people to handle them.

Combination Contracts. Based on my experience, I prefer not to combine audiovisual phases in a single contract, particularly script and production. I find it better, if time permits, to contract for each phase separately and at different times.

Speculative Script Treatments. Another major point to discuss is the widespread and abhorrent practice of soliciting speculative script treatments for a proposed audiovisual production. Most often these are induced by vague assurances of having an edge in getting the scripting or production contract. Such a scheme is conniving balderdash.

Critically acclaimed author and information film producer Molly Gregory dubs the practices as ". . . immoral, unethical, and unreasonable, and . . . not in the best interests of anyone." (4)

Though these comments may be ancillary to the main thrust of this paper, they are sufficiently germane to discuss in this section and to spotlight professional reaction.

Speciality Service Contracts. Also to be considered, if the client has audiovisual producing skills, is the concept of using specialty service contracts to augment the in-house group, or to develop a total audiovisual package (including distribution).

Again, from my perspective as a manager of an in-house film production group, I've found the use of specialty service contracts to be very productive and cost effective.

The Contract Specialist. Translating the communication analysis, distribution plan, and contracting scheme into procurement specifications requires the skills of a contract specialist. Ideally, working together, the audiovisual manager and the specialist can develop a statement of the requirements that is contractable—namely, the specifications are definitive and well-written (with minimum use of audiovisual and procurement jargon) and are legally viable. The attendant contract is easy to understand, easy to administer, and easy to comply with.

If that were only true! Unfortunately these ideal goals rarely are achieved. Most governmental audiovisual contracts I've seen (and private sector ones, for that matter) reflect the bureaucracies from which they come. They are tedious, ambiguous, voluminous, and encumbered with inconsequential "boiler plate." They are usually legal, however!

PROCUREMENT PLANNING

With the requirements clearly defined, the audiovisual manager and the contract specialist decide on the procurement method, type of contract, schedule, assignment of responsibilities, and a host of other details to ensure that a viable contract is let (5). A milestone chart is often used to monitor progress and forestall delays.

Procurement Methods

Basically, procurements (contracts) are classified into two broad categories, noncompetitive and competitive, with each category having numerous subsets.

Noncompetitive. Of primary importance in this category are

1. Single Source. In a competitive market, purchasing from only one of the sources may be advantageous because of that source's experience, ongoing contract work, or other features.

2. Sole Source. Only one contractor has the unique capability that is essential to meet the requirements.

3. Small Purchase. Basic Purchase Agreements placed with local vendors may be the best source for low-cost, off-the-shelf items.

Competitive. Of the many variations in the competitive category, the two most common and appropriate to audiovisual contracting are formal advertising and negotiated procurement.

Formal Advertising. When the requirements have been defined with pinpoint precision to encourage maximum competition, formal advertising is used to solicit bids. Contract award is based on the lowest bid from the "responsible" and "responsive" bidders. Inherent in this process is sufficient time for the formalities of preparing the Invitation To Bid, evaluating the bids, and determining the responsibility and responsiveness of the lowest bidder.

Negotiated Procurement. Maximum competition, flexibility, selectivity, and speed in dealing with contractors are the salient points of a negotiated procurement. A Request for Proposal, which details the requirements and establishes the criteria for evaluation, is used to solicit offers. The client attempts to get the best possible terms by negotiating with the contractors in the "competitive range" (client-established estimated cost range). Contract award is made to the contractor whose price, technical proposal, and other factors appear to offer the most advantages (6).

• In summary, a sound procurement plan is a requisite for a timely and viable contract award, professional business conduct, and responsible contract administration (7).

PREPARATION OF THE SOLICITATION DOCUMENT

Information gathered and codified during the definition of requirements and procurement planning is used to prepare the solicitation document (Invitation To Bid or Request for Proposal). The solicitation document must include enough data (terms, conditions, responsibilities, schedules, and other pertinent information) to constitute a definitive contract when it is signed and dated by the contractor and the client. The statement of work in the solicitation document tells the offeror what must be accomplished as an end product. It may be as simple as a one-line statement for an off-the-shelf item or as complex as a complete plan for an audiovisual production.

An audiovisual plan would include data from the requirements statement such as communication goals, target audience, detailed and approved script, storyboard, complete production details, and the distribution plan. Client-furnished technical advice, equipment and operators, and other details are committed. Contingencies are specified (weather delays and unavailability of equipment, for example), and locations are pinpointed (8).

Specifications in the plan detail physical aspects of the production; for example, color, 16mm, sound (synchronous and stock music), and length—all the data needed to describe the production.

The solicitation document must also detail all the information a contractor should provide so that the offer (bid) can be evaluated properly in terms of costs, technical approach, and contractor qualifications (9).

Finally, included in the solicitation document (as appropriate) are the general provisions ("boiler plate"), inspection criteria, acceptance procedures, payment schedule, and other contract administration details.

SELECTION OF CONTRACT TYPE

There are as many types of contracts as contract specialists can devise to meet specific situations of risk, responsibility, legal liability, and products and services to be delivered. To discuss them in more than general terms is beyond the scope of this paper. However, some of the more salient features of each type can be highlighted.

Contract Form

Form defines what the contractor is to deliver, either a definitive end product or some level of effort in man-hours and materials.

Contract Family Groups

Contract family groups define fiscal risk and responsibility of the contractor and the client: *Cost reimbursement contracts* specify the contractor's legal obligation to "try" to accomplish the work specified—risk is low for contractor and high for the client. Conversely, *fixed-price contracts* detail the contractor's obligation to "guarantee" to deliver the work specified—risk is high for contractor and low for client. (Of course, for the client to realize the low risk, the contractor must stay in business and deliver satisfactorily on time.)

Each of these two family groups is subdivided into a number of specific contract types. In decreasing order of client's fiscal risk, they are

- Cost Reimbursement
  - Cost Plus Fixed Fee
  - Cost Plus Award Fee
  - Cost Plus Incentive Fee
  - Time and Materials
- Fixed Price
  - Fixed Price Redeterminable
  - Fixed Price Incentive
  - Fixed Price With Economic Price Adjustment
  - Fixed Price Indefinite Quantity
  - Firm Fixed Price (10)

Cost Reimbursement. In cost reimbursement contracts, an estimated cost is set during the negotiation process as a ceiling to attain a specified end product or some level of effort. The contractor may exceed this ceiling only with risk. Particulars of the contractor's fee also are agreed on during negotiations. Incidentally, statutes forbid contracts in which the fee is a direct function of cost; that is, the fee increases as costs are incurred. In general, the client is obligated to reimburse the contractor for all the allocatable, allowable, and reasonable costs incurred, including approved cost overruns.

These contracts are useful particularly when flexibility is needed to redirect the contractor's effort to meet changes in requirements that fall within the scope of the contract and when specifications cannot be detailed with pinpoint accuracy.

Cost Plus Fixed Fee. During negotiations the contractor's fee is set at a specific dollar amount. The fee does not vary with actual costs, including cost overruns. Since this contract has minimum incentive for the contractor to manage effectively and tends to increase client costs, its use should be limited.

If the scope of the work is broadened significantly, the fee may be increased proportionally. Conversely, if actual costs for an end product or level of effort are significantly less than estimated, or if the schedule is not met, the fee should be decreased to reflect the reduced scope of the contract.

Appropriate for this kind of contract would be production of a technical film report of an ongoing research and development program for which unknown factors and technical delays cannot be predicted.

Cost Plus Award Fee. The fee paid to the contractor is composed of two parts:

1. A fixed amount, or basic fee, which does not vary regardless of performance
2. An award fee determined by the client's unilateral and subjective evaluation of the contractor's performance in meeting requirements, schedule, and cost

Because the motivation for contractor excellence is strong in award fee contracts, they are used widely to procure technical, administrative, and housekeeping services; the audiovisual distribution phase, for instance.

Cost Plus Incentive Fee. When the uncertainties in the work statement and specifications are moderate (firmer than in award fee and fixed fee contracts), the incentive fee contract is appropriate, as the risks are shared about equally between client and contractor. Target costs and target fee are set in the negotiations.

If the contractor reduces costs below the estimate, a bonus fee, based on a percentage of cost saved, is awarded. By its nature, the incentive fee contract exerts exceptional pressure on the contractor to reduce cost, improve performance, and meet schedules. It is appropriate for the operation of a company-owned, contractor-operated audiovisual facility—a film laboratory, for example.

Time and Materials. Though not usually classified by the purist as a cost reimbursement contract, I'm including this hybrid here because it has many similar features and fits at this point on the risk scale. Time and materials contracts are appropriate when neither the extent, duration, or overall cost of the work can be estimated with any degree of certainty. With a ceiling price established, the task is accomplished by contractor-furnished labor and materials. A fixed hourly (or daily) rate, including profit and overhead, is established for each labor category to be used on the task. Materials are purchased at contractor "cost"—usually cost plus 15%. Time and materials contracts are suited to buying engineering or design services (special effects, for example), and to maintenance, repair, and overhaul functions.

A slight variation of the time and materials contract is the labor hour contract, which requires no materials. It is used to buy film scripts or storyboards, for instance.

Fixed Price. A fixed price contract is an agreement to pay a specific price (sometimes with adjustments) upon delivery and acceptance of the goods or services contracted for. The contractor's risk ranges from moderate to very high. The fixed price contract, by its nature, encourages maximum competition and is generally the most cost effective contract. It is, therefore, much preferred in industrial practices.

Fixed Price Redeterminable. This contract is used to buy quantity production items when adequate estimates of labor and material cannot be made because the specifications are somewhat vague, the initial buy is very small (or very large) and not reflective of realistic production costs, or the delivery schedule is short. Fiscal risk, because of uncertainties, is shared about equally between client and contractor. The contract price is adjusted (up or down) during the life of the contract as specifications are firmed or other variables are controlled. This contract could be used to buy, for example, special laboratory processing of large quantities of film to be exposed under unusual conditions and delivered to the laboratory in varying quantities over an extended period.

Fixed Price Incentive. When the contingencies relating to cost, performance, and schedule are less than associated with a redeterminable contract, an incentive contract is appropriate. Fiscal risk is shifted toward the contractor because of more certainty in the contingencies. Cost reduction, performance improvement goals, and the formula for determining profit are agreed on during negotiations. The incentive is strong for the contractor to meet target goals because profit is a direct function of overall performance. This contract might have application in the procurement of a complete audiovisual training package where student performance (goal) can be objectively measured.

Fixed Price With Economic Price Adjustment. Special contingencies clauses in this fixed price contract allow for price adjustment (up or down) to protect the contractor and client from major fluctuations in labor or material costs. Price adjustments are limited to contingencies the contractor does not control. Because of the wide variance in the price of silver, this contract would be useful for purchasing raw stock (film) and for routine laboratory processing and printing.

Fixed Price Indefinite Quantity. Several factors make this contract especially attractive to client and contractor. A fixed price is set for goods and services as line items in a contract—a catalogue, as it were. Goods and services may be purchased piecemeal to meet varying client requirements over the life of the contract. The client guarantees to buy a stated minimum of each line item. (Sometimes a maximum purchase limitation is established to protect the contractor.) Frequently the services are categories of highly skilled labor plus the attendant specialized equipment. This contract is suited very well to purchase of specialized audiovisual functions such as art and animation, sound services, and laboratory services.

Firm Fixed Price. In a firm fixed price contract, the contractor is obligated legally to deliver an end product that meets specifications, on time and at the agreed price regardless of the actual cost, or be liable for breach of contract. Profit is the difference between contract price and contractor's costs. Accordingly, this contract places a maximum incentive on the contractor for cost control. Sometimes, however, quality can suffer. A firm fixed price contract should be used only for procurements that have few or no uncertainties in cost, end product description, or schedule. It is the simplest contract to structure and to administer, and is ideal for buying a complete audiovisual production whose script is completely detailed and approved and for which contingencies are inconsequential.

• As can be seen, many factors must be evaluated in the selection of contract type: complexity of goods or services needed, schedule, duration, cost and performance evaluation, prior experience with the contractor, subcontracting plan, and the client risk. There are no easy answers and no formula contracts for easy solutions. Nonetheless, most of these factors ought to be resolved with a high level of satisfaction during the next two basic contracting functions, resulting in contract award.

ANNOUNCEMENT OF PROCUREMENT

To have maximum competition from responsible contractors, the proposed procurement should be publicized extensively through informal advertising, and formal advertising if appropriate. Publicity usually is handled by the purchasing (supply) department to ensure that legal requirements are met.

As a first step, the audiovisual manager prepares a list of prospective contractors known to have the capability to accomplish the work. Also, the organization's records are screened to uncover others. For large procurements a presolicitation conference may be held with prospective contractors to discuss the work and to peak their interest.

In negotiated procurements, the "letter of interest" is useful to solicit interest. A caution, however: the letter should not become an informal request for proposal (11).

EVALUATION OF PROPOSALS

The evaluation process is used in negotiated procurements to discern which offeror has the best prospect of satisfactorily completing the work on schedule and within cost. As part of the procurement package, the solicitation document must state all criteria used to evaluate and rate a prospective contractor's proposal. The criteria should be comprehensive and detailed enough for a critical assessment of all relevant technical and managerial capabilities needed to accomplish the tasks as delineated in the statement of work (12).

The key criteria used to evaluate contractor capabilities are past performance, client continuity, integrity, credit rating, financial resources, management, organization, quality control procedures, access to classified information (if appropriate), plant or office location, facilities (space and equipment), personnel (staff and free-lance associates), and subcontractor access (13).

The audiovisual manager plays an active and important role in the evaluation process. The manager's specialized background is a requisite for proper analysis and critique of the technical aspects of the proposals.

Eventually, all proposals must be rated either acceptable, unacceptable, or capable of being made acceptable with additional information.

In formally advertised procurements, evaluation criteria may also be used. However, the criteria must be highly specific and essential for contract performance. Since no communication may be had with the bidders, after evaluation their bids may be rated only as acceptable or unacceptable.

The preaward survey, a detailed evaluation of all aspects of an organization's operations, is a particularly valuable instrument in the evaluation process. It helps to ferret out the true capabilities of a prospective contractor. A visit to the contractor's place of business is made to inspect the facilities, equipment, and procedures, and to interview personnel. (Oftentimes, it is easier to write fiction in a proposal or bid than it is to produce the facts during a survey. In the years I've been in the audiovisual business, I've seen some very elaborate fiction, indeed—always uncovered in the preaward survey for the fraud it is.)

Rigorous preaward surveys should be conducted on any proposed contractor with whom the client has no experience and whenever any doubt exists about the contractor's managerial, financial, or professional capabilities.

CONTRACT AWARD

In negotiated procurements, the client dickers with the offerors rated acceptable in the "competitive range" to get the best possible terms. All factors of the proposal may be explored. The contractor's technical approach, resource allocation, and cost are of particular importance. (For instance, exactly how is the contractor going to produce the information film and what equipment and personnel are to be used?) Award is made to the responsible and responsive offeror whose proposal has the most advantages and the least risk for the client.

In formally advertised procurements, prospective contractors submit sealed bids. Bids are opened publicly, and after evaluation the award is made to the lowest responsible and responsive bidder. Because award is based on price, the contract, by definition, must be either firm fixed price or fixed price with escalation clauses (14).

CONTRACT ADMINISTRATION AND PERFORMANCE MONITORING

Efficient administration and performance monitoring are critical to satisfactory contract progress and completion. These functions are accomplished through active cooperation between technical, purchasing, and legal personnel.

The gamut of contract administration and performance monitoring specifics is almost endless. Most of them can be handled with common sense, some are legally complex, and others involve business ethics:

• Frequently contracts must be modified to meet changing circumstances. If the modification affects a material provision of the contract, a bilateral agreement needs to be completed (a minicontract, in effect). Changes in the work statement or specifications would constitute a material change and require some agreement between client and contractor. Examples are cancellation of extensive location photography or doubling the release print order. In most organizations, bilateral contract modifications may be negotiated only by the contract specialist.

Unilateral actions provided for in the original contract, such as administrative changes or exercising of options, require only written notification from the client to be effective.

• Problems relating to technical performance, cost, and schedule are to be expected. However, close monitoring and expeditious reporting will identify potential problems and permit timely corrective actions to be taken. If a problem persists, the contractor may become delinquent by reason of nonperformance. Several remedies are available to the client (and to the contractor if the client becomes delinquent). The easiest solution, though not necessarily the most practical, is to terminate the contract, either with a settlement or at no cost to the client. Another remedy is to terminate for default—an ugly business. In defaults both parties almost always lose. To preclude these pernicious events, a viable contract between a responsible contractor and a responsible client is imperative.

Lastly, delivery and acceptance constitutes contract completion. Usually a certification of some sort is submitted to conclude formally the business (15). Practically, however, cashing the final payment check is The End.

CONCLUSION

Though no one contract will fit all situations (or even most of them for that matter), the Standard Motion Picture Production Contract developed by the International Quorum of Motion Picture Producers (IQ) is an excellent example of a firm fixed price contract for an audiovisual production (16). As a client there are several clauses in it that I find unsatisfactory, but they can be struck.

In this brief discussion of the intricate topic of audiovisual contracting, only a broad-brush treatment could be developed to highlight a few cogent points. Although the treatment has a governmental ring to it, it is sufficiently general to be applicable to all facets of audiovisual contracting and for audiovisual managers to gain a keener appreciation of the complexities and pitfalls involved.

As a last thought, expert counsel is a "must."

REFERENCES

1. Sobel, Stanford, "Paradox," *Business Screen*, September/October 1975, p. 10.
2. Shelton, S. Martin, "Do You Need a Film?" *Proceedings, 22nd International Technical Communication Conference*, May 14-17, 1975, Anaheim, California. Washington, D.C.: Society for Technical Communication, 1975, pp. 229-234.
3. Ibid., p. 229.
4. Gregory, Molly, *Making Films Your Business*. New York: Schocken Books, 1979, p. 41.
5. Killoran, J. D., *NWC Procurement Guidelines*. China Lake, California: Naval Weapons Center, 1977, pp. 1-3.
6. MacFarlan, Greg, and Edward Pulsifer, *Navy Research and Development Contracting*. Washington, D.C.: Sterling Institute (DAC), 1976, p. 20.
7. Gaisor, Charles, *Defense Procurement Management*. Boston: Harbridge House, 1976, p. I-11.
8. Swanson, Robert, "What You Need To Know When Buying a Company Movie," *Business Screen*, June 1980.
9. Gaisor, op. cit., pp. I-1—I-53.
10. Killoran, op. cit., p. H-3.
11. Gaisor, op. cit., p. II-25.
12. Clement, H. P., "Selection of Contractual Services for Major Defense Systems," Department of Defense Directive No. 41052. Washington, D.C.: Government Printing Office, 1976.
13. MacFarlan and Pulsifer, op. cit., p. 34.
14. Gaisor, op. cit., pp. II-1—II-75.
15. Gaisor, op. cit., pp. III-1—III-82.
16. International Quorum of Motion Picture Producers, Standard Motion Picture Production Contract, Oakton, Virginia, IQ. (This copyrighted contract is available from IQ, International Headquarters, P.O. Box 395, Oakton, VA 22124, for $10 a copy.)

TECHNICAL POSTER FABRICATION

J. T. WHITE
Oak Ridge National Laboratory

J. T. WHITE

Mr. White is a designer in the Graphic Arts Department of the Oak Ridge National Laboratory, which is operated for the U.S. Department of Energy by Union Carbide Nuclear Division.

He began his graphics career in 1941 as a draftsman with the Tennessee Valley Authority. He worked as a cartographer for the U.S. Geological Service, and joined Union Carbide in Oak Ridge as a technical illustrator, rendering pictorials for a variety of publications, including reports, brochures, and books. He presently designs industrial exhibits and scientific displays for the Oak Ridge National Laboratory.

The poster group at the Oak Ridge National Laboratory is an important part of the system for dispensing information. The graphics group provides quality visuals that are needed to explain a technical process or a complete laboratory program. The work done includes the drafting of illustrations and condensing text and tables. The visuals must simplify highly technical information and language, so that the poster is self-explanatory or understandable with only a short oral explanation.

The graphic arts section of the Oak Ridge National Laboratory (ORNL) has used a standard format for several years and finds it easy to set up. We lack the space for an ideal session, but we manage to have effective poster sessions even in crowded situations. The ideal situation would be to have in front of each display a conversation area for at least 12 people.

Our graphics group has made posters for many years but never in such large numbers as in the last several years. A poster session of 30 to 40 posters puts a burden on a small graphics group unless you have well qualified people to fabricate them.

Our graphic arts department has 23 people with poster work actually assigned to only six in this group. These six have become quite proficient in this medium and will do most of the art work. Some typing and POS camera shots will be made by at least two other people, so we use about one-fourth of our staff on poster work.

Many steps are involved in preparing each poster display for a poster session. It is vital that each step be scheduled appropriately to allow the author to interact with the artist as needed. Such a schedule begins with a meeting between myself and all the technical people who plan to use the poster method to present their papers. This is usually about three months before the meeting at which the posters will be presented.

We discuss the visual medium of posters and the many "why's" and "why nots" of this medium. I show color slides of good and bad examples of posters and offer to discuss any problems they may have in condensing their material-many will be "first timers" and to others it will be "old hat."

The poster session chairman sends some helpful comments on poster preparation to each person compiling information for a poster:

1. Designers can use two or three panels (32 x 40 in. each) per poster topic.
2. Prepare sketch of each panel on 8 1/2 x 11 sheet and label each item on that panel.
3. Develop individual items each on a separate page and assemble in a folder. Type all material and submit to the chairman. Attach charge numbers, names of designers, and telephone numbers.
4. Keep the poster design simple and clear. Do not crowd or use excessive words.
5. Poster can include photographs, graphs, tables, and hardware; the hardware can be mounted on the panels or be on a table in front of the display.
6. Include only enough text to allow complete examination of the poster set in 5 to 7 minutes.
7. Specify color schemes only if relevant to the poster topic. Otherwise leave the color selection to the graphic artist.
8. Major steps (and their scheduling) in the preparation of posters include:

MAJOR STEPS

	weeks before poster needed
Poster outline completed	9
Poster outline reviewed	8
Art work begins	7
All materials to graphic artist	6
Posters complete	1
Poster rehearsals	poster must be available by this date

To better establish the schedule in the minds of all involved in the poster work and to be sure everything gets done in the order and time in which it is needed, an action plan is sent to graphic arts and to all the technical people involved:

TIME SCHEDULE

ACTIVITY	WEEKS BEFORE FINAL DATE	PERSON RESPONSIBLE
1. Instruction session by artist for authors	12	Division directors and each author
2. Poster outlines completed	9	Author
3. Poster plans reviewed	8	Author and PS chairman
4. Start art work (order photos, drawings)	7	Artist
5. All poster plans done and submitted to graphic artist	6	Author
6. Review of preliminary layout and checking as work progresses	5-6	Author
7. Posters checked and completed	4	Artist and author
8. Photos of posters completed (for reference)	·3	Artist and photography
9. Posters displayed	3	Authors
10. Poster summary books	2	PS chairman
11. Poster rehearsals	2	Division director and authors
12. Formal poster session (to technical audience)		Division director and PS chairman
13. Informal presentation (to public)		Authors

The schedule for the action plan is designed to allow some slack time, every effort is made to meet the schedule in order to meet the session deadline.

EQUIPMENT

The ORNL graphics section uses good material and equipment, but we are always trying to upgrade our methods. Below is a listing of the materials and equipment we use in preparing posters:

POS-ONE CAMERA, model 320 (Varigraph Corporation). This camera allows quick reduction or enlargement of copy. Negative paper prints can be enhanced with the addition of color.

COMPUGRAPHIC MODEL 7200 and IBM FREESTANDING COMPOSER. These two pieces of equipment gives us a good selection of type and a quick way to do text and tables.

I.n.T. (an image and transfer system by 3M Company). This is a dry method for transfering a page of text or an illutration by a rub-on to the poster. It uses a negative film from POS-ONE camera. Colors are black, red, blue, and white.

LETRASET TYPE (other brands available). An adhesive-back dry transfer lettering for special types to use for headings and small amounts of text.

ZIPATONE (other brands available). An adhesive-back transparent acetate for blocking-in large areas of color.

MAGIC MAARKERS. A color marker to accent curves on graphs–especially useful on negative-positive prints.

TAPE (Scotch brand, double-sided foam tape). This tape is strong and comes in 1/16 to 1-in. thickness. It is excellent for floating some panels from the background.

EMBOSOGRAPH (a lettering press made by EMBOSOGRAPH CORP. OF AMERICA). Title boards are made with this machine. Metal letters press an embossed letter into poster board.

CRESCENT MAT BOARD (a picture framing mat board of single thickness in 32 x 40-in. size). This board comes in a variety of pastel colors and will not fade as quickly as some poster boards. We use these for the background on all posters.

FOME-COR BOARDS (a thick lightweight board, also called GATOR FOAM). This lightweight board comes with white and brown paper bcking; the poster is mounted on FOME-COR with double-sided foam tape. The poster is then protected and can be used in a variety of ways. FOME-COR comes in 4 x 8-ft sheets; it can be cut with an EXACTO knife.

Poster sessions are made of many people presenting many posters; obviously the quality of the session as a whole depends on the quality of the individual presentions. Careful planning, close cooperation, and skillful graphics are all essential to producing quality posters and it is with quality posters that effective presentations start.

W
Writing and Editing

Stem Managers

Lois Kathryn Thuss
Technical Writer and Editor
The Johns Hopkins University Applied Physics Laboratory
Laurel, Maryland 20810

Member, Washington, DC Chapter, STC
Member, IEEE
Member, IEEE Professional Communication Society
Member, Toastmasters International

Lacy R. Martin
Project Manager
Procon, Inc.
El Monte, California 91731

Senior Member, STC
Member, IEEE Professional Communication Society
Administrative Committee

WRITING AND EDITING STEM

The 28th ITCC's Writing and Editing Stem has a new twist that offers writers and editors innovative ideas in many areas of communications.

To whet the appetite and set the scene, the stem speaker, a communications consultant, will expound on his many years' experiences in all facets of communications.

For the main course, authorities from the following eight fields — Advertising and Public Relations, Engineering, Journals and Tradepress, Legal, Magazines and Newspapers, Medical, Movies and Audiovisuals, and Scientific — have messages to deliver about communicating in their specialized areas. Leaders from each field will explain their session and format in an "Introduction to the Fields Meeting" prior to the start of the program.

Now for the dessert! The following workshops offer participants and attendees alike the chance to exchange new and dynamic ideas:

Student Workshop. A panel of enthusiastic students will explore new horizons in the communications world as they see it from their vantage. This three-hour workshop should provide us "old hands" with some young insights.

Shortcut to Proposal Writing. This workshop will explain a method that helps authors organize their thoughts, align their approaches with that of other authors, and reach agreement with the proposal manager on the message and content — all before starting to write.

A Plain-English Workshop. This workshop offers no magic formula — only good advice, long preached by clear-writing advocates — and should help us drop bad writing habits and remain with the basics.

Corporate Communications and the Press. In an excess of so-called "investigative reporting," many have been putting business in bondage. Balanced news coverage, in many cases, has given way to biased personal opinions, and objectivity has been lost. This must be stopped. Our workshop will focus on such problems and explain how they can be resolved.

That's the menu. Bon Appetit.

MEDICAL WRITING: SATISFYING YOUR AUDIENCE, YOUR CONSCIENCE, YOUR HUMAN RESEARCH COMMITTEE, AND THE FEDS

Lawrence Bachorik
U.S. Food and Drug Administration

Lawrence Bachorik
Food and Drug Admin.
Writer-Editor
5600 Fishers Lane (HFJ-1)
Rockville, MD 20857

(301) 443-5004

As medicine becomes inevitably more complex, many forces are likely to interfere with the production of clear and straightforward medical writing. These forces may combine to produce a "bureaucratic imperative" that has implications for all medical writers. After defining the bureaucratic imperative and discussing its potential effects, this article examines how FDA's review of patient package insert proposals altered the final guidelines. The review shows that even in the most stringently-controlled tasks medical writers can shift tone, emphasis, and meaning in highly significant ways--but it also suggests that the review process does not always improve the product.

As 1984 grows ever nearer, I may be vainly attempting to stem the Orwellian tide. But as medical writers face the task of conveying inevitably more complex information, they will simultaneously be working under greater constraints. These constraints may come from editors, from universities, from the government, or from sources yet unknown. Some of them may be designed to encourage clear communication, but in practice they may have the opposite effect. Medical writers may be forced to choose between satisfying some of those demands and remaining true to their work. My purpose in these remarks, therefore, is to shed light on these conflicting pressures by (1) describing the "bureaucratic imperative," (2) discussing its implications for several types of medical writing, and (3) examining how the review of a recently-published Federal Register document affected its final form.

My opening reference to George Orwell is apt, because Nineteen Eighty-Four deals with the difficulty of communication in the modern totalitarian state. In the work of its main character, Winston Smith, this chilling novel also illustrates the fullest expression of the "bureaucratic imperative." Employed in the Records Department of the Ministry of Truth, Smith reworks old articles from the Times; his daily task is to reaffirm Big Brother's infallibility by revising the leader's predictions, bringing past accounts of battles into line with current military alliances and disputes. In Orwell's brave new imaginary world, history as we know it no longer exists. The process of writing history has been so debased that, instead of chronicling events to fix their significance, history in Nineteen Eighty-Four means daily tinkering with back issues of the Times to produce new and consistent (but ever-changing) "daily" newspapers that justify the present.

This terrifying prediction provides the best example I know of the bureaucratic imperative, the process by which the means for accomplishing a specific end gradually becomes the end in itself. It consists of three stages. As form replaces function, the initial procedure is restricted. The outcome of this procedure becomes correspondingly restricted. Finally, in an attempt to improve the diminished outcome, people tinker with the procedure. This tinkering, initially viewed as a way of improving an outcome, becomes in itself the evidence of progess.

The Bureaucratic Imperative and Medical Writing

Few would argue today that the world of Nineteen Eighty-Four is upon us, and yet some of the Orwellian features of Winston Smith's world loom much nearer than they must have in 1949 when the novel was published. While the last thing I want to suggest is that some form of "Thought Police" is making the rounds of Washington, D.C., Palo Alto, or northern New Jersey, I do want to explore whether medical writing faces a danger from this bureaucratic imperative.

In articulating this problem, I owe a debt to Bradford Gray and his excellent work on informed consent and the vexing issues it raises. Although he does not use the phrase "bureaucratic imperative," he does warn against the dangers of "approaching informed consent as if it were nothing more than a set of procedures to be followed" (1). Gray lucidly describes the process by which this process transforms an ethical concern into a legal one.

In the pharmaceutical industry, for example, a piece of patient information or physician information (to use the current jargon) may be emended subtly yet significantly as it moves from the original drafter to the legal and marketing departments, on its way through the chain of approval. The original emphasis on compliance with a drug regimen (to cite one possibility) may be displaced by emphasis on compliance with a regulation.

Likewise writers and investigators at university research centers may find themselves facing gentle encouragement to alter the phrasing of crucial passages. This process may produce useful changes in research protocols, but tinkering with the process (to include the presence of certain old reliable words or "catchy" phrases) may displace the original outcome of humane experimentation.

Finally, the production of a document in any large organization is fraught with the potential for running afoul of the bureaucratic imperative. The steps by which conventional wisdom acts to produce a better document--review at many levels, concurrent review, meetings to reconcile differences of policy--tend to consume the product, so that the process itself begins to take on an importance greater than and ultimately independent of the product. And what remains to be released, published, or sent out--produced at great pains by following all the established procedures--is vague, "mushy," or misleading. This product, the result of following all the approved procedures, may conform to a standard--but the standard is no longer the original one of clear communication.

The questions facing medical writers in this context, then, are these: are the means we apply (layers of review, editorial guidelines, company procedure) becoming ends in themselves, and are they tending to displace the results--clarity, forthrightness, coherence--that we value in good medical writing?

The Bureaucratic Imperative at Work

Consider a couple of examples. The first involves the broader case of scientific writing, and it suggests that stricter controls and review mechanisms may be in store. In June 1980, Science reported the astounding news that a Jordanian researcher working in the U.S. had published five papers subsequently shown to be plagiarisms. This discovery naturally cast into doubt the authenticity of the more than 55 others he had published in a meteoric and perhaps ephemeral rise (2). It also cast considerable doubt on the review and publication practices of many prestigious journals. Although this episode may point to the need for more (not less) bureaucracy, it also illustrates perhaps a different bureaucratic imperative: one in which the original goals of good work and better understanding have been betrayed by the very great pressure to get one's work into print.

A second instance deserves mention. If current economic woes persist, another outlet for medical writers may begin to dry up. The myriad controlled circulation journals may refuse articles with unfavorable findings about certain products or procedures; here, the ostensible goal of communicating may become clouded by the desire not to offend sponsors. And it may also mean a shift in subject matter, as a journal of medical economics may ask for yet another article about what furniture to buy for setting up private practice instead of one that examines (for example) the structure, effectiveness, and viability of health maintenance organizations.

The Review of PPI Guidelines: One Example

Having identified the bureaucratic imperative and suggested a number of its potential effects, I want to sketch out the significance of this phenomenon for medical writers of all sorts. What I would like to do briefly is to compare a few selected passages from draft patient package insert guidelines published in September 1980 (3) with the final guidelines, published in November 1980 and in January 1981 (4,5).

The question to keep in mind for this comparison is whether getting the guidelines past the reviewers replaced the original goal of making available the information that allows patients to make informed decisions about their own health care. I am especially interested (1) in observing how a large organization, working under a strict publication deadline (and perhaps other pressures as well) alters its product and (2) asking whether these alterations are significant.

Both the proposed and final versions of the PPI for ampicillin prominently feature a brief section titled "Uses of Ampicillin" directly after the introductory "Summary" section. After two sentences explaining that ampicillin is a penicillin antibiotic and listing its uses against infections, this paragraph in the September proposal included the statement that "Ampicillin kills bacteria but not viruses. It should not be used to treat the common cold" (3).

While there is nothing incorrect about these final sentences, they assume that a reader will first recognize the distinction between bacteria and viruses and, next, understand that colds are caused by viruses. In the final version, one sentence has replaced two: "Ampicillin and related antibiotics have no effect on infections caused by viruses, such as the common cold" (5). The change is noteworthy for two reasons. It removes the distinction between bacterial and viral infections, a potentially meaningless distinction for large segments of the population, reserving "bacterial" as an adjective to modify the infections listed in the preceeding sentence. More important, it explicitly makes the significant point for patients: that ampicillin is worthless ("has no effect") in treating the common cold. Nothing much has changed factually, but the change can be very significant--the significance depending presumably on which side of the waiting room, pharmacy counter, or regulatory "fence" one occupies.

Another fairly significant, though seemingly minor, change occurred in the PPI guideline for clofibrate, a drug used to lower serum cholesterol levels. (The idea behind this class of medication is to decrease the incidence and seriousness of heart attacks by lowering blood cholesterol levels.) Because two major studies have suggested reasons why clofibrate should not be used extensively for this purpose, the guideline prominently features a discussion under the heading, "Risks of Taking Clofibrate." The second paragraph in both versions outlines the results of one study. That paragraph in the September proposal begins, "One of the studies suggested a 30% increased risk of getting cancer." After discussing the numbers of people with cancer in the control group and explaining the number of increased carcinomas that occurred in people taking clofibrate, the paragraph concludes thus: "Mice and rats that have been given clofibrate at five to eight times the human dose showed an increased risk of getting liver tumors, some of which were cancerous" (3).

The final, November, version of the paragraph differs primarily from the proposal in order of presentation. It begins without citing the percentage, noting only that "One of the studies also suggested that people who take clofibrate have an increased risk of getting cancer." After a simplified discussion of how many people developed tumors during the trials, it explains that "this is about a 30% increase above the expected cancer rate." The previous version's final sentence becomes, "Mice and rats given clofibrate at five to eight times the human dose showed an incresed number of liver tumors, some of which were cancerous." Here, "an increased risk of getting liver tumors" has become "an incresed number of liver tumors"--a more concrete observation. The final sentence in the November version echoes the opening statement: "This further supports the evidence that clofibrate may produce cancer in humans" (4).

It is difficult to estimate the effect of these changes on the ways patients will understand the information, and it would probably be even more challenging a task to measure the differences. But this example illustrates precisely how meaning can be manipulated by rearranging order and by shifting emphasis.

The third and final example is briefer. It concerns the section headed "Dependence" in the guideline package insert for propoxyphene, a drug used to treat pain. The first version begins, "You may become dependent on propoxyphene"; the final version replaces "may become" with "can become" (3,4). The purpose of such an emendation remains a mystery to me, unless it is merely for correctness of usage. (Even so this change is puzzling, for the "may-can" distinction is probably lost on many contemporary readers anyway.) The original paragraph concludes, "Dependence has occurred to people who have taken larger than recommended doses of propoxyphene over a long period of time." This becomes, "Dependence may occur in people. . ." (emphasis mine), a change that seems to make less definite the possibility of growing dependent on the substance (3,4).

I don't want to insist on the importance of these emendations in themselves, for I chose PPI's primarily because of my familiarity with the subject. I do not claim to know precisely why certain changes occurred, although I can report that the proposed guidelines generated considerable public comment, and that professionals from a number of disciplines inside FDA reviewed both the proposals and the comments received on them before revising the guidelines.

I do want to emphasize, though, that these two versions passed through the same organizational review within three months. The words before and after this review are nearly the same, but the effect of the changes varies considerably. Although the guidelines survived the bureaucratic review, and although many would probably agree that the comment period and the revisions improved the product, it would be unwise to claim that the emendations invariably improved the clarity, readability, or information of the final guidelines. Whether the entire process of requiring this information with prescriptions will become another bureaucratic imperative--that is, whether the preparation, distribution, and dispensing of PPI's will replace their having any effect--is an issue that is receiving close scrutiny. Besides, that is the matter for another tale.

Implications for Medical Writers

The message for medical writers is mixed. That there is no great difference between nominal conformity to a standard and essential conformity is hardly a novel observation. But recognizing once again that slight changes in syntax, vocablulary, and emphasis can markedly alter meaning should prove great solace to professional communicators everywhere. And being aware of the bureaucratic imperative is doubtless the best defense against it.

What all of this means for us, for people who seek the twin goals of clarity and honesty of expression, remains unclear in an age when the debasement of speech (forseen by Orwell) and the ascendence of the bureaucratic imperative threaten intelligent and useful discourse. But in closing, I would simply like to urge that, as professionals in communications, we can work together toward those goals. First, in setting up review cycles and clearance procedures, we should always take care to exclude the bureaucratic imperative wherever possible, so that the process does not devour the product. Second, as authors, editors, and reviewers, we can be active--even agressive--in advocating forthright communication. We can make a difference; fighting for the correct word, the appropriate emphasis, and the relevant topic are not mere matters of authorial whim or editorial predilection.

Finally, as professional writers, editors, and reviewers, we can invoke and rely on professional standards. We must continue to be aware of the fragile nature of communication, and of the human community it serves and unites. The continuing effectiveness of discourse depends on a consensus about language, and it should be our constant care to preserve--and not debase--the linguistic element which sustains and permits informed discussion. I do not intend to suggest that we have reached the Orwellian point where language is so devalued that there is no longer any contradiction in Big Brother's notorious slogans: "War is Peace. Freedom is Slavery. Ignorance is Strength." But plain dealing, attention to meaning, and some wariness of the review process will help us satisfy ourselves. This satisfaction holds the key to satisfying editors, human research committees, and the feds. That may be increasingly difficult to do, with 1984 only three years away.

References

1. Gray, Bradford H. "Complexities of Informed Consent." Annals of the American Academy of Political and Social Science 437:41-44, May 1978.

2. Broad, William J. "Would-Be Academician Pirates Papers." Science 208:1438-40, 1980.

3. "Prescription Drug Products; Patient Package Inserts Requirements." Federal Register 45:60754-817, 1980.

4. "Prescription Drug Products That Require Patient Package Inserts; Cimetidine, Clofibrate, and Propoxyphene." Federal Register 45:78514-22.

5. "Prescription Drug Products That Require Patient Package Inserts: Ampicillin and Phenytoin." Federal Register 46: No. 1, January 2, 1981.

QUALITY SOFTWARE AND THE TECHNICAL WRITER

John P. Brinegar and B. Dale Farrar

Honeywell Process Management Systems Division

John P. Brinegar
Honeywell PMSD

Sr. Documentation
& Support Analyst

16404 N. Black Canyon
Highway, Phoenix AZ 85023

(602)863-5263

B. Dale Farrar
Honeywell PMSD

Manager, Software
Methodology

16404 N. Black Canyon
Highway, Phoenix AZ 85023

(602) 863-5521

ABSTRACT

Usually, the most management-visible part of sofware development, the coding effort, is given the greatest attention. It is now being recognized that effort spent in design activity is more effective in meeting software quality goals.

User documents should be produced early in the development, so that they drive the design and implementation. Moving the preparation of these documents to the design phase causes more thought about user problems during the design. Technical communicators should contribute to the design, helping designers cover concepts they often miss, and avoiding delays due to rework.

The technical writer should develop an empathic relationship with the software developers. Mutual understanding of needs, resources, and objectives during a software development aids each in doing a better job by improving the documentation which drives the development.

WHAT IS QUALITY SOFTWARE?

How does a technical writer aid in the production of quality software? To answer this question, we must define quality software and delve into the software development process.

Quality Software is documented software, meeting established requirements, produced on schedule and within budget, which is correct, reliable, and maintainable over the life of the product.

Documentation is the key to quality software, and quality software *requires* a documentation-driven development process.

THAT'S THE WAY WE'VE ALWAYS DONE IT!

Traditional software development has seemed more of a magical art than a methodical application of science. Software systems were developed without much consideration for documentation. The production of program code was often the primary measure of productivity. System performance was the primary measure of quality - often the only measure.

Because productivity was measured in lines of code, documentation was often not considered until the programs were fully devleoped. With the documentation phase tacked onto the end of the development process, the documentation effort was measured in pages-per-day, due to the need to start on the next development program. Professional technical writers were seldom involved in software documentation except as editors. Usually, there was not enough time and the need was not apparent.

The results of this process were pretty good because the software and the systems it supported were not as large and complex as they are today. Generally, the customer base was limited and fairly sophisticated. Documentation wasn't considered such an important contributor to quality as it is today. And now, technology is both forcing and facilitating a new approach: **document driven software development**.

TECHNOLOGY'S IMPACT

The biggest impact of technology on documentation is that there is now so much of it - so much technology and so much documentation.

Recognition is growing that documentation should not be just a byproduct of a development program but is vital to successful development, use, and maintenance of the developed product or system. Significant research in technical communication is being accomplished. Universities are now preparing professional technical communicators. A larger portion of research and development budgets is being spent on technical documentation. And technology, itself, is offering more efficient ways to write and publish.

Larger Product Populations; Fewer Specials

Modern, microprocessor-based systems can pack a lot of functions into relatively inexpensive packages that can be made to support a broad range of applications. New technology, including modern human interfaces and high-level programming languages, makes it possible to move application engineering out of the manufacturer's plant and into the customer's hands.

Those who apply and use modern computers and their software systems are usually not involved in the inner workings of the hardware and software, but see the computer system as an interactive friend, that understands them and speaks their language.

Complexity

When integrated circuits first appeared, it looked like devices and systems would be simplified and there would be less need for documentation and fewer technical writers. Fortunately, because the hardware costs were significantly lower, and because so much more could be packed into a small package, electronic systems became much more versatile and complex.

While the cost of the hardware dropped, the cost of the software sky-rocketed. The hardware cost per unit is so reasonable that a computer in every home is a possibility, but development program costs are huge.

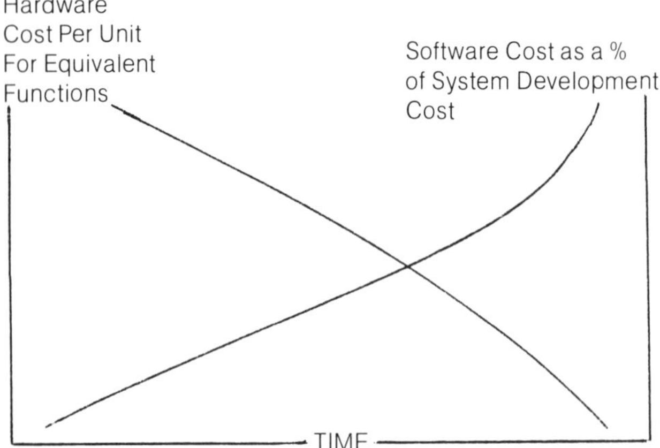

Documentation Technology

As this paper is being prepared, it has not yet appeared on paper. We are preparing the original draft at a word processing terminal. The first time this paper will appear on paper will be when the draft is printed for review and approval. The paper copy is necessary only for reviewers who do not have access to one of the word processor's terminals. Once the draft is approved, it will be sent directly to a phototypesetter which will set the type as it is to appear in the *Proceedings* of the 28th ITCC.

Similar terminals are used by the software developers for document and program preparation. Transfer of text and line drawing images from the software development system to the publications development system is possible. In fact, while not used in this instance, some facilities provide the software development system and the publications development system as the same system.

Conclusions

1. Technology is advancing so rapidly that new products must get to the market place quickly or the competition will beat you.

2. The development of most new software systems is expensive, and many must be sold to amortize the development cost. A few somewhat dissatisfied customers are usually no problem. Many somewhat dissatisfied customers are a big problem. And a few really irate customers can ruin you.

3. While incredibly complex, modern systems appear relatively simple to the user, and the user expects correct operation in all situations. Since there are many users, it is unreasonable to expect each user to understand the functional descriptions and theory in order to use the sysem, so the user publications must provide fool-proof procedures.

4. Quality user documentation must be available as soon as the first deliveries are made. If the new software systems are put on the shelf while the publications are prepared, competition will win over the waiting customers.

So, both quality software and quality documentation must be produced. Since the qualitity of each contributes to the quality of the other, success is more probable if their development is merged.

A DOCUMENT-DRIVEN DEVELOPMENT PROCESS

Documentation By Design

The documentation package must be carefully designed. It must not be a byproduct of the development process but the foundation for it. The design documentation should include a definition of the user and support documentation package. All design documents should be prepared with recognition that much of the information they contain is to be transferred to the user and support documentation.

The Technical Writer's Role

Qualified technical writers should be applied to each development program from the planning stages through field trials. Serving as an advocate for users and support personnel, the techncial writer brings important skills and knowledge to the development process. As a member of the development team, the technical writer contributes to good internal communication and gains knowledge necessary for the preparation of the published user and support documents.

Initially, the developers may resist this concept. Most everyone thinks he or she communicates fairly well and no one wants criticism from an outsider. If technical writers are integrated into development programs with preparation and care, they should turn out to be more helpful than critical.

The Developer's Role

The software developers take the lead in satisfying the users' needs. They prepare specifications and other documents needed to define the software structure and its solutions of the users' problems. In the preparation of this documentation and the program listings, they consider that these documents are the source of the procedural and technical information that will be in the user and support publications.

Design and Implementation

All modern software development methodologies are based on design followed by implementation. First the user requirements, then the structure and content of the software are defined. The design is then implemented by writing the programs and verifying proper operation.

The publications are prepared by a similar procedure. The requirements of the publications users are defined first. Then the design is implmented by preparing the publications and verifying proper operation. The functional and procedural information should be derived from the software design documents. In fact, it's best to use preliminary drafts of the publications as software design and test documents. Thus both the software design and the accuracy of publications may be verified at the same time.

Development Methodology

Modern software development programs should use an approved methodology that defines:

- Required internal documents and their content.
- Development phases and significant milestones.
- Development techniques and tools.
- Publications users and their needs.
- Publications structure and content.
- Publications development phases and significant milestones.
- Document preparation and publishing techniques.

This methodology should be defined by a joint team including the developers, marketing personnel as representatives of the users, and the technical communicators. With such a methodology, all members of the development team can work toward common objectives, with a mutual understanding of the development requirements and techniques.

TRADITIONAL PROCEESS

Developers:

User Requirements	Implement	Verify, Field Trial	Introduction

Technical Communicators:

Draft, Edit	Verify, Field Trial	Publish

DOCUMENT-DRIVEN PROCESS

Developers:

User Requirements	Design	Implementation	Verify, Field Trial	Introduction

C o o r d i n a t i o n & C o o p e r a t i o n

Technical Communicators:

User Requirements	Pubs. Design	Implementation	Verify, Field Trial	Publish

A bibliography is provided with this paper, listing considerable material devoted to "a better way" for software development. All of the publications listed advocate the document-driven process.

Analysis of Publications Users

Effective publications for computer system users and support people must present information in the order it is needed, with adequate detail, and in a form that can be readily understood.

Most technical writers keep up a continuing, informal user survey by seizing every opportunity to make contact with their readers. Many, if not most technical publishing organizations print reader comment forms in every publication. But the value of formal, direct user surveys is seldom recognized.

Surveys should include interviews with users and support personnel by key technical communicators. Visits to typical customer sites by the technical writers are well worth the expense. Other opportunities for formulating user profiles and determining information needs include:

- Joining studies by marketing personnel in defining new product requirements.
- Participation by the technical writers in training classes that include users and support personnel.
- Distribution of questionnaires to publications users. The value of information received this way is inhanced if the task is handled by a professional polling organization, skilled in the preparation of questions and the analysis of responses.

User Documentation and Human Interfaces

The best user publications are of little help if the human interfaces are poor. The design of the human interfaces and the design of the user publications should be interrelated. Technical communicators should review and comment on the human interface design documentation, The designers and implementers of the human interfaces should contribute to and review the user manuals. Reviews should indicate opportunities for improvement of both the human interfaces and the publications.

Traditional user and support publications have required each reader to analyze functions and theory to develop individual procedures. It's far more effective to develop standard procedures and present them in clear, unambiguous documents. These "cookbooks" won't necessarily omit theory, but what theory they have will be presented in the order it is needed.

Publications Implementation

This is draft preparation by the technical writers. They use the publications design documents to prepare text and illustrations. Ideally, the development and publications methodologies would be so well integrated that information needed would become available in the developmemet documents in time to be transferred to the publications drafts - documentation system to documentation system, or within the same documentation system.

In any case, preliminary drafts of the publications should be used by the developers for verification of that the software meets the users' needs, and verification of the accuracy and adequacy of the user manuals.

SUMMARY

Relatively haphazard methods for software development have been a moderate success. Recognizing that software development is a mental process and attitude, objective proofs that one process is better than another is difficult, if not impossible. Experience and observation indicate that close association between technical writers and software developers, using a document-driven software development process, results in a well documented, quality software product.

WHAT DOES THE SOFTWARE DEVELOPMENT TEAM REALLY PRODUCE?

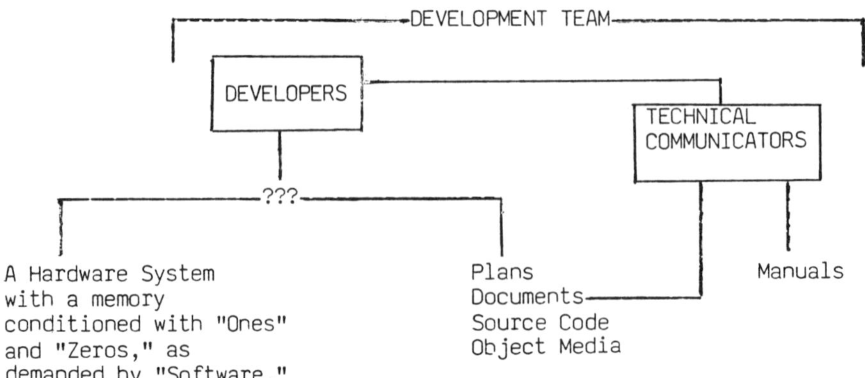

Bibliography

1. "Corporate Level Software Management;" John D. Cooper; IEEE TRANSACTIONS ON SOFTWARE ENGINEERING, Vol. SE-4, No. 4; July 1978.

2. "Estimating Software Costs;" Lawrence H:. Putnam and Ann Fitzsimmons; DATAMATION, Vol. 25, Nos. 10, 11, 12; Sept./Oct./Nov. 1979.

3. ETHNOTECHNICAL REVIEW HANDBOOK, 2nd ED.; Daniel P. Freedman, Gerald Weinberg; Ethnotech, Inc.; 1979.

4. FACTORS IN SOFTWARE QUALITY - CONCEPT AND DEFINITIONS OF SOFTWARE QUALITY; Jim A. McCall, Paul K. Richards, Gene F. Walters; Rome Air Development Center RADC-TR-77-369 Volume I, 1977.

5. FACTORS IN SOFTWARE QUALITY - METRIC DATA COLLECTION AND VALIDATION; Jim A. McCall, Paul K. Richards, Gene F. Walters; Rome Air Development Center RADC-TR-77-369 Volume II, 1977.

6. FACTORS IN SOFTWARE QUALITY - PRELIMINARY HANDBOOK ON SOFTWARE QUALITY FOR AN ACQUISITION MANAGER; Jim A. McCall, Paul K. Richards, Gene F. Walters; Rome Air Development Center RADC-TR-77-369 Volume III, 1977.

7. MANAGING A PROGRAMMING PROJECT; Philip W. Metzger, Prentice-Hall Inc.; 1973.

8. MANAGING THE SYSTEMS DEVELOMENT PROCESS: Charles L. Biggs, Evan G. Birks, William Atkins; Prentice-Hall Inc.; 1980.

9. RELIABLE SOFTWARE THROUGH COMPOSITE DESIGN; Glenford J. Meyers; Petrocelli/Charter; 1975.

10. SOFTWARE METRICS; Tom Gilb; Winthrop Publishers; 1977.

11. SOFTWARE QUALITY MANAGEMENT; John D. Cooper and Mathew J. Fisher, eds.; Petroceli Books; 1979.

12. SOFTWARE RELIABILITY GUIDEBOOK: Robert L. Glass; Prentice-Hall Inc.; 1979.

13. SOFTWARE RELIABILITY PRINCIPLES & PRACTICES: Glenford J. Meyers; John Wiley & Sons; 1976.

14. SPECIAL ISSUE ON COMPUTER MANUALS; Technical Communication, Journal of the Society for Technical Communication, Vol. 27, No. 4; 4th Quarter, 1980.

15. "Structured Documentation as a Design and Maintenance Tool;" B. D. Farrar; PROCEEDINGS 27TH INTERNATIONAL TECHNICAL COMMUNICATIONS CONFERENCE (STC), Vol. 1; 1980.

16. THE ART OF SOFTWARE TESTING: Glenford J. Meyers; John Wiley & Sons; 1979.

17. "The Cost of Developing Large-Scale Software;" Ray W. Wolverton; IEEE TRANSACTIONS ON COMPUTERS, Vol. C-23, No.6; June 1974.

18. THE MYTHICAL MANMONTH; Frederick P. Brooks, Jr.; Addison-Wesley Publishing Co.; 1975.

19. "We're Doing It All Wrong;" Robert L. Glass; DATAMATION, Vol. 25, No. 12; November 1979.

PREVENTION OF DONKEYISM: THE ROLE OF THE MEDICAL AUTHOR'S EDITOR

Judith Gunn Bronson, M.S.
University of Minnesota College of Health Sciences

Judith Gunn Bronson, M.S.

Department of Urologic Surgery
University of Minnesota
420 Delaware St., S.E.
Minneapolis, MN 55455

Senior Editor

Many of the faults in medical manuscripts are failures of logic or argumentation. These usually result from the author's lack of training in these areas and from his or her closeness to the work. One of the most valuable services a medical author's editor performs is calling the author's attention to these failings, which look foolish if they appear in print. Some of the most common errors are described, accompanied by suggestions for dealing with them and helping authors avoid them in the future.

INTRODUCTION

(Metric) tons of ink have flowed to tell physicians of their written crimes against the English language, so perhaps they can be forgiven for thinking that the near-impenetrable thickets of polysyllabic weeds that spring up behind their pens are the worst offenses they commit in their articles.

However, incorrect grammar and poor word choices often are not the worst of a manuscript's faults. As medicine became more scientific, it increasingly adopted the standards of science, demanding that authors adduce cogent evidence for each conclusion, present it logically, and deal satisfactorily with conflicting data. These are things physician-authors are not always prepared to do, because their training usually does not include the intensive drill in the scientific method, techniques of reasoning, and argumentation that scientists receive routinely. Some of them also are handicapped by the failure of their schools to teach them how to search the medical literature systematically. In other words, the failures usually stem from inadequate training, not from failures of intellect; and once the problem has been pointed out, the author can correct it. But someone has to point it out.

Peer review at the journals is supposed to stop shaky or unworthy manuscripts, but, as any regular reader of medical literature can tell you, too often it does not. Perhaps the reviewer is too busy to give the manuscript close attention or already agrees with the conclusion and thus is disinclined to give the argument the necessary scrutiny. In other cases, review is not sufficiently rigorous because the editor is afraid of being perceived as suppressing controversial findings (1). Whatever the reason, the shaky manuscript slips into print, where it is perused by thousands of beady-eyed readers who will make their dissatisfaction known in letters to the editor and in their own articles. The author is left wailing, "Why didn't I see that? Why didn't someone point that out to me?" and feeling like an--er--donkey.

Whenever anyone asks me to describe the responsibilities of a medical author's editor, I summarize them by saying that it is to keep your authors from making donkeys of themselves in print. The most important single part of this is helping them find and correct failures of logic or argumentation. To this task, you must bring your knowledge of biological science and its conventions, of medical publishing and its conventions, and of rhetoric as well as all of your negotiating skills. Usually, the author recognizes the problem when it is pointed out, but occasionally he or she will see your efforts as contrary to his or her interests. In the long run, of course, they are not; authors do not benefit from the publication of faulty manuscripts. It is helpful to have a file of published letters and editorials showing what happens when an author is caught violating the law.

What, then, are the most common material fallacies one encounters in medical manuscripts? I offer a sample.

"THE BITE OF SCHOLARS MAY BE AS SHARP AS THAT OF A FOX" (2)

Ad hominem arguments are alleged to be a uniquely feminine way of coping with problems--attack the person, not the problem--but I find men are no less prone to them. Of course, a great deal of fun would vanish from the medical literature if Surgeons A and B or Radiation Therapists C and D stopped burying those delicately poisoned darts for each other in their papers for the cognoscente to enjoy, but the author who relies on *ad hominem* arguments will always be suspected of having nothing better in his arsenal and is guilty of the fallacy of irrelevant reason. Doctor Doe may indeed be a certifiable blithering idiot, but that does not mean that his treatment for hangman's toenail isn't the greatest thing since Lazarus was raised from the dead. In explaining the weakness of an *ad hominem* argument to younger authors, it often helps to point out that if the personal qualities of an author, such as academic rank (or ancestry, whether human or otherwise) really were important in determining the value of a diagnostic method or treatment, no one would be interested in reading a paper by a resident.

FORTUITOUS AND PRODIGIOUS FALLACIES

These usually result from a failure to search the literature thoroughly.

In the fortuitous fallacy, the author puts together everything he or she knows about a subject and writes a manuscript without a systematic search for other (especially contradictory) data, with the result that an important area of the subject is ignored. For example, an author might describe the diagnosis and treatment of gonorrhea without mentioning that some strains of *Neisseria gonorrhoeae* are resistant to penicillin or that some cases of what appear to be gonorrhea are in fact nongonococcal urethritis. Few things can make an author look as foolish as does the comission of a fortuitous fallacy. The editor must check to make certain the material used in the manuscript is a reasonable sample of what has been written on that subject. University editors have an advantage here, because departments tend to write many of their articles on the same few diseases. By the time you've edited 51 articles on testicular cancer (as I have), the only literature you do not know is that in the journals that were published today.

In the prodigious fallacy, the unusual is mistaken for the significant. A young gynecologist related to the *New England Journal of Medicine* that her department had treated a patient for a condition none of them had seen before, and it was suggested that the case would make an interesting paper. Fortunately, the gynecologist began by searching the literature, where she found reports of nearly 500 cases of the same condition. Thus she saved herself the writing of a paper that an expert reviewer would have returned with a comment that said, in effect, "Yawn." The physician who treats such a zebra (from the admonition to interns, "When you hear hoofbeats, don't think of zebras") may be bitterly disappointed that the *New England Journal* or *JAMA* will not be interested in the case, but it can be pointed out that a state journal may be and that this is a good place for a fledgling writer to start.

BEGGING THE QUESTION

As Doctor John Dirckx has pointed out in his delightful book, some form of question begging appears in medical articles far more often than it should. He concocted a marvelous satire on the most flagrant kind, namely sheer impudence: "After listing several objections to his theory without even bothering to refute them, the writer returns to the charge by stating, with a total absence of logic and only the merest pretense at rhetoric, '_The fact remains_ that rye bread cures viral nostalgia'" (3).

In my experience, this list-and-ignore error is more often the result of exhaustion than of impudence, however. After carefully presenting each objection, taking pains to present it dispassionately, the author thinks, "I've done a lot of work on that; that should take care of it!" and moves on to something else, forgetting to marshall the opposing arguments. The author usually has those arguments in his or her mind, but readers read papers not minds.

The prototype of the question-begging argument is one in which the conclusion is included in the premises. Silly as this sounds, it's easy to do; the study has been completed and the conclusion is demonstrated to the author's satisfaction. To take a nonmedical example that came my way recently, the "scientific creationists" wish to prove that the first chapter of Genesis is a literal description of Creation. They start convinced that it is, and thus all scientific data must be interpreted to conform with that idea. Not surprisingly, those who make the argument convince themselves. Also not surprisingly, they convince no one who was not inclined to believe them in the first place.

In a question-begging argument like this, the truth or falsity of the conclusion is not the issue; the structure of the argument is. The things in medicine that everyone agrees on are rarely written about. Somewhere in that audience of beady-eyed readers will be someone who does not believe the conclusion and will look hard for reasons not to change his or her mind. If the author has used the conclusion as a premise, that someone will not have to look very far.

HONESTY AND OTHERWISE

"Dishonesty in print consists less often in barefaced lying than in the author's arrogant assumption of authority" (4). Dirckx might have added carelessness, laziness, and overenthusiasm. For example, it is easy to slip into statements about things that will "never be done." The great surgeon Billroth proclaimed that no one would ever operate on the human heart--only a few years before it was done (5). Ravitch reviewed the history of the introduction of Listerism (antiseptic surgery), showing how many famous surgeons said it was foolishness (5). Such arrogant claims to know the future are best left to the newspaper columnists, who are rarely held responsible when they turn out to be entirely wrong. It also is easy to proclaim that "the cornerstone of diagnosis (of some common disease) is the CAT scan," but chances are the author doesn't practice that way and shouldn't pretend to. "Be cautious in teaching others, for even an unintentional error in teaching is tantamount to a deliberate transgression" (6).

And what about that mention of someone else's work that is referenced to one of the author's papers in which the original work is cited? Is that honest? Of course not; it is taking credit for another's work. The failing here is usually laziness. The author knows the citation of his or her reference to that original paper but would have to dig in the files to find that original article. However, if the author's referenced paper collects data from many articles, especially if they were published in places like the _Outer Mongolian and Inner Tasmanian Journal of Transcendental Proctology_, the secondary reference is both useful and economical; but the text should make it clear that the data are not the author's. "The righteous scholar...quotes a lesson in the name of its author" (7).

Some of the stickiest situations an editor gets into concern authorship. Lancet recently published several letters and an editorial stemming from an author's removal of the name of another author from the galley proofs, without the knowledge of that author. The journal, caught in the cross-fire, was understandably unhappy. One wonders what effect this will have on the offending author's career, now that his unethical behavior has been spread across the pages of one of the world's most widely read medical journals (8).

Far more common, in my experience, is the inclusion on a manuscript of authors who should not be there or who don't want to be there. No one should be listed who did not contribute intellectually to the work or who does not agree with the conclusions. Young authors often think everyone wants as many publications as possible and so put several names on the manuscript of persons who do not want to be there. Getting those names off without offending the author who put them there can challenge the negotiating skills of a professional diplomat.

A final word of the subject of honesty and authorship. Physicians who have been led gently from a rough imitation of a manuscript to an article accepted for publication often want to express their gratitude to the editor by putting her or his name on as an author. This must be resisted firmly. Authorship in medicine and science is not like authorship in literary writing. In the latter, the words and the way they are put together is at least as important as the content, and the writer is the author. In the former, authorship is a statement of authority; and no matter how many articles I have written in part or entirely on testicular cancer, I am not an authority and thus must not be an author. An acknowledgment is proper; authorship is not. As Susan Eastwood said, "No one suggested that Hemingway's Nobel Prize in Literature should be shared with Maxwell Perkins, least of all Perkins himself" (9). Nearly all the medical author's editors I have talked to believe strongly, as I do, that it is unethical for the editor's name to appear in the list of authors. Besides, don't too many medical papers already have too many authors? As Bronson's Law has it, "Almost every medical manuscript will be improved if you cross off the first paragraph, all nontechnical adjectives, the word 'very' wherever it appears, and the names of three of the authors."

CONCLUSION

Editors are supposed to be guardians of grammar, word usage, and topic sentences, and certainly we must be that, too. However, it is foolish to spend hours polishing an apple that is rotten at its core. Help your authors refine their logic and argumentation before you go after their language. Then they will not look like donkeys in print and they will never say of you what is said of some editors: "Editors are a low form of life--inferior to the viruses and only slightly above academic deans."

REFERENCES

1. Comar C: "Bad science and social penalties." Science 200, 16 June 1978 (no page number)

2. Babylonian Talmud, Tractate Pirkeh Avot, II, 15

3. Dirckx JH: Dx + Rx: A Physician's Guide to Medical Writing. Boston: G. K. Hall, 1977, p 107

4. Ibid., p 100

5. Ravitch MM: "The danger of stating an opinion or the hazards of going on record." Surgery, Gynecology & Obstetrics 151:810-816, 1980

6. Pirkeh Avot, op. cit., IV, 16

7. Ibid., VI, 6

8. Heath DA: "Authorship of paper on primary hyperparathyroidism in Lancet of June 21." Lancet 2:30, 5 July 1980

9. Eastwood S: "It's turtles all the way down: The editorial office in a university or research center." Workshop, American Medical Writers Association 40th Annual Meeting, Atlanta, GA, 2 October 1980

TECHNICAL WRITER ON STAGE
THE ILLUSTRATED SPEECH

Kathleen Gail Browne
American Telephone & Telegraph Co.

Kathleen Gail Browne
A.T.&T. Long Lines Dept.
Supervisor
Room 3A150
Bedminster, N.J. 07921
201-234-7131

A WRITER RESPONSIBLE FOR AN ILLUSTRATED PRESENTATION FACES NEW CHALLENGES WHEN HE OR SHE IS UNUNINITIATED IN AREAS OF ART. A SPEECH CAN BE ACCOMPANIED BY RESPECTABLE VISUALS WHETHER THE WRITER WORKS WITH AN ARTIST OR NOT.

When the technical writer becomes the speech preparer, getting to the point, isn't what it was.

In a smaller company (or even a division of a large business), when a speech is needed, a technical writer is often considered to be best able to take the responsibility. Whether the technical writer is asked to prepare and present the talk, or to prepare a script for someone else, it will be assumed that any writer can produce a speech. This writer may be facing some new challenges.

In composing the text of the speech, a writer draws on skills which are used in all writing efforts.

Things for the speech writer to keep in mind (or to remind the speaker of) are available in many references. Some are listed at the end of this paper.

There is, however, another area that will help or hinder in making a successful presentation.

Generally, the writer will also be expected to produce visual aids for use with the text. It behooves any writer to have a rudimentary awareness of the value and use of visuals; but, many writers have not had an opportunity to hone skills in the area of visual communication. Yet the visuals prepared by our writer will be seen by audiences that have extremely sophisticated tastes. No matter how many poorly prepared business or technical presentations the audience has been witness to, they have been trained (by years of exposure to the commerical media) to know what good visuals are. The criteria of acceptance will be based on comparison with the highest rather than the lowest quality to which that audience has previously been exposed. Therefore, this paper introduces the technical writer-cum-script producer to some fundamentals.

BEFUDDLE OR BEGUILE

Consider yourself as a member of an audience. Do you remember an unhappy experience of sitting in a darkened room looking up at a screen and feeling as if you were being asked to decipher hieroglyphics without benefit of a Rosetta Stone?

When you see poor visuals is it a good production? Is the presentation, as a whole, good communication?

Now you are responsible for the visuals. How do you overcome this problem? (This paper will refer to 35mm slides; however, most of the pointers also apply to other visuals such as overheads, and in a smaller room, flipchards and easels).

Would your first choice be to collaborate with a professional technical artist who has experience in preparing illustrations for a speech? If you cannot arrange such a collaboration, can you prepare acceptable camera ready illustrations and have them made into slides by a professional shop? Probably!

But what goes on the slides? Generally, the speech should be written first and then the art designed to correspond with the main areas of the talk and to illustrate specific points. A great artist may be able to convey an entire philosophy or technology in one slide, but no audience wants to see a treatise on a screen. A picture can be worth a thousand words, but a picture of a thousand words isn't worth much. The speaker who wants the audience to see a complete copy of the text should arrange for printed copies to be available. If a speech is so poor that an audience would read such a slide, it will probably hypnotize them. Good writers should not be writing boring speeches and novice speakers need the help of interesting material, both verbal and visual.

Often speakers feel there must be something, anything, on the screen while they are speaking. Why? Well, sitting in a dark room with a black slide for a longish time is condusive to sleep, and a bright white light from no slide is blinding. But you can insert holding slides, like the speech logo, the title of the section just covered, or the title of the section being brought up, as appropriate. Repeat - as appropriate! Just as, ideally, each word should serve a purpose; so, each illustration should serve. The key is: visuals should illustrate the talk. They should enlighten, clarify, and demonstrate.

Supportive slides have pictures or main words that, like a theme sentence, express the core of the idea. Also, although details have a way of changing at the last minute, when it's too late to get a new slide made, the main idea is usually unaffected. For instance, if each of six regions report a six figure number, all of which increased over the previous month, show the "trend" in the slide, but leave the exact figures to be quoted by the speaker.

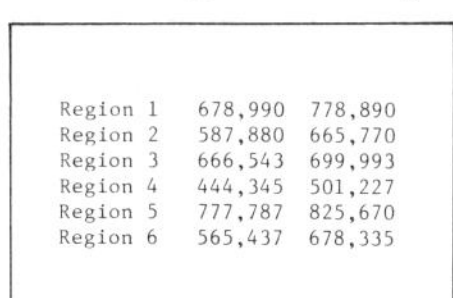

Region 1	678,990	778,890
Region 2	587,880	665,770
Region 3	666,543	699,993
Region 4	444,345	501,227
Region 5	777,787	825,670
Region 6	565,437	678,335

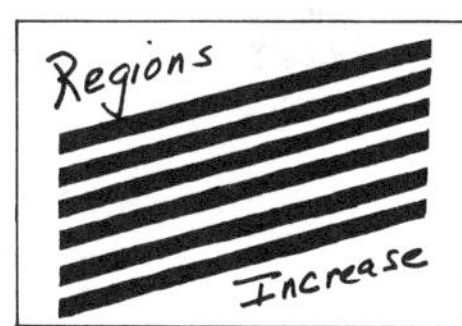

Figure 1 - Details vs. Main Ideas

To create interesting visuals, keep the words brief and bold, the pictures simple and relevant. Someone may be thinking "What about a schematic diagram?" (Or pharmaceutical description, or parts list, etc.) Such a visual aid is very useful to the technician, but can this person read or use it from the back of the room? Is your speaker going to be using this diagram in its entirety? Is the whole speech about it? If so, wouldn't it be kinder to allow everyone in the audience to trace his or her own way through an individual copy, making important notes along the way?

If there is some particular part of a complex drawing that is being referred to, could we focus on that, mask the exteraneous material, eliminate the clutter, get to the point?

LEGIBILITY

Legibility is not readability. Writers know about readability - making sentences flow together and having the text make sense. Legibility is what third grade teachers try to have their students achieve when they learn cursive letters and handwriting. Desiring legibility, we have our products typeset instead of offsetting the authors handwritten manuscript.

Slides should be legible, whether words or pictures. The audience should be able to tell if it says nun or run, whether it shows the Liberty Bell or a bell shaped curve.

Figure 2 - What vs. Ah ha!

Unless you're planning a great career change, probably none of us will be producing great works of art, but speech visuals are part of the art world. The artist, like the writer, must know and understand some fundamentals.

Whether or not you get your visual masters done by your art or graphics department, as I strongly suggest, you should keep these basics in mind:

- Use large letters and bold lines.
- Present one main idea at a time (so that it can be quickly understood).
- Condense numerical data (show only the sum, the average, the bottom line, the trends, etc.).
- Use color for impact.
- Keep the size of the slide in mind.

This last point can use further illustration. Referring back to our poor soul without a Rosetta Stone, many overinflated slides come about when someone tries to get an 8½ x 11" typed page made into a slide.

To put things in perspective, draw and reproduce "slide size boxes." (Actually I use boxes of 1½ X 1 inch). These can show pictures, hand printing, or typed words as they will be perceived by the audience in the back of a banquet room. If the boxes are legible at twelve inches, the projected slides should be legible at the back of the room. Using this size box makes it difficult to put too much material on it.

THE WRITERS ART

Do you know what supplies you'll need? How to use them? The guidelines for preparing a camera ready master?

When the writer has the task of preparing the camera ready art, more guidelines are needed. The first step is to visit a drafting or art supply store and acquire a supply of:

- 15" x 12" paper (or slightly larger)
- Rub-on Type - two sizes
  - ½ inch caps for titles
  - ¼ inch caps for body
- Two transparent, ruled straight edges
- A light blue pencil (or two)
- Masking tape
- As much free advice as they'll give

When ready to begin, use the masking tape to tack the corners of the paper to your desk or table. In the center of the paper, using the light blue pencil and a straight edge, lightly draw a box 6 inches high by 9 inches wide. This is your critical field. (See Fig. 3) The information on the slide will be within the critical field. However, the camera will take in slightly more area (to avoid missing some of your art); so, draw another box around the one in the center. The second box should be 10 inches high by 13 inches wide, this represents the camera field. (See Fig. 3)

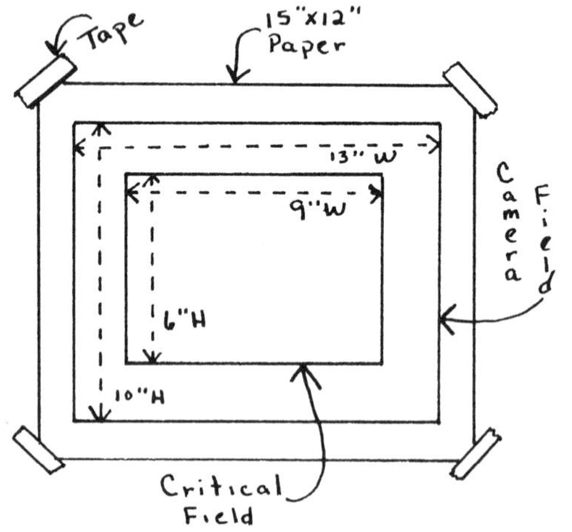

Figure 3 - Critical Field

Now, you are ready to do a rough draft, or pencil copy of your slide. The pencil copy will help you to line up and center your work. You will need to plan your spacing. (Two lines of ½ inch letters take more than 1 inch of space. There should be a blank line of 3/8 to ½ inch.)

If you do your own hand lettering, it is best to try for an informal look. If you try to have it look "printed" it will look like a poor imitation. If you use stencil or "rub-on" lettering, achieving correct spacing takes some practice. Start from the center and work toward the edges. (The vertical center should allow for the spaces between lines). Use your light blue pencil to draw a base for each line or the end result may cause seasickness. (Erase the base line, and all other lines when you've finished.)

The space between letters can be tricky. The easiest way to deal with this is to always allow the same amount of space between the right side of the first letter and the left side of the next. Another way is to place to "M"s side by side, measure the space between their centers and always use that space for measuring from the center of one letter to the center of the next. However, the human eye comprehends the <u>area</u> rather than the <u>length</u> of the white space between letters, so with some experience, you may want to adjust the space between letters to allow for this visual effect. (See Fig. 4)

1/4" 1/4" 3/16"

Figure 4 - Spacing

If you do hand lettering, stencil lettering or drawings, you may wish to double the size when drawing the original and then have it photographically reduced by 50% before having slides made. Reductions make the lines, letters, and drawings look smoother by diminishing the irregularities.

STAGE IMAGE

But now that you've prepared respectable masters and had them made into transparencies, do you know that the audience can still be deprived of seeing the illustrations?

When determining that "Slide size boxes" will represent the legibility of the projected image from the rear of the audience, the placement of the projector is assumed to be correct. If the projector is placed too close to the screen, the speaker may be able to see the slides while the audience cannot.

If there is no projection booth, the speaker should arrange to review the slides while seated in various places in the room. Failing this, placing the projector in a center aisle, approximately three quarters of the way back (or more) from the first row of the audience is a good approximation. A Vu-Graph proector should be checked to see that it is far enough back too, any visual has to be seen.

The production is ready! It may not be Broadway, but the audience should get the point.

ENCORE

The first time a writer becomes a speech producer, he or she may have little background, but if it happened once, will it happen again?

When the writer is often asked to be the speech producer, reference works will be wanted for this new challenge.

Some, for example are:

Artwork

Nelms, Henning. Thinking With A Pencil. 6th Printing, 1968; New York: Barnes & Nobel, Inc., 1968.

Visual Aids

Hare, Carol R. "Preparation of Slides For Oral Presentations." Proceedings: 21st ITCC Washington, D.C.: Society For Technical Communication, 1974, 7-8.

Huff, Darell. How To Lie With Statistics. New York: W. W. Norton & Company, Inc. 1954.

Johnsen, Richard A. "Six For The Price Of One." Proceedings: 20th ITCC. Washington, D. C.: Society For Technical Communications, 1973, 171-174.

Lutz, R. R. Graphic Presentation Simplified. New York: Funk & Wagnalls Company, 1949.

Perlmutter, Jerome H. "The Federal Graphics Improvement Program." Proceedings: 20th ITCC. Washington, D.C.: Society For Technical Communications, 1973, 155-156.

Selby, Peter H. Using Graphs And Tables. New York: John Wiley & Sons, Inc. 1979.

Speaking

Higgenson, Margaret V., et al. The Ambitious Woman's Guide To A Successful Career. New York: AMACOM, 1980.

MacKay, Colin Neil. Speak For Yourself. Cambridge: Brooke Crutchley, University Printer, 1971.

Sager, Arthur W. Speak Your Way To Success. New York: McGraw-Hill Book Company, 1968.

Weiss, Harold, et.al. Technically Speaking. USA: McGraw Hill Book Company, 1963.

A Picture of a Thousand Words Isn't Worth Much.

DESIGNING USER MANUALS FOR APPLICATIONS SOFTWARE SYSTEMS: AN INTEGRATED APPROACH

John D. Browne
Boeing Computer Services Co.

John D. Browne
Boeing Computer Services
Mgr., Financial Services
Publications
PO Box 24346 M/S 1E-61
Seattle, WA 98124
(206) 575-7263
Member, Puget Sound Chapter, STC; Former Chairman, Orlando Chapter, STC.

Historically, user documentation of in-house applications software systems has fallen into two categories. The first category has technical personnel such as programmers preparing documentation. Structure, in this instance, is largely a function of system design, with most of the documentation consisting of module descriptions. The second category has in-house procedures personnel preparing documentation. These manuals generally ignore system considerations in favor of step by step methods and procedures. This paper examines the advantages and disadvantages of each method, and then explores an integrated approach which answers more of the key objectives for documentation without the disadvantages associated with the two historical methods.

Many companies are developing their own application software these days, and sooner or later they are faced with the same problem: documenting that software. Unfortunately, small shops with a handful of programmers usually don't have qualified technical writers on hand. As a consequence, they don't know how to go about designing the structure of the documentation, which is one of the most important ingredients in a useful manual. What do they do? Historically they have approached the problem in one of two directions -- neither one particularly effective. In this paper we will examine those two ways, and then look at a third, more effective way of designing documentation for applications software.

Before we get started, however, we need to define our terms. For the purposes of this paper, I want to make the following definitions:

> Application software: Software that does work supporting a company's business activities, such as an inventory, payroll, or general ledger system.
>
> In-house: Doing something by yourself. In this case, we are refering to companies who, rather than purchase a system from a vendor or buy time on a system from a time-sharing company, elect to design and develop software by themselves, using their own programmers.
>
> User Documentation The documentation which supports an existing system. May include descriptions of programs, files, flowcharts, report narratives, instructions on completing input forms, etc. Does not include the documentation supporting the design and development of the system, such as problem definition, functional specifications, etc.

CONVENTIONAL APPROACHES

I'm going to label the two historical methods of developing documentation as the systems approach and the procedures approach. Each is characteristic of the types of people who prepare the documentation -- each is an evolution of the ways in which these people perceive software systems.

Robert W. Pirsig, in his book Zen and the Art of Motorcycle Maintenance (New York: William Morrow, 1974), divides perceivers of technology into classical and romantic reasoners. People who use classical reasoning tend to view technology in terms of its underlying form, while those who use romantic reasoning tend to view technology in terms of its apparent (or surface) form. Technologists tend to exercise classical reasoning; non-technologists tend to use romantic reasoning. For our purposes a "technologist" is a programmer who develops and maintains a piece of applications software. A "non-technologist" is a user of that software -- a clerk, manager, accountant, or whatever.

These two groups of people have perceived software in two different ways, and have thus developed documentation according to their perceptions. The programmer, using classical reasoning, views the system in terms of its underlying form; that is, as a collection of programs or modules. The user, on the other hand, views the system at a surface level; that is, as a collection of human-related activities. For example, when discussing a payroll system, a programmer might speak in terms of program PAY150 or PAY200, while a user might speak in terms of filling out the Payroll Adjustment Form, or balancing a particular report.

Now that we've covered this background material, we can see why documentation developed by programmers has traditionally been a collection of program or module descriptions. This is what I call the "Systems Approach." Documentation developed by users has traditionally been a collection of human activities -- the "Procedures Approach." As we discuss each in detail, keep in mind what we have said so far about the types of reasoning and the perceptions of the people involved.

The Systems Approach

As we have already said, the key aspect of the systems approach is that it views the system as a collection of modules, rather than human activities. This is characteristic of field service manuals for pieces of hardware. The technician is expected to be able to identify which module or component the problem is related to, and then uses the documentation to further troubleshoot the component. The documentation, of course, is structured around those components.

You know you have a software manual structured like this when it begins:

> "This system is made up of three modules: the entry/edit module, the posting module, and the reporting module. The entry/edit module is made up of the Transaction Capture Program, the Transaction Edit Program, and the Edited Transaction Sort Program. The posting module is made up of ..."

At this point the manual tells you everything you could possibly want to know about these three modules, and all the programs that make them up, and anything else the writer could think of, relevant or not, about the system.

This type of documentation is useful for programmers who already are familiar with the system and only need to find the occasional detail which has slipped out of their memory, but it is useless for the new programmer who doesn't know what module a particular problem relates to. More importantly, this method doesn't show relationships within the system.

Summing up, this method is favored by programmers writing documentation, and can be useful for those people familiar with the system, but is lacking in other, critical areas.

The Procedures Approach

This is the one that methods and procedures people are fond of when they write documentation. This, like the other approach, is characteristic of the way in which they perceive documentation.

Documentation written using this approach will consist of a collection of procedures: how to enter a new account, how to balance the Daily Activity Recap, how to set up an Account Exception Record, and so forth. Frequently the procedures tie system-related activities (those dependent only on the software) to organization-related activities(those dependent only on the internal workings of the organization). For example, the reports produced by a Payroll System is a function purely of the software; the names of the people who receive those reports is a function purely of the organization. Documentation written with the Procedures Approach would tend to tie those two pieces of information together. This in effect doubles the revision problem, as either software changes or organization changes will impact the documentation.

There are several problems with this type of documentation. First, it is hard to know when to quit. If one searches hard enough, one can find an infinite number of procedures in any organization which could be documented. The more procedures, the more revisions to the manual necessary. Secondly, this kind of documentation really sees the software as a kind of black box, with no interest in how the system functions. This makes it difficult (or downright impossible) for a reader to understand *why* the system does what it does, knowledge one needs to handle the unexpected problem or error. The chief advantage of this approach is that many companies already have procedures personnel in-place, and they can then just pick up this function.

THE INTEGRATED APPROACH

Here's where we depart from conventional thinking and explore an integration of the two accepted ways of doing things. I think this approach combines the advantages of the other two without the corresponding disadvantages.

Before we can explore the integrated approach, however, we must first talk about something I call the "Three Part Principle."

The Three Part Principle

The Three Part Principle says that all documentation, whether for a software system, a missile, or a refrigerator, is comprised of three parts: Description, Operation, and Troubleshooting. Let's use the owner's manual for an automobile as a sample document, and find the three parts. In the beginning of the owner's manual we find an illustration of the dashboard with all the controls identified. That is Description. (In addition, all the specifications in the back for fluid capacities and lubricants are also description.) Next in this manual we find instructions on how to start the car, how to set the air conditioning, and so forth. This is Operation. Finally we find information on what to do if the car won't start, what to do if the oil light comes on, what to do if you have a flat tire. This is Troubleshooting.

If you're thinking ahead, you will have noted that the Description part of our manual uses the systems approach, and the Operation and Troubleshooting parts use the procedures approach. Mixing these two according to the Three Part Principle is what gives us the integrated approach. We can define the integrated approach, then, as follows:

> <u>The integrated approach to documentation uses the Three Part Principle to integrate human activities with system design considerations.</u>

When you use the integrated approach to user documentation, you find that you need descriptive information, such as program descriptions, record descriptions, flowcharts, descriptions of input forms and reports, and so forth. You will need operational information, such as instructions on how to fill out input forms, use input/maintenance screens, balance reports, set up accounts, perform file maintenance, and like activities. (Note that this operational information would include operating instructions for the computer operators who actually run the system, for activities like daily processing, month-end processing, special report production, and so forth.) You will finally need troubleshooting information such as how to correct errors, how to recover the system, restart procedures, etc.

Experience tells me that a logical way to break down the total documentation required for a software product is into three manuals:

1. User's Guide
2. Systems Manual (or Programmer's Guide)
3. Operator's Manual

We apply the Three Part Principle to each of these manuals, and find that they are best written according to the integrated approach.

Experience with this approach has taught me that <u>documentation design is a function of the product line, not the individual product.</u> This means that normally a single overall design will work for documenting different applications in one shop. That way we get standardized documentation, and we do not have to reinvent the wheel each time we produce a manual.

An example of the way manuals might be designed (using the integrated approach) for a shop documenting batch applications software products might be as follows:

User's Guide

- Section 1 - Introduction
- Section 2 - System Overview
- Section 3 - Form Descriptions
- Section 4 - "How To" Procedures
- Section 5 - Report Descriptions
- Section 6 - File Maintenance and Transaction Codes
- Section 7 - Troubleshooting
- Glossary
- Index

Systems Manual

- Section 1 - Technical Overview
- Section 2 - Program Descriptions
- Section 3 - Record Descriptions
- Section 4 - System Flowcharts

Operator's Manual

- Section 1 - Run Instructions
- Section 2 - Console Messages
- Section 3 - Data Entry Instructions
- Section 4 - Restart and Recovery

Note that each section represents one of the three parts we discussed earlier:

either Description, Operation, or Troubleshooting. Note also that the integrated approach has been applied: the Systems manual, for example, is not structured around individual programs, but rather around types of information, thus showing relationships between parts. This carries through for all the documentation -- it is based on functional units rather than solely modules or activities. By integrating the two, we have achieved better documentation.

The advantages to the integrated approach are many. For one thing, the relationships between human activities and the design of the system are shown. The document becomes useful as both a training and reference tool. Because it is organized more logically, it becomes easier to update. Because the design is not peculiar to one product, standardization of documentation becomes possible and practical.

The disadvantages of this approach are those of any good method of documentation: it is not the cheapest way, nor can it be done with untrained people. It does indeed require professional technical writers, not just "programmers with a flair for writing." It requires a good understanding of the system, the programming involved, and the tasks of the users. It requires research and time and a commitment.

If it costs more, why bother? Because it is cost effective. More logical documentation means that users and programmers can find the answers to questions quicker. This saves their time searching for information, which in turn saves the company money. Better documentation means people understand how to use the software better, which means fewer errors with their corresponding expense. When users can find answers to system questions by looking in the manual, they will be less likely to call a programmer everytime they are puzzled, and thus programmers can spend more time programming, and less time answering elementary questions over the phone. All these things (and there are others - think about it) mean savings to the company.

Don't think of good, integrated, documentation as an expense: think of it as an investment.

THE EFFECTIVENESS OF INDUSTRIAL (IN-HOUSE) MAGAZINES

Janie J. Cambridge

IBM
Federal Systems Division

Janie J. Cambridge
IBM
Comm. Svcs. Mgr.
Bodle Hill Road
Owego, NY 13827

(607) 751-2608

A company publication functions as the official and primary source of obtaining management's position on company issues and the facts concerning company actions. It puts the varied pieces of the organization together so that each employee can get the "big picture." Company publications are the only medium of employee communications that regularly reaches spouses and families as well as the employees.

Unlike conventional newspapers and magazines, most industrial publications operate with a small staff. In order to have an effective in-house publication, it is essential to have a competent, well organized editor/writer, and a well qualified layout artist, sometimes called a designer.

INTRODUCTION

American industry spends well over $100 million a year on company publications as their major channel for getting information to its many audiences: employees, customers, stockholders, community and civic leaders. Most firms have some form of publication, generally a newsletter or newspaper. The Bureau of National Affairs states that 80% of manufacturers publish in-house magazines or newspapers, 65% of non-manufacturing firms, 65% of large companies, and 85% of small companies. Almost half publish monthly, 20% bimonthly, 20% quarterly, and the remainder weekly, biweekly or other.

The frequency of the publication depends on a number of factors such as how much news there is to report, size of the publication, budget and staff. Frequency, however, should not be so high that its impact on the production cycle reduces the quality of the publication. But given the proper staffing, sufficient budget and significant material, there is no reason for greater frequency to mean lesser quality.

Timeliness is important on any major news events, such as major re-organizations and announcements of new products and employee benefit plans. Monthly or semi-monthly publications should not be the announcement vehicle for this type news. Instead, special issues or bulletin board postings should be used.

ADVANTAGES OF INDUSTRIAL PUBLICATIONS

Employees want to feel they are a part of the company and their work efforts are appreciated. In-house publications are ideal for creating that feeling. It is also ideal for solving other communications problems: maintaining high participation in employee benefits plans; gaining goodwill by pointing out the company's contributions to those plans; promoting health and safety; reducing accidents and waste; increasing productivity; developing teamwork; and creating a family feeling among the workers.

The publication can be used to show why the company is a good place to work. Acquaint employees with conditions of employment, help them become familiar with their place in the organization, show them the advantages of long service, answer their questions, inform them about future plans and new policies, and show that the company is a good corporate citizen and an asset to the community. It is also a good place for management to get across its own point of view.

STAFF

In order to produce an effective in-house publication, it is necessary to have a competent, well organized editor/writer, and a knowledgeable designer. These people should have good communication skills, including the ability to write clearly and in an appropriate style, a feeling for graphics, and an ability to meet deadlines.

The editor/writer is responsible for maintaining high standards of taste in the publication. The demands of good taste reflect the company's standards and in no way interfere with the publication's communication objectives.

No one in the reading audience should ever be offended by an item or photo in an in-house publication. Therefore, judgement on good taste requires considerable sensitivity; it must be applied to everything published. At times, this may seem to be a form of censorship, but it is actually just the exercise of good manners.

Undertaking the job of editor/writer of a company magazine or newspaper can be an exciting challenge and a rewarding experience. The job will certainly provide exposure to a broad spectrum of functions and activities which include all the planning, organizing, directing, and controlling of each publication. And there is a great sense of personal accomplishment in producing the finished product.

The layout attracts the readers, leading them to stop, look and read. This is the job of the designer. Good layout does these things without making the reader aware of the design. It is simple, direct and orderly, and is never layout merely for the sake of layout.

The publication should be designed for easy reading. The designer should never dazzle the reader with strange and arty configurations. For each issue, the designer should make up a dummy, a kind of blueprint for each page.

Pictures should be found in all parts of the publication, but instead of portraits and posed shots, photos should show action or people involved in their work.

It would be ideal for each company to employ a professional photographer, however, in many cases it is not economical or feasible. To replace this talent, the designer can be taught to take 35mm or polaroid shots. This method requires additional training but there is a plus to it. The designer knows the publication's pictorial needs. In many instances, he or she has worked with the editor/writer for a long time and knows the purpose of the publication. Therefore, the photographer needs to know more than shutter snapping and lens openings; he or she needs to know how to compose the picture to make it eye appealing and "catch" the reader.

GATHERING NEWS

In researching his material, deciding which articles are right for each issue and the most appealing layout, the editor/writer determines the effect of the publication. If he or she supports only management, the appearance of fairness will be destroyed, and with it, the purpose of the publication itself. Direct quotes from management and employees in news articles add merit to any publication.

An annual schedule of planned major articles should be outlined by the editor/writer. These projected, major milestones can be obtained from each department head. Keeping in mind that these are projected events, the editor/writer should keep a somewhat firm quarterly schedule (based on the frequency of the publication) with built-in flexibility that would accommodate fast breaking news. For example, if a reorganization is announced that was not scheduled, the editor/writer should have at least one article that could be held for the next issue.

A simple way to gather news is by taking informal tours of the plant or laboratory weekly or bi-weekly. A photographer should accompany the editor on these tours because many interesting pictures and articles can be discovered through this method.

If the company has plants, laboratories or offices throughout the country, valuable material may be obtained by means of requesting articles from each of the company's geographical locations. By establishing a contact at each location, a news bureau can be set up. According to the frequency of the publication, a phone call, weekly, bi-weekly or monthly, should be made to these contacts soliciting their inputs.

A popular way to gather news for your publications is by soliciting material from employees by means of articles or letters to the editor. When employees see their fellow workers getting into print, they will realize that the publication gives them direct access to everyone in the company.

Finally, Suggestion, Cost Effectiveness, Outstanding Contribution and Quality Awareness Awards are some more popular news stories that give "solid" recognition to employees. This type of news also publicizes the company's Awards Program.

While jokes, funny stories and cartoons are often found in company publications, they are costly, and there is no indication that they increase readership. Stick to the primary purpose which is information, not entertainment.

Strive for consistency in the publication so that employees can anticipate and watch for the next issues. Deadlines must be met. The more frequency the publication, the more difficult it is to meet deadlines.

Publishing a newsletter or magazine does not by itself solve an organization's communication problems. In fact, it can be counter-productive unless it motivates and gives recognition to the work force.

DESIGN AND LAYOUT

Any publication should be designed so that it is easy to read. Exaggerated type styles may look good in a word or a line, but they make paragraphs and pages hard to read. Colored ink is also hard to read. Stick to simple, black letters and use bold type sparingly. Keep headlines simple and large.

A sound knowledge of typography and typographic standards is only part of the story. The designer will benefit by being constantly aware of who the reader is and of his motivation and his reading habits. Young designers, for instance, often forget that an eight-point letter looks a lot larger to them than to the reader who is over forty. Above all, words are to be read.

A publication designer who reads the words he works with will not only improve his relations with the editor, but will be in a better position to find meaningful solutions to design problems.

For each issue of the publication, a dummy layout, a kind of blueprint of each page, should be made up using the grid sheets.

Basically, there are two types of dummy layouts. One is a rough layout. When the photographs have been completed and the headlines written, a plan is drawn to fit the actual space. This serves as the "blueprint" for the printed pages. Little time is devoted to rendering the photographic areas, but careful attention should be given to space, typography, proportion, and photographic cropping.

The second type is the comprehensive layout which translates the rough into a reasonable facsimile of the printed page. Photostats or photoprints of the pictures are mounted in position. Photographs should be positioned so that small ones do not lose their impact and then plan the copy to fit around them. Dummy type is used for the typed areas and headlines are either set in type or carefully rendered.

All publications may not be able to execute the comprehensive layout as outlined. Time and cost limitations may require shortcuts, but this procedure is the best way to achieve a polished and coordinated publication design.

PRODUCTION

There is much to be learned of the various printing techniques and processes. Even the page size can affect economies and the weight and kind of paper used can have considerable bearing on the finished appearance of the work. A glossy page can be much more appealing than a dull, rough page and will greatly enhance the finished product.

While it is not essential that a publication's designer be skilled in all of the techniques of printing and production, it is valuable for them to know the limitations, and more important, the potential of the process which they use. This means working closely with the Publications Department or vendor responsible for the final product.

SUMMARY

The objectives of industrial magazines and newspapers are to inform, educate, recognize and motivate. In meeting these goals, the publication serves to reduce speculation, rumors, and misinformation by providing the employees with facts. It gives employees a better understanding of what is expected of them and why, and helps create a sense of community.

But all of this can be academic unless the publication has good readership. In order to obtain this, the news coverage must be balanced and well written and presented in an attractive, readable and inviting package. Balance means a mixture of news such as business coverage, editorials, individual and promotion recognition, and community involvement activities.

Publishing a company magazine or newspaper is not easy. There can be particularly upsetting moments. These may range from misspelled names to the omission or mention of people who may have merited inclusion. At such times, one must have the fortitude to shift into second gear and carry on regardless. If you persevere, the overall publication cannot fail to be an astounding success and a source of great pride.

Notes and References

1. Allen Hurlburt, Publication Design Van Nostrand Reinhold Co., a Division of Litton Educational Publishing, Inc. 1976

2. Charles B. Moore and William F. Blue, Jr., Editing and Layout Techniques for the Company Editor, Ink Arts Publications, 1979

3. La Rae H. Wales, A Practical Guide to Newsletter Editing and Design. Iowa State University Press, 1976

4. Linda Nanfria, How to Publish an Organizational Newsletter, Creative Book Company, 1976

5. Thomas G. Hargreaves, C.A.M. LL.B, "The Manager As Editor", Management World, Administrative Management Society, Volume 7, Number 6, June 1978

6. Alan Unwin, "Communicating Company Information", Industrial and Commercial Training (UK) Volume 10, Number 10, October 1978

7. Small Business Report, CA, "Company Publications", Small Business Monitoring and Research Co. Inc., March 1980

ASTRONOMY FOR BIOLOGISTS (AND VICE VERSA)

R. B. Crawford
Lawrence Livermore National Laboratory

Richard B. Crawford
Lawrence Livermore National Laboratory
Technical Editor/Writer
P.O. Box 808 L-52
Livermore, CA 94550
(415)422-3711

In our monthly reports we try to make the many and varied projects at LLNL intelligible to readers from diverse technical backgrounds. Since our readers are busy, we make it easy for them to skim. Each article has a stand-alone abstract and full figure captions that reiterate information from the article. We make a special effort to seize and hold a reader's attention, using attractive design, functional organization, crisp writing, and concrete images. Science writing is a fascinating specialty and a rewarding career. It requires more preparation and dedication than other kinds of technical writing, but it also permits more creativity and provides more satisfaction.

"Work performed under the auspices of the U.S. Department of Energy by the Lawrence Livermore National Laboratory under contract number W-7405-ENG-48."

INTRODUCTION

Energy and Technology Review, the Lawrence Livermore National Laboratory's monthly report of its unclassified activities, is a slick-paper periodical of 24 to 40 pages. It has a circulation of about 6,000, including college libraries throughout the country. Its primary audience, however, consists of the managers and administrators (drawn from diverse technical backgrounds) who oversee our work for the DOE. Its main function is to tell these administrators about the variety and quality of work being done at LLNL.

The pure and applied research of the Laboratory encompasses an enormous range, from astronomy and pure mathematics to such down-to-earth concerns as medicine, metallurgy, and meteorology, from inertial- and magnetic-fusion energy, next-generation high-speed computers, and nuclear weapons to nondestructive testing, cell counting, holography, and laser-fluorescence chemical analysis. No member of our audience, from whatever technical background, can be expected to be familiar with all these fields. Hence we must make a special effort to avoid specialized vocabularies and explain our methods and results in common terms intelligible to all.

The material submitted to us comes from authors who are used to writing reports and technical articles for professional journals. Many of them communicate very well, on that level. Their audience already knows a lot about their specialized field, and doesn't need to be filled in on every little detail. Hence they can concentrate on presenting what's new and different, and more or less assume that anyone who can't keep up with the discussion probably isn't that interested, anyway.

Furthermore, such an author can assume that his readers will stick with him through the hard parts because they really want to get the message. Most readers of a professional journal are working in the same field. They have a built-in motivation to keep trying until they figure out anything obscure. They need the information it contains to help them perform more effectively.

Our readers, on the other hand, are mostly busy with a lot of different concerns, only a few of which have anything to do with our efforts. We can't assume that there is any built-in need for them to read our articles. If we want them to stay with us, we must catch and hold them any way we can.

DESIGN

One of our tools for doing this is attractive design. Graphic design is the conscious selection and placement of visual elements to form a harmonious whole. We have chosen a type face that is attractive and easy to read. We try to provide enough illustrations to break up the monotony of solid blocks of type. We use color, both in photographs and in diagrams, to enliven the image and spark interest.

Design is the first thing you see when you open a book or magazine, but it should be the last thing you notice. Good design is subtle, not blatant. A good design is like the background music in a movie, which succeeds by unobtrusively conveying the movie's mood. You'd miss it if it weren't there, but you don't want to be conscious of it.

Good design, like good writing, is based on respect for the rules. Just as in writing, however, respect for the rules is not enough. It is all too easy to spout nonsense in grammatically correct sentences, or to produce dull, uninspired designs that follow all the rules. The rules provide the starting place, and the real creative work takes off from there.

Just as nonwriters can learn to recognize bad writing, however, nondesigners can learn to spot inadequate designs. They may not be able to put their finger on the exact flaw, much less suggest how to fix it, but they can at least register their disapproval. All they need do is to cultivate a seeing eye.

The product of design is a format, which is to layout as the outline is to writing. It is the blueprint from which a series of related stuctures can be built. There is no perfect format for all purposes, nor even one perfect format for each purpose. Design is, first and foremost, an exercise in taste.

It is possible, however, to isolate certain elements of design just as it is possible to define parts of speech. In general, everything that goes onto the printed page, including blank space, is an element of design. This includes the type face, of course, but it extends to the column width, the margins, the placement and size of headlines and headings, illustrations, color, and the organization of all these elements into a harmonious whole.

A full discussion of all aspects of our format would fill a whole article, all by itself, and leave no time for the other facets of our communication strategy. In any case, a brief side-by-side comparison between our periodical and others will quickly reveal most of our tricks, if not the reasons behind our choices. There is time, however, to mention two of the less obvious design features.

We use a ragged right format, for example, that eliminates the hordes of end-of-line hyphens we would need with constant-spaced text if we insisted on right-justified columns. But since our text is loaded with scientific terms, which tend to be long words, a fully ragged right column as narrow as ours would look jarring. We therefore use a modified ragged right, slipping in a few end-of-line hyphens here and there where they will leave the column less formal than justified type, but not as undisciplined as a wholly ragged format would be.

We also use a ragged bottom, giving the layout artist the freedom to adjust column lengths to avoid layout problems. The bottoms are not randomly ragged, however. Within the ragged bottom format we chose to place the longest columns inside, near the gutter, and the shortest columns outside, at the page edge. Thus each spread, even if it lacks illustrations, has something visually different (but not outrageous) to please the eye.

ORGANIZATION

Another tool is organization, the aspect of our format that shows most clearly our concern for the reading patterns of our potential audience. Since some of our readers barely have time to flip through the pages, on the contents page we give them capsule two-line descriptions of each article. This at least gives them the main topics covered and some notion of what we have accomplished. It also tells them where to find those articles that relate directly to their interests.

For those with a little more time, we provide an abstract (in boldface) and full figure captions. A reader who absorbs just the abstract, the illustrations, and the figure captions should get a reasonable approximation of the article's main points.

Full figure captions have their disadvantages from the editorial standpoint. Every last-minute change in the text must be double checked in the figure captions as well, to prevent conflicting statements. We consider the end result worth the extra trouble, however, in improved communication.

In the text itself, we find that most articles need section headings to divide them into logical segments. Here again, tastes differ. Some readers want to know how the experiment was conducted, others are more interested in the results, still others want the theoretical background. Section headings help such readers zero in on their main interest.

Another useful device is the side-bar, or box. In some articles we need to go into more detail for some of our readers, although the rest already know the background. Isolating such a brief tutorial in a box puts it within the reach of all, but out of the way of those who don't need it.

WRITING

Since few of our authors are adept at writing for a general audience, often we must rewrite extensively to convert a specialist's draft into the kind of text we can use. In doing this it helps to keep in mind the composition and habits of audience we are trying to reach.

What do our readers want to know, and what do we want to tell them? In Energy and Technology Review we want to present as complete and coherent a picture as possible of the breadth and depth of research activities at LLNL. To do this we adopt certain conventions that involve some risks, but also help us achieve better communication.

In the first place, we abandon the passive voice so strongly favored by most journal authors and write almost exclusively in the first person plural. Energy and Technology Review is our report on what our laboratory has accomplished. This comes across most effectively if we say so.

This means that we must make certain who did what and give credit where it is due. An author who writes that such and such a result "was obtained" may easily forget to specify who did it. In joint efforts with a cooperating laboratory, therefore, we need to define clearly the contributions of each and claim for ourselves only those we deserve. Having made the effort to define these boundaries, however, we can then express them clearly and unequivocally so as not to leave our readers guessing.

In the second place, we need to restrict ourselves as much as possible to a generally understood vocabulary. Every specialty has its well accepted jargon and acronyms, a verbal shorthand that serves well among those who understand it. What many authors forget, however, is that these same catch words can baffle anyone outside the inner circle. We need to find common words that express the same meaning or, failing that, to define each special usage clearly for those who may be seeing it for the first time.

In expressing dimensions and numerical results we use the SI metric system exclusively. To some this may seem to contradict our policy of using common words. How many of our readers grew up with SI and use it in preference to the other systems of units in their daily lives? But in a larger sense, we see this as a compliment to our readers. The metric system is the one consistent system of units to which all technically trained readers can relate. Over 90% of the world already uses it, and it seems logical that we will some day follow. In this we prefer to be among the leaders rather than the foot-draggers.

In trying to make advanced research results intelligible to a wide range of nonspecialists, it helps to have a broad general background in science. No one could possibly be an expert in all fields, but it is important at least to master the vocabulary. One way is to make science a hobby, and to read regularly such publications as Science and Scientific American.

It also helps to tell the author frankly when you need help. Most of the authors I deal with enjoy explaining their specialty to an appreciative audience (they probably get few such chances). Just be sure to ask intelligent questions, and to record the answers accurately.

And how do I find intelligent questions to ask? First I read the draft several times. By the third reading, usually, things are beginning to make sense. Then I start to jot down passages that are still ambiguous. Another reading or two will sometimes turn up answers to a few of these questions. The rest go on my list to the author.

ATTITUDE

Behind all this, directing and informing everything we do, is an attitude of appreciation and enjoyment of science as a worthwhile human endeavor. Most scientists work at science because they like it, and they can't understand anyone who doesn't. To communicate with scientists at all, especially when the subject is outside their own specialty, we must share at least some of the excitement they feel over new discoveries, new techniques, new ways of understanding.

An appreciation of science is not something that can be turned on like a faucet when needed. If you don't feel it you can't fake it, and maybe you should find some other branch of technical writing. But if you do feel it, then maybe interpretive science writing is for you.

Technical Writing For The Market Place – Why Not?

Daniel W. Focht,
Supelco, Inc., Bellefonte, PA

Daniel W. Focht
Supelco, Inc.
Writer/Editor
Supelco Park
Bellefonte, PA 16823
(814) 359-2732

This paper was prepared while Mr. Focht was employed by Supelco, Inc. Mr. Focht is now with HRB-Singer, Inc., P.O. Box 60, State College, Pa. (814) 238-4311.

Many technical writers yearn to exercise their imagination and creativity by writing promotion material. What's involved? – – Can you, as a technical writer, make the change? Many technical writers will tell you it's impossible. I believe, however, that a technical writer is the logical choice to promote a technical product.

Although there are many attributes of technical writing which are invaluable in promotional writing, there are many other habits which must be changed. These changes are required for one basic reason: your new audience does not need to read, or necessarily want to read, your copy. The audience you wrote to as a technical writer was usually looking for information and you were its source. They had to read your copy to obtain information. Your new audience has other priorities, and now you are competing with those priorities.

How many technical writers are involved in some way with marketing? Judging from the articles published by STC, I have to conclude that there are not very many. The vast majority of material coming from STC seems to deal with writing manuals, reports, test procedures, test reports, or one of the multitude of data items required on government programs, and very little pertains to promotional writing. A prime example was last year's ITCC in Minneapolis where very little was said about promotional writing. I too spent many years writing according to government specifications, but then I wanted an opportunity to show that I had an imagination – – to show that I could do something creative. I wanted to try writing something else, such as promotional material, and I believe there may be many more writers in STC who feel the same way. It is for those writers that I'm addressing the subject of promotional writing.

First of all, is it feasible for a technical writer to try to make the transition to promotional writing? Many STC members seem to believe that technical writers should not attempt to write promotional material. I submit, however, that a technical background is required to write promotional material on a technical product, therefore a technical writer is the logical choice when you want to promote a technical product. In a technical industry, I believe that it is definitely possible for a technical writer to succeed as a promotional writer. It should, in fact, be viewed as a logical progression. I went from writing about electronic systems for structured government programs to writing promotional copy for a chemical firm with a highly technical product line. I do not pretend to have all of the answers, but I believe I can now offer a few thoughts that will help other writers make the same transition. Even if you never choose to write promotional copy full time, you may be required to occasionally write some type of promotional material. You will find that you have to alter your thinking in some ways, but you will also find that many of the qualities you developed as a technical writer will serve you well when writing to the market place.

The following are some of the qualities of technical writers which I find helpful in my present position.

1. *The ability to give attention to detail* – Probably the most important question any technical writer needs to answer is "Have I left anything unanswered?" This attention to minute detail is often the most difficult trait to acquire as a beginning writer. If you are now a technical writer, this attention to detail is one of your most important qualities. Do not forget it when you begin writing promotional copy.

"But his hair is parted on the wrong side"

2. *The ability to look at your writing through the eyes of your reader* – As a technical writer, you probably ask yourself what information or background your prospective reader has. In promotional writing you must ask the same questions: "What does the reader know?", and "Can I expect the reader to understand what I am saying?".
3. *The ability to determine the ultimate objective before you start writing.* You must know why you are writing each paragraph – each sentence. Establish what you want to accomplish by your writing, and then work toward that goal.

You can probably think of many more traits which you believe are necessary, but I believe these to be the most essential.

Now let us consider some aspects of promotional writing which are not necessarily a part of technical writing – new ideas which you will need to develop.

- *Create interest in your material.* You must remember that the prospective readers of promotional material do not need to read your material – or want to read it. In technical writing, you normally communicate to an audience who has a need to know. They have referred to your material because it is their source for information. Your primary objective is to provide that information in detail. In contrast, your first job as a promotional writer is to entice prospective readers to read your material. To do this, you must first capture their attention, and then convince them that what you have to say is worth their time. If you fail on this, everything else is a failure.

begun to read your material, you must maintain that interest.

- *Avoid all wordiness.* If you add a little extra verbage to your technical documents, the readers will probably wade through it, just as long as the important facts are included. In many instances, technical writers intentionally use extra words, or sentences to restate something just to be sure everything is covered.

  In promotional writing, however, you must avoid all extra words. They are added noise to the communication process. A large block of text will immediately turn away many readers because they surmise it will take time to read it, and they often are not willing to give you that time. This sometimes requires that you make hard decisions about what is *really* important, and what isn't. *Remember*, you must avoid anything which might cause the reader to refrain from reading your copy.

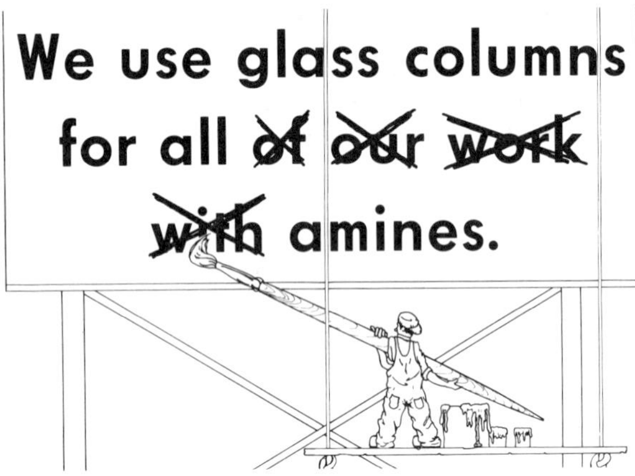

- *Always speak of benefits to the reader.* If you talk of features, as we often do in technical reports or manuals, you force the readers to translate those features into benefits, and such translations are not always obvious.
  Do not just say a hand calculator is 3/8" x 2 3/4" x 5". Instead say it fits easily into your pocket, or it fits the palm of your hand!
  Do not just say a transmitter/receiver has 10 watts of power, instead say it transmits up to 2 miles, or 3 miles, or some other distance.

  Do not make the readers figure out how a
  tell them how!
  This presentation of benefits is often forbidden in te
  material, but is mandatory in promotional writing.

- *Make the important points noticeable.* Put the significa important, words up front where the reader sees them immediately – without reading half of the copy. Most of your readers are busy, and if the first sentence or two does not strike them as worthwhile, they will move on to something else. If your message is to show your readers that you can help them do an existing job better, easier, quicker, or cheaper, make sure that point is clearly made in the first sentence. Then don't cloud your message by adding unnecessary words. If you can tell the whole story in a few sentences, do so!

- *Anticipate the questions your readers might have.* Answer probable questions as soon as possible. If your readers are preoccupied by some question while they are reading, you will only have part of their attention for the remainder of your message. For example, if your product operates on 120 *or* 240 volts, tell them immediately. Don't state that it operates on 120, then, as an added thought at the end, that it also is capable of operation on 240 volts.

- *Make your message flow logically.* Do not make the reader figure out how everything you say fits together. You understand what you are saying, and you may be able to follow your own logic, but will another reader be able to fit it together? Here it is important to see your material through the reader's eyes. Uncertainty may cause your reader to stop reading or even to abandon your material. This point is appropriate regardless of what you are writing, but it is especially critical in promotional writing.

- *Use familiar words which the typical reader easily understands.* If you make the reader study a word, or compel him or her to look it up to find the meaning, you may never get that reader to come back to your material. Use of familiar words is a practice which I feel should be exercised in all writing, but which is especially critical in promotional writing.

To summarize, you as a technical writer already have the following qualities which are desirable in promotional writing – especially when a technical product is being promoted.

- Your attention to detail.
- Your consideration for who the audience is so that you can see your material through their eyes.
- Your consideration of what the final objective is, before you start writing.

In addition, there are requirements in promotional writing which are quite different from technical writing. In promotional writing, you must:

- Create and maintain interest in your material. In technical writing, this is not a large concern, but in promotion, it is your largest concern.
- Keep your message as brief as possible. There are exceptions, but this is a good general rule.
- Promote the benefits – not the features.
- Make sure the important message stands out.
- Make your message as easy to read as possible. This means don't introduce questions without providing answers, make your message flow logically, and use familiar words.

Most of these are good practices regardless of what form of writing you do, but they are especially critical in promotional writing. *Always remember, your readers do not need to read your material.*

AVOIDING CLIENT INTERFACE PROBLEMS

Nancy Lee Guiland
B-K Dynamics, Inc.

Nancy Lee Guiland
B-K Dynamics, Inc.
Senior Technical Writer
15825 Shady Grove Road
Rockville, Maryland 20850
(301) 948-0650

Produces long-range strategic plans for government ADP organization

The paper discusses communication problems arising in the requirements analysis, development, and implementation of Automated Data Processing (ADP) systems. Three relationships (user-analyst, client-contractor, and manager-developer) converge to create a great potential for major communication gaps, thereby resulting in major problems in the software development process. Out of the client interviews upon which the paper is based, the author identifies two generic problems in the three relationships: lack of communication and lack of professionalism. After discussing the effects of these problems, the author proposes Ten Commandments for Corporate Communication. Technical writers in the area of software development are encouraged to pursue a broader role in corporate communication and to function as a bridge between user and developer.

The Information Systems (IS) Division of B-K Dynamics, Inc. (BKD) specializes in the requirements analysis, development, and implementation of Automated Data Processing (ADP) systems, primarily for government agencies. As a successful, eighteen-year old research and development company, BKD has always emphasized the professional development and excellence of its primary resource -- people. As part of its staff support, the company sponsors a monthly lecture series on information-exchange and technical training. Speakers include both experts from other fields as well as in-house professionals. In my career of technical writing, I had often been interested and concerned with the special hindrances to communication arising in the software development process. These obstructions seen to be taken for granted as organic, unavoidable occurrences by both parties involved in system design: the user-manager-client and the analyst-developer-contractor. Occasionally, this relationship involved only two people with three different functions; other times, as many as six persons filled these roles. In the particular area and agency in which BKD operates, we had enjoyed a reputation for minimizing endemic communication problems in the software life-cycle.

As the Senior Technical Writer for the IS Division, I became aware not only of these problems with client communication, but also of an undirected desire on the part of our technical professionals to further rectify the situation. As they knew, without strong and healthy client communications, a research and development consulting firm can expect a series of disheartening and financially threatening events: increasing schisms between user requirements and systems development, cost overruns, growing client dissatisfaction, and eventual loss of those clients. The material presented here resulted from a series of interviews with IS clients and was first presented at the BKD Lecture Series. During the interview process, the clients addressed themselves to other contracting companies as well as BKD, looking at the relationship generically, rather than specifically. Both the clients and I assumed a basic technical expertise on the part of everyone involved; the communication problems discussed here arose even though everyone was quite expert in their field. The interview itself was relatively unstructured, following the client's area of concern as much as possible.

At the center of the communication problem lay the well-documented rift between ADP system users on the one hand, and analysts responsible for developing systems to meet their requirements on the other. In a well-known text on software development which addresses this problem, the authors "emphasize the need for effective communications in the development of software and the disastrous effects of communication failures in the design process."(1) They later restate that "communication is an important aspect of the design process, second perhaps only to clear thinking. A design that cannot be communicated is worthless as a design."(2) The impact poor communications have on project success, and resulting corporate image, is dramatic. Dave Mollen, of the IBM Systems Science Institute, cites an IBM study which suggested a possible 400% increase in productivity due to improved communication; in the same article, 70% to 80% of the DP managers asked to identify their major problem overwhelmingly replied: "lack of proper user involvement."(3)

The second factor discouraging open communication grew out of the contractual relationship of client and contractor which was superimposed over that of user/analyst. The mass of confused and confusing federal regulations governing federal procurement compounded this already troubled relationship. In the early 1970's, the Congressionally-mandated Commission on Government Procurement identified the following problems:

- A void in Executive Branch Policy leadership;
- A fragmented, inconsistent, and outdated statutory base for contradicting regulations;
- A massive labyrinth of uncoordinated regulations within the Federal agencies;
- An unqualified, untrained work force in disarray;
- An absence of data, research, and management attention to assess how well Federal procurement was working.(4)

Ultimately, the legal aspects of the relationship introduced not only contractual difficulties but a third player, so to speak, in the Contracting Officer's Technical Representative (COTR). The COTR attempted to carry out his duties as government liason between the contractor and contracting officer with little, if any, guidance from the quagmire of federal regulations. Further, the COTR acted as a middle man, imposing a further impediment in the user/analyst line of communication.

Acting as the third factor in the deterioration of client/contractor communication was the antithetical focus of the client-manager and contractor-developer. Viewed from the black-box paradigm of

managerial concerns focus on the input-output, or what is produced and for how much. Further, scarce-resource allocations dominate managerial decision-making. Usually, the only slack variable a manager had was the process or end-result of the project. In-house personnel quality and resource constraints added a further frustrating factor to the government manager's job. The contractor, on the other hand, viewed his system design with excitement and resented the imposition on the design which resulted when a manager pulled in his belt (witness President Reagan's hiring freeze and call for a 5% reduction in contracting). The developer/contractor also felt a sense of isolation since he took no part in the decision on spending cuts which affected "his" system's development.

It was little surprise that trouble with communication resulted when three difficult relationships -- user/analyst, client/contractor, and manager/developer -- overlayed each other. After several interviews with our clients, it became obvious that two central and generic tendencies dominated others as the keystone to success or failure in all three relationships. Clients unanimously identified communication problems as the most damaging, followed by problems centering around a lack of professionalism. Lack of communication caused the manager:

- A loss of trust and confidence in the analyst,
- A sense of losing his money,
- A lost feeling of ownership in the project,
- A feeling that the analyst misunderstood his problem,
- An insecure sense of the project's progression,
- A feeling of isolation and not being allowed to input, and ultimately, a sense of superfluity.

Compared to the isolation caused by no communication, misuse of the proper chains of communication caused outright hostility. Actions such as going over the client's head, submitting deliverables to a higher or different agency, or not dealing with the COTR when contract boundaries were changed consistently resulted in severe damage to the relationship. Finally, and quite frankly, clients underscored the problems they had with abrasive personalities. In such cases, basic mutual lack of perception of each other's capabilities marked the inception of a contractual relationship; when the contractor injected an abrasive or caustic element through personality, the prospects for a good working relationship were doomed.

When the question of professionalism arose, the problems appeared much more clear and straightforward than the nebulous sense of disaffected isolation experienced with a lack of communication. Clients expressed anger at overselling or overextending of a company's ability, whether through outright dishonesty or misguided optimism. They also pointed out the contractor's occasional lack of sensitivity about committing dollars and manpower in ways not specified in the contract. A surprising comment pinpointed the lack of business savvy sometimes seen in the narrow-focussed technological types. Clients expected consultants to know the corporate capabilities of their company, not merely the advisability of installing one data base management system over another. Finally, clients expressed concern over developers' adherence to industry-standards rather than the requirements of the user's environment. Especially in situations locked-in to old equipment, state-of-the-art was often highly inappropriate as well as unresponsive to the user.

Faced with the mounting problems of the client/contractor relationship, it was with equal parts of grim determination and enforced optimism that I extrapolated a set of guidelines for our professionals. The Ten Commandments of Corporate Communication follow.

1) Maintain direct and close communication with the client; periodic "sitreps" are inadequate, and only being there with the manager reveals his problems.

2) Emphasize your experience with and knowledge of his problem; if you need information, formulate questions and prepare for them before the meeting.

3) Never fail to report as specified by the contract, regardless of how close, warm, etc. you feel the relationship to be.

4) Use working papers, decision points, and presentations to increase the investment of the government personnel in the project, particularly during lengthy projects.

5) Ask the manager for input, not solutions, and keep him appraised of the project's progress.

6) Never submit a deliverable to any but the tasking agency, no matter how convenient, interested, etc. such an agency might be.

7) Primary loyalties should always be with your end-user; have him mediate with his boss, even though you might know his boss better.

8) Be aware of the need for an equalizing time-span to know your user, and make every effort to increase communication.

9) Build bridging functions which aid your user in understanding, such as metaphor, or any tool which helps you both to better comprehend each other.

10) Maintain the highest level of professionalism in terms of knowing your own company and meeting your user's requirements.

Without exception, our clients expressed positive reaction to the interview and our interest in fostering better communication. One of the most gratifying moments of this research came when two of our clients asked, "What are your problems with us?"

That last question also points to where the technical writer fits into the garbled and difficult relationship described in this paper. Many of the suggestions for increasing communication outlined above might seem obvious at best. Without question, it is our very skill at communication which makes these observations unnecessary, and conversely, increases our value in the area of corporate communications. In my career as a technical writer/editor, much of my work involved acting as a go-between, or translator, for the user and developer. The area of corporate communication -- whether for industry, government, or consulting firms -- offers a new and renumerative avenue for technical writers. It is my contention that not only will we gain by applying skills we use in documentation to communication, but the technology and personnel which we support will likewise benefit.

---

(1) Jensen, Randall W. and Tonies, Charles C., Software Engineering (Prentice-Hall, Inc.; Englewood Cliffs, N.J.) p.13.

(2) Jensen, p. 96.

(3) Mollen, Dave, "Narrowing the Gap," Datamation (November 1980).

(4) Hall, Robert, "Procurement: Government's Newest Non-System," Government Executive (November 1980), p. 34.

PROJECT MANAGEMENT—AN EXPANDED ROLE FOR THE TECHNICAL COMMUNICATOR

JoAnn T. Hackos
Colorado School of Mines

Joann T. Hackos
Colorado School of Mines
Director of Writing Programs
Golden, Colorado 80401
(303)279-0300

Technical communicators need to take an expanded role in project management in the engineering firm. The communicator is often best informed about the needs of the multiple complex audiences that are the target of much technical reporting. Too often in the past, the communicator's role has come at project end, well after the report drafts have been through several technical reviews. The communicator has little or no time to make the major changes needed to make the report effective. However, if he or she is involved in a project from the start, he or she can play a integral part in keeping the project team focused on their research's end product–the report. And, by providing a clear view of reporting requirements, the communicator can help keep a project on track.

In a speech to District 5 of the International Association of Business Communicator's Keystone, Colorado conference in September, 1980, Chuck Palmer, of Sun Gas, Dallas, Texas, stated that "acceptance of communication in business is growing, but if communicators are going to meet management's needs in the future, we're going to have to change."

Palmer said that one reason communicators have failed to do as good a job as they could be doing was that the communicators perceived their jobs too narrowly. Palmer was talking about public affairs communication, but the same holds true for technical communication. Because both communicators and management have had a narrow conception of what communication is about, less than effective communication persists. That persistance is especially the case in the minerals industry where management and the engineering staff itself have traditionally viewed communication skills as secondary and communicators as poorly-paid grammarians. For a communicator to secure a significant role within a company, a role that can enhance both the company's image and its productivity, management must be made to appreciate how valuable a professional communicator can be.

What can the technical communicator's role be? In engineering and in the minerals industry in particular, the communicator can contribute to several activities essential to a company's mission—technical report writing and proposal writing. Many engineering companies function as consultants. In addition, many exploration, research, computer, and similar divisions of larger nonconsulting companies function as in-house consultants to management. The primary product of such consultants is the report. Professional engineers and scientists in these divisions do not produce gasoline or heating oil; they do not mine gold, copper, coal, or moly; they produce reports, stacks of paperwork distilling mountains of data from the field and from the laboratory. Yet, most of the people responsible for producing this paperwork have had little training, formal or informal, in communication.

Such project engineers and managers can profit immensely from working closely with a trained professional communicator who can help them appreciate the need for good reporting, understand what the components of good reporting are, and help them write and edit their reports.

Because their education has been primarily technical, engineers, geologists, and other scientists have often never taken a single course in technical writing. The report writing model provided by many of their college instructors was most likely the academic journal article or the doctoral dissertation, a model often 30 or 40 years out of date.

If people trained in accordance to this model manage to produce readable and informative reports, they do so through native talent, a good mentor in a first job, and a great deal of hard work and personal dedication. Unfortunately, those individuals are few and far between. More often, young engineers first entering an industrial setting have little knowledge of the business side of technical communication and little understanding of the role communication plays in the decision-making process. The technical communicator can provide a formal structure, through project management, for this learning to take place.

To provide the communication structure necessary for good reporting to take place, the communicator needs to develop several different, although interconnected, roles. The primary role, as I see it, is to remind the engineer that an audience exists for his work and must be recognized and analyzed from the start. Many of the writers with whom we deal lack an audience awareness; they fail to recognize the significant selling component in all the writing they do for their companies.

I once had a young engineering graduate tell me when I asked him to identify the audiences for the reports he was writing, that he never knew because his supervisor and the secretary put the distribution list on his reports after he had completed them. Neither had he been told nor had he thought to inquire about who would be reading his reports. He was writing in a complete vacuum; no wonder that his management complained that they could not understand what he was writing.

In a similar case, a bright and highly competent computer systems analyst was asked to write a memo to a group of his company's clients about some significant changes in the way they would have to handle their major software package. He looked at the distribution list for the proposed memo and noted, accurately, that only 10% of the people on the list would be able to understand on his level just what was being changed. He decided to write the memo to that 10%! When he asked me to read the memo before it went out, I was happy that I had the opportunity to review it before it was sent. The very first paragraph read as follows:

> Within the next few weeks, we will update the system software on the 66/20 from release 3/I of GCOS to release 4S (also known as 4JS). Most commands that work under the current release will work under the new release; however, there are some differences which will be noticeable to you.

You should realize that 90% of the audience for this memo has almost no technical data processing knowledge. They are primarily administrators, often accountants. The memo went on for three pages in a similarly cryptic style. If it were released in this state, it would only reinforce an already strong belief that this company was talking over everyone's head.

The second most significant role is to help focus the information in the report so that each piece of data clearly supports the conclusions. Many technical people have great difficulty organizing their thoughts in writing. Their tendency is to overwhelm the reader with detail, without explaining exactly how that detail is significant. They are unable to focus their information and discussion toward a conclusion. If asked to explain what is most important about a subject, they have difficulty making themselves clear.

A third role for the communicator is to help the project engineers and scientists appreciate the importance of integrating the reporting process with the research process. An executive of a large engineering company explained the problem this way:

> There is a tendency to spend five of the available six weeks of a project investigating all facets of a subject and then try to write a report, and have it typed and edited, in the remaining week. Some engineers—very few in my experience—can produce good work in this manner. The majority of us will tend to conduct the investigation in an inefficient manner, going off on tangents or not fitting effort to work scope, and then write a hurried report that may be less than satisfactory.

We call reports produced under these conditions "file cabinet reports" because that is where they usually end up—unread.

I find that two interrelated circumstances of concern to the technical communicator are at least partially responsible for the reporting process described above. First, most engineers are more interested in investigation, design, and research than in writing. In fact, a few student engineers have told me that one of the reasons they chose science as a career was that they wanted to avoid writing. Second, in their engineering education, students are rarely taught the close relationship between engineering design and report writing. Consequently, they do not understand that they must design and build a report like any other product they put out.

If a professional communicator is on hand, is skilled in report design, and is given the proper job responsibility, he or she can emphasize the importance of audience analysis to effective communication and help the writer perform that analysis, aid the writer in unifying and focusing the report as a whole, and provide direction to achieve good report design.

Unfortunately, many technical communicators in engineering-related firms find that their advice and direction come too late. In many cases they have little opportunity to review material until the last minute, after the poor badgered author has already been through a large number of content reviews. The author probably has revised her copy five or six times, trying to accommodate the shifting stylistic demands of colleagues and supervisors. By the time the technical communicator gets his turn, the author has no patience remaining for further changes beyond a few commas and an occasional misspelled word.

Under such circumstances, the communicator who already is being paid only half the salary the young engineer can command, has little time or opportunity to make the substantive changes to the text that it needs. The story remains "Too late and too little."

The solution is to involve the professional communicator directly in project management from the start.

PROJECT MANAGEMENT

The quotation above from the engineering company executive illustrates one problem with project management—keeping the project on track. The director of a scientific and engineering research laboratory once complained that many of his staff professionals would go off on their own special interests rather than concentrating on what their clients wanted from a project. One source of this "tracking" problem comes from a general confusion about the direction of the research itself and from an initial lack of clarification about the requirements for the final reporting. When a project team does not know precisely who will use the information developed and how they will use it to reach some decision, the team will have great difficulty keeping their investigations from going off on interesting but unproductive and expensive tangents.

Another source of the tracking problem is the relative inexperience of some project managers. They may have had little training in project management and may never have experienced a well-managed project. Whatever knowledge and skill they may have accumulated derives from trial and error.

Other managers do, by virtue of their experience in the company, have a sound understanding of the company's mission and its relationship with its clients and with its management. In fact, such project managers know at first hand most of the people who receive and review the reports produced under their supervision. Unfortunately, they often do not communicate this knowledge to their staff members.

One new employee of an engineering research lab explained that he included everything he had discovered in his reports because no one had had taken the time to inform him who was going to read his reports or what they were going to do with the information. It had never occurred to him that his project manager might have this information. And, it had never occurred to the project manager to tell him.

How does the professional technical communicator fit into this quagmire? The communicator may be one of a few individuals in a technical company who is closely interested in reporting requirements and audience needs. By virtue of this interest and expertise, the communicator can fulfill a very important function in project management—looking at the project from the outside. By this I mean that the communicator should be able to keep reporting requirements as they occur at frequent stages in a project life cycle, strongly visible to the researchers. She can continue to urge that the product of the investigation, the report, be planned from the outset.

But to play an effective role in project management, the communicator must have the status of a in-house consultant, a status equivalent to that of the engineers involved in project management. Without the "clout" that accompanies such status, the communicator can achieve little of importance.

Unfortunate as it is for the profession as a whole, many technical communicators are hired by engineering and scientific firms, particularly smaller ones, at roughly the same level and status as a secretaries. The job titles may be different but the salaries are the same. And, little as we may like it, salaries convey prestige in most firms. At this level, even the most qualified writer can do little because he is seen as a punctuator, someone who knows little of the real business of the firm, someone concerned only with trivia rather than important technical and scientific subjects.

As long as this view dominates, and as long as we put up with it, communicators will not have much impact on the engineering industry. Yet, if we maintain our professionalism and work toward a more significant role, we can change this situation.

MAKING REPORTING AN INTEGRAL PART OF A RESEARCH PROJECT

As part of the research project staff, the technical communicator can perform several valuable functions:

1. Detailed audience analysis, including, if possible, direct interviews with potential audience members.

2. Clarification, in writing, of the initial assignment. In appropriate situations, the communicator could prepare a document outlining the audience's requirements for a research project or engineering study.

3. Review of the client's or of management's response to the requirements document. The communicator could then, in consultation with the technical project manager, revise the requirements document.

4. Preparation of a detailed thematic outline. While most engineers outline a report only after the research or study has been completed, for most projects an outline could be prepared at project initiation.

Frequently, the engineering staff has a pretty fair idea of how a study is going to come out, even if they don't know the detail. Under these circumstances, a detailed thematic outline developed at project initiation can be written. The existence of such an outline can become a major tool for keeping a project on track. For example, the research manager at a metallurgy laboratory was very excited when we described this DISCUSSION OUTLINE method, our name for a thematic outline which addresses itself both to the audience's needs and to the focus required of the technical information. He said that he frequently had difficulty keeping his scientists and engineers from spending a lot of the lab's money on interesting tangents. In that context, he immediately saw the utility of the discussion outline developed at the beginning of the research project, as a project management tool. The outline, once established, could always be revised to reflect new directions for the project.

5. Establish sound feedback mechanisms. Part of the problem with project management is the absence of adequate written review of what is being accomplished. The verbal and informal progress reviews which ordinarily occur are frequently inexact and misunderstood. Only after the final report has been submitted for comment does a project engineer discover that he has not done what management wanted.

I have encountered many cases in which an engineer will have worked six months or more on a project, submitted a final report (usually three days before it is due at headquarters 2000 miles away), and then be told that the report does not meet requirements. The morale problems that occur under these circumstances are enormous.

Feedback can include a variety of methods. Some of the more effective are clarification memos at project initiation, requirements or feasibility studies, discussion outlines prepared at an initial brainstorming sessions and revised if the scope and direction of the project changes, and conclusion highlights when the project is nearing completion. These highlights, which stress preliminary conclusions developed by the project team, can be used by management to determine before a final report is written if the project has indeed gone in the direction intended. Of course, if sound project management techniques have been used throughout, few problems should be encountered at this stage. However, as the conclusions of a project or study emerge, individuals in management may perceive applications or problems which are not perceived by the people involved in the project. At the highlighting or preliminary conclusion stage, the technical communicator can help the project team discover if any unforeseen approaches should be addressed in the final report.

6. Develop an organization and style for the final report. If the communicator as project manager has done a good job of managing the reporting part of a project, she should have less work to do when the final report has to be drafted and edited. In fact, the number of drafts required and the corrections needed should be minimal. To keep the number of drafts under control, the communicator should consider limiting the reviewers comments to substantive points. Writers become frustrated and editing becomes interminable when reviewers begin to take apart an individual style and substitute their own.

HOW DOES A COMMUNICATOR BECOME A PROJECT MANAGER?

More writing needs to be done in most companies than the professional engineering or research staff can possibly handle. If an experienced communicator can offer to help with report writing, while at the same time explaining that those reports will be easier to write if she takes part in project planning and review sessions, she is already on her way to a responsible position. All that is needed is initiative and a degree of job flexibility.

Unfortunately, acquiring a project manager role is difficult for a communicator who is hired in at a salary equivalent to a beginning secretary. The job market for inexperienced people is tight enough that many take what they can get. But for an experienced professional communicator, a position which will clearly have influence on a company's writing policies, can be developed from the outset.

By asking tactful but pointed questions about audience, direction, conclusions, etc., the communicator can keep project team members aware of reporting requirements. By volunteering to create some of the feedback mechanisms, by collecting significant audience profiles, by helping with outlines and in brainstorming sessions, the communicator can quietly keep a project on track. And once he has established his expertise, he may be asked back.

A more direct route is to obtain management's help. If the communicator's ideas for project design are approved by one or more company executives, they can become policy. But such approval occurs only if the communicator has demonstrated his ability to learn quickly about the company's technical direction, the process management uses to make decisions, and the way management perceives the company's interaction with its clients, internal or external.

An individual who is neither a quick study nor has an apptitude for technical material will not be trusted to handle technical communication problems. Only by close attention to detail and a willingness to engage in background study will the communicator succeed in convincing management that his abilities should be fully utilized.

To participate in project management, the technical communicator must acquire knowledge and skill in the technical fields of her company, through self-study, continuing education courses, or the like. She must establish a role that goes beyond editing, although editing will always remain important. Since she has considerable training and expertise in organizing the reporting part of a design or research project, she can provide a service to management in the company's interest. Sound project management and effective reporting ultimately save money. If the communicator can save her company money, then she has justified her salary in a very visible manner.

EDITORIAL DIALOGUE: AN ALTERNATIVE WRITER-EDITOR RELATIONSHIP

Mary S. Hageman-Louise M. Vest-Patrick M. Kelley
New Mexico State University

Mary S. Hageman

Information Services
New Mexico State University
Box 3-K
Las Cruces, NM 88003

Publications Assistant Editor; Consultant, Merck-Sigurdson; Graduate Student in Technical and Professional Writing

Louise M. Vest

Department of English
New Mexico State University
Box 3-E
Las Cruces, NM 88003

Instructor in Technical Writing; Consultant, Merck-Sigurdson; Graduate Student in Technical and Professional Writing

Writers and editors often regard each other as adversaries. We propose Editorial Dialogue (ED) as an alternative writer-editor relationship. The theory for ED is based on Martin Buber's philosophy of dialogue, as applied by Richard L. Johannesen to communication and by Nelson A. Briggs to editing. The practice of ED is based on Johannesen's six major components of dialogue, as applied by us to the writer-editor relationship.

Patrick M. Kelley

Department of English
New Mexico State University
Box 3-E
Las Cruces, NM 88003
(505) 646-3931

Coordinator of Technical Writing; Senior Member, STC; Charter Member, Sigma Tau Chi

INTRODUCTION

As writers and editors, we all know too well that scientists, engineers, and other technical people who must report on their work in writing and then submit their writing for editing often regard editors as adversaries. At best, these writers regard editors as an impersonal _them_ with whom they must contend in an _us_-versus-_them_ relationship; at worst, these writers regard editors as an alien force with which they must contend unequally--mere humans that they are--in an _us_-versus-_It_ relationship. So regarded, editors too--mere humans that they are--often regard writers as adversaries.

These alienating, adversarial writer-editor relationships, characterized by _us_-versus-_them_ and _us_-versus-_It_, are monologic, in a term from the modern philosopher Martin Buber. In less philosophical and more practical terms, we believe that monologic writer-editor relationships are inefficient in the editing process and ineffective in producing written products.

As the basis for an alternative writer-editor relationship--a caring, supportive relationship characterized by Buber's _I-Thou_ that is dialogic--we propose Editorial Dialogue (ED). We believe that a dialogic writer-editor relationship is more efficient and effective than monologic relationships, not to mention more appropriate to the very human writers and editors engaged in the enterprise that we all know as human communication.

THEORY

The theory for our ED is based on Martin Buber's philosophy of dialogue. Buber (1) defined _dialogue_ as "the essential or inner action of man as he turns toward the other and is concerned with the wholeness of the person." The emphasis in his discussion of dialogue is on our recognition of each other's humanity and the responsibility that each of us has in any encounter to affirm the humanity of the other. Buber differentiated dialogue from monologue with chilling contrast: a dialogic encounter is "I-Thou"; a monologic encounter is "I-it."

Richardson L. Johannesen (2) has brought the concept of communication as dialogue to communication theory. Beyond that, by describing the six major components of dialogue--genuineness, accurate empathic understanding, unconditional positive regard, presentness, spirit of mutual equality, and supportive psychological climate--in terms that make it possible to practice dialogue in everyday life, Johannesen has brought dialogue from the airy realms of philosophy and theory toward the earthly world of practical application.

And Nelson A. Briggs (3), before us, has brought dialogue to that extraordinarily practical part of the world where technical writers and editors work. Briggs has gone far in applying Buber's philosophy of dialogue to a set of practical, common-sense principles of editing. He emphasizes that to edit by dialogue, instead of by monologic fiat, editors must remember that with each draft to be edited comes a human being--the writer.

Although Briggs acknowledges that there are times when "tough-minded" editing is necessary--as Buber acknowledged that there are times in life when monologue is unavoidable--he encourages editors to remain flexible enough for dialogue at other times. We go further by maintaining that dialogue can be practiced even during those times of "tough-minded" editing, _if_ the editor has established a caring, supportive human relationship with the writer.

Dialogue is not the denial of the self, but the affirmation of the other. It is not a false unanimity, but a true mutuality achieved through honest agreement and disagreement with another human being. Therefore, we do not believe that the writer and the editor always need to agree during the editing process. In fact, we believe that disagreement is sometimes necessary to produce an effective written product efficiently. We also believe that the editor who is working with the writer in an established dialogic relationship can make decisions contrary to the writer's desires without damaging their relationship. A writer-editor relationship that is a caring, supportive human relationship is also a functional, durable relationship.

This is true, however, only if the writer-editor relationship is genuinely dialogic. The editor who uses monologue under the guise of dialogue is manipulating the writer. And artificial dialogue--which is really monologue--is lethal to human communication and liable to Buber's definitive term for monologue, "Lucifer."

PRACTICE

The practice of our ED is based on Johannesen's six major components of dialogue. We have developed general guidelines for six practices, which are based on these components, but which apply specifically to the writer-editor relationship. Johannesen might be skeptical of our guidelines, believing that they destroy spontaneity. We believe, however, that our guidelines foster dialogue. Furthermore, we do not believe that they are manipulative.

Following are Johannesen's six major components of dialogue and our guidelines for fostering a dialogic writer-editor relationship:

1. _Genuineness_. The editor should "impart himself as he really is," instead of playing the role of editor in a conventional writer-editor relationship; he should expect the writer to impart herself as she really is, instead of playing the role of edited writer.

2. _Accurate Empathic Understanding_. The editor should gain experience as the writer in the writer-editor relationship so that he can understand the relationship "from the side of the other as well as from [his] own side"; he should help the writer become her own editor so that she can understand the relationship from his side.

3. _Unconditional Positive Regard_. The editor should affirm the writer as "a partner in dialogue" and confirm her "in [her] right to [her] individuality" even when--or especially when--"[she] opposes him" on an editorial decision; he should hope that the writer affirms and confirms him when he opposes her.

4. _Presentness_. The editor should "demonstrate willingness to become fully involved" by taking time, making space, avoiding distractions, listening actively, responding readily, and risking at least professional attachment within the writer-editor relationship; he should insist on nothing less from the writer.

5. Spirit of Mutual Equality. The editor should respect the writer as his equal with each of them equal "as persons" within the writer-editor relationship and with her knowledge of content the equal of his ability with language; he should earn the writer's respect for him as her equal.

6. Supportive Psychological Climate. The editor should "encourage the [writer] to communicate" during the editing process, listening to her "without anticipating, interfering, competing, refuting, or warping meanings into preconceived interpretations" as he helps her communicate in conference whatever she did not communicate in writing; he should respond to the writer's support for him during the editing process by producing a written product that is hers--not hers as it was in the draft, which was not hers as she meant it to be, but hers as she means it to be--the ultimate support that the editor can provide for the writer.

These general guidelines form the basis for a series of workshops that we offer in which we teach participants specific practices for implementing dialogue in the writer-editor relationship.

During our practice of ED, we have learned a bit about dialogue. Immediately, we learned of the difference between understanding dialogue intellectually and integrating that understanding so that one is dialogic. Gradually, we have learned that a dialogic relationship is an ideal to aspire toward--at least for us with our very human limitations. And although we are still learning about dialogue, what we have learned of it and from it already makes us eager to propose it as the basis for an alternative writer-editor relationship.

CONCLUSION

We believe that a dialogic writer-editor relationship is more efficient and effective than monologic relationships. Unlike alienating, adversarial monologic relationships, a dialogic relationship is truly a mutuality achieved through honest agreement and disagreement between two human beings with each affirming the other. We believe in the potential of a writer and an editor, each caring for the other, each supporting the other, and both working mutually--even during times of "tough-minded" editing--to produce an effective written product efficiently.

The dialogic writer-editor relationship that we propose is a professional relationship, we emphasize finally. It is based not on friendship or love, necessarily, but on human communication. With human communication as its basis, we believe strongly in such a relationship.

REFERENCES

1. Buber, Martin. Between Man and Man. New York: MacMillan, 1965.

2. Johannesen, Richard L. "The Emerging Concept of Communication as Dialogue." The Quarterly Journal of Speech, Vol. 57, December 1971.

3. Briggs, Nelson A. "Editing by Dialogue." Technical Communication, Vol. 16, Second Quarter, 1969.

JOURNALISM IN TECHNICAL ARTICLES
Or: It was all right for Einstein ...

Størk Halstensen
The Ship Research Institute of Norway

Størk Halstensen
Ship Res. Inst. of Norway
Editor
Rostengrenda 159
7075 Tiller - Norway

The paper claims that there are no technical writers, no technical language, only good or bad writers, good or bad language. We must take a completely new attitude towards writing and the education of writers. This implies that we give the writing profession status as such, and forget the word technical. It also discusses a few aspects of clear writing and how to reach the readers. The journalistic attitude towards writing must be adopted by the technical field, and so must the journalistic technique.

$E = m C^2$

I once landed in an heated argument with a scientist about language and form of expression. As all scientists he thought that his academic background also qualified him to voice an opinion on this matter. Our views were sharply opposed to each other, and almost in desperation I at last asked him to name a few persons, who in his opinion had given a clear expression for a complicated problem. He mentioned Moses and the ten commandments, Albert Schweitzer and some others before he threw down the trump - EINSTEIN.

$E = m C^2$ - The formula on which the theory of relativity is based.

I could have found no one better myself. Short, simple and clear. Another story is that he spent much of his remaining life explaining to others what it really ment.

Years of experience have taught me that the ability of man to comprehend the information he is recieving, is very low. Lower than most sceptics are willing to admit, and lower than society is demanding of us. In this respect it is a wonder that humanity has progressed at all.

There are several studies that support this. Some years ago in Sweden, a report was published on "Reading and writing abilities among grown ups" (Hans U. Grundin 1977). We are used to regarding the Swedes as a fairly well educated people, and with a school system that takes care of everybody. Grundin concludes, however, that among people in the age group of 25 years and with good educational background, 7 % are not functionall literate. If we move on to the 35 year olds, who have a somewhat lower degree of education, 13 % are not functionally literate.

More than 40 % of the participants in the survey were dissatisfied with their own ability to read and write, and one in four found their skills inadequate for their occupational reading and writing. I have no illusions whatsoever that the situation is better in any other country, my own included.

DISMAL RESULTS

The figures are discouraging, but let us look at another example. A Norwegian research institution made in the last half of 1980 a sorvey of how the users were recieving information from R & D activities (Kari Moxnes Osmundsen 1981). It showed quite clearly that their main source of information were mass media like TV, newspapers and magazines. Reports and scientific magazines interested only those who were more than average engaged in their work.

At one stage they talked to a group of potential users about a specific work, got them interested, and mailed them the final report. A check some time after disclosed a dismal situation. About 70 % of them had either thrown away the report directly, or put it aside without reading it. Their reason was that the expression and the wording in the report was so difficult, that they could not understand it. Not to mention put it into use.

A READABLE MESSAGE

This in short, is the background on which we must base the communication of technical information. Not only do we have to make the message readable, but we also have to make it so interesting that it will be read. In our society nobody can afford to bear over with sloppy or inadequate writing. The consequences can be both economical and legal liability. One solution is to adopt a journalistic, in the best sense of the word, attitude. Therefor let us take a look at what journalism is.

MARKING TARGETS

A somewhat sarcastic definition says that a journalist is an expert at separating the wheat from the tares - and printing the tares. This may be hard to swallow for an old journalist, but it states an important issue. A valuation has taken place. The important facts have been separated from the unimportant. When this is done, and not before, can we consentrate on why we are writing - if we have a message, something new and important, that we want someone else to know about.

Before we start, we must also consider what we are trying to communicate and to whom. In marketing the expression "Target Groups" is vital. It is just as important when we are writing an article for a newspaper or a scientific magazine. One of the best examples I have found of structuring the message to a specific group of people, was in the American outdoor magazine "Field and Stream". It was a review of a book most of us know, at least by reputation:

" Although written many years ago, "Lady Catterley's Lover" has just been re-issued by Grove Press, and this fictional account of the day-to day life of an English gamekeeper is still of considerable interest to outdoor-minded readers, as it contains many passages on pheasant raising the apprehending of poachers, ways to control vermin and other chores and duties of the professional gamekeeper.

Unfortunately,one is obliged to wade through many pages of extraneous material in order to discover and savour these sidelights on the management of a Midland shooting estate, and in this reviewers opinion the book cannot take the place of J.R. Miller's "Practical Gamekeeping".

Let it be that the piece violates many of the rules of styling. The important thing is that the reviewer had found what he thaught would be important to his readers and structured the piece accordingly. And, his message is clear Another discussion can be if this was the true message og the novel, but we must leave that.

This is also a long step towards good journalism. A journalist must allways keep in mind that his message is the most important thing. It is his only obligation.

SIMPLE and CLEAR

Remembering people's ability to take and comprehend information, the message must be carried across as simple and clear as possible. Journalists are the only professionals I know at this. They have made it a living and a style og life.

Journalism is also a question of expression and style of writing. It is short, clear and convincing. The fact that a problem is treated in a light and readable style, is not the same as treating it lightheartedly. I risk the statemnt, that behind the expression "Technical Writing" is hidden oceans of idiomatic insecurity, inability and shere negligense. The worst failiure any writer can do, is to think that some words and expressions are more scientific than others, and finer than others. Sooner or later we are bound to use the words wrong, and then we are really on thin ice. Let it be that the meaning may be lost, but the whole affair can be turned completely upside down.

A LABOUR OF LOGIC

A clear and easy expression does not come by itself. Writing is a continous process of logic, and it must be kept to the bitter end. This is not easy. Anyone who has tried to write simple and straight forward will agree that it is far more difficult than writing in a roundabout style with a series of reservations.

Writing on technical matters is demanding even more of clarity and presition, than writing on everyday matters. it may be at litle difficult to see the immediate similarity between a users manual and an article on a technical question. But, they have something important in common. A manual must be short and clear and give exactly the right information at the right time, one point at a time. Exactly the same aplies to an article from a journalistic point of view. The only difference is, that in an article we have the possibility of going deeper and give the points a more interesting wrapping.

BROKEN RULES

There are certain basic rules for good writing. For some reason or other it seems necessary to learn them again and again in every generation. To take som at random:

- Sentences must be short and presise
- Expression must be active and concrete, not passive and abstract, even when treating abstracts.
- The message, the important points, must be put forward.
- The language must siut the reader, not the writer.

Most people will readily agree to this, and maybe say that this is selfevident. Still, journalists seem to be the only ones who dear put them into use. To them they are vital if they want to reach their readers at all. What surprises me, is that people writing about technical problems seem reluctant to do the same. They, more than others have a difficult message to present.

Let us have a look at these random points.

LOOK TO THE VIKINGS

During the 1930'ies people started investigating in readability. Every survey up to now has concluded that short, presise sentenses are more easy to understand than long, intricate ones. Usually a sentense should not contain more than about 20 words. After that we very soon loose sight of what the writer is trying to tell us. This is no set formula, but when the sentenses rave on and on, jumping from one point to another, even the most consentrated is falling of the cart.

We then also loose sight of the message.

I know of no other prose that illustrates this better than the old norse sagas. In short, consise sentenses they chisel out drama and tender story. We never loose sight of what they are conveying to us. This is not only good prose - it is also exellent journalism!

There is also no use in hiding behind a passive and causious style. It does not free the writer from responsibility for what is written. Active language may often even stall chritsism. It also creates an impression that we know what we are writing about, and in addition that we know how to write it. The last is important when we think about convincing the readers that we are right.

UP and FORWARD

A precise and active language also makes it easier to point out the meaning of the article. It is a good journalistic trick to put the message up front where it will be seen - and remembered. Unfortunately the case is often the opposite when treating technical matters. In stead of a clear and understandable message, we find, as in "Lady Chatterley's Lover", pages of uninteresting history and discussion of methods. The results, or the message, the reason for writing the article, has tragically drowned in an ocean of indifference.

SUBORDINATION

What differs an article with a technical theme from a usual discription or protrayal, is how the writer must subordinate himself to the message. His or her personality must never dominate! Towards the message we must be absolutely true and objective to the last sentence. But there the difference ends. In presenting the contents we may use every trick in the book, tricks that novelists and journalists use dayly - variation in length of sentense, painting with large strokes in the general and fine intimation in detail, porsion the information in exact doses. The language and our personal style are the means we use, and they must be crystal clear without leaving a shadow of doubt. An old journalist once gave me a sentense to ponder: "A foggy expression is normally the result of foggy thinking". This is also true in conveying technical information.

ACCEPT the WRITER

To write is a profession. The sooner we accept that, the sooner we may do something to the seemingly sorry situation we are in. Writing is not somet-ing that should be left to a technician as a hobby. Talent and a wish to write are also not enough. Much can and must be learned, and knowledge in technical matters are more often than not a hinderance. Not because knowledge gets in the way, but because a person with a technical background has lost the necessary openness during his eduaction. Engineers and technicians, it is said, have often both feet firmly planted on the ground, or in what for them is reality. They are then subject to the law that says: "A person with both feet on the ground is standing still!"

We can do quite a lot to improve the informational value of technical articles. I have allready mentioned one point; start regarding writing as a profession similar to others. It must be learned, and knowledge in technical problems, and the ability to solve them, is not the same as writing ability. An immediate solution is to hire journalists to do it. After all they are allready professionals. Unfortunately the education of journalists does not take technical information into consideration. It has a low priority compared to new in general.

For a long term solution we must reconsider the whole educational system. No educational system, that I know of, is trying to develop the ability of free presentation. This is very important in treating both technical matters and other. Alvin Toffler is voicing the same thaughts in his book "The Third Wave". He maintains that we for generations have educated youngsters to click into specific roles in society, and frightened them from crossing borders.

This pertains not only to our western world. We find the same pattern in every industrialised country, and also in the countries that are taking the step into industrialised society. We educate our students to punctuality, obedience and a life of repetitive toil. This is more evident the higher the education, especially in the technical field. The more technical the subject, the less is tolerated of individual thinking. Not to mention expressing themselves in a free and individual style.

Here we also find another fenomenon, and maybe an excuse. Almost every textbook is written in a language that has litle to do with clear and simple expression. And that is the language they learn! The schools and universities demand that they use the same lumbering way of expressing themselves. When they eventually have to communicate with others, they loose allready before they are at the starting gate.

INDUSTRY SUPPORTS IT

Industry and society support this situation by hiering people more by diplomas than ability. Such diplomas are actually of litle value. They only tell that the person has a certain ammount of knowledge. This he has conveyed to another person with exactly the same knowledge. In such a case presentation and form is not important. They still understand each other. The problem arises when he or she wants to communicate with a third person. What is more, the dimploma says nothing about using the knowledge, only that it is repeated correctly.

The whole educational system subordinates to this demand. And as long as industry does not ask for ability of expression, nothing is going to change.

It will be a long and difficult process altering this. I still believe it will be worth while.

The technique of presenting something in writing can be learned, and very often must be learned. Some universities and sentres of higher education are getting aware of the problem and are taking the consequences. Unfortunately, as of yet they are a minority.

FORGET "TECHNICAL"

I think there is only one proven way to achieve a better structure in technical articles. This we find in the education of journalists. It is a combination of practical work and theory. In addition we must forget the word "TECHNICAL" when talking about writing. There is no difference whatsoever in writing an article on a technical problem and about a dairy farm.

At FORUM 80 in Lillehammer in Norway in 1980, the two american researchers Julie Stensrud Held and Margaret Goodman Burns presented a work on writing and education. Their conclusions harmonise with other, similar work. They maintain that traditional teaching in language does not improve writing ability. Neither does a repeated composition writing. Their conclusion was that the work shop format was the best and most fruitful way. There articles and other written material can be discussed and commented. This, they said, would benefit the student most.

I tend to disagree with the two ladies. Even Cicero was sceptical to formal education in expression. In Rome rethorics was a highly estimated profession. In Cicero's time, however, it had sunken to be a formal discussion about words. The eloquent speaker detested that.

During the Renaissance a man was judged by his ability to express himself and his ideas. People then had three geuidelines for education:

Away with all formalism
The use of a language acceptable to society
Repeated practical training

WRONG STARTING POINT

What is wrong, is that everybody is starting with a totally wrong belief. They are locked in the superstition that to treat technical matters in writing, you have to be technically educated. Writing is something you do on the side. Nothing could be more wrong. Starting from such a platform, the only thing a work shop will do, is to destroy whatever tendency towards personal style the student has.

The education of journalists has a different view. First of all it regards writing as a profession. Secondly it takes on students who allready are using the language as their tool. Most of the education, as I know it, is concentrating on techniques of using the language, and the technique of expression. Those are matters that can be taught. Form and style are more diffuse and have to be added from our own personality. Another important point is that they are also taught a host of subjects aimed at broadening their spiritual background.

For journalists practical training is vital. No one who has ever worked under a choleric city editor will forget his lesson. These Devils Advocates have no compromise in their demand for brevity, clarity and eloquence. They have broken the dreams of m any a romantic. But, they have also matured others into real artists.

WRITING EXPERIENCE

No one is submitted to any higher degree of journalistic education wothout a minimum of practical experience. This implies that they know what they are up against before they start. This is unfortunately not allways the case when educating writers for the technical profession. It is also one of the reasons why Dr. emeritus Earl Britton of the University of Michigan could say after 40 years of experience in educating "Technical Writers": I recommend to all of my students, that they go out and get at least some journalistic experience before starting in their profession".

At last there is one point, which I consider very important. There is nothing more useful in improving ones writing than reading widely, particularely good novels. They contain everything we are trying to achieve.

Not all of us are of the same greatness as Einstein, and not all of us can hope to reach his simple way of expressing a complicated problem. Still, he was a person with a certain humor and a mind for the unexpected and bizarre. I like to think that he would appreciate my own personal interpretation of his formula:

$$E = m\,C^2$$

E = Eloquence
m = message
C = clarity

The last item is so important that it has to be squared.

My conclusion will therefore be:
There are no Technical Writers
There is no technical language
There are only good or bad writers, good or bad language.

PUBLIC RELATIONS FOR ANY COMPANY

James F. Harroun
International Business Machines Corporation

James F. Harroun, IBM
Mgr. of Communications,
Systems Integration
10215 Fernwood Road
Bethesda, Maryland 20034

There appears, in the eyes of John Q. Public, that a big vise is moving in on the American consumer and from the viewer's perspective, the greedy American business-person is at the controls of these relent-lessly moving jaws. If he looks in one direction, John Q. sees, or is told, the businessperson is polluting his atmosphere, ripping his good green earth to shreds trying to find coal, is foisting unproven drugs and chemicals onto the market in consumer prod-ucts, and is generally at fault for what's wrong with John Q's quality of life.

In the other direction, John Q. sees an economy of rising prices, shoddy merchandise and shortages all the fault of the greedy businessperson who is only interested in profits, profits, and more profits.

And where does John Q. get this perspec-tive? The most likely place is this nation's mass communications industry, which with hours and hours to fill, has brought the good and bad (mostly bad) news about the American industrialist into the living rooms.

What can business people do to change this image? There are no simple solutions. Factors causing today's problems are complex. They are interwoven with laws and govern-mental agency regulations, and are influenced by John Q. Public's desire to know more about what goes into and what affects the environ-ment around him.

While Public Relations will not alter the attention business people are getting from the media, increasing the amount of time and expertise going into this effort by business people will help explain to the consuming public the causes of a deteriorating economy and what he, the businessperson, is doing to improve the quality of life.

But before any recommendations, first a look at what has caused the businessperson to become the object of such intense media interest.

THE PHENOMENA OF MASS COMMUNICATIONS

At the beginning of the '30's, America was made up of homogeneous regions: the industrialized north and northeast; the south, where king cotton rules; southwestern states with their long-horned cattle and painted deserts, the farm belt, and onto the west through the Rockies to the Salmon fishing and snow-capped mountains of Oregon and Washington, down the coast to California where the mere mention of the state's name conjured visions of tinsel, sequins and the exciting life of Hollywood.

Washington at that time was not an exciting place. Three events unified America and made Washington the focal point: the development of radio, the introduction of the airplane, and the election of Franklin D. Roosevelt. Under Roosevelt, Washington became the hub of the American Wheel, determining how people worked, how much they made, what they ate, and where they lived. Before he arrived, government was smaller and more timid; by the time he died it had reached everywhere and was the source of power to make things change. As David Halberstram points out in his book, THE POWERS THAT BE, Roose-velt's use of mass communications became the textbook for future politicians.

What politicians learned about television and its impact, television executives had known for a long time, and, as good businessmen and women, they began to exploit what they had wrought.

Television was not the only thing to explode. Throughout industry, technol-ogy, business acumen, new tax laws, faster depreciation, new plants and machinery provided both businessmen and women with more profit and the employee with more free time. Over the past thirty years, the work week has shrunk from 48 to 40 hours, and in some industries, it is now 35. Flex-time has been introduced to U.S. industry; unions and the collective bargaining process have become stronger, demanding and getting more benefits, more money and more free time for members. With time on his hands, the American worker did two things: he expanded his recreational time, and when not fishing, hunting, golfing or working in his garden, he or she turned to television for news and commentary.

Printed media declined. No need to list the national magazines that have folded, nor the PM newspapers that have ceased to publish. How did TV become the primary news source? Two events meshed to push video past the print media. First, the launching of the Communications Satellite, which shrunk the world to the size of a living room television screen, and second, the TV networks expanded national news coverage. The fifteen minute national news broadcast at six p.m., local time, was expanded to thirty, and local TV stations began running from thirty minutes to an hour of local news before the evening edition. Radio also was keeping pace. National network news every hour on the hour from six a.m. to midnight became standard, with local stations providing news inserts on the half-hour.

While John Q. Public was being enticed to the tube with an amazing increase in sports and cultural programs, his view of the business world was one of negativism. Examples of TV and newspaper reporting on how business activity was (and is) causing problems with John Q.'s quality of life are abundant: the chemical industry polluting Love Canal in upstate New York forcing people to leave their homes forever, and oil companies forcing up prices making driving and heating homes more expensive. If protests at nuclear power plant construction failed to bring the dangers of this controversial subject into sharper focus, TV's reporting of the Three Mile Island incident outlined "Nuke Power" with crystal clarity, and, finally, thanks to TV and newspapers, toxic shock syndrome and the hint that the feminine TAMPON was the cause is as familiar as the words bread, butter and milk. Thus, a dichotomy: the good news in the form of entertainment and cultural shows are standard TV fare, while negative news about business, and what business does that results in damage to the quality of life, are going through together.

Couple this with the reporting of Watergate and ABSCAM; is there any wonder that John Q. Public has a poor image of business and politicians.

And, this kind of TV coverage of the American scene is not going to diminish; expansion is the word for tomorrow.

Cable television and its wiring of America, industry analysts predict, will completely alter our lives. Already, Ted Turner, the sports tycoon from Atlanta, Georgia has the super television station that saturates the Southern part of the U.S. Also, Turner has started the first, seven-day-a-week, 24-hours-a-day news television network. Launched in July, 1980, Cable News Network has about 4 million subscribers, and business news is a big part of the network's format. And, in December, '80, the Federal Communications Commission approved a COMSAT Corporation request to continue its study of developing a backyard dish antenna so that buyers (or renters) will be able to pick up satellite signals from around the world for themselves.

The expansion of the mass communications industry has only just begun, and with all that information flowing, it is obvious that the businessmen and women will be the object of much reporting.

THE ECONOMIC DILEMMA

On the opposite jaw of this vise is America's deteriorating economy. Again, business people are in a "Catch-22" position: as wages go up, so do expenses; since stockholders want a return on investment, the business manager raises prices to cover cost. The consumer is squeezed and begins to yell about "the high cost of living," and soon news media personnel are turning to business people to get the answers to "why," "how come," and "what are you going to do to fix the problem?"

"To judge by newspaper editorials, opinion polls and comments from competitors, American business is caught in a trap of its own creation: American goods do not meet the test of quality in the market, and American business people have not understood that the future belongs to those who sell to the world. Blessed for years with a large domestic market, many American firms have neglected to look overseas. Conversely, look in stores for an American-made radio or tape recorder; you will probably not find one. One half of all the cars that leave California showrooms this year will be imports, mainly Japanese. More than half of our exports to Japan are food, fuel, crude material, and more than two thirds of imports from Japan are finished, high-technology, manufactured goods.

"Better morale and teamwork produce improved quality and high productivity, so the saying goes. While morale and teamwork may be necessary for productivity and quality, business people are quick to add that capital investment is the other necessary condition. For the businessperson, it is only one step from the observation that lack of capital investment is what is at the heart of his or her problem to the conclusion that the government is ultimately to blame. Some observe that government has been a drag on business' flexibility and its capital resources. The government has built inflation into the economy, they say. Meanwhile, government officials claim it is not rules and regulations that inhibit businesses, but industry's method of using its own resources -- and especially its inability to plan for the long run." **

Whatever side you might choose, there is, as James Fallows wrote last year in THE ATLANTIC Magazine, a piquant chicken-and-egg side to these phenomena.

> "Because inflation is so bad, business can't raise money to invest. Because they don't invest, productivity doesn't rise. Because productivity is bad, inflation gets worse. Because inflation is worse, the dollar is weak, OPEC (which is paid in dollars) raises its prices. Because oil prices go up, so does inflation. And because they want to be re-elected, presidents pump up the economy every four years and move inflation up to a new platform."

The attempt in this paper has been to point out two unusual circumstances that have crashed together to make life uncomfortable for business people. As was pointed out, the problems are complex: regulations are interwoven with the consumer's desire (and right in some cases) to know more about what is affecting his or her life. Business people are relectant to give anyone just any kind of information for fear of improper use. And, some governmental bodies are asking for proprietary information.

Public Relations will not solve these problems, but wisely put to use within and without the company, PR will go a long way to help explain the whys and wherefores of business action. PR supports the goals and principles of the corporation and the function's and duties range from publicizing products and services to assuring that company positions are presented accurately to reporters. PR does not create a corporate image, that is fashioned by the performance of the product or service in the marketplace along with the actions of employees and executives in the community. Whether it is favorable or unfavorable publicity, it is PR's job to explain the company's position to its publics and to make that position believable and palatable.

There are numerous ways in which this can be done, since a company interfaces with its publics in many ways: the press release, internal or external publications, technical papers, attendance at conferences and through community relations. First, and most important for any PR program, the company must maintain high standards of conduct. Second, the firm should be fair to all reporters. It is dangerous to "leak" a favorable story to just one reporter; that can hurt in the long run.

The selection of a person or persons to interface with the press should be well thought out. The spokesperson(s) should be easily identified in the company, be thoroughly knowledgeable about the business, in an organizational alignment that clearly shows he or she speaks for the chief executive officer, be cool under pressure and personable enough to talk convincingly to a group of strangers. These requirements are necessary because today's reporters are far from movie stereotypes. They, too, are articulate, most of them are college-trained in such fields as business, economics, medicine, history, philosophy or government. Many are specialists, concentrating on a few key industries, like stock analysts, seeking answers to their readers' (listeners) questions.

It is PR's job to understand the needs and deadlines of reporters and to research, write and clear answers to their questions. PR people should be courteous and as responsive as possible to reporters, but the interests of the company must be first. Most journalists respect this if the PR department is not unreasonably restrictive in the name of company interest. PR should try to provide as much useful information as possible, as long as it does not contain proprietary material, infringe on the privacy of employees or otherwise violate basic corporate principles.

Most of the time, reporters call the PR Department to set up interviews or ask questions. It is not unusual, however, for reporters to call executives on the telephone or approach them at a conference or other public gathering. With due respect to reporters, the rule of thumb should be caution. The executive probably does not know the publication; and may not know precisely what can and should be said. Unless the executive is sure, the best advice is to have him or her limit his or her comments to his or her own area of expertise, or give no comment at all. It is good practice and perfectly acceptable to get the reporter's name, publication and deadline and promise a return call.

News interviews are important to a company's image. Interviews offer advantages that a press release or a written answer cannot. They can be more convincing; they may bring to light misconceptions of the reporter; they should result in a better article. But they involve risks as well as advantages.

When the decision has been reached, the PR professional should help the executive. There are many things the professional knows about the reporter, the publication, or about press activities in general that would not be apparent to the executive. Before the interview, the PR professional should cover several points:

What does the reporter want? General areas of discussion should be made clear. What points does the executive want to make? What have others in the industry said, and what are the ground-rules for the interview? The PR professional should cover these points with the reporter ahead of time, and they may include:

- A time limit. Thirty to 45 minutes is usually enough.

- Everything on the record? This is questionable. Probably the best rule of thumb is to tell the reporter nothing that cannot be printed.

- Will the executive see copy? Most reporters will not review copy with the PR professional or the interviewee. However, either of the company employees can and should ask to check direct quotes and key facts and statistics over the phone before the story goes to press.

When answering a reporter's questions, there are five main points to bear in mind:

1. The primary concern should be for the reputation and security of the company, the dignity of its people, the best interests of its customers, and tell the truth.

2. The executive should stay within his area of competence.

3. Consider that every word said will appear in print.

4. There is no obligation to answer every question.

5. The executive should seize the opportunity to make the points he wants to make.

Appearances on TV news programs or documentaries, or statements for TV news, require special attention. If the reporter is asking a question for a TV news program, the answer will have to be distilled to no more than two or three sentences. That is all TV newscasts usually carry. If it is an on-camera interview, the executive should get a thorough and detailed description of the thrust of the news story. Again, as in the interview, the PR professional can and should study the situation and recommend the best course of action.

Finally, some additional suggestions: Deny firmly; reporters are suspicious of hedged denials. Do not provide "inside" anecdotes about the company or employees that could later prove embarrassing. Treat "no comment" cautiously. Improperly used, "no comment" can sound like a confirmation. Do not speak for others in the company or for the industry. Be careful of colorful language; do not tell a reporter to call someone else and be careful of answering a reporter's comment, "I have heard..." He may be fishing for something bigger.

Again, Public Relations will not make the businessperson instantly loved, but used properly, PR is one of the best tools that businessmen and women have today to counter the poor image that rests in the mind of John Q. Public.

**Excerpted from American Industry by James Fallows, Atlantic Magazine, 1980.

EDITING FOR THE FIRST HALF SECOND: THE PERCEPTUAL PROCESS AND THE TECHNICAL EDITOR

Arnold C. Henderson
Princeton Aqua Science

Arnold C. Henderson, PhD

Princeton Aqua Science
Editor/
Communications Specialist
P.O. Box 151
New Brunswick, NJ 08902

(201)846-8800

Editor of environmental reports; has also published literary criticism and photography.

Modern studies of perception stress the active role of the mind in its rapid roughing out of partial meaning moments before we decide what it is our eyes have seen. When slides, charts, and texts fail to communicate, the fault may lie in the gropings of that first half second. A quick-sketching exercise used at Princeton Aqua Science helps editors and graphic artists rapidly anticipate an audience's most likely misreadings and frustrations.

Modern studies in perception, by and large, emphasize the brain and the actively shaping powers of the mind, not the eye and the passively receiving screen of the retina. While psychologists probe with renewed interest the workings of optical illusions, and art historians led by E.H. Gombrich rethink the perceptual premises of visual stylistics, technical editors and graphic artists, too, can take fuller account of the perceptual distortions that arise as our audiences grope toward the meaning of our reports and audio-visuals. The eye perhaps detects the line on the page, but only the mind knows the figure the line makes.

Seeing is Believing–and Vice Versa

Now the mind is a sly and unpredictable beast. It makes mistakes. It persists in those mistakes despite all logic. It laughs at the writers, editors, and graphic artists who try so hard to make at least some things perfectly clear. The mind--the audience's mind--achieves its mistakes despite our care, and the amazing thing is, it does it so quickly. Show us a page and we can misread it in less than a second. And there's the clue.

Find a way to freeze that first fraction of a second, and you'll have seized on the first act of the mind, the first act that leads to each succeeding groping for meaning and form, the act that initiates the whole chain of error or of communication. As editor at Princeton Aqua Science, a firm with a special concern for the immediate clarity of our reports and audio-visuals, I have been exploring a quick-sketching technique that helps our graphic artists and writers quickly test some basic ways that a given page or slide is likely to be perceived or misperceived.

Princeton Aqua Science is an environmental consulting firm whose environmental assessments must--by Federal law--be intelligible to both technically trained professionals and interested citizens. Any good writer knows to write with the audience in mind, but it is not easy to visualize so mixed an audience; their potential misunderstandings are hard to anticipate. But trained or untrained, every member of the audience shares in the basic perceptual process, and we could anticipate at least some part of their response if we could only know what happens in that basic perceptual process.

To get at those first moments of perception, show your chart, map, photograph, or slide to a few colleagues, giving them perhaps 5 seconds to look. Then have them sketch it--from memory. You get marvelous mistakes that way. It can be fun. But try that same page on several people, and you start seeing amazing agreement about not only what *is* there on your page, but about what *isn't*.

Different people will make the same mistakes, mistakes that come neither purely from the memory nor purely from perception, but which represent what goes on in the first moments of the whole confusing process of communication.

Perhaps we are not surprised that people will sometimes draw things that weren't there to be seen in the original; it is harder to anticipate that several different people will draw the *same* nonexistent thing. Since seeing is believing, I present below a photograph by Robert Frank that nearly a hundred people have sketched for me. (Don't look yet--if you want to test your own response, you'll have to look for only 2-3 seconds, then look away and sketch what you can remember.) I use a photograph as my first illustration rather than a technical document both because I happen to have seen what happens when dozens of people try to sketch this particular photograph, and also because a photograph looks "real" and should require less technical training to read than, say, a flow chart or resource capability map.

Most people will, in a 5-second glance at a distance of several feet from the full-sized photograph, realize that they need to sketch two windows and a flag. Somewhat fewer will realize that they need to put people in the windows.

The editor or graphic artist who wants a slide or page to register quickly will learn from such quick sketches that the eye typically picks up objects before the mind puts them into order, and that the mind apparently tries out a few wrong orders before settling on a right--or at least satisfying--one. Trying the quick-sketch experiment with your latest flow chart may well bring you sketches of unrelated boxes. If so--particularly if you are making a slide that must be seen in a fixed time--you'll be glad of the chance to go back to the drawing board to fatten up the arrows between your boxes.

But the amazing mistake in sketching this particular photograph is just the sort of mistake we might have thought wouldn't happen, a mistake in which several people agree in drawing the same nonexistent thing. Look at the figure in the right-hand window. In sketch after sketch (perhaps up to a third or fourth of the total number), this figure has a head--not surprising until you realize that there was no head in the photograph. Look more carefully at the photograph and you will see that any head that figure may have had in reality is obliterated by the whipping flag. With no head to be seen, person after person will nonetheless draw one in. Amazing.

So sure are my viewers that there is a head in the window that, rather than cover the head, many will refuse to draw (or see) the overlap of flag and window. They draw the flag above or between the windows.

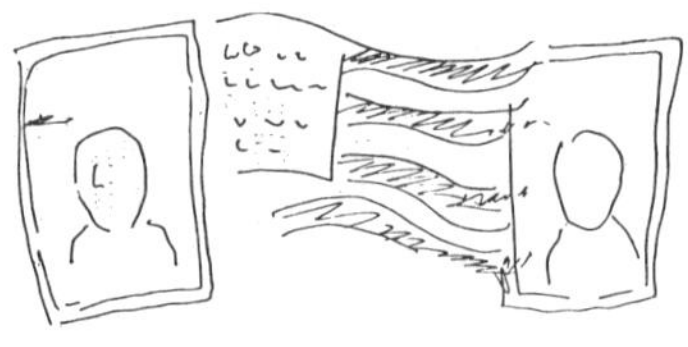

Others will correctly draw the overlap and then (o height of obstinacy!) go right on to draw the head that the overlap ought to conceal.

For these viewers, concept conquers sensory data: it's a person; people have heads; *draw* a head!

What my test audiences believed they saw has overridden what they did see. Or did I say that right? If person after person draws the same thing, isn't that pretty good evidence that they *did* see it? Even if it didn't exist?

This amazing agreement on something that wasn't there is good evidence that what we at first believe we see is in fact what we do see--for that moment. We may later, given a few more moments to look, revise our opinions, but cutting our unconscious testing off at an early stage, as in this sketching exercise, preserves the fleeting mistakes we are all making at every moment as we look about us.

How does this tentative and error-rich groping for form and meaning affect our perception of things other than photographs, our perception of technical graphics?

The Mountains and the Molehills of Technical Communication

Unlike a photograph, a flow chart or technical illustration generally has some words right inside it. One would think these words would guide the viewer away from seeing heads where the flags are, or making mountains out of the molehills. But in that first half second when the slide flashes up on the screen or when the reader first opens the page and the words have yet to be made out, some highly annoying mistakes may start happening. Once a mistake gets ingrained at this first stage, the audience may or may not have the time to correct it. What correcting does get done has taken time and effort, has necessitated changing one's mind--and that can be annoying.

I am convinced that much of the frustration readers and viewers feel at "badly written" reports or "confusing" slide shows is essentially the frustration of having had to change their minds so often, the frustration of looking for clues too difficult to find on a badly laid-out page.

Below on the left is a flow chart as I showed it quickly to our graphics designer; on the right, her sketch.

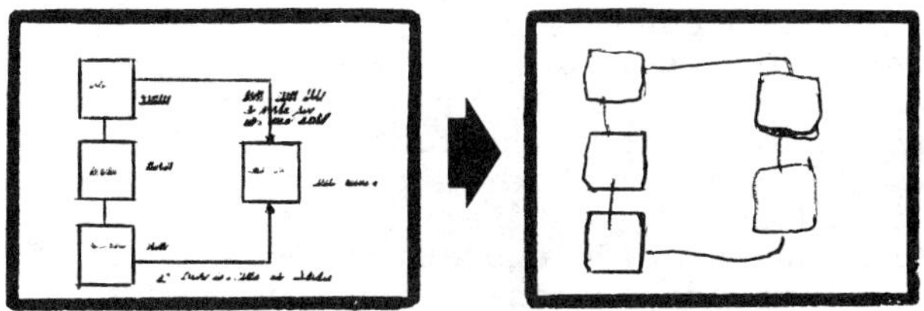

On the left, three boxes flow to one; on the right, to two. She mistook a block of *words* for another box. The error is the sort we can call typically right-brain, one where concepts have been ignored but shape noted. Visual thinking is what makes this woman such an outstanding graphics designer, but it has its flaws. To her companion in the experiment, a cartographer with a more conceptual, left-brain mode of thinking, words were words--he even managed to catch a few in his few seconds--and his three boxes correctly flowed to one. Real audiences contain both types and confuse technical editors horribly.

Moments later, the cartographer made his typically left-brain misperception. Given a map with, of course, some words on it, he was so used to seeing words as names of towns on the maps he made, that he created for his sketch a nonexistent city to go with the word he took for its name, though it happened to mark a forest preserve. To each his own misperception, but we all have them.

We all misperceive because we must: misperception is the road to perception. Without a tentative order, there is no final order.

Such confusions of the first glance may never get corrected. Time lacks; slide shows zip along; readers skim. Worse, they skim most when they think they understand best--and it is just this *impression* of understanding that we have seen leading to all those extraneous heads and fictive cities.
Once mistakes have occurred, it is too late for the editor to save the day completely with an explanation three pages later. Getting the facts finally straight doesn't unkink the gut response left over from having had to work to get them straight. Readers resent it. They blame the people who made up the document. They say, why do engineers have to write like that?

A transient mistake, as modern percept-ual studies show, satisfies the mind just as well, while it lasts, as a true perception does. That is why optical illusions work, why psycholo-gists can construct special rooms (Ames rooms) that have crazy proportions but fool viewers--who assume normally rect-angular walls--into thinking them nor-mal rooms even as every other piece of sense data screams that something is askew. That is why my sketchers were so ready to move the flag out of the way to make room for a head they be-lieved in. In our perceptions, the flag steps aside and anyone walking in the Ames room seems to loom and shrink, but we prefer those obvious mistakes to denying our convictions that peo-ple have heads and rooms have right angles. Believing is seeing, and it lasts.

Studies of visual illusions all show how stubborn we are, how very hard we find it to deny a fixed interpretation.

This often-studied "ambiguous figure" may be a duck or it may be a rabbit, but the one thing it is not (in our minds) is both-at-once. Somehow we admit both are inherently there, but at the instant of our perceiving a duck, our rabbit vanishes. The lunar craters we see in photographs pop from crater to dome as our minds hesitate in deciding the direction of the light: without contextual clues, we wouldn't know a hill from a hole in the ground.

There are principles one can isolate from errors in such sketches. Below are some I have seen often enough to be sure they are worth watching for in technical documents and audio-visuals.

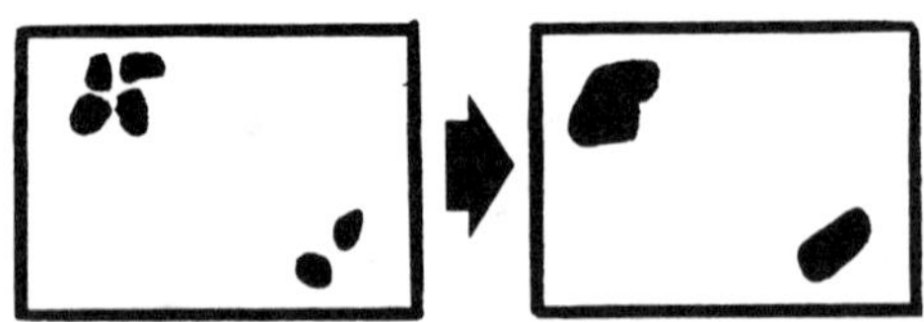

In this pair, things in proximity to each other in the left-hand drawing tend to get lumped together in our per-ceptions, memories, and sketches (right). Below, incomplete things, like a circle with a gap--or a body without a head--tend to get completed by our minds.

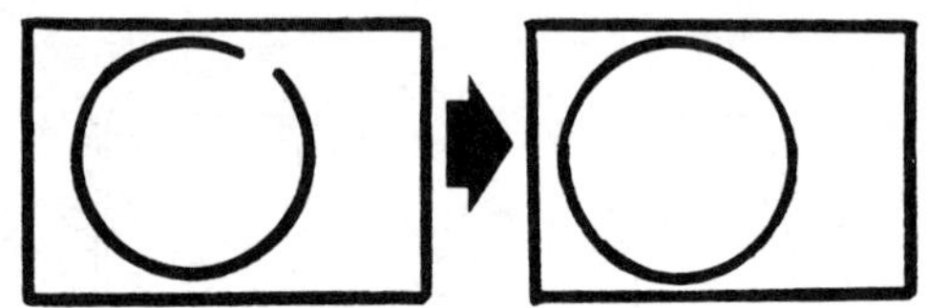

Other trends commonly seen involve grouping similar shapes together, or simplifying and distorting complex shapes to approach shapes we prefer, perhaps simpler geometrical ones or ones especially familiar to us. The Gestalt description of this as the "good figure" principle doesn't explain much beyond our simply preferring the figure for some reason; the psychologists using information theory may be more subtle in their references to boundaries and to the probability that a given signal will be distinguished from background noise.

What we find useful at Princeton Aqua Science, in our daily designing and ed-iting of environmental documents, is not naming the phenomena by one theory or another, but being alert for moments when they may arise. For fostering a-lertness, the quick-sketching exercise has an advantage over the old but still helpful tricks of squinting at your il-lustrations or stepping back: with the sketches, you can relatively objectively gather the responses of several people and not have to trust your own squint-eyed hunches.

No technical editor or graphic artist should expect to keep the mind from mak-ing its tentative misreadings--groping is now generally recognized among psych-ologists as something like the very es-sence of perception. You can edit what is on the page, but not what is *not* on the page, the heads and cities and extra boxes that someone will sooner or later perceive. All you can do is be wary. The quick-sketching method helps: by probing beneath the conscious trained activity of our minds, it can help the differently trained scientific, graphics, and editorial staffs communicate to each other their perceptions of the material at hand, and it can help researchers and communicators alike to remember the needs of those ultimate perceivers, the final audience.

Readings on Perception

1. Gombrich, Ernst H. *Art and Il-lusion: A Study in the Psychology of Pictorial Presentation*. Princeton, N.J.: Princeton University Press, 2d edn., 1961.
2. Gregory, R. L. *Eye and Brain: The Psychology of Seeing*. New York: McGraw-Hill, 2d edn. 1973.

ESTABLISHING THE IN-HOUSE NEWSPAPER
Some Guidelines for Success*

C. W. (Bill) Jensen   Ann P. Silverstein

Lawrence Livermore National Laboratory
Electronics Engineering Department

C. W. (Bill) Jensen
Consulting Editor, ENGINEERING NEWS, an LLNL in-house publication with distribution of 4500.
Co-author with E. R. Fisher, *PET and the IEEE 488 Bus (GPIB)*, Osborne/McGraw-Hill, 1980, and *IEEE 488 Interface Parts Handbook*, Osborne/McGraw-Hill, 1981.
Educated at Brooklyn Polytechnic Institute; UC Berkeley.
Member, Pacifica Chapter, STC.

Ann P. Silverstein
Technical Editor/Writer for the Electronics Engineering Department and Reporter/Photographer/Assistant Editor for ENGINEERING NEWS.
Educated at UC Berkeley; received B.A. in Physics/English Literature in 1977.
Member: Pacifica Chapter, STC; International Association of Business Communicators (IABC).

In February 1979, the Associate Director for Engineering at Lawrence Livermore National Laboratory (LLNL) asked to have a proposal submitted to him that would be the basis for a new in-house newspaper. The resulting publication, ENGINEERING NEWS, is delivered bimonthly to over 4500 people including employees of the Electronics, Mechanical and Plant Engineering Departments plus retirees and selected universities/colleges. The proposal for ENGINEERING NEWS addresses the following topics: audience, purpose, contents, style, format, paper stock/ink, production schedules, staffing/organization, and operating costs.

While such a proposal can help in obtaining management's approval and in publishing the first several issues, other problems crop up that, unless solved quickly, can result in loss of readership.

Introduction

So you're it! You, and perhaps a couple of your fellow writers, have been directed by management to start an in-house newspaper: "Nothing fancy...just something to keep everyone informed. You can fit it in with the rest of your work!"

Why you? Well, you're a technical writer, right? Never mind that the last time you wrote in a journalistic style was in a college creative-writing class or that you've been buried in writing technical manuals and specifications since you came to work for your company three years ago. You are now, by management edict, reporter/photographer/editor/(and maybe) graphic artist. So, with tape recorder and camera in hand, you sally forth to interview the president of the company...but wait!

Before you do that, you may want to find out what we, the staff of LLNL's ENGINEERING NEWS, went through after management assigned us to our roles as part-time editors/writers/photographers of a then yet-to-be-defined in-house newspaper. We hope you will learn from our mistakes, and start out on top!

The Proposal, A First Step

In the beginning there was the proposal...and management looked upon it and saw that it was good.

Whether management asks for a proposal, or you take the initiative by seeking management's approval to publish a new in-house newspaper, it's a good idea to write a proposal and have the basics well defined before you meet face-to-face with those who must bless the project. Your proposal should not only provide a clear understanding of the paper's policies but should also address management's concerns. A little thinking ahead of time may save some embarrassing moments of silence as you grope for words in answer to the vice president's question, "What do you anticipate to be the cost per page?"

Know Your Audience

Basic as this statement may seem, no dissertation on the subject of communication would be complete without a few words on the subject of audience. Knowing who your readers are will help you zero in on subjects of interest to them. Plumbers, electricians, gardeners and various crafts people will never be seen absorbed in reading your news article—the one you sweated over 16 hours to beat the deadline—when it's about management's efforts to recruit talented new engineers.

Our audience has pretty much been defined for us through the initial management edict. It includes *all* Engineering employees: engineers, technicians, drafters, secretaries, custodians, painters ...*everyone*. We also include retirees, certain Lab personnel outside Engineering, selected university/college professors, and Personnel's recruiters. And, the mailing list continues to grow. After every issue we get letters and phone calls asking us to add names to the distribution, which is now more than 4500.

Knowing full well that all articles will not be of interest to everyone, we try in each publication to strike a balance among

* Work performed under the auspices of the U.S. Department of Energy by the UCLLNL under contract number W-7405-ENG-48.

our three Engineering Departments. Management, we have found, is particularly sensitive to our "coverage"; if one issue happens to feature one department more than others, we can always count on a "blast" from the two under-represented departments. "At least," we tell ourselves, "some people out there are reading it all."

Define Your Purpose

Why do it? This is a question that elicits an outpouring of lofty ideals and objectives. The verbage can run the gamut: "To act as a motivational tool... to reinforce the mission of the ABC division... to develop awareness of support efforts... to promote cooperation... to create a direct communication link... to enhance visibility of the XYZ division... to serve as a public relations tool...." The list goes on and, as one of our staff blurted out near the end of one of our planning meetings, "What we really want to do is tell everybody what the hell is going on!"

We do not intend to demean the subject of "purpose." Quite the contrary. Unless you can honestly answer the question, "Why do it?" and write each article with these definite purposes in mind, your publication will emerge as a mishmash of ineffective communication. In our case, we found that our purposes expanded with each issue as we tried to build more human interest into our articles. In summary, our purposes evolved over the months into a single slogan now appearing over the masthead: "Highlighting achievements and activities of Engineering's people." The slogan serves to remind us of the answer to the question "Why do it?" each time we write our articles.

Outline The Paper's Contents

Once you have the purpose of your upcoming paper well in mind, you can consider the contents.

We knew we wanted three kinds of articles in the paper. The first was "Project Feature Articles," which publicized how different Engineering groups work together on a single effort. Intended to promote cooperation among various Engineering groups, these feature articles were, originally, a major thrust of our paper. The second was an "Engineering Corner," which took the place of the usual "Message from the President" found in other company publications. The Corner communicates information from top Engineering management to all levels of Engineering employees. The third was a "Welcome Aboard" column covering Engineering's new hires: their backgrounds, their hobbies, their current positions. (Interestingly enough, this column has a wide readership; it's popular among both new hires and old-timers as they are all seeking acquaintances with common interests.)

Since those early days, we've added several different areas, most of them our attempts to bring a personal touch to the newspaper. New features include a "Spotlight" of a person for a technical achievement and a "Community Services Corner" highlighting individual achievements outside the Laboratory; we've also added "Newsbriefs" for small articles that we feel rate only a paragraph or so.

Having definite headings and examples for sections of your upcoming newspaper will lend a detailed touch to your proposal and demonstrate to your management that you have "done your homework."

Determine Style

If most of your experience has been technical writing, you may have problems adapting to a journalistic style—that is, if you decide to go that way. Adopting a particular style of writing is a decision to face early-on in your planning phase. Some factors to be considered here (there are others) include the frequency of your publication, and the size and the background (type) of your audience.

Frequency of publication is important. If your newspaper is to be a "weekly," chances are you'll want to adopt the true journalistic style. The reason? Your paper will probably cover *current* company news (such as a ground-breaking ceremony for the next multi-million-dollar project) and report on it right away. This is when you must get all the typical who, what, where, when, why, and how answers up front. Paragraphs should be short, with the least important facts at the last. (This prevents bloodshed when your paste-up artist insists that some of the text must go if you want photos on the page as well.) On the other hand, if your newspaper is published once a month, chances are that by the time you "hit the street" the local town paper has printed the picture, run the story and—it's no longer news! In this case, you can "relax" and, by judicious planning, write articles of interest—not necessarily news— on subjects where you won't be scooped.

Audience size and background are determining factors when you decide on style. If you're writing to only 100 people about their activities, you may want to adopt a folksy style ("Well, Linda finally gave birth to her 13-lb baby girl..."). Such a style just isn't appropriate for a distribution of 2000. In either case, people enjoy reading articles that come alive with active verbs and short, to-the-point sentences. Just because you write to engineers and scientists, don't think your language has to be complex and written in the passive voice. There's no quicker way to lose your readers!

You may find, as we did, that your style will evolve over time. The important thing is to grab a pencil (or a typewriter) and start in. (One suggestion is to pattern your style after a respected publication. Buy some copies, study them, and go to work.) As the months went by, we noticed our beginnings had more punch. Articles shrunk in size as we learned to be brief. At present we're trying for a style reflecting human interest; we now concentrate on the "who" and "why" and "what," rather than emphasizing the technical "how."

Mock-up the Format

As an appendix to your proposal, you'll probably want to give your management an example of how the newspaper will look. Here, your graphics artist can help you sell your ideas. The important thing is to have the proposed masthead, the regular features such as "The President's Message," and any special columns (like our "Welcome Aboard") prominently identified. While giving this mock-up your best efforts, the final product in the proposal need not be elaborate. A pencil or freehand ink sketch will do for the masthead and headlines, while the text can be wavy lines or pasted-up copy from a suitable source. Remember, you're just after a visual effect. In fact, we found it best to keep this part of the proposal quite rough so that the people who must approve the proposal could add or sketch in their ideas.

The present format of ENGINEERING NEWS has changed radically from the format of its first issues. From four pages to eight, from buff paper stock/brown ink to white paper stock/black ink, from small photos surrounded by columns of text to large photos and judicious use of white space...we've been told we've come a long way. (See Figures 1, 2, and 3.)

The reason we made these changes (more or less all at once) was to increase the readability of our paper. The easier it is to read, the more chance our audience will read at least part of the paper before consigning it to the circular file. Two changes were especially important in making the newspaper more attractive to readers:

- A change to eight pages gives us a center spread for our feature article and allows us to jump the gutter with photos or text.
- A change in stock and ink makes the paper easier to read. Black and white provide more contrast than buff and brown, and the combination is better for photos, too. (More on photos later.)

Figure 1. The ENGINEERING NEWS original masthead has given way to the one below. The new masthead includes the paper's slogan.

A hard day's night no longer — custodians and the switch

Caluwé outstanding apprentice award goes to ME's Ashley Long

TURN OFF

ENGINEERING NEWS STAFF
Editor: C.W. (Bill) Jensen
Assistant Editors/Reporters: Donna Lee Paquette, Ann Silverstein
Secretarial/Layout: Roberta Mackey/Liz Caires

Figure 3. The no-longer-used format of ENGINEERING NEWS. Gone are the tiny photos, cramped columns.

Dauphin

By C. W. (Bill) Jensen

"... weight ... only 383,000 pounds!"

... a complex experiment ... a cast of 'thousands'

... devotion to duty is not at all uncommon.

"It's the human element out here that makes it work."

New prototype board offers unlimited applications

Figure 2. The eight-page format provides the area for the feature story with large photos.

Evaluate Paper Stock/Ink Color

From "newsprint" to "Madison Avenue" glossy-type stock—somewhere in between you will probably find the paper stock suitable for your publication. Zellerbach Paper Company,[1] for example, produces a full line for you to look at. Offering more than 800 types, this company manufactures a variety of paper stocks and colors under the following headings: Book Papers and Matching Covers, Texts and Covers, Covers and Bristols, Business Communication Papers, and Simpson Lee Text and Cover Papers. Not that you need to evaluate all these; within the Business Communication line are paper stocks recommended for offset printing. Other companies for your consideration are Blake, Moffit and Towne, Simpson, Carpenter, Noland, and Strathmore.

We suggest you contact a local representative of one or more paper companies and get prices and availability of a group of paper colors and types. In your survey, get assurance that the paper mill (many are in Canada) manufacturing your choice of stock will (1) not be threatened by severe strikes in the near term, (2) not simply go out of business (some have recently), and (3) not discontinue the particular line of paper of your choice.

Due to the three factors just mentioned, printers often stock large supplies of certain paper types. Check with them; you might just get a deal. Sometimes they have a large supply of end cuts left over from a large order. If the page size of your paper is smaller than 11 by 17 inches, here again you may save some money. However, if you intend to maintain continuity of your paper stock and color, give careful consideration to the long-term availability of the product. For instance, many colors available as recently as 1978 are now discontinued.

Our current paper stock:

> Paper stock: White Matte Coded Offset Book
> Weight: 120 lbs per 1000 sheets, 25 by 38 in.
> Cost per 1000 sheets: $5.70 (for our Nov/Dec 1980 issue)

Once you have decided on the paper stock, arrange to have the printer store the paper; don't store it in your own plant and transport it to the printer for each publication. One company supplying its own paper to the printer recently avoided, by chance, a disaster. The company's forklift driver accidentally dropped the entire paper supply on a gravel walkway; the wrapping broke and gravel ended up between the sheets. The paper was delivered to the printer and loaded into the press. Moments before the operator pushed the start button he discovered, to his horror, gravel sifting from the paper supply! Disaster was averted. Had it not been discovered, we leave the consequences to your imagination.

In picking an ink color, remember that you want a good contrast between the color of the paper stock and the ink. Black ink on white paper is still your best bet, although you may think it a trifle mundane. If you are a government agency or under contract with a government agency, chances are that you will be limited to certain standard colors of ink. GPO (Government Printing Office) policies do not provide for mixing colors. Consult the USGPO Government Paper Specifications.

Work Up A Production Schedule

Although you may not want to include a sample schedule as an appendix to your proposal, it's a good idea to rough one out for your own planning and peace of mind. Try to estimate how much time is needed to put out each issue. Include the efforts of everyone involved: editor, reporters/photographers, word-processor operators, layout/paste-up artists, clerical support and the printer. Never mind that the schedule may wax and wane once you get into production. Again, the important thing is to get down (preferably on a large calendar with squares big enough to write in) all overlapping and end-to-end or serialized tasks that need to be accomplished. Such a planning chart might display the days set aside for (1) interviewing/writing/photography, (2) typing/editing, (3) rough draft approval by those interviewed, (4) typesetting, (5) proofing galleys, (6) total layout using photocopies of galleys/photos, (7) final review by patent/classification/management, (8) typesetting of management changes and galley errors, (9) printing, and finally, (10) distribution.

In our case, we begin each issue by sitting down and putting together a schedule, usually as the previous issue is going through its final approvals. *Everyone* involved gets a copy of the schedule: writers, secretaries, artists, and printer. That way if there are any conflicts (artist having other projects, etc.) we find out right away. For our available time, it takes about 1½ months to put out one eight-page issue.

While putting together the schedule, the editor and reporters discuss topics that might make good articles. Topics come from management, as well as from other employees. In the beginning, we went directly to department heads for topics; we received lots of technical article suggestions, but not too many topics of human interest. Most good "human interest" articles come from the employees' suggestions. In addition, reporters are always on the lookout for possibilities. You will find that as your paper becomes better known you'll get more suggestions for articles.

Decide On Staff/Organization/Equipment/Workspace

Even though there may be only a few people directly involved in the production of your paper, it's important to define their duties and to spell out who reports to whom. You might find, as we did, that a simple organization chart will clarify the amount of staffing effort your newspaper will require.

Some other factors to be considered are equipment and work space. We have found our darkroom—a photostat camera complete with developing trays and drying facilities—invaluable in meeting production schedules.

Estimate Operating Costs

Having estimated the amount of employee effort, consider two major production costs: phototypesetting and printing.

Costs for typesetting can vary widely and can be determined best after getting bids from several type houses in your area. In getting bids, be sure to include transporation costs: with today's gasoline prices, it's well to crank in a cost factor of how your text gets to the typesetter and how the galleys get back. If the type house is not close by, don't be surprised if the pick-up and delivery costs run as high as $50.

Our typesetting costs have varied as follows:

Phototypesetter A
- $600 - An issue of six pages (photos included).
- $100 - Cost per page (includes finished pages with photos).

Phototypesetter B
- $950 - A typical issue of eight pages (photos included).
- $118.75 - Cost per page (includes finished pages with photos).

Printing:
- $125 - Average cost per page.

Proposal Accepted, You're In Business What are the Problems?

Plenty!

While the production problems can be some multiple of the people and processes involved, and your problems may be somewhat different than ours, let's list common problems that occur and then go on to ways of improving in-house publications.

Some of the common problems that arise for in-house publications (so we are told) are:

- Management does not grant final approval on time.
- Managers, not knowing the publication business, impose their ideas on the style/format of the paper.
- Management limits communication to one way—out, but not back.
- Articles do not come alive with human interest.
- Newspaper staffs have problems gaining credibility with management.
- The typesetter doesn't produce on schedule; when it's delivered, it's full of errors.
- The printer delivers the finished product—all 5000 copies—and there are ample grounds for rejection: the photos are too dark (screened holes plugged from too much ink), copies too light (not enough ink), edges trimmed out of square, etc. Despite the mess, you have to distribute, because the articles are time-sensitive!
- Photos: uninteresting, small, out of focus, underexposed or overexposed.

Of these problems, one that can "make or break" you is the one of photos.

Writers, being "verbal" rather than "visual" people, find it easy to rationalize the poorly taken photograph by exclaiming, "But the story is so *fantastic*. With that neat headline it doesn't matter what the photo looks like. I'll just make the picture small and no one will notice it's out-of-focus." We've all seen such results: unrecognizable, postage-stamp-size photos drowning in an ocean of text. The truth is that, no matter how well written your story is, a large, clear, well executed photo will improve it.

Two problems crop up regularly in photography: the poorly executed photo (out-of-focus, underdeveloped, etc.) and the boring photo that says nothing (the handshake from the president, the stone-faced group snapshot, etc.). First, poor execution can be solved by *getting to know your camera system*. Practice with it; become an expert on controlling the depth-of-field and the shutter speed; learn about the various kinds of film and how to take photos in different situations and lighting. When operating your camera becomes second nature, you're then free to concentrate on the second problem: creating photos that will grab your readers' attention and get them to read your wonderfully written story. A few suggestions follow.

Move in Close

Don't be afraid to move in close to your subject. Figure 4 gives an interesting perspective to the violinist, eyes closed in deep concentration. Figure 5 illustrates the photographer's position to get the shot. Figure 6 is a snapshot of the violinist that conveys little meaning to the reader.

Try for the unusual angle. Photographers should be ready to crouch, kneel, even lie on their backs to capture the angle that will grab the reader's attention. Such are the photos in Figures 7 and 8.

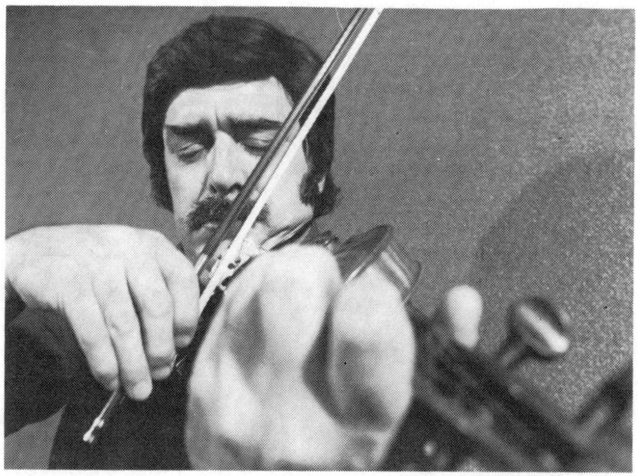

Figure 4. Some unusual effects of focus/soft focus can be achieved by bringing the camera close to the subject.

Figure 5. Close in, the camera caught this concert-violinist-turned-computer scientist as he fiddled his way through the Mendelssohn Concerto.

Figure 6. A nice friendly snapshot that conveys little meaning. Compare this with Figure 4.

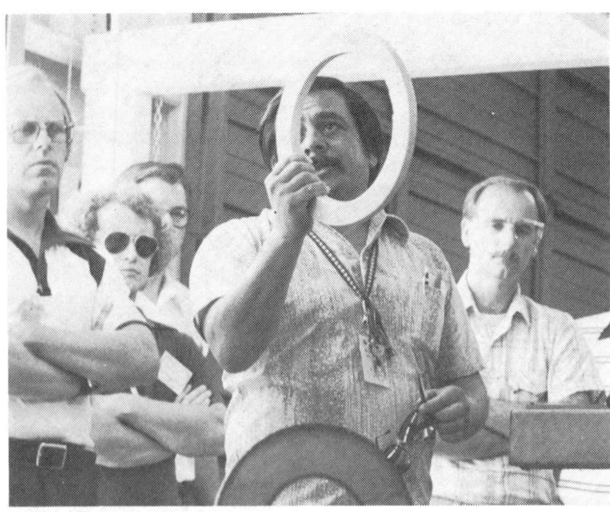

Figure 7. To capture this angle, the photographer sat on the cement floor, his legs cramped under a low iron table.

Figure 8. Another view of how the reporter/photographer captures the story in a picture.

Avoid the Stand-Up Award

Avoid, if possible, the standard stand-up-award shot unless you can make the photo come alive with facial expressions other than "ho-hum" or stage fright. Such a photo is Figure 9. Compare this with Figure 10. Here is an award that not only sent the audience at the presentation into fits of laughter, but will elicit some chuckles from the readers.

Figure 9. Typical of many award-type shots, this lacks reader interest. Compare this with the award being received in Figure 10.

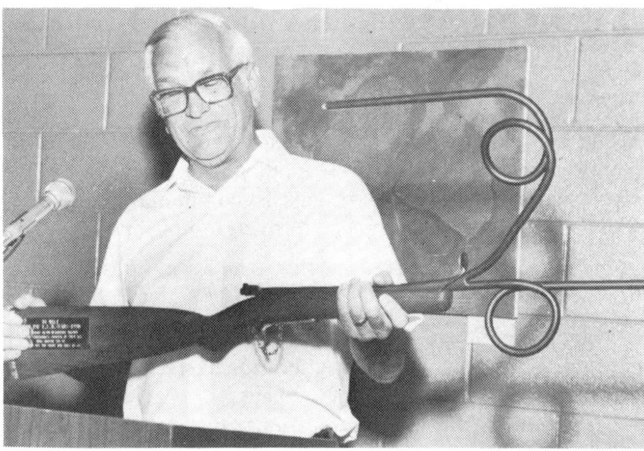

Figure 10. Not only the award, which is in itself unusual, but the look of mixed reactions (skepticism and amusement) on the face of the winner, make this photo worthy of publication.

Look For Human Interest

Sometimes reluctant to shoot up an entire roll of film, photographers using 35mm cameras often miss that special moment. Remember, film is cheap. Don't feel badly if you "blow" 32 exposures out of 36; of the four that are usable, you may get one like the 25-year celebration photo in Figure 11.

Figure 11. Try to bring photos alive with human interest. At an award ceremony, the camera caught this winner (r) in an emotional moment as his wife came from the audience to grasp him around the neck in an affectionate embrace.

Build Incongruity Into Your Photographs

Lately, we've sought to bring incongruity into our ENGINEERING NEWS photos. Incongruity brings tension into a photograph; it grabs the reader's attention and at the same time helps to tell the story. Cover photos for our recent issues illustrate this principle. Figure 12 is the cover from an issue featuring an article on magnetic mirror machines for fusion research. Focusing on the idea of the *mirror* machine, we posed people (involved with building the magnet) atop and in front of the huge magnet. We then achieved a mirror image by printing the photo backward and affixing it to the bottom of the original photo.

Figure 12. To bring incongruity into a group photograph, we printed the original photo backward and affixed it to the bottom of the original to achieve the mirror effect.

In Conclusion

Putting together an in-house newspaper is an experience where you learn as you go. The proposal is a good place to start; it will help you visualize what your publication's thrust will be. Then go to it! In getting started, you'll probably quake at taking photos, agonize over headlines, struggle with the text and differ with your management. But when you see those issues roll off the press, and your supervisor wanders in a few days later to say, "Hey, I've got a great idea for an article...," you'll know your hard work has been recognized.

[1] Reference to a company or product name does not imply approval or recommendation of the product by the University of California or the U.S. Department of Energy to the exclusion of others that may be suitable.

ETHICS AND MEDICAL WRITING: A PROLEGOMENON

Anne Hudson Jones
University of Texas Medical Branch

Anne Hudson Jones
Inst. for the Med.Humanities
Assistant Professor
Univ. of Texas Med. Branch
Galveston, Texas 77550
(713) 765-2376

Ph. D., University of
North Carolina

Recent concern among members of the Society for Technical Communication about a professional code of ethics for technical communicators prompts this paper. My purpose is to examine those complexities inherent in the profession of medical writing that will complicate efforts to formulate a code of ethics for medical writers. I distinguish among different groups of medical writers, each with its own specific ethical dilemmas and professional societies, as necessary groundwork for discussion of a code of ethics for medical writers in general. This introduction to ethics and medical writing may be helpful in discussions of professional codes for other technical communicators.

INTRODUCTION

A 1980 issue of Technical Communication, the journal of the Society for Technical Communication (STC), features a special section on ethics (1). In her introduction to the section, Mary M. Schaefer, an associate editor of the journal, traces the background of the Society's concern with ethics. She reports that as early as 1958 the Society adopted "Canons of Ethics," but that gradually distribution of the Canons was discontinued and eventually the Society's committee on ethics was disbanded. Only after Watergate and the ensuing scandals in some professional organizations did STC decide to appoint an ethics committee again. The STC "Code for Communicators," the product of that post-Watergate ethics committee, was adopted in 1978. But because the 1978 Code does not give specific guidance for communicators who confront ethical dilemmas as they practice their profession, some members of the Society want a new STC code (2). The three articles that follow Schaefer's introduction discuss ethical problems of technical communicators and call for more discussion of professional ethics, especially at STC program meetings.

I would like to think of this paper as one answer to those calls and as part of what I hope will be continuing discussion of ethical problems in technical communication. My focus will be on ethics and medical writing, a subset of technical writing. The current national preoccupation with ethics--the "Great National Ethics Debate," as Frank Radez calls it (3)--is even more intense for medicine than for most professions, with corresponding fallout for medical writing. For example, in 1979-80 alone, The Hastings Center Report published an article about the ethical responsibilities of biomedical journal editors (4); Medical Communications, the official publication of the American Medical Writers Association, published an article on morals and medical writing (5); and CBE Views, the journal of the Council of Biology Editors, published an article about ethics and publication (6).

My purpose in this paper is to outline and briefly examine those aspects of the emerging profession known as medical writing that will complicate attempts to formulate any uniform code of ethics for medical writers. Medical writing is not what it used to be, and medical writers are no longer exclusively physicians who write. The enormous growth in the health care industry in the past few decades has produced several new corollary professions, among them medical writers as well as physician's assistants and nurse practitioners. I will try to distinguish among significantly different groups of medical writers; point to the most important ethical dilemmas each group faces; and identify professional societies of each group. This scrutiny is essential groundwork for serious discussion of professional ethics for medical writers.

MEDICAL EDITORS

At one end of the spectrum of medical writers are medical editors. These may be journal editors or authors' editors; they may be physicians, more likely so in the case of journal editors, or nonphysicians, usually the case of authors' editors. To those who would argue that medical editors are not medical writers, I would respond that most editors also write. And I would point out that one of the five divisions of the American Medical Writers Association (AMWA) is the Editors Section, and one of the four divisions of the 28th International Technical Communication Conference is the Editing and Writing stem. Medical editors belong to both AMWA and STC and are as bound by any professional codes of those organizations as the rest of the members are. But in addition to these two organizations, medical editors have one of their own, the Council of Biology Editors (CBE), which admits to membership only those medical writers whose principal responsibility is editing manuscripts to be published in biological journals. In the past, CBE has issued recommendations for its members about important ethical issues.

Journal Editors

In her book The Scientific Journal (7), Lois DeBakey outlines responsibilities of editors, reviewers, and authors. But references under "Ethics," "Ethical," and "Ethically" in the index of that book are almost all to one chapter, "Ethical experimentation and the editor." That chapter focuses on the responsibility of editors of biomedical journals to evaluate articles submitted to them according to accepted ethical standards of the biomedical community. As the proportionate emphasis in DeBakey's book suggests, whether or not articles reporting unethical research should be suppressed is currently the most important ethical question for editors of biomedical journals.

The question is already many years old. An attempt by CBE to resolve it by publishing recommendations for CBE members to follow was not entirely successful. That attempt may be instructive, though, for those interested in professional codes that would offer specific guidelines to resolve ethical dilemmas. The Committee on Editorial Policy of CBE first published its recommendations in 1970 (8). The most basic of those recommendations were that editors subject manuscripts to ethical review and that editors require authors to submit proof that an institutional review board (IRB) had approved their research. In 1980, Yvonne Brackbill and André E. Hellegers reported on a 1977 survey designed to determine how many of 138 medical journals followed the 1970 CBE recommendations. Of seventy-five editors who responded to the survey, only twenty-seven said they instructed reviewing editors to evaluate articles on the basis of ethical criteria and only sixteen said they required authors to submit proof of IRB approval of their research. Yet fifty-eight agreed that editors should refuse to publish unethical research. There is, as Brackbill and Hellegers point out, a troublesome discrepancy between editors' attitudes and their behavior (9).

Journal editors set standards for other medical writers because, as gatekeepers, they control access to the publication that is so necessary for their colleagues' professional success. Thus, their inability to resolve their own major ethical dilemma may not be reassuring to other medical writers who look to them for guidance. But the journal editors have shown no lack of concern, and their reluctance to act as censors--even of unethical research and even at the behest of their professional organization--is itself reassuring.

Authors' Editors

Authors' editors, one of the fastest growing groups of medical writers, may serve as editor for a single author (in which case they are author's editors) or as editor for a group of authors, often in a university department. They wield far less power than journal editors, and correspondingly little attention has been given in print to their specific ethical problems. Judith Gunn Bronson's article and presentation for this session is an attempt to fill that breach (10). Because authors' editors are in an intermediary position between the author and the journal editor, resolution of their dilemmas often requires more negotiating skill than philosophical insight.

Although eligible for membership in CBE as well as in STC, authors' editors are more likely to belong to AMWA, where their numbers increase every year. Organized in 1940 as an association of physician-writers, AMWA now includes more nonphysicians (70%) than physicians (11). AMWA's nonphysician members are beginning to think of themselves as the new professional group known as medical writers, but unlike AMWA's other members, they are not bound by the physicians' professional code.

PHARMACEUTICAL COMPANY WRITERS

By far the largest concentration of medical writers working in business are the pharmaceutical company writers. The most important ethical issue these writers face may be the most difficult and complex of all dilemmas for medical writers. That issue has been identified by Tabitha M. Powledge, an associate for biosocial studies at The Hastings Center, as the double-agent problem: "a real problem faced by those who write with two aims: one to impart information and another to sell the product" (12). In short, pharmaceutical company writers serve at least two masters: those whom they would inform, such as physicians, patients, and the general public; and the company for whom they work, whose goal must ultimately be to sell products. In addition, these writers are often by necessity writing for government and, thus, are further pressured by what Lawrence Bachorik, in his paper for this session, has called the "bureaucratic imperative" (13).

Although pharmaceutical company writers experience the double-agent dilemma more acutely than other groups of medical writers do, the dilemma itself is not unique to them. Many of the ethical problems Harley Sachs poses for STC members in his article "Ethics and the Technical Communicator" (14) are, at base, double-agent problems. It might appear, then, that the ethical dilemmas of these medical writers do not significantly differ from those of other technical writers. But Powledge raises an important question: whether or not medical writers "have moral obligations that transcend those of other writers" (15). If they do--and they do when what they write about has more serious potential consequences for people's lives than what other writers write about does--then codes of ethics for medical writers would have to reflect those greater moral obligations.

Most pharmaceutical company writers probably do not qualify for membership in CBE. Some may belong to STC, but most join AMWA, which is sustained by corporate memberships of pharmaceutical companies. The companies also help support the AMWA annual meetings and supply the funds for their writers to travel to the meetings. The double-agent dilemma may even extend to AMWA itself. If the organization were to establish a code of ethics that placed its members in explicit conflict with their companies, the companies might withdraw their support. Like authors' editors, pharmaceutical company writers have increased in remarkable numbers over the past few decades. Most are nonphysicians and are trained on the job, by their companies. Guidelines of one company may vary from those of another--for ethical considerations as well as for stylistic ones. Whether a professional code could resolve the double-agent problem for these writers is highly questionable.

MEDICAL JOURNALISTS

As should by now be obvious, the term "medical writer" is not an exact analogy to the term "science writer." By "science writer" most of us mean someone, usually a nonscientist, who writes about science for the general public. Some science writers are reporters for newspapers; others write for magazines, either as staff writers or free lancers. By "medical journalist" I mean those medical writers who are analogous to science writers.

Medical journalists, like science writers, are translators or intermediaries between physicians or biomedical researchers and the general public. Although their ethical problems correspond more closely to those of science writers and journalists than to those of medical editors or pharmaceutical company writers, medical journalists have greater moral obligations than other journalists do unless those other journalists write about subjects that have equally serious effects on people's lives and health. Lois DeBakey's and Selma DeBakey's article on media and medicine, in the Encyclopedia of Bioethics, reviews the development of ethical concerns in journalism and points out the special ethical problems of medical reporters. Premature publication, invasion of privacy, and sensationalism are the most prominent (16). Of the three, premature publication is the one that is currently receiving the most attention. Whether or not new medical information should be published for the general public before it has been published in professional journals is the question most debated. That debate pits one group of medical writers against another: journalists against journal editors.

Like other journalists, medical journalists have been given some ethical guidelines in the "Code of Ethics or Canons of Journalism" adopted by the American Society of Newspaper Editors in 1923. If they belong to the National Association of Science Writers, medical journalists have additional guidance in that association's 1960 code, "Science Writer Ethics" (17). Many free-lance medical journalists belong to AMWA, and some may belong to STC. Potentially, then, medical journalists might find themselves with four professional societies' codes of ethics to follow.

CONCLUSION

Within the broader classification of "technical communicator," "medical writer" appears to be a small, homogeneous subset. The prospect of drafting a professional code of ethics for medical writers sounds correspondingly less awesome than the prospect of drafting such a code for the larger group, technical communicators. Yet examining the subset "medical writers" quickly reveals several significantly different component groups, each with its own specific ethical dilemmas and professional societies. Optimism about drafting a code of ethics that would be useful to all medical writers fades quickly. But as discouraging as it may at first seem, this introduction to the complexities inherent in considerations of ethics and medical writing is intended as the basis for further discussion. I hope the groundwork is useful, not only for medical writers but also for the larger group of technical communicators to whom medical writers belong.

REFERENCES

1. "Special Section on Ethics." Technical Communication, 27: 3,4-12 (1980).
2. Schaefer, Mary M. "Introduction." Technical Communication, 27: 3,4 (1980).
3. Radez, Frank. "STC and the Professional Ethic." Technical Communication, 27: 3,5-7 (1980).
4. Brackbill, Yvonne, and André E. Hellegers. "Ethics and Editors." The Hastings Center Report, 10: 2, 20-22 (1980).
5. Powledge, Tabitha M. "Morals and Medical Writing." Medical Communications, 8: 1, 1-10 (1980).
6. Renkin, Barbara Z. "Ethics and Publication: The Journal as Mentor." CBE Views, 2: 4, 2-7 (1979).
7. DeBakey, Lois. The Scientific Journal: Editorial policies and practices. St. Louis: C.V. Mosby, 1976.
8. Committee on Editorial Policy, Council of Biology Editors. "Ethical Experimentation and the Editor." Council of Biology Editors Newsletter, pp. 17-19 (July 1970).
9. Brackbill and Hellegers, p. 21.
10. Bronson, Judith Gunn. "Prevention of Donkeyism: The Role of the Author's Editor." Proceedings, 28th ITCC, Pittsburgh, May 20-23, 1981.
11. Manheimer, Ann. "American Medical Writers Association." Association & Society Manager. 11: 2, 44-47 (1979).
12. Powledge, pp. 6-7.
13. Bachorik, Lawrence. "Medical Writing: Satisfying Your Audience, Your Conscience, Your Human Research Committee, and the Feds." Proceedings, 28th ITCC, Pittsburgh, May 20-23, 1981.
14. Sachs, Harley. "Ethics and the Technical Communicator." Technical Communication. 27: 3, 7-10 (1980).
15. Powledge, p. 4.
16. DeBakey, Lois, and Selma DeBakey. "Media and Medicine." In Encyclopedia of Bioethics, vol. 1. Ed. Warren T. Reich. New York: Free Press, 1978, pp. 180-188.
17. DeBakey and DeBakey, pp. 180-181.

MUSINGS OF AN ENGINEERING PROFESSOR ON LEAVE AS A TECHNICAL EDITOR

Hugh F. Keedy
Lawrence Livermore National Laboratory
Livermore, CA*

On leave 1980-81 from Vanderbilt University,
Nashville, TN

Hugh F. Keedy

Lawrence Livermore
National Laboratory
Livermore, CA 94550

Technical Editor

East Tennessee Chapter member

Properly planned arrangements in which a faculty member spends a year working in a research or industrial editorial group can prove to be advantageous to both the individual and the employer. While the faculty member is learning by being part of a working group using current methods, the employer can use talents of the faculty member to develop or upgrade selected parts of training programs, or otherwise enhance the operation of the employer's business. This paper describes one arrangement of this type, and includes personal reflections of an engineering faculty member on leave as a technical editor.

INTRODUCTION

At the time of this writing, I have completed a little over half of a sabbatical year leave from Vanderbilt University. This paper focuses on how arrangements were made with the Lab for spending a year in a technical editing group, and describes some of the assignments and opportunities of the first six months. Although some of my personal reflections are included here, overall appraisals of the year's experience will be delayed until the presentation at the annual meeting in May, at which time the year at the Lab will nearly be complete.

In considering a leave such as the one that will be described, several questions must be answered by the individual as well as the employer. Can a professor of engineering or science find growth opportunities and rewards in a sabbatical year as a technical editor? What motivations are there to work in an editorial division of a large research laboratory or industrial group? Can both the professor and the employer profit from an arrangement of one year? What understandings and obligations should each party to such a relationship have? What is expected of each party?

After a half year as a technical editor, I can say that the experience has proven to be personally interesting and useful, and I believe the same can be said on behalf of the Laboratory.

PERSONAL BACKGROUND

My personal interest in technical communications might be viewed as a response to student needs. After 20 years as an engineering faculty member at Vanderbilt University teaching a variety of mathematics and engineering courses, I became involved several years ago as a planner and later as coordinator of the freshman year program for engineering students. Part of the design course that we developed required a team effort that terminated in a project report. As the course evolved, local engineers and managers who acted as judges of the team reports increasingly emphasized the need for better written communication skills. Efforts were made to increase the writing instruction and requirements in the design course.

After six years, student pressures were building for a junior-senior level course in technical communication. Students, often as a result of summer employment in engineering related jobs, recognized the need for more instruction in communications of the type found in industry and management. However, there was no course in technical communication available on campus, and the English department was not interested in developing a suitable course.

Four years ago, primarily motivated by a sense of need more than personal expertise, I developed and began co-teaching a course in technical communications for upperclassmen. The course has been offered once each year since, with classes limited to 25 students.

*Work performed under the auspices of the U.S. Department of Energy by the Lawrence Livermore National Laboratory under contract No. W-74095-ENG-48.

Although the course has gone well each time, my own confidence in the "reality" of what we were doing in the course was low, since I had had no actual experience in writing for industry or government. Coupled with the fact that I had not had a leave from teaching for some years, I felt that a year in an industrial or research writing environment would be helpful to me as a teacher of both engineering and communications.

HOW ARRANGEMENTS WERE MADE

At a conference sponsored by the East Tennessee Chapter of the STC, I met a speaker from the Lawrence Livermore National Laboratory's editing staff. As a result of our conversation, I wrote to the manager of the editorial division about the possibility of spending a year at the Lab. Fortunately, my idea fitted well with some thinking the editing organization had been giving to increasing their contacts with academia, and with engineering schools in particular. There was a feeling that editing should also be a teaching process, and that part of maintaining a good editing group is getting new ideas. My background as an engineer and teacher with editing interests matched well. With mutual interests established, planning for my year at the Lab proceeded rather rapidly. A visit to the Lab and the Technical Information Department (TID) was made in January 1980, and the process of obtaining a security clearance was begun shortly thereafter.

A temporary appointment of one year was arranged. The Lab was insistent on a one year limit because one employed for more than a year must be classified as permanent. From my point of view, for tax purposes my appointment could not exceed one year.

Because of extra expenses of moving and maintaining two residences for a year, financial arrangements were discussed carefully. Final arrangements were for the Lab to support travel and limited moving expense both to and from Livermore, to pay an agreed salary, and to extend standard sick leave and vacation benefits for the year. Attempts to work out a method whereby the Lab would pay Vanderbilt directly, with my salary to continue through them, did not prove to be feasible. Such a plan does have certain advantages if it can be worked out. Hospitilization and retirement benefits were continued through Vanderbilt. Two conferences related to engineering education that I was previously committed to attend (as part of my participation as an officer in a division of the American Society for Engineering Education) were supported by the Lab with travel expenses and compensated time.

WORK ASSIGNMENTS

When I arrived at the Lab in early July, I was assigned to a group of about eight editors who edit various types of reports from several areas of basic science. The group leader provided me with references and standards of editing used at the Lab, and with a minimum of additional orientation gave me a manuscript on computer graphics to begin editing. As I began this first editing task, I must say there was a certain level of apprehension about how well my editing would fit the standards of TID. Fortunately, the author was great to work with and all went well between us. At the Lab, after a report is printed it is reviewed by the group leader and feedback is supplied to the editor about the quality of editing. My confidence level was raised considerably by feedback on this first report. Although there were some comments and suggestions, I felt a sense of relief to know that my teaching practices were both workable and valuable under actual editing conditions.

Rather than being assigned to longer reports that would have confined my attention to one subject or to one specific group within the Lab for possibly the whole year, efforts were made for me to work with a variety of reports in a wide range of subjects. Subjects have included topics in machine technology, earthquake reports, geology reports, environmental impact reports, a modeling technique reported in a Lab journal, and reports on brittle fracture and nuclear plant safety to be submitted to the Nuclear Regulatory Commission. Three articles have been edited for submission to professional journals for publication.

The variety of topics and formats has provided valuable examples for use in both my engineering and communications courses when I return to teaching. At the same time, my engineering and teaching background allowed me to make contributions to several of the reports that few of the other editors could. In several of the reports, I raised technical questions or pointed out errors in technical content. In some cases, changes that resulted were both necessary and significant.

TRAINING OPPORTUNITIES

The opportunity for participation in training programs has been excellent, and should be a part of any leave arrangement with an employer. In coming from an academic atmosphere to a government funded environment, I found myself amid a vast range of research efforts about which I knew little. Fortunately, a number of short tours of research facilities are offered as a part of the program for summer employees. I was able to go on tours of research facilities in laser fusion, magnetic fusion, solar research and other areas. These tours added directly to my overall understanding of the programs at the Lab and indirectly to my ability to prepare better reports after having seen the actual facilities. In fact, before beginning actual editing, I find it very helpful to ask the author to give me a short orientation tour of any facility or equipment that is featured in a report.

I quickly found that opportunities for training or increased knowledge in a number of areas far exceeded the time I could set aside for them. Numerous college level courses in technical and technically related subjects are available in class format, on closed circuit TV, and on videotape. Also, approximately 15-20 seminars or lectures each week are sponsored by divisions within the Lab and are open to most Lab employees. Topics vary widely and interest levels range from highly technical discussions to project reports given in lay language. Unfortunately, I have not taken advantage of as many seminars and courses as I probably should have.

One of my major objectives of working with an editing group was to gain experience in word processing using modern equipment. While I was familiar with word processing on the computer at Vanderbilt, my experience with it was rather limited. When I reached the Lab, I immediately began using a localized system consisting of ten terminals, a CPU, and two printers. With my prior exposure to word processing and some reading of the system's reference guide, I soon felt comfortable with basic manipulations of text. Unfortunately, it was several weeks before I was able to attend a two-morning class on the use of the system. While much of the instruction was at a beginning level, I did pick up a number of valuable features of the system that have been useful since.

TID offers two courses in preparing visual aids for use with oral reports, another course in oral presentation, and also a course in written communication. Courses, which range in length up to five half-day sessions, are available to all Lab employees. As soon as possible, I attended the two courses on preparing visual aids. While the basic principles are similar, each course offers a different approach to the actual preparation of aids, and to organization and presentation of material. In fact, these courses made me reconsider some of my own ways of presenting material to classes. For example, my personal tendency of putting too many key points on one overhead projector slide was brought to light.

CONTRIBUTIONS TO THE LAB

Partly through personal interest in the content and operation of the course in written communication, and partly as a resource person, I attended the course the last time it was offered. As it happened, many of the 20 participants expressed interest in memo writing, a topic not formerly included in the syllabus. Having taught sessions on memo preparation in my course at Vanderbilt, I was able to quickly put together and present a 20 minute discussion of some principles of memo writing. I later wrote some notes that course instructors might use if interest in memos is expressed when the course is offered again. Also, in a debriefing session of instructors following the course, I was able to contribute some feedback and suggestions.

An engineer working as an editor can make definite contributions to editors with less background in engineering. My combined engineering and teaching background was a new resource within our editing group. The group leader and I discussed ways that I could help improve editing of technical material. One result has been a series of "Engineering Notes" that is distributed throughout TID. Content focuses on terminology and principles fundamental to engineering, and points out relationships that editors may not be familiar with. Each of the ten sets of notes prepared thus far consists of two single spaced pages. Topics discussed have included dimensions and units, stress and strain, forces, engineering design principles, dimensionality and dimensionless parameters, measurement techniques, principles of wave motion, types of experimental data, memo preparation, and fundamental equations frequently encountered in engineering. Editors are encouraged to suggest topics for additional sets of notes, and to pass on to me terms that they encounter and are not familiar with.

Another positive outcome of my work as an editor has been good relationships with the authors. Authors at times feel defensive about having changes made in their manuscripts other than corrections in grammar and possibly organization. It appears that my technical background and experience as a teacher add a dimension that helps establish good relationships with authors. In working with some authors, I initially feel that they are somewhat aloof and are expecting to take an adversarial position to the editing I will be doing. In such cases, I take care to establish technical points of contact that can be discussed easily and which make the author feel that I have his best interest in mind when I make suggestions--which are always presented as suggestions rather than decrees for change. Negative attitudes vanish as it becomes evident that an editor might be helpful not only in suggesting improvements in organization and clarity but in expressing and evaluating technical content. The value of a sound background in technical terminology and basic engineering principles is quite obvious in many cases, even though my knowledge of the specific subject may at times be almost nonexistent. Needless to say, the feeling of being able to discuss content of an unfamiliar paper on a reasonably sound technical basis has been uplifting to my personal self-esteem as an engineering educator.

SUMMARY

As I review the first half of this sabbatical year, I find that the experience has already been rewarding to both the Lab and myself, but that there are still some areas where we can be of further benefit to each other before the end of the year. One area for investigation may be the close similarity between teaching and working with authors in such a way that an author's current editing experiences may be carried over to improve his or her future manuscripts. The Lab is well pleased with my editing abilities, contributions to their training efforts via the sets of engineering notes, and availability as a resource person to the written communications course. Personally, I have been refreshed by working with engineers and scientists on real frontiers in technical areas. Ideas for improving my courses at Vanderbilt have been numerous. My experience in editing has brought out many points I never could have gotten from texts, and the confidence I have gained is immeasurable.

From this experience, I can strongly recommend similar arrangements between faculty members and employers as being of mutual benefit. Just be sure:

1. To start negotiations early.
2. To clearly define expectations and financial arrangements.
3. To carefully lay plans that will both use the faculty member's talents to the best advantage of the employer and provide the faculty member with the experience, new perspectives, and a break from the academic world that a sabbatical is intended to provide.

If properly executed, the result is a win-win situation in which the employer and the faculty member, as well as the faculty member's home institution, all come out ahead.

HELPING STUDENT WRITERS BE THEIR OWN EDITORS

Bob Mandino
Purdue University

Bob Mandino

Purdue University
West Lafayette, IN 47906

Technical Writing Instructor

Any writing fails when the writer neglects his audience and ceases to communicate. Helping student writers realize that they are in fact communicating real thoughts to real persons -- and not merely generating words in a vacuum -- is the end of technical writing instruction in the college classroom. This paper, based on my work in Purdue University's Technical Writing Program, addresses several important pedagogical problems unique to the classroom, an environment where students with a wide variety of abilities, backgrounds, and interests are brought together. Some of the methods found successful for sharpening self-reflexive editing in students are detailed herein.

THE PAPER

Let me start by stopping to set some limits for this paper.

First, I consider it to be a case study of sorts. It describes the work of one instructor, me, in a particular educational backdrop, the Technical Writing Program at Purdue University. But it also addresses a composition problem common to any type of writing, albeit especially egregious to the technical document: the lack of thorough editing and consideration of the audience.

Second, my concerns lie with the basics of grammar, far more than with formats, in paper writing. Too many upperclassmen write too poorly, and can't be expected to run when they cannot walk.

I will be outlining some straightforward teaching strategies that have proven successful in my classes, hoping they might aid those of you facing similar teaching situations in the university, the high school, or the corporation.

THE COURSE

Students in Purdue's "English 421: Technical Writing" meet three times per week. Two of the meetings are large lectures, called General Assembly Sessions (GASs). Program Director Mrs. Juanita Dudley and a doctoral student cover material in Mrs. Dudley's Text-Workbook For English 421. The third meeting is a Critique and Question Session (CQS). The approximately three-hundred students enrolled are divided into groups of roughly twenty students; each of these groups is assigned a graduate teaching assistant. These instructors handle anywhere from twenty to sixty students per semester, depending on their employment status with Purdue.

Both the GAS and CQS meet for the first eight weeks of the semester, and once about a month later to check the final report status of each student. (The final report is an extensive paper on a topic of the student's choosing, obeying outlined formats and utilizing the writing skills taught in CQS. It is worth fifty percent of their final grade in the course.) The remainder of the course is given to conferencing individually with students.

Mrs. Dudley allows a comfortable degree of latitude in the way her instructors chose to cover the necessary materials in CQS. In my three sections, we correct anonymous copies of each paper assigned in the first eight weeks. There are some special handouts on writing style which we work with, but the concern is largely with criticising papers. And during conferences, I discuss with students their progress on the final report, paying particular attention to grammar and style.

These technical writing classes have a diverse make-up. Majors range from building construction technology to chemical engineering. Some students have had more writing practice than others. Some have a natural bent towards writing — a few will go so far as to admit they enjoy it — while others would much rather run a computer program than correct their dangling modifiers.

Even though the students do share an analytical frame of mind, technical writing classes at Purdue present quite a mix of writing abilities and proclivities.

"EGOCENTRIC" WRITERS

The problem is a unique one. Here are twenty students in a class, all having science and technology backgrounds, with no one common, topical technical or scientific interest to single out and teach around. If I were to concentrate on senior design projects for electrical engineering technologists, for example, the biochemists wishing to learn about journal article writing would be irate — or at least bored silly.

Of necessity I have to dig deeper to some shared roots: they all write in English, and few of them can do it well. That's the common ground I work, trying to remove the stones, and boulders, that block the way to effective communication.

One focus of technical writing instruction in my classes can be put this way: any writing fails to do its job when the writer forgets his audience and thus ceases to communicate. It is this "egocentric writing" which underlies so many writing problems. Students invariably begin the semester with papers that assault their reader. They hurl jargon about, continually skip definitions and cause-and-effect details, and often drop the reader _in medias res_, to name a few of their literary offenses.

But before we can write for an audience, we have to define one. It certainly is not a pat definition, and it sometimes varies between the papers. But as a rule, students aim to write for the layman-executive reader, that imagined person interested in what the student has to say but who, unlike the scientist or technician, does not already know a great deal about the subject. It's a good premise to teach audience awareness and editing from. It demands that students reread, rethink, and rewrite their papers with an eye to English 421's ABCs: accuracy, brevity, and clarity.

Most of my students intend to begin their careers in the mainstream of hard sciences. I try to make the course more relevant by getting them to see why they practice writing for this layman-executive audience. I make it imperative that they realize that if ever asked to draft a report for an employer, they cannot plan to use jargon and hope to communicate with management. A researcher could not write a status report on his work in plasma physics studies, for example, encumbering it throughout with non-transliterated equations and esoteric terminology, and have managerial readers make a decision on the feasibility of continued funding for the project.

That is not to say that all parts of all papers are written for this audience. The comprehensive final reports can be any of the following: a case history or status report, a journal article, a presentation of a design or process, a manual of instructions, a physical research report, or a feasibility study. And there are, quite properly, those sections in each paper class that deal with complex, technical materials. These sections are written especially for experts in the field. We instructors make it clear that these sections are crucial, and demand thorough, accurate recording of data under established formats.

Secondary elements, the various formats, and other mechanical conventions are not taught at any great length. All the necessary details are delimited in the _Text-Workbook_ and are followed by the students in cookbook fashion. What is stressed is the idea that these details cannot be overlooked.

More importantly, we focus on abstracts, introductions, conclusions, and recommendations. It is here that the layman-executive audience needs hard data translated and interpreted in readable prose.

RAPPORT AS GROUNDWORK

Helping student writers become aware that they are in fact communicating, and not generating words in a vacuum, is the end of technical writing in the classroom. I'd like to explain some of the ways I've found successful for engendering editorial insight in this hodge-podge of science and technology majors.

One way to begin is, simply, by establishing rapport. It's a great groundwork asset that we should not slight. From my experience on both sides of the gradebook, the instructors who have given their students most when the course is over are those who diminish, though not remove, the teacher-student barrier. Students seem to benefit more from teachers who establish interpersonal relationships, who can somehow make mundane subjects more interesting, immediate, and even humorous. This is the kind of verbal rapport developed verbally in classroom dialogue when an instructor opens to all questions, and comments, and criticism.

In the evaluation of papers is another chance for creating rapport with students. Their writings are quite often, and quite frankly, inadequate pieces of prose. And they obviously need their errors signalled. But just as necessary as fault-finding is _positive_ feedback, both via the student models critiqued orally in class and in the written comments on their papers. When I "bleed" red ink on their work, as one student described my grading, I take care to include negative criticism _and_ praise. Teachers must be B. F. Skinners. You can't open eyes to obtuse writing by sighting only faults.

An instructor's worst failing is to shut off communication with the students he seeks to help. It's an ironic situation that probably happens too often. Yet by encouraging new writing awareness through rewarding their improvements with verbal and written recognition, learning — rather than a return to poor composition patterns — will occur.

Let me elaborate on the CQS at this point. Each week I select one or two student papers from the sixty I receive, and have copies made for all the students. They are chosen from various competencies. That is, I pick poor, excellent, and mediocre ones. Too many poor papers and most students suddenly start thinking that they write damn well. Too many excellent papers and most students become intimidated. Too many mediocre papers cause boredom. The use of one excellent paper as a model and one poor or average paper as a foil seems to make for the most interesting, lively, and beneficial class discussions.

The CQSs are also helpful for establishing rapport because of the very nature of the critique. Students and their instructor join forces, finding good and bad aspects of the papers under scrutiny. The net result in terms of rapport is that "team grading" gets students on your side.

Some students might argue the judgment calls regarding style and usage, but by getting them to pick out the more obvious errors, they get in tune with similar offenses they might be committing against their readers. They come to realize that the red marks peppering their papers are, believe it or not, there for a purpose other than ego-bruising. They see that they are doing more than sterile homework assignments. And they are more aware of their dual role as audience and editor.

The CQS critique has proven quite successful for a simple reason: examples before one's eyes are easier to learn from than an isolated, hard-and-fast list of dos and dónts. No matter how thoroughly and with how much conviction a teacher stands before a class and lectures on grammar, an example is worth a thousand more words. And its anti-boredom value is high. With any sense of humor and any teacher-student rapport, these paper editing sessions invariably have light moments.

The critique also helps students sharpen an important editorial tool, the ear. They learn to hear sentences that don't read well or make sense and they become especially proficient at punctuating, condensing, catching vagueness, and reorganizing logically. It's a weaning process whose end is enabling student writers to edit by sound and then to revise their papers for the ABCs without my help.

THE MIXED-BAG PREDICAMENT

Corporations, research foundations, the military — all such institutions employing professional technical writers have standards. Manuals delineate how all documents are to be prepared. And the personnel working on the literature are all trained, competent writers.

But how can you find any regularity in the classroom? How do you set standards to follow and goals to achieve by the end of one semester? It isn't that easy.

The main impediment is the mixed bag: a variety of student interests and a variety of writing abilities in one class. Some students cannot seem to remember when to use "its" and not "it's," or how to use a colon, or where to begin a new paragraph. Other students are more advanced and compose papers that are mechanically quite good — they need advice on stylistics.

Getting most students to recognize common mechanical errors can be an accomplishment for one semester. But if you work always at this level, the more skilled students will leave the course no better than they enrolled. You must attend to their more advanced needs.

AN ANSWER: THE SELECTIVE CRITIQUE

Once again, the very nature of the CQS can offer at least a partial solution to the problem.

Because the class is entirely given to orally dissecting papers, I can focus on the selective critique, a homespun term that means tailoring questions to suit the different needs of different students. After a couple of weeks, the abilities of students start to become apparent, and I begin working in class on a variety of levels. In this way, nobody idles or gets frustrated for very long. Every student must offer criticism.

There is an unskirtable problem, however. When you ask Average Al about the use of a comma, Bill Skill nods off. And when you ask Bill about the advantages of a certain organization for a certain paper, Al gets lost. There seems to be no way around this classroom crisis.

A teacher must not succumb to temptation and concentrate on either polishing the advanced writers' skills or on changing a 400-level course into a remedial one. Likewise, neither extreme can be sacrificed in the interest of catering to mediocrity. All abilities have to be tapped by discussing both style and mechanics in the classroom.

ANOTHER ANSWER: CONFERENCING

Each student's unique needs can best be filled in the conference. The four meetings I schedule, lasting from fifteen to thirty minutes each, necessitate individualized teaching. These meetings are designed to keep tabs on the progress of the final report. In some conferences, we review segments of papers previously turned in and corrected. In others, the student brings in new pieces, about two-hundred words each, and we correct them together.

Most students get more out of these conferences than from crowded lecture halls. One-to-one work on individual problems is a rare opportunity for students in a university the size of Purdue.

They come prepared with questions on their writing, and they have to answer mine — it's an intensive half-hour.

In keeping with the nature of the course, questions on format are kept to a minimum. I simply warn them about the necessity of plugging into the primary and secondary element formulas in Mrs. Dudley's Text-Workbook. These conferences are, above all, further opportunities to work on the universals of writing mechanics and on editing for the audience.

BEYOND THE CLASSROOM

"You are enrolled in 'English 421: Technical Writing.' That's ENGLISH, and that's FOUR-twenty-one. And while you will leave Purdue and probably never use identical formats when preparing technical documents for your future employers, you can't leave here and change languages. English is the common denominator in technical writing, and it has basic rules. Which you will learn in here."

That's the preface I deliver to students at the first CQS.

From Day One, then, I impress upon students that 421 is a practical course. Whenever they compose documents for their employers, they'll use what they learn in the classroom. If they remember that they are writing for real, live people, the result will be self-reflexive editing and a more successful composition process.

And when their final reports come in at the end of the semester, they are usually quite good. (There are always "The Untouchables" who seem to refuse to learn from their mistakes and whose final papers show it. Luckily, for my own self-esteem, they are always few in number.) The steady improvement shown in nearly all students culminates in the final report.

Rough drafts have been edited. And the audience, me, has been spared mounds of confusion.

ORGANIZING AND PREPARING A DATA QUALITY PLAN

Michael E. Manni
Communication Consultant

Michael E. Manni

Communication Consultant
Route 2, Box 182
Hudson, WI 54016
(715) 386-5980

B.A., Business and Economics, University of Minnesota, Duluth, 1965.
Publications: "Quality Assurance; People Meeting Requirements Together," 1979 ITCC
Senior Member, Twin Cities Chapter, STC

The Department of Defense is increasing its emphasis on quality requirements for technical data. Amendment 5 to MIL-M-38784A is a good example. It invokes readability standards upon technical manuals. New specifications have also been written for software. A guiding force through the myriad of specifications is a data quality program. This program assures that all technical data delivered meets contractural requirements. This paper describes how to develop a data quality program in consonance with the contractor's other administrative and technical programs. The interrelationship between the procedures and methodologies of data preparation is explained by text and illustration.

A data quality plan is a useful management tool for any contractor required to prepare and deliver technical data. Applicable technical data includes hardware/software manuals, provisioning technical documentation, test reports, maintainability analyses, microfilm, and associated data. The emphasis on such data is increasing within the Department of Defense, with the subsequent challenge falling upon the contractor to prepare acceptable technical data in accordance with schedule requirements.

This paper considers several typical elements of technical data quality plan. They are as follows:

* *Organizational Chart*
* *DD Form 1423, Contract Data Requirements List*
* *Recordkeeping*
* *Corrective Action*
* *Validation or Verification*
* *Modifications - Changes or Revisions*
* *Applicable Documents*

ORGANIZATIONAL CHART

If a company decides to document its technical data quality planning, the next decision that must be made is the format and content. The format must be designed to fit the individual company needs. No one is better suited or more qualified to write a data quality plan than the people actually preparing the data. Various levels of approval and control can be applied, but the major design work must emanate from the developers of the data. Another key factor in a data quality plan is the assignment of responsibility. This can be done through text and standard organizational charts. Figure 1 shows a functional and organizational relationship for technical manual development. In this example, quality assurance is shown as a separate and independent function.

The debate over the independence of quality assurance remains unresolved. Size of the company is a major determinant, with small companies often unable to support an independent quality assurance function. It should be noted, however, that federal contracts require two considerations:

1) Quality assurance must be independent of manufacturing, and
2) Quality assurance must report to upper management.

Organizational considerations not withstanding, few people dispute that final acceptance of technical data is a quality assurance function.

As previously mentioned, the data quality plan identifies organizational responsibilities and authorities and outlines events critical to the plan's implementation. That becomes a matter of communication. Many of the technical data quality problems stem from a lack of communication. It is difficult to prepare acceptable technical data without a full understanding of the requirements. The requirements must also be passed along in an easily understood form. A

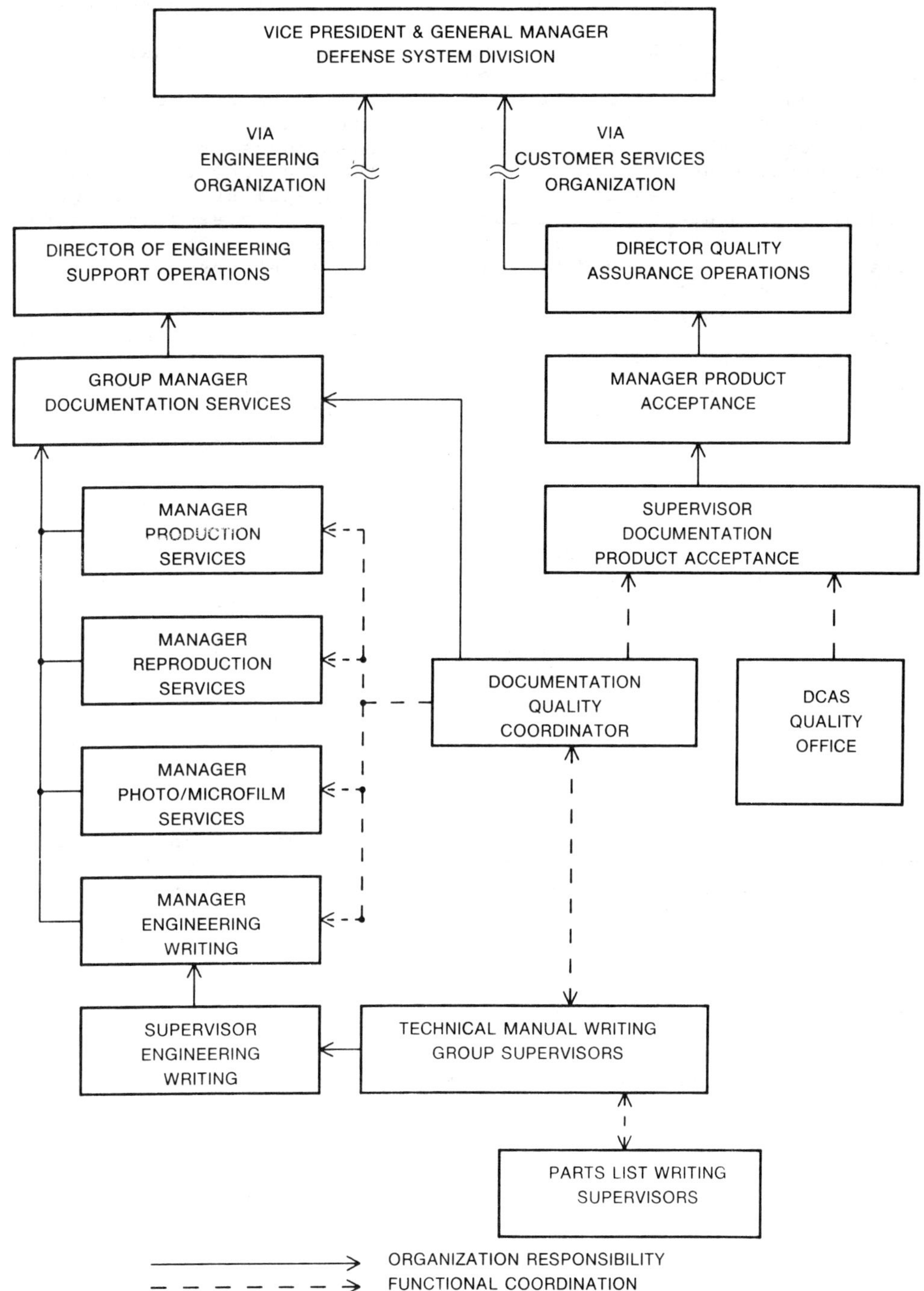

Figure 1. Technical Publications and Quality Assurance Organizational Chart

technical data quality plan describes this system for transmitting information.

CONTRACT DATA REQUIREMENTS LIST

A key item to reference in the data quality plan is the DD Form 1423, Contract Data Requirements List (CDRL). The CDRL, or authorized equivalent, is the sole contractural document listing kinds and amounts of data required. This information, shown on the CDRL, includes the format of the data, type and number of copies, degree of detail required, delivery schedules, and the purpose of submission. Refer to Armed Forces Procurement Regulation (ASPR) Section F200, 1423 for the official description. Of major concern to quality assurance personnel on the CDRL are the blocks which indicate the governing specifications and work statement tasks. This is the nucleus of the CDRL for quality control purposes.

RECORDKEEPING

The body of the data quality plan in a separate section, should identify practices, procedures, and any applicable instruments needed for inspection. The procedures must be unique to the specific data and referenced on quality records. A separate section could also be devoted to the important function of recordkeeping. Figure 2 shows a general-use form, Data Inspection Record, containing key information needed for quality records. Figure 3, the Request for Data Inspection, demonstrates the type of form needed when final acceptance of the data is performed by the customer, These forms become part of departmental historical records providing excellent traceability in future reviews and serve as valuable back-up information for cost proposals and audits. It should also be noted that few better tools exist for instructing new employees than previous inspection records. The entire history of the data can be contained in one folder with the use of typical forms, such as the examples shown here.

CORRECTIVE ACTION

The data quality plan also documents the corrective action program. The plan references procedures which assure the prompt detection, documentation, and correction of technical data problems and deficiencies. Procedures include information such as:

* Identifying and describing problems or conditions which may adversely affect data quality or costs.

* Formally requesting action be taken by cognizant individuals to correct significant problems or deficiencies

* Documenting and reporting to management any problems which may seriously impact cost or schedule.

* Review of corrective measures for follow-up reporting.

VALIDATION OR VERIFICATION

The data quality plan defines validation and verification. Validation is the process by which the contractor tests the accuracy and adequacy of the technical data against the hardware. In verification, this exercise is performed by the Government. The data quality plan outlines responsibility for validation/verification.

MODIFICATIONS - CHANGES OR REVISIONS

Modifications consist of changes or revisions to existing technical data. The data quality plan describes the configuration control of changes and revisions. Key elements of this configuration control are as follows:

* Technical data revision or changes are accurately identified.
* Ensure that no unauthorized modifications are made.
* Verify that all approved modifications are properly incorporated.
* Certify that technical data submitted for validation/verification is the correct version.

APPLICABLE DOCUMENTS

A section on applicable documents forms a part of the data quality plan. Publishing this list of referenced specifications promotes coordination between the responsible technical groups. This section would include documents such as:

MIL-Q-9858	"Quality Program Requirements"
MIL-S-52779A	"Software Quality Assurance Program Requirements"
MIL-M-15071	"Manuals, Technical: Equipments and Systems Content Requirements for"
MIL-STD-480	"Configuration Control-Engineering Changes, Deviations and Waivers"
MIL-STD-831	"Test Reports, Preparation of"
MIL-STD-105	"Sampling Procedures and Tables for Inspection by Attributes"
MIL-M-38784A	"Manuals Technical: General Type and Format Requirements"

The above list is an example of typical reference documents in a data quality plan. The point to be made here

DATA INSPECTION RECORD

1. PRODUCT DESCRIPTION

2. CONTRACT OR PURCHASE ORDER NUMBER

3. SIGNATURE & DATE

4. LOT SIZE

5. SAMPLE SIZE

6. APPLICABLE DOCUMENTS

	8. ITEM IDENTIFIER																7. REMARKS
1.																	
2.																	
3.																	
4.																	
5.																	
6.																	
7.																	
8.																	
9.																	
10.																	
11.																	
12.																	
13.																	
14.																	
15.																	
16.																	
17.																	
18.																	

FIGURE 2. GENERAL INSPECTION RECORD

REQUEST FOR DATA INSPECTION

1. PRODUCT DESCRIPTION	3. SIGNATURE & DATE	6. APPLICABLE DOCUMENTS
	4. LOT SIZE	
2. CONTRACT NUMBER	5. SAMPLE SIZE	

INSPECTION SEQUENCE	SPERRY UNIVAC	CUSTOMER REP.

VISUAL DEFECTS	EVENT	DISPOSITION	INSP.

APPROVED FOR SHIPMENT

DATE	CUSTOMER REPRESENTATIVE	DATE

DISTRIBUTION: 1 - WHITE → PROD. ACCEPTANCE 2. YELLOW → CUSTOMER REP.

FIGURE 3. CUSTOMER ACCEPTANCE FORM

is that publishing your list puts personnel on notice to become knowledgeable in their respective area. It serves as an excellent reminder of upcoming requirements.

Software plays a continually larger role in Government contracts. This shifting emphasis places additional requirements on the contractor for technical data. As previously mentioned, Amendment 5 to MIL-M-38784A illustrates this. In accordance with this amendment, the procuring activity determines a Reading Grade Level (RGL) that is compatible with the reading skills of the target audience. The contractor then must prepare technical manuals that meet the Reading Grade Level. To determine compliance, the manuals are validated using inspection sampling. From the samples, raw data is collected by counting the number of words, sentences, and syllables. An Overall Grade Level (OGL) and the sample Grade Level (GL) are calculated using formulas. The OGL and the sample GL then cannot exceed the predetermined RGL beyond specified limits. If the manuals don't meet these requirements, they must be rewritten to meet the readability standards. Obviously a close working relationship between quality assurance and the preparing activity is necessary to exercise a policy of defect prevention.

The intent of this paper is not to attempt to describe all aspects of or the philosophy behind a data quality plan. It is, rather, to demonstrate the importance of such a plan in the context of software development. The plan recognizes that it is people who implement any plan. A data quality plan, as described here, provides the means for combining the creative abilities of software personnel and management. The end result, for the benefit of the customer, is an excellent product.

PRODUCING MASS VOLUME DOCUMENTATION WITHOUT TECHNICAL WRITERS

Tatsuro Matsuyama and Yasuhiko Kobayashi
Fujitsu Limited

Tatsuro Matsuyama
Manager of Technical
Documentation Department

Yasuhiko Kobayashi
Technical Editor

Fujitsu Limited
17-25, Shinkamata, 1-chome, Ohta-ku,
Tokyo 144, Japan

This paper describes a method for producing mass volume documentation without technical writers.
Three services were implemented:

- printing quality control service
- document editing service
- writer training service

The result was quite successful. Our documentation is now considered to be second only to that of the world's largest manufacturer. Not only was there great progress in quality, but the additional manpower cost is equivalently zero.

INTRODUCTION

Although professional technical writers are considered to be absolutely necessary for good documentation, not every company is in a position to use such writers.

This paper describes a method for producing mass volume documentation for companies who do not have access to technical writers.

Over 64,000 pages of Japanese Language computer user manuals were produced by Fujitsu Limited last year, and that volume is increasing every year. Almost 56,000 pages of these were written by a total of 2,700 different software engineers. They are not professional technical writers. Grammar and style problems are common. It is very hard for them as engineers to write computer manuals which nonengineers (the ordinary user) can understand. This has caused many problems in the past.

Seven years ago, however, the technical documentation department was established, and we found one way to solve that problem without using technical writers.

The result was quite successful. Our documentation is now considered to be second only to that of the world's largest manufacturer. Not only was there great progress in quality, but also the additional manpower cost is equivalently zero.

Three services were implemented in solving this problem:

- Printing quality control service
- Document editing service
- Writer training service

This paper describes the problems, the solutions (three services), and future plans.

PROBLEMS

Development of large scale software systems requires much manpower over long periods of time. But since software is an intangible product, documents are essential for communication between the developers throughout all stages of the development process.

Today we have reached the point where development of software is severely limited by our ability to write development documentation. When those same developers must also write the documentation for the end user, it is, of course, almost impossible to produce good documentation.

The problems are as follows:

- The manuscripts do not suggest the best usage techniques for the computer users.

- The manuscripts do not have correct sentence structures or suitable logic flow to make them easier to read.

- The quality of the illustrations (line printer forms, photographs, etc.) is not good enough for printing. And graphs and charts are drawn so badly that professional illustrators can not correct them.

- Free rewriting by the developers during the printing stage delays schedules and caused cost overruns.

- The variation in workload for the editing and printing processes is too great to be manageable.

SOLUTIONS

Planning, writing, making camera-ready copy and printing were the stages in our publishing process. For each stage we applied the following measures:

- Instituting a printing quality control service for the standardization of camera-ready copy and printing.

- Inserting an editing stage (document editing service) between the writing stage and the stage for making camera-ready copy.

- Instituting a writer training service for the planning and writing stages.

<u>Printing Quality Control Service</u>

The first area we tried to improve was the printing quality. We instituted a service to standardize the camera-ready copy and printing into an easier-to-read format, hoping thereby to standardize the whole documentation process.

We used three standards to control printing quality. First, we established the formatting standards. Secondly, we established standards for general terms and usage in technical documents. Thirdly, we created the Fujitsu Information Processing Terminology standards.

Writers write their manuscripts using the general and technical standards. Editors check general terms and usage with the general standard before typing. Printing companies make galley proofs using the formatting standards.

The resulting higher quality appearance make reading easier and thus even content errors become easier to find and correct. (As an incidental result we have reduced the annual printing cost by more than the manpower cost of our whole department.)

<u>Document Editing Service (reading and checking manuscripts service)</u>

The second improvement we implemented was the establishment of an editing stage. In this stage, customer support engineers (SEs) read every page of the manuscripts, and add comments on the content from a user's point of view, then return the documents to be rewritten.

The major check points are as follows:

- Systematicalness check
- Readability check
- User's point of view check

Systematicalness check

By pretending we were novices in the area being described, we found many problems in the manuals. In particular we found that the manuals were not organized systematically. So we established the following two checks.

- Checking the manuals position in the manual system for that product

- Checking cross references in a manuscript and between mutually interrelated manuals.

Readability check

Since we have been trained as customer support engineers (SEs), we can recognize if the technical content is clear. In particular we check for definition of technical terms, readability of sentences, and clear logic flow. This check often enables us also to discover errors in the technical descriptions.

User's point of view check

We (the documentation department) and representatives of users (customer support engineers) check the content from the point of views of user's system engineers, programmers, and operators. In particular we check for effective examples and for multiple entry points to the information to allow readers with different needs to access their information quickly.

Writer Training Service

Our third major service was to provide technical writing education. The purpose of this service is to reduce the time required for document editing service by eliminating common mistakes made by all the writers. In effect, we teach to writers what we learned from our printing quality control service and the reading and checking manuscripts service.

Japanese enterprises encourage in-service training of employees by other employees. Thus our course, originally a biannual event, is now held monthly by popular demand. In the last five years we have trained one-third of the software engineers of our company.

FUTURE PLANS

To further improve our documentation services we will have to implement computer processing of the documents. But standardization of manuscripts is a precondition of EDP (Electronic Data Processing) and we have already created fairly good standards.

We are now testing a method of producing manuals by the Japanese Processing Extended Feature, a software product of Fujitsu. Our ultimate goal is office automation, to create a paperless office by using computers.

But Japanese language processing has a big problem in the areas of data input, updating, and printing. For example, the size of the Japanese character set (kanji, kana-characters and the alphabet) is more than 14,000 characters. It is very difficult to process it by computers, but we feel that our company will be able to surmount this technical problem in the near future. And we believe this effort will also be useful in the future for processing other languages throughout the world.

SUMMARY

There are no professional technical writers in Japan. Using technical writers may be the best solution to our problem, but it is not possible for us at this moment. Therefore we developed the method for producing mass volume documentation without technical writers as described in this paper. And this method is quite effective as we had expected.

ERROR PATTERN ANALYSIS APPLIED TO TECHNICAL WRITING: AN EDITOR'S GUIDE FOR WRITERS

E. Brette Monagle
New Mexico State University

E. Brette Monagle

Dept. of English
New Mexico State Univ.
Box 3-E
Las Cruces, NM 88003

Instructor in Technical Writing; Freelance Editor; Graduate Student in Technical and Professional Writing

This paper uses error pattern analysis to reduce the time and money spent on editing and correcting manuscripts. Error pattern analysis involves noting, classifying, and keeping a frequency count on errors. Once this pattern is identified, an editor can assign a number to each error. Lengthy manuscript notations are thus eliminated because errors are circled and given a number. The writer, upon receiving his edited manuscript, can refer to a Master List of Most Common Errors, which lists each numbered error and the rules for correcting it. Results from error pattern analysis can save writers and editors both time and money.

AN OPTIMISTIC BEGINNING

As the newly-employed graduate school editor, I looked forward to a rather easy time of editing. I knew I had time limitations, of course. Once submitted, theses and dissertations had to be checked and returned within ten working days. Further, there were particularly busy periods. The month I had been hired, over thirty theses and dissertations had required editing in an eighty-hour period. Mentally, I did a rapid calculation. If each thesis and dissertation had a minimum of 50 pages, an editor would have to edit 1,500 pages and possibly 3,000 pages if all the pages had to be rechecked. But even with these figures, I wasn't alarmed.

After all, weren't all the writers advanced students? Some of them had as many as five years of graduate schooling. Besides the editorial guidelines had no section on mechanics and style, thus further supporting my belief that the writers didn't require any assistance in these areas. (I found information on such points as prepage format, proper bibliographical form, documentation in the text, correct pagination, and so forth in the editorial guidelines.) Surely a section would have been included on the most common mechanical errors, if many of the writers were making them.

A QUICK AWAKENING

But I found mechanical errors and, in some cases, a large number of them. I tackled the job with enthusiasm, circling each error and citing the rule it broke (as briefly as possible, of course.) But the system was cumbersome because the edited copy had to be returned to the author and a second copy had to be made for the editor's files. I slipped into the desperate feeling of racing against the clock and knowing I was losing.

SOME POSSIBLE ALTERNATIVES

How could I deal with the problem? Since I was working many hours past the time I was being paid for, I began to search for alternative solutions. Perhaps I could "nod with Homer" and overlook some of the more minor errors. But when faced with that choice, I rejected it. Since these papers were going to be published, it was important not to "let go." Could I simply circle an error and correct it without explanation? No, that wasn't the way either. Certainly it would save time, but frequently writers objected instinctively to such high-handed ways of dealing with their "babies." What if I just circled the error and returned it to the author for corrections? Before I had even finished framing this alternative, I could see its flaw. Would an author have written an error, if he had known what rule he was breaking?

AN EMERGING PATTERN

I began to notice certain errors were being repeated over and over, and an idea began to form in my mind. To test this idea, I ran an error pattern analysis, which meant simply circling, noting, and classifying errors under their respective categories. Even I was surprised when I had finished. I had expected to find a small number of types of errors, but not so few as I actually did. In all, there were approximately 18 classes, which can be classified as follows:

- Awkward Syntax
- Choice of Word
- Lack of Colon for Lists and Definitions
- Lack of Conditional Verb
- Lack of Hyphen in Compound Nouns/Adjectives
- Lack of Parallelism in Infinitives
- Lack of Parallelism in Sentences
- Misplaced Modifier
- Not Spelling Out Numbers Ten and Below
- Ommitted/Incorrectly Placed Comma
- Omitted Word/Words
- Overcapitalization/and Undercapitalization
- Subject-Verb Agreement
- Typographical Error*
- Unidiomatic Preposition
- Use of Adverb for Pronoun (Where for Which)
- Use of Verb as a Noun (Affect/Effect)
  Use of Noun as a Verb (Effect/Affect)
- Use of Verb for Adjective (the Return Check)

In addition, within these classes of errors, the majority of errors, 80 percent, were in the placement of commas, the lack of hyphens with compound nouns and adjectives, and the failure of spelling out numbers of ten and below. And approximately 50 percent of all errors were in the category of comma usage alone.

The results of this analysis supported what I had intuitively sensed. <u>While it is possible for authors to make a wide variety of errors, they actually made only a few, and these they continued to repeat</u>.

*I broadened the category of typos to include misspelled words which is discretionary, of course. Some editors might not want to call a Jack of Hearts the King of Clubs.

PUTTING THE FINDINGS TO USE

Now that I had put my idea to the test, I needed a way to put it to practical use, especially to a time-saving use. What if I assigned a number to each of these classes of errors, such as (1) subject-verb agreement, (2) lack of hyphens with compound nouns/and adjectives, (3) omitted word/words, (4) use of verb for adjective, and so forth? In one quick stroke, I could cut out wordy, written notations. I could also do away with some of the editorial symbols, many of which were unfamiliar to authors.

Simultaneously, this method provided me with a substitute for the cumbersome record-keeping used formerly. Now all that was needed was to transfer the numbered error and its page number to another page for the editor's files. Finally and most important, I realized I had found a way to eliminate the flashpoint of confrontation that can occasionally arise between a writer and editor. By introducing a "third party," a nameless and faceless expert whose grammatical rules and references could be cited on the Master Sheet of Most Common Errors, I could do away with the difficult editor role which few editors relish.

DOING AN ERROR PATTERN ANALYSIS

<u>Excerpt from Student X's Dissertation</u>

Folklore, misinformation, religious beliefs, and emotional outlook may influence sexual appetities in the pregnant woman far more than anatomic and physiological variables, which will be described later. In the early part of pregnancy women who report less sexual tension than they did before becoming pregnant may be fearful of harming the fetus, feel insecure in their changing relationship with their husband, or have new personal anxieties or concerns.

From these and other pages, categories of errors were set up, numbers assigned to them, and tallies kept of errors.

SOME COMMA PROBLEMS

However, I hadn't gone very far before I could see commas were going to be a special problem. It wasn't enough to assign a number to them, cite a rule, and give a brief example because within the comma category were large numbers of sub-categories and rules governing them.

Moreover, the question of length had to be considered. A typical grammar book, such as Warriner's English Grammar and Composition (Third Course), spent as many as 20 pages on comma rules, explanations, and examples. To reproduce such a mass of material would be self-defeating. Would anyone be likely to read through so many pages? Even if they did, many of the terms would be difficult to understand. This problem can best be seen by citing, in a condensed version, the seven basic categories for comma use from Warriner's grammar text.

1. Use commas to separate items in a series.
   Example: Students, teachers, parents, and visitors attended the picnic.
2. Use a comma to separate two or more adjectives preceding a noun.
   Example: This is a rough, narrow, dangerous road.
3. Use a comma before and, but, or, nor, for, and yet when you have two independent clauses (two complete thoughts on either side of your conjunction).
   Example: He did not come to my birthday party, nor did he even bother to answer the invitation.

So far, so good. The definitions are simply stated, and the terms easy enough to understand, particularly when accompanied by an example. However, the remaining four rules have sub-categories and use some grammatically technical language.

4. Use a comma to set off non-essential clauses and non-essential participial phrases.
   Example: The National Bank, which was a firetrap, was torn down.
5. Use a comma after certain introductory elements.
   a. Use a comma after such words as well, yes, no, why, etc. when they begin a sentence.
      Example: No, I have not answered her letter.
   b. Use a comma after an introductory participial phrase.
      Example: Pausing for a moment in the doorway, the teacher smiled.
   c. Use a comma after a succession of introductory prepositional phrases.
      Example: After Alexander had hit the ball over the fence, the crowd gave him a standing ovation.
6. Use commas to set off elements that interrupt the sentence.
   a. Appositives and appositive phrases are usually set off by commas.
      Example: Everyone, even his enemies, respects him.
   b. Words used in direct address are set off by commas.
      Example: That program, Florence, has been changed.
   c. Parenthetical expressions are set off by commas.
      Example: He did not, however, keep his promise.
7. Use a comma in certain conventional situations.
   a. Items in dates and addresses are set off by commas.
      Example: My family moved to Knoxville, Tennessee, on Monday, May 4, 1964.
   b. The salutation from a friend's letter and the closing of a letter both use commas.
      Example: Dear Aunt Edith,
      Sincerely,
   c. Names followed by Jr., Sr., M.D., etc. are followed by commas.
      Example: Allen Davis, Jr.

The problem becomes clear when rules four through seven are cited; the terms, non-essential clauses, non-essential participial phrases, introductory participial phrases, a succession of introductory prepositional phrases, and so forth, are too technical. In addition the sub-categories and sub-definitions were too numerous. I knew I would lose the writer's attention if I didn't condense this material.

The question was, could I simplify the terms and categories without sacrificing accuracy? After studying the types of errors writers actually made, I thought I could.

Rule six could easily be combined into number four to read as follows: use commas with elements that interrupt the flow of a sentence or separate the subject and verb.
   Example: The National Bank, which was a firetrap, was torn down.
   Senator Morgan, hoping for a compromise, began a filibuster.

Rule five can be restated in simpler language yet maintain its meaning: use commas after introductory phrases and clauses and such adverbs as however, therefore, and nevertheless.

Example: While driving the car in the wintertime, I often forget to check the antifreeze level. Nevertheless, I've always been lucky and never damaged a car yet.

Rule seven would become rule six and, since it is not stated too technically, it would stay the same.

HOW ERROR PATTERN ANALYSIS CAN WORK FOR YOU

You, as an editor, would take a typical page of writing and circle each error as was done with Student X's paper. After you had done a sufficiently large number of pages to identify an error pattern, you would arrange these errors in categories, alphabetically, and assign numbers to the categories as follows: (1) Awkward Syntax, (2) Choice of Word, (3) Lack of Colon for Lists and Definitions, and so forth. With commas, of course, you would need sub-categories such as (4a), (4b), (4c), etc. Once you had established your pattern and numbering system, you can then simply circle an error and place a number above it. The authors would then refer to the rules cited on the Master Sheet of the Most Common Errors, which accompanies their papers.

For example, a writer might find on page 3, line 6, a word circled with (3) above it. He would refer to the Master Sheet, which gives all the common errors and the rules for correcting them. He would find, by referring to (3), that he had left out a colon for a list and that it contained the rules for colons.

A Master Sheet with all the most frequent errors categorized and numbered simplifies bookkeeping. It allows you to reduce your record of an edited paper to a few pages because you can simply list errors by page, line, and number [p 1, line 16, (3)].

Finally, it is often helpful to send a prospective author an advance copy of this Master Sheet of Most Common Errors and the rules for them. I found authors frequently read the editorial guidelines carefully; they were anxious to avoid pitfalls and had this list been drawn up, I believe the error frequency would have gone down sharply.

The most important aspect of error pattern analysis is that it reduces what seems to be a thicket of possible errors to a relatively small list of those people _actually make_. Your list might not look exactly like mine. You might also disregard some of the errors, such as parallel infinitives, as being relatively unimportant.

Though doing an error pattern analysis for your writers takes a little time and work in the beginning, it saves a lot of time and work in the end.

TEACHING TECHNICAL WRITING: A STUDENT'S PERSPECTIVE

Margaret J. Moore
Marshall University

Margaret J. Moore
Marshall University
Graduate Assistant
English Department
Huntington, WV 25701

The problem with grouping students into one course entitled "Technical Writing" is discussed in relationship to my past experience as a student in such a course. From a student's perspective, I view my previous experience as three-fold: as a student with no formal training in technical writing; as a student with past experience or "hands-on" training; as a student who is a professional technical writer. However, this paper does not attempt to center around the argument of what is or what is not common in a technical writing class. On the other hand, it does attempt to zero in on the problems arising from such a varied audience. Although instructors should and must be aware of such varied backgrounds in their students, some are not aware that any problem exists. Therefore, suggested methods for interfacing the expectations and goals of the instructor with the expectations and goals of the students are noted.

INTRODUCTION

Several colleges and universities offer degree programs in technical writing. However, the majority of higher educational institutions do not offer technical writing curriculums. They, instead, offer a single course entitled Technical Writing, Report Writing, or Scientific and Technical Writing. Often there are no prerequisites for taking such a course, therefore, all students are grouped into a single course. This, indeed, creates a problem for the instructor as well as the student.

From a student's perspective, I view my previous experience in such a course as three-fold:

- First, as a student with no formal training in technical writing;
- Second, as a student with past experience or "hands-on" training;
- Third, as a student who is a professional technical writer.

Even though it is not unusual to find students who fall into any one of the above groups, similarly, a student (like myself), who can relate to all three groups, is also not uncommon. However, this paper does not attempt to center around the argument of what is or what is not common in a technical writing class. On the other hand, it does attempt to zero in on the problems arising from such a varied audience.

THE STUDENT'S STANDPOINT

A Student with no Formal Training

Several colleges allow two-year degree students to substitute a course in technical writing for a course in English composition. Consequently, a student without experience in technical writing or any experience in composition is placed in a class with a student with experience. The student feels insecure or that he must "prove" himself as good as the student with experience. Often the terminology is more difficult for him to grasp, he becomes bored and frustrated. Nonetheless, he is doomed! If he drops the class, he looses an education he is entitled to; if he remains in the course, he must work twice as hard as the student with experience. He expects to acquire a general understanding of technical writing. His goal is difficult to define or predict. Hypothetically, he does not plan to be a professional technical writer, but obtain a two or four year degree in a technical or technical related field.

Past Experience or "Hands-On" Training

This student is usually a "returning" student. He is generally in a degree program or hopes to refresh his already acquired skills. He has written before or has worked with writers. The understanding of the variety of reports, proposals, and nomenclature which help to define technical writing is not a problem for him as it might be for the previous student. The expectation of a student with past experience in technical writing differs greatly from the one without experience. He expects a deeper understanding of technical writing and to become more specialized in the writing of technical reports and manuals.

Professional Technical Writer

A professional technical writer is, for the most part, in such a course because his supervisor erged him to take the class. It is good for his "promotional record." Such a course in technical writing is not advanced enough for him. However, he does attend regularly, participates, and has no problem understanding the assignments. He also has an advantage the other students because he has ready access to copiers, binders, typesetters, word processors, and or some type of editing equipment. This particular student's expectation corresponds to the student's who has past experience in technical writing. His goal may be to become more expertise in the writing field.

INSTRUCTOR'S STANDPOINT

Not only does such a course create problems for the student, but it creates equal problems for the instructor. Most instructors teaching a basic course in technical writing do not actually have a technical background. The instructor generally is not expertise in teaching a technical writing class. Therefore, he tends to resort to his former techniques used in teaching composition. Consequently, the basic technical writing class becomes a basic composition course. A basic composition class may be worthwhile for the student who has no previous technical writing experience. However, the questions of what happens to the more advanced students or how does the instructor contend with and motivate the student who may even possess a degree in English still remain.

Now, the dispute really begins! The students with varied backgrounds clash with each other and the entire class colides with the instructor. The instructor is left with the major decision of dealing with such a varied audience and to solve the problems arising from it.

One answer to the problem would be to have two or three classes in technical writing offered on different levels. Idealistically, the instructor would have only those students in his class who possesses the qualified prerequisites to to take the class. This might be the simplest answer to the problem, but it definitely would not be the most realistic. Financing, as well as other reasons, would not permit it. Therefore, in order to help solve some of the problems, without the addition of more technical writing classes, methods for interfacing the expectations and goals of the instructor with the expectations and goals of the students are suggested.

SUGGESTIONS

- Both instructor and student define their expectations and goals of the course.
- Students should write a short abstract of their past experience in technical writing in relationship to what they plan to gain from the course, and how they plan to use the knowledge they obtain from the course.
- Even though instructors generally do not like to give separate assignments to individual students; it is feasible (at least for the first few weeks) that individual assignments should be given to fit each student's past experience.
- Workshop time should be scheduled in order for all students to interface their own experience with each other.

Again, it is unrealistic to believe that if an instructor follows the above suggestions that he will have the perfict class. However, following these suggestions will help to ease the burden of the instructor as well as the student. The instructor will have a clearer understanding of the students' expectations and goals of the course and can work in order to meet those expectations and goals. Through the use of the abstract suggestion, the student informs the instructor of his experience, if any, in technical writing. And at the same time, it gives the instructor an example of each student's writing ability. The separate assignment exercise eliminates some of the competition presented by the more advanced students and helps the student overcome any writer's block he may have. Workshop time is one of the most effective techniques for solving the problems arising from a varied audience. The students are grouped into sets of three to six students (all with varied backgrounds) depending on the amount of students in the class. Each group is given an assignment (generally the identical assignment) which requires each student in the group to participate. The students discuss the assignment and must all agree on the most effective method for completing the assignment. The more experienced students in the group help explain any vagueness in terminology or the assignment that the less experienced students may have. The students conversationally exchange ideas allowing for a more relaxed environment. Each student feels that he adds to the class lessening his frustration. By photocopying, binding, or editing any part of the assignment which might require it, the student who has access to such equipment helps the student who does not. In turn, no single student feels at a disadvantage, and the instructor as well as the students give positive reinforcement to the class.

MANAGING THE SOFTWARE DOCUMENTATION MANAGEMENT REVIEW

Robert A. Schorle
Gilbert Associates, Inc.

Robert A. Schorle
Gilbert Associates, Inc.
P.O. Box 1498
Reading, PA 19603

(215) 775-2600

Software Documentation
Supervisor/Analyst

The software documentation management review must be managed by the writer to ensure that review objectives are achieved in an efficient manner. Good communication between the writer and the reviewers is necessary. To manage the review, the writer must help reviewers understand their responsibilities, the objectives of the review, the manner in which they should review the documentation, and the type of comments they should make. This understanding increases productivity for both the writer and the reviewers.

NECESSITY OF REVIEW PROCESS

The documentation produced by most software technical writers is subject to technical and executive management reviews in addition to editorial reviews conducted by a writing group editor, supervisor, or senior writer. The emphasis of this paper is to provide the writer with review guidelines that encourage a more effective management review of the documentation and more useful and comprehensive review comments.

This paper will explain why review management is necessary and describe how to implement it. This information will allow the writer to manage the review for the reviewers, and thus ensure the efficient use of management time and the receipt of valid review comments from a thorough review.

IMPORTANCE OF AN EFFICIENT REVIEW

Review management is necessary for our technical publications because of the limited time usually allocated for the review process and the requirement that software documentation must be verified. If the software documentation is insufficient, many computer users may be unable to reliably use the computer system resources.

The software documentation produced at Gilbert/Commonwealth consists of program user guides, computer system reference manuals, policy manuals, procedure manuals, and a monthly technical newsletter. This documentation is distributed to approximately 1000 consulting engineers, support people, and managers who use the computer resources. The non-editorial reviewers typically consist of systems programmers, business and engineering applications programmers, data processing executives, corporate executives, and selected users of the documentation.

The management review process must be efficient because it requires the time and knowledge of expensive people, and because it is an important examination for the accuracy and completeness of technical documentation. A management review conducted in an inefficient manner wastes the time of many people, reduces overall productivity, discourages the timely cooperation of the reviewers, and causes manuals to be published with extraneous information and technical errors. It is logical to conclude that a company would find it too costly to produce technical documentation that was not efficiently reviewed.

UNDEFINED RESPONSIBILITY

Obtaining review copies of the software documentation usually represents a small victory for the writer. However, the writer's task is not complete until the documentation is reviewed and the comments are incorporated. This process is usually considered unpleasant and difficult by everyone involved.

For the writer, the first problem may be deciding who should review the documentation. The second problem may be convincing the chosen reviewers to cooperate. This is the heart of the documentation review problem. The task of reviewing software technical documentation most likely is not the responsibility of anyone other than the software documentation group supervisor or editor. Few people, however, would recommend that software documentation be published without a comprehensive review. The writer's responsibility includes writing the documentation and securing an adequate group of interested reviewers. The review process may be marred for the writer when he must sift through vague and extraneous review comments and resolve conflicting questions concerning technical accuracy, corporate policy, and style.

For the reviewer, the task is often unpleasant because he does not consider the review of software documentation to be a primary part of his job, and he does not understand his responsibility as a reviewer. These problems are compounded when the management commitment to a comprehensive review of software documentation is expressed in a statement such as: "Try to take a quick look at this when you have some free time."

More often than not, the quick look probably occurs after the writer calls to inquire why the documentation was not returned by the requested date. Now the reviewer probably feels pressure and resentment in addition to his earlier feelings of confusion. The situation does not encourage a comprehensive review, and it is doubtful that such a review occurs. Instead, the reviewer may briefly look at the documentation and may even attempt to find something to comment on as proof to the writer, upper management, and himself that he performed as requested. Spelling and punctuation errors are an easy mark. He may look at the technical information and be angered that a favorite descriptive phrase was not included. He may decide that the presentation seems a little confusing, but does not bother to comment because he understands it. The executive management reviewer may indicate that a paragraph should be removed without realizing that some commitment or direction must be given to allow users to complete a specified action. Finally, the technician may spend considerable time looking at policies, and the executive may spend considerable time examining the technical information. Overall, this paragraph describes an expensive waste of time and skills. The loss of productivity is increased when the writer must resolve the confusing review comments.

Fortunately, there are some dedicated people who understand the reasons behind the software documentation management review and accept the responsibility of participating as a reviewer.

WRITER'S RESPONSIBILITY

Good management reviewers certainly make the writer's job easier and more productive. The writer, however, can also do much to make the review easier and more productive. The writer should provide each reviewer with detailed objectives for the review. It is the writer's responsibility as the professional communicator to improve the communication between the management reviewers and himself.

The writer should exercise review management in order to make the review a productive step in the process of producing good software documentation. If the writer does not receive useful comments from reviewers, it is probably because he did not adequately define the responsibilities of each reviewer. To ensure that each non-editorial reviewer knows how to review the documentation, the writer must:

- Define who should provide the technical, executive, and editorial reviews.
- Define the objectives of the technical, executive, and editorial reviews.
- Define how technical and executive reviewers should study the documentation to be effective and efficient.
- Define what types of technical and executive review comments are useful.

In essence, the review is part of the writing assignment, and the writer is the one who must manage the review.

DEFINE REVIEWERS

The first step in review management is to identify the reviewers so that they understand and accept their responsibilities. Reviewing software documentation is a difficult but important activity. Each reviewer is on the front line when it comes to ensuring the accuracy of a technical publication. Misunderstood responsibilities or a lack of enthusiasm can allow inaccuracies to be published.

Technical Reviewer

The technical management reviewer must know the technical information well. If the documentation concerns a specific software product the technical reviewer could be the same person who initially supplied the technical input. However, an inferior technical review can occur in this situation, because the technical expert may be so close to the information that he may overlook errors. The supervisor of the technical expert might be a more objective technical reviewer. In some instances, selected user representatives may also be able to provide an effective technical review of the software documentation.

Executive Reviewer

The executive management reviewer is probably in the corporate management structure and is therefore easier for the writer to identify. While this reviewer recognizes his responsibility, it is unlikely that he has considerable time to review software documentation. Executive reviewers should be able to identify established departmental and corporate practices and objectives.

Editorial Reviewer

The editorial reviewer is (or should be) an expert in the field of technical communications who displays excellent language use skills and has considerable experience in the software documentation field. Some companies employ editors, while many other companies rely on the next available level of talent which is usually the writing group supervisor or senior writer. To be effective, the editor must be able to provide the writer with an objective appraisal of the literary reference value of the software documentation.

DEFINE REVIEW OBJECTIVES

When the reviewers are identified, the writer should inform the technical and executive management reviewers of the objectives associated with each type of review. This communication will increase the ability of each reviewer to correctly examine the software documentation and identify problems within their area of responsibility.

Some technical and executive management reviewers may find it helpful to be aware of the objectives of the editorial review. It is important for the success of the review, however, that they understand that editorial review objectives are not their responsibility.

Technical Review Objectives

The technical management review is the official opportunity to verify the technical accuracy and completeness of the software documentation. The review should not be a re-analysis of the documentation objective, scope, content, or appearance, as this information was reviewed and finalized in an earlier phase of the documentation project. The review should be a careful examination of technical details in the text, lists, tables, illustrations, examples, procedures, instructions, references, and definitions. Technical approval is necessary for publication.

Executive Review Objectives

The executive management review is the official opportunity to ensure that corporate activities and publications are in compliance with established departmental and corporate practices and objectives. This review should consist of a general examination of the entire document and a careful examination of statements which describe policies, objectives, practices, guidelines, services, commitments, responsibilities, and actions related to the purpose of the software documentation. Executive approval is necessary for publication.

Editorial Review Objectives

The editorial reviews for quality, organization, and completeness should occur before the documentation is distributed for the management review and after all management review comments are resolved and incorporated. These reviews ensure that management review copies and final approval/publication copies of the documentation are of the highest editorial quality possible. The editorial review should consist of several passes through the documentation to examine:

- The structure and organization of the documentation.
- The paragraph, sentence, word, and punctuation use.
- The overall consistency of style.

Editorial comments typically must be incorporated by the writer. Editorial approval is necessary for publication.

DEFINE REVIEW METHOD

When the objectives of the review are understood by the reviewers, the writer should suggest how they should review the documentation. The manner in which the documentation is reviewed is very important in terms of efficient use of time and effective discovery of errors. Some differences exist between the technical and executive management review methods. These differences result from the emphasis of the review changing from details to overall scope and applicability.

Technical Review Method

The technical management review is primarily detail-oriented. If the technical information is to be effectively reviewed, the reviewer must check every last technical detail for accuracy and completeness. The writer can assist the technical management reviewer by pointing out those parts of the documentation which contain technical details and by reminding the reviewer that editors typically examine the documentation several times while looking at separate characteristics in each pass.

While this may initially appear to create more work for the reviewer, the writer should remind him that he is reviewing the documentation for two characteristics: accuracy and completeness.

To begin the review, the technical management reviewer should briefly recall on paper the intended function of the software documentation. Outlines developed in the Documentation Analysis or Design Phases could also be used. The reviewer should recall the intended function of the software documentation and be able to examine the entire document for completeness. One review pass should be used to determine that all relevant technical information was included. Typically this review requires very little time and can provide an accurate determination of the completeness, applicability, and timeliness of the documentation.

The accuracy of the technical details is the second characteristic of the documentation to be checked by the technical management reviewer. Verifying this information is a very tedious task and should therefore be his only concern. It is an inefficient use of time for the reviewer to stop examining technical details in order to check if related technical information is also included in the documentation.

The technical management reviewer should verify the accuracy of all technical details and the manner in which they are used in the documentation. These details may span the entire spectrum of computer system technology, applications, and use. The reviewer should also carefully check terms, abbreviations, tradenames, acronyms, statements, examples, and illustrations.

Executive Review Method

The executive management review is an examination of the software documentation for problems in the scope, applicability, content, timeliness, and appearance. This review is usually less complicated than the technical management review but is of equal importance because of intercompany politics and the need for management approval of corporate and departmental activities.

To be most effective, this review also should be completed in two review passes. As in the technical review method, the reviewer should understand the intended function of the software documentation before beginning the review. The writer can assist the executive management reviewer by pointing out those parts of the documentation which pertain to established departmental and corporate practices and objectives. The first review pass should be used to determine the format and contents of the documentation. The second review pass should be an examination of all statements for which the departmental or corporate management is responsible.

DEFINE USEFUL COMMENTS

The final information to be defined by the writer for the management reviewers concerns their review comments. By defining what type of review comments are meaningful, the writer can ensure that the comments he receives will be useful and the communication between the reviewer and himself will be good.

The writer should request that technical and executive management reviewers provide comments that specifically resolve the problems in the documentation. Reviewers should not make comments which are unanswered or unresolved questions because the writer is probably less capable of resolving the issue than the reviewer. If information from sources other than the reviewer is required, the reviewer should list the necessary references. Review comments should be specific directions to the writer, rather than puzzles, and all comments should specifically pertain to the quality of the documentation being reviewed. Unrelated or theoretical comments and suggestions should not be included with review comments.

RESOLVE ALL COMMENTS

The manner in which the writer handles review comments can significantly affect the cooperation and enthusiasm of the reviewers. The writer should explain that all review comments which are not entirely incorporated into the documentation will be resolved with the reviewer. This practice is important because reviewers expect the writer to use the comments they were asked to provide, and they deserve the respect of consultation when comments must be omitted.

PUBLICATION AND USE

Despite the tedious nature of the software documentation review, final approval and subsequent printing is eventually reached. This point should be recognized as publication rather than perfection. Given the dynamic nature of the computer industry, most software documentation could be reviewed and enhanced almost indefinitely and still not be complete or perfect. Another characteristic of software documentation is that it usually requires updating at least once a year. The concept to be remembered by the writer and the reviewers is that software documentation is worthwhile only when it can be used. The management review is the polishing stage of documentation developed for a software project. The final measure of success for software documentation is whether the intended user finds it useful.

SUMMARY

The information provided in this paper was developed to improve the efficiency and the effectiveness of the software documentation management review within the Gilbert/Commonwealth offices. This paper does not represent new developments in the documentation field. Instead, it contains a thoughtful analysis of the documentation review participants, objectives, activities, and results in order to establish guidelines by which the review can be made more productive. When this information is communicated to the reviewers the writer will be able to successfully manage the review.

HERE, EDIT THIS!

Patricia N. Smith
Rensselaer Polytechnic Institute

Patricia N. Smith

Graduate Student
Technical Writing

RPI
Troy, New York 12181

First-time professional editors can soon become respected editors by following two rules: always deliver good work and always deliver on time. This paper specifically suggests how new editors can meet these major goals, even with their very first editorial assignments.

Finally, after years of academic preparation, you have been hired to do what you do best: write and edit. Your first day at work, your supervisor hands you a sheaf of papers and says, "Here, edit this." On the one hand, you're anxious to tackle the job; on the other hand, you're afraid to begin. How can you demonstrate from the very start that you are a capable and competent editor? This paper discusses two time-tested rules to stand you in good stead.

THE TWO STRATEGIES

Respected editors deliver good work on time. You should try to follow their example.

Deliver A Good Job. Good editorial work requires that you:

- understand the task before you start,
- pinpoint your problems, and
- resolve your difficulties.

But your efforts may go unnoticed unless you submit your work within the deadline. Your supervisor will not be impressed by the report you provide on Friday if she needed it at 9 A.M. on Thursday.

Deliver On Time. To meet established timelines, you must:

- know the timelines before you begin, and
- adjust your style to meet the demands.

Observe each of these points and your first editorial assignment will undoubtedly be a successful one.

WHAT DO YOU NEED TO KNOW?

Before you can edit your paper, you must understand its purpose: why was the document written? Ask your supervisor or a knowledgeable source about the paper's readers: who are they? Material prepared for in-house distribution will be different from materials written for out-of-house use. Knowing the paper's objectives will help you clarify your task.

Scan the paper for its contents: what is it about? You may need some background knowledge to do your job, especially if you will be extensively editing and rewriting. Be prepared to contact resource people or to investigate the company or public library. Don't underestimate your own abilities, however. Your common sense and critical mind may be the two most important tools you apply to the task.

You should also be aware of any restrictions that could affect your editorial decisions. Don't wait until you have reorganized the piece to discover that the original format is the only acceptable format. Find out if a house style manual (or other style guide) is used and use it.

When you have all this information at hand, you are ready to pinpoint your problems.

WHAT PROBLEMS DO YOU FIND?

Editors approach materials in various ways. Some go through a piece section by section, revising and reworking as they proceed. Others tackle the work as a whole, making changes only after they have acquired a firm sense of the entire paper. In either case, editors evaluate the following characteristics:

- organization,
- diction,
- style,
- content, and
- layout and design.

After thoroughly reading the paper, decide whether its organization is logical. If, on the first review, you lost the author's train of thought and struggled to locate it again, no doubt the reader will also flounder. A reorganization of sentences, the deletion or addition of material, or a refinement of the author's focus may be necessary.

Since you know its intended audience, determine whether the paper's diction is appropriate. An article written for highly trained microbiologists would not incorporate the same vocabulary as a selection geared for tenth-grade biology students. A report directed at the chairman of the board would require more formal diction than a long memo written to a good friend and colleague. Whoever its readers are, however, the material should be written in clear and precise language.

Style is just as important as diction; in fact, the two cannot easily be separated. Diction directly affects an author's style. Colloquial phrases contribute to a "friendly" style; technical terms create a "scientific" style. When considering a paper's style, ask yourself these questions: Does the style suit the paper's purpose and audience? Does the style effectively convey the paper's information? Is the style consistent throughout the paper? What is the overall effect?

You thought about general content when you first accepted the editorial job. Now you must examine the paper closely. If you are familiar with the paper's subject matter, you might be able to detect inaccuracies, inconsistencies, and omissions. If you are new to the topic, you can apply your analytical skills to determine whether the material flows logically; what questions remain unanswered; and where confusion exists. Be certain that your revisions are not misleading or inaccurate. When in doubt, check with a reputable source. Verification adds time to the editorial process, but your efforts will pay off for both yourself and your employer.

Finally, pay attention to the paper's layout and design. You may be unaccustomed to this criterion, but eye appeal does play an important role in readability. Consider substituting graphics for prose, or breaking dense stretches of text with heads and bullets in order to create aesthetically pleasing material. But be sure that all your revisions conform to house policy.

After you've completed your evaluation of organization, diction, style, content, and layout and design, you should have isolated the problems.

HOW DO YOU RESOLVE THE DIFFICULTIES?

Now comes the hard part--solving the problems you have just identified. It's one thing to pinpoint trouble spots, but it's another thing to fix them. Basically, you have two choices: lightly editing the work or heavily revising the material. Your decision will depend on at least three factors:

- the original quality of the paper,
- the type of rework necessary, and
- the time available.

Analyze how much work the paper needs before it will be acceptable. Then, consider what kind of work is involved: attending to grammar and punctuation; adding transitions; tightening lengthy, redundant prose; reorganizing a section; redirecting the focus. Your answers to the questions of "how much" and "what kind" will largely determine the amount of time demanded by the task. Conversely, the time you can devote to the assignment affects "how much" and "what kind" of editing you do. You want to submit good work; and remember, you want to submit it on time. A few words about meeting deadlines are now in order.

Recognizing The Timelines. Before you begin any editorial work, you should know how long you have to complete the task. If you realize that you cannot finish the assignment within that time frame, tell your supervisor immediately. While you may feel uncomfortable saying, "I've looked over the paper, and, as you can see from the extent of my suggestions, I will need until Friday to edit the material," such a statement serves two purposes. First, it indicates the types of revisions you have in mind. Your supervisor may feel that such changes aren't warranted. On the other hand, she may readily agree to an extension now that she knows what you have planned. Second, it indicates that you are a responsible editor who is conscious and respectful of deadlines.

Realizing Limitations. Some deadlines are impossible to change. If that's your case, decide just what you can produce under those conditions. Don't try to edit, reorganize, and reformat a 100-page document in two days in the same way that you would rework it in three weeks. Instead, assess what you can do--and make it your best effort.

The decision to lightly or heavily edit the paper will ultimately depend on your circumstances. Some materials are drastically improved with few revisions. The following example, for instance, demonstrates how simple editorial changes can make an appreciable difference.

> On July 1, the lowered bus fares were made effective. The weekly Wednesday head count was continued as before with the bus drivers using calculators to count all paying riders. Mr. O'Bannon pointed out a need to gather this weekly information, unstable as it might be. For he wants to keep abreast of the situation; that is, he wants to see the increase (or decrease) as it is happening.

Other materials require extensive work to be made comprehensible. The next excerpt illustrates this situation.

> Advertising witnessed its first beginnings centuries ago, prior to the age of printing, when shops used signs to announce their store of goods; from thence forward it has evolved to its present stage and format. The development of the advertising world in the 20th Century parallels those changes found virtually everywhere today--for just as any other social institution, it symbolizes the common values of society, and admittedly, these values have undergone transformation and reorientation.

It is your responsibility to use the method that will produce the best job within your time restrictions.

YOU'VE MADE IT!

So you've chosen the best approach for your task, and you've completed the job within the prescribed deadlines. Now you can sit back and relax before the next assignment. Your supervisor probably won't let you rest for very long. You are, after all, an editor who delivers a good job and delivers on time--a valuable addition to any organization.

ELECTRONIC COMPONENT DOCUMENTATION IN A RESEARCH-LABORATORY ENVIRONMENT

Jack W. Spencer
Lawrence Livermore National Laboratory

Jack W. Spencer
Lawrence Livermore National Laboratory
P. O. Box 5504, L-121
Livermore, CA 94550
(415) 422-9082

Supervising Engineer, EE
Technical Publications

Documenting electronic components and assemblies used by design personnel at a scientific research laboratory presents some unusual challenges. We must provide up-to-date documentation which is complete enough to provide designers the information they need for unusual applications. Yet, because of the relatively small quantities of components used, the documentation must be inexpensive to produce and distribute. The system we have evolved depends on close interaction between technical publications editor/writers and components engineers, and results in components documentation manuals closely tailored to the needs of our design personnel. This paper discuses the philosophy underlying our efforts, as well as the organization, production, and distribution of our component documentation.

INTRODUCTION

In one form or another, documentation of the items from which we make up our technology is of vital interest to almost everyone. Adequate documentation of all of the bits and pieces is an absolute necessity, if designers are to adequately design, users properly use, and service personnel properly service and maintain, equipment.

Documentation of electronic components and assemblies is particularly difficult, both because of the complexity of many of the items and because of the rapid introduction of new items and obsolescence of old ones. Devising ways to see that the "paper" keeps up with this continually-changing mix is a real challenge.

Work performed under the auspices of the U.S. Department of Energy under contract No. W-7405-Eng-48.

This paper discusses electronic component documentation in a rather unusual environment: a large scientific laboratory. Lawrence Livermore National Laboratory is a major scientific research complex, operated for the U.S. Department of Energy by the University of California. The Laboratory was established in 1952 to do nuclear weapons research for DOE's forerunner, the U.S. Atomic Energy Commission. From a few hundred people then, LLNL has grown into one of the most modern and well-equipped research facilities in the world, with more than 7,000 employees and a budget of more than $450 million. The scope of our work has also expanded greatly: today, in addition to theoretical and applied research in nuclear explosives technology, we have major, long-term research programs in laser fusion, magnetic fusion, advanced isotope separation, non-nuclear energy (such as underground coal gassification, wind and solar energy systems, metal-air power cells, and alternative fuels) nuclear systems safety, and biomedical and environmental research. The more than 1,000 employees of the LLNL Electronics Engineering (EE) Department are an integral part of all these research programs, providing the electronic systems necessary for a tremendous range of experimental applications.

The Laboratory stocks nearly 50,000 different items of an electronic nature for use by our designers. One of the responsibilities of the group to which I belong is documenting those items: providing designers with the information needed to use them properly.

OUR SYSTEM

Some of the situations leading to our documentation problems are rather different from those of a typical publications group. First, few of the Laboratory's designs are replicated more than once or twice. Second, the demanding nature of control and instrumentation electronics for energy research means that a significant amount of the equipment designed and constructed at LLNL is unusual in design or application and is often approaching the state-of-the-art.

We have three principal challenges: first, to provide up-to-date documentation, complete enough to provide our designers information needed for unusual applications; second, to rapidly disseminate the documentation for new items which may be added to stock; and third, do it all inexpensively since small-quantity use cannot justify the cost of an elaborate documentation system.

Our system depends on close interaction between two groups: EE Components Engineering, responsible for evaluation and selection of items for Laboratory stock, and ourselves, EE Technical Publications. (Components Engineering consists of a group of engineers, each with responsibility for being the Department's "expert" in one or more areas of electronic technology. Working closely with engineers assigned to various research programs, they are responsible for seeing that the majority of the items needed for design are available from stock.)

When Components Engineering adds a new item or group of items to stock, they immediately publish a Components Bulletin--distributed to Department design personnel--which announces the addition, and which may provide some minimum design information. At the same time, they will begin working with an editor/writer to prepare a complete engineering-design data sheet, which will be distributed to the EE Standards Manual.

THE EE STANDARDS MANUAL

The EE Standards Manual (Figure 1) is intended to be a complete, single source of design information for all of the electronic items in LLNL stock. As of now, the Standards Manual consists of five loose-leaf binders, more than 2500 pages of material. The contents have changed radically over the years but the basic philosophy behind the Manual and its overall organization have remained remarkably constant. We've found that the best compromise between usability (by design engineers) and produceability (by us, at a reasonable cost, and in a reasonable time) results in a manual that is:

- Loose-leaf, for easy revision
- Sectionalized by component type, with large, titled, tabbed dividers.
- Made up of of separate documents--one or more sheets--rather than totally of individual sheets.

The data included is usually rather typical of what one would expect to find in a manual of this nature (Figure 2). However, there are many cases where--because of the unusual uses to which our designers put things--we include rather specialized material. One example might be "pulse" ratings for components usually rated only for steady-state operation. Such information is often valuable at LLNL, and impossible to obtain from the usual manufacturers' data sheets. (One of the component engineer's jobs is to develop or collect information of this nature. See Figure 2 for an example.)

Each document in the manual is numbered; this is necessary to remove, insert, or file it. We have experimented with several numbering systems and have finally invented a system which seems to work for us, and which allows us to be very flexible in arrangement and subordination of material. The number looks like this:

LED SS-GG-SG-TT

"LED" is a prefix which identifies a document from our group, relating to stores-stock items. The number itself is composed of:

SS = Standards Manual Section	00 to 99
GG = Group Number	00 to 99
SG = Sub Group Number	00 to 99
TT = Item Number	A1 to Z9

Although unwieldly, the numbering seems to work better than any of the others we know of, and arrangement and subordination are evident to almost everyone. Using such numbers, it is easy to group components into "families", of as many levels as appropriate for a particular situation.

There are some additional features of the manual which we've discovered make it more useful for its intended purpose. Three of the most important are:

Table(s) of Contents. Although there is an alphabetical index for the manual as a whole, we also include a complete table of contents as the first document in each individual section. We include the document number, title, and last revision date of each document in the section. These are handy for looking items up, and particularly handy for checking to see if the manual is up-to-date and complete.

Section Summaries. For sections which include items where the most useful information is whether an item is in stock at all (and our stock number), rather than detailed information, we provide a Stock Summary immediately behind the Table of Contents for that section. These summaries have proved to be very useful, and are often pinned up on walls for quick reference, etc. Figure 3 is a typical example.

Photographs. Having discovered that elaborate indexes are often a waste of time (because most people look in the index only after they can't find an item by "thumbing through") we head each data sheet with a photograph or drawing of the item(s) described. Because the items in the manual are grouped in sections with all similar items, quickly flipping through will usually locate the item you want; even if you don't know its exact description, you know what it looks like.

PUBLISHING THE STANDARDS MANUAL

In most publications groups, the biggest problem is the lack of people--and time--to do the job the way it really should be done. Our component documentation work is no exception.

On an average, we publish 100 to 150 documents (500 to 600 new or revised pages) per year, in four quarterly revisions. We manage that with an editor spending about half time, and the equivalent of one clerk-typist (although the work may be spread among several people).

We are able to do as well as we do for three main reasons: skilled people, cooperative effort, and standards of quality relaxed as far as we feel we can and still produce a product which reflects credit on the group: quick, but professional.

The most important contributor to our success is close cooperation between our editor and the components engineers. The editor must have a broad knowledge of electronic components; genuine hardware familiarity. Thus the editor helps the engineer produce a clean draft on the first try, which saves much retyping. Often, the editor drafts data sheets for the engineer's approval. The editor, as a single individual familiar with the entire manual, sees to consistency in style, proper cross-referencing and similar tasks, and is responsible for keeping the index and tables of contents up to date. The editor also oversees production, deals with the printers (a genuine headache!) and monitors the distribution of material. Each revision we send out includes a "comment" sheet, and these come directly back to the editor--immediate feedback when people who hold the manuals have questions or spot errors. (We reply to all comment sheets sent in, thanking the authors and letting them know what action was taken. This has proved to be an excellent way of developing good relations with our "customers".)

For production, we use standard quick-pasteup techniques. Because we have elected not to be concerned about such niceties as type matched throughout a document, we are free to borrow--without retyping or redrawing--all the usable material we can from vendor's data sheets. This saves us an immense amount of work. Of course, we always get permission to borrow material, and give credit as needed. (Getting this permission is a job for the components engineer, at the time a new item is being added to stock.)

Surprisingly, we make very little use of word processors in our component documentation work. There is a good reason for this. Our experience in the past has been that we must use a variety of formats, with a relatively small amount of type scattered among a lot of pasted-up "borrowed" material. Attempting to use a word processor for that sort of work is frustrating and wasteful: it's much quicker to correct a small error with a quick paste-up. However, we do make good use of word processors for documents such as the Section Summaries mentioned previously. As word processors become more sophisticated and their memories get larger we plan to make more use of them. There are additional comments on the subject of word processors under "THE FUTURE", below.

We cannot hire compositors or technical illustrators to do our paste-up work: all our work is done by clerk-typists who "learn by doing", after some minimum amount of instruction. This works surprisingly well, and has advantages on both sides: people develop skills useful in other areas of our publications work, and useful to themselves should they decide to move on--or up--either into more conventional office jobs, or into jobs that make more use of the skills they've learned. (For example, three clerks trained in our group have gone on to become technical illustrators.) The biggest problem is that paste-up work can become deadly dull after a while, so we try to mix jobs as much as possible.

For each quarterly Standards Manual revision, we make up a complete set of revision instructions. We've found that this is very important: sketchy or misleading instructions quickly result in manuals with incorrect, obsolete, or improperly organized contents. We've found it important to give step-by-step instructions for the revision, <u>and</u>, for each deleted or revised document, to list the reason (as specifically as we can) why the change was made. This prevents a surprising number of telephone calls from people who, at a quick glance, think "the new sheet looks just like the old one."

THE FUTURE

Attempting to predict the future is always a chancy business: people's memories may be long enough to prove you wrong. Nevertheless, based on trends over the last few years, I'll venture two predictions: one a problem, and one a very hopeful situation.

The biggest problem we are going to face will result from the increased complexity of the items we have to document. A NAND gate, for example, is relatively easy to define rather completely on a data sheet. A one-chip microprocessor is another matter: often a whole book may not be enough to provide complete information. Our approach will be to provide selected, consolidated, basic information, and references to locally-available sources of complete information (such as our Laboratory's Small Computer User's Library). We are also going to borrow even more information from vendor's published data, and this may eventually lead to a rather unusual situation: vendors of a particular item being selected on the basis of the adequacy of the documentation they provide, rather than on price, availability, or other more usual grounds. Another consideration will be finding or developing editors with enough technical expertise to confidently deal with selecting and consolidating such material: this may be a real challenge to many of us.

The hopeful situation I foresee is based on the steadily-increasing power and sophistication of word-processing systems, and the developing possibility of making them perform functions previously available only in the realm of data processing. The merging of word- and data-processing functions in relatively small-scale, inexpensive systems has the potential to provide us with radically new capabilities. For example, all our Standards Manual material in a single data base from which we can extract indexes, tables of contents, section summaries, and any other material needed. The system could also do the cross-referencing, the filing, the keeping track of the latest revision, and so forth; it could even tell us the status of each individual document. Large users have, of course, had these sorts of capabilities for some time, using computer-based text-processing systems of many types. The overall effect on most of us has, however, been minimal so far. This situation is going to change. When the small publications group can begin to acquire these capabilities for a reasonable investment--something which it seems is just beginning to happen--we will see some really significant changes in how we do our jobs.

Figure 1. The EE Standards Manual. The Logic System Designer's Manual, also shown, is a selected subset of material from the five-volume manual; it is published and distributed separately for designers who do not need the full set.

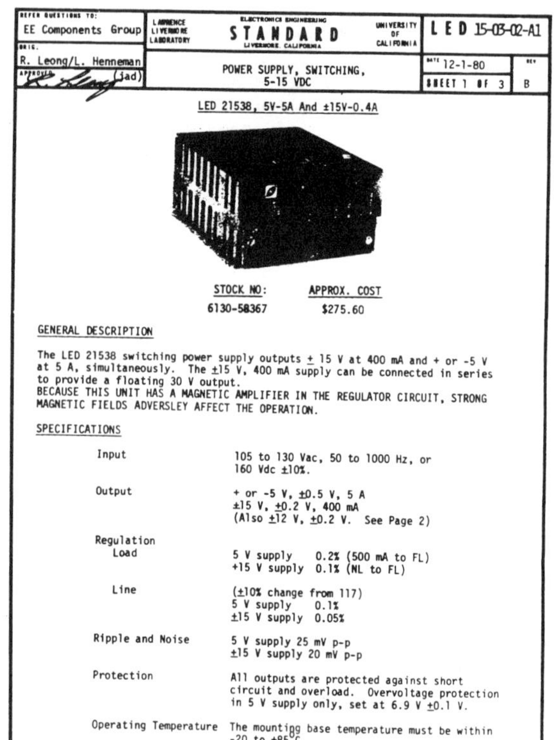

REFER QUESTIONS TO: EE Components Group | LAWRENCE LIVERMORE LABORATORY | ELECTRONICS ENGINEERING STANDARD LIVERMORE, CALIFORNIA | UNIVERSITY OF CALIFORNIA | LED 15-03-02-A1

ORIG.: R. Leong/L. Henneman | POWER SUPPLY, SWITCHING, 5-15 VDC | DATE 12-1-80 | SHEET 1 OF 3 | REV B

APPROVED: R. Leong (jad)

LED 21538, 5V-5A And ±15V-0.4A

STOCK NO: 6130-58367 APPROX. COST $275.60

GENERAL DESCRIPTION

The LED 21538 switching power supply outputs ± 15 V at 400 mA and + or -5 V at 5 A, simultaneously. The ±15 V, 400 mA supply can be connected in series to provide a floating 30 V output.
BECAUSE THIS UNIT HAS A MAGNETIC AMPLIFIER IN THE REGULATOR CIRCUIT, STRONG MAGNETIC FIELDS ADVERSLEY AFFECT THE OPERATION.

SPECIFICATIONS

Input	105 to 130 Vac, 50 to 1000 Hz, or 160 Vdc ±10%.
Output	+ or -5 V, ±0.5 V, 5 A ±15 V, ±0.2 V, 400 mA (Also ±12 V, ±0.2 V. See Page 2)
Regulation	
Load	5 V supply 0.2% (500 mA to FL) +15 V supply 0.1% (NL to FL)
Line	(±10% change from 117) 5 V supply 0.1% ±15 V supply 0.05%
Ripple and Noise	5 V supply 25 mV p-p ±15 V supply 20 mV p-p
Protection	All outputs are protected against short circuit and overload. Overvoltage protection in 5 V supply only, set at 6.9 V ±0.1 V.
Operating Temperature	The mountigg base temperature must be within -20 to +85°C.

Figure 2. The first page of a typical document. The caution about strong magnetic fields is something the Laboratory learned the hard way--not a part of typical vendor information.

REFER QUESTIONS TO: EE COMPONENTS GRP. | LAWRENCE LIVERMORE LABORATORY | ELECTRONICS ENGINEERING STANDARD LIVERMORE, CALIFORNIA | UNIVERSITY OF CALIFORNIA | LED 11-00-02-A1

ORIG.: D.Peschel/L.Henneman | INTEGRATED CIRCUITS SUMMARY OF LIVERMORE STOCK | DATE 12/1/80 | SHEET 1 OF 16 | REV I

APPROVED: R. Leong

1453F

INTEGRATED CIRCUITS, SUMMARY OF LIVERMORE STOCK

NOTE: THE TYPE NUMBER LISTED IN COLUMN ONE REFERS TO THE NUMBER BY WHICH AN INTEGRATED CIRCUIT IS MOST LIKELY TO BE KNOWN. THE TYPE NUMBERS ARE ARRANGED IN ASCENDING ORDER BY FAMILIES (I.E., DTL, TTL, CMOS, AMPLIFIERS, REGULATORS, ETC.).

THE PRIME TYPE NUMBER REFERS TO THE NUMBER ASSIGNED TO A DEVICE AS A SPECIFICATION FOR PROCUREMENT AND QUALITY ASSURANCE.

KEY TO PACKAGING CASE CODE

D	TO-3	K	DIP, 16-PIN
E	CN18, 12-PIN	L	DIP, 6-PIN
F	TO-99, 8-PIN	M	DIP, 24-PIN
G	TO-100, 10-PIN	N	DIP, 20-PIN
GA	TO-220, 3-TERMINAL		
H	DIP, 8-PIN (DUAL-INLINE-PACKAGE)	R	DIP, 18-PIN
J	DIP, 14-PIN	X	DIP, 40-PIN

TYPE	DEVICE	DOCUMENT LED11-	PRIME TYPE	CASE	STOCK NO. 5961-	APPROX. COST
DIGITAL INTEGRATED CIRCUITS						
DTL (DIODE-TRANSISTOR) LOGIC						
* 631A	QUAD 2-INPUT EXPANDER	01-02-A1	LED 21139/631A	J	51585	$ 1.25
* 930	DUAL 4-INPUT NAND GATE	01-02-A1	LED 21088/930	J	50813	0.50
* 932	DUAL 4-INPUT NAND BUFFER	01-04-A1	LED 21089/932	J	50814	0.75
* 936	HEX INVERTER	01-03-A1	LED 21137/936	J	51542	0.50
* 946	QUAD 2-INPUT NAND GATE	01-02-A1	LED 21090/946	J	50815	0.50
* 948	J-K FLIP-FLOP	01-05-A1	LED 21091/948	J	50816	1.50
* 951	MONOSTABLE MULTIVIBRATOR	01-08-A1	LED 21092/951	J	50817	2.25
* 962	TRIPLE 3-INPUT NAND GATE	01-02-A1	LED 21138/962	J	51543	0.50
*15835	HEX INVERTER W/O INPUT DIODES	01-03-A1	SN15835N	J	57633	0.50

*NOT FOR NEW DESIGN

Figure 3. A typical Section Summary. These summaries are maintained as word-processor files, which makes them easy and quick to update.

COMMUNICATING THE RESULTS OF SCIENTIFIC WORK TO THE LAITY: AN APPRAISAL OF THE PROCESS

Donald C. Spencer
Sandia National Laboratories, Livermore

Donald C. Spencer
Sandia National Laboratories
East Avenue
Livermore, CA 94550
(415)422-3255

Technical Writer, Film-Maker

Ph.D. Speech (Theatre),
University of Wisconsin

Vital to the health and support of science and to the needs of the public is effective communication of the results and implications of scientific work to the laity. This process has improved substantially but is still inadequate.

The human mind has created quite unexpectedly a set of conditions so dangerous and so complex that if we're to survive them, all intelligent and responsible people may, for the first time in history, have to make a common effort to understand just what we've created and how we're to deal with it. The mental process originally responsible for most of these novel and threatening conditions is, of course, science, the highly specialized occupation of fewer than one percent of our kind. Now as we, the laity, turn to look at the problem, we realize dramatically that few of us have more that a smattering of scientific knowledge, and even that had to be translated for us from a language that might as well be Venutian. We have no clear idea of how science works or how the management of its results may help or harm us in the future.

How could such a state of ignorance have come about? Obviously, because communication from the scientific community to the rest of us has been grossly inadequate in the past. Is it any better now? To answer this question, let me treat it as three questions: I. Why communicate at all? II. How well is the communication going? III. How can scientific communication be improved?

To help me answer these questions, I wrote to scientists, heads of scientific societies, publishers, editors, broadcasters, and science writers; also, I read a little. There was nothing statistical about the effort; I only wanted some authoritative views to correct for my limited experience. Responses were extremely interesting and helpful, though in no way surprising.

I. Why Communicate at All?

Why should scientists try to communicate with the laity, when the effort is so great and the result is so often just one more misunderstanding? After all, scientists can go about their business quite well in isolation from the public, can't they? Public relations doesn't further research; it only takes up precious time that could--and should--be spent in the laboratory. Let reporters sniff out "events" and spoon up a little pablum, or let the Sagans and the Erlichs go on television,[1] and leave other scientists free to labor in their sanctuaries, associating with their own kind and speaking in tongues. Let scientists discharge their societal duties by doing an occasional stint in Washington or by serving on various councils--activities in keeping with the image of scholar and expert.[1]

This aloof attitude, common before World War II, still persists in many quarters. But the bomb did its work, not only bringing the war to an abrupt end, but marking the beginning of the end for the genteel isolation of science from issues that were troubling the rest of society. We could say that the bomb gave science a conscience,[1] and things could never be the same again.

Moreover, WW II also made government the main source of research funding,[2] and since then the cost of science and technology has risen to about $50 billion annually, shared between government and major private industry,[3] because only they can afford it. Yet science affects everybody's daily life in a multitude of ways; so it's everybody's business. Consider military and space research. Both involve billions in tax dollars, raise complex questions of priorities, and have strong impact on world peace.

There are other crucial areas, too, wherein we look to science for help. There are the serious remaining diseases; the deteriorating environment; the dwindling natural resources; and, of course, the questions of food supply and population control to be grappled with. In each area, science will contribute greatly, though none is the concern of science alone. We'll also need help from ethics, law, engineering, and education, among others, if we're to deal satisfactorily with any of them. We need daily to discuss and decide--often by vote--the most involved questions of policy and action, questions which demand a knowledge and appreciation of science. A paragraph on the ballot simply won't do!

And it's not just the citizen; our leaders, too, are uninformed.[4] Government cannot function without platoons of scientist-advisors and

Work supported by the U.S. Department of Energy under Contract DE-AC04-76DP00789.

officials at every level; yet the information is still not getting through:[3,5] terrifying when you consider that more than half the bills before Congress have a scientific or technological basis.[1] Sagan puts it nicely: "We have a society which... uses science in every one of the interstices of national life, and in which the public, the executive, the legistlative and the judicial have very little understanding of what science is about. That is a clear disaster signal."[5]

To try to solve this problem by following the unalloyed advice of scientists, innocent of worldly affairs, is unsound.[6] Moreover, as Hogben warns, "No society is safe in the hands of so few clever people".[7]

Why, then, should scientists bother to communicate with the laity? Because they cannot responsibly avoid it; it's their humane obligation to take us into the sanctuary.

II. How Well Is the Communication Going?

The answer depends on who you talk to, and ranges from an occasional "excellent"[8] to a more frequent "O.k., but ...." There is certainly no groundswell of encomiums. To be specific, social scientists (except psychologists) rightly feel neglected and misunderstood[9]--and there are reasons.[10] Scientists in the "hard sciences," too, feel they're being ill-served, especially by commercial television and the daily press; though many praise the science programing on PBS.[11,12]

These scientists have a point; reporting has too often been scandalous. Yet scientists' perceptions in this regard may be a direct result of their own unwillingness to meet the problem with anything better than complaints about details or withdrawal from the arena. Further, the very incompetence they ascribe to the press is the predictable symptom of a nonfunctional communication channel.

Now let's look at the several media.

Public Television (PBS). Public television does an outstanding job. Series such as The Ascent of Man and Cosmos combine a comprehensive view of the history of science with the broader context of a history of ideas and make the whole business palatable to a broad assortment of viewers. Programs like Odyssey and Nova, together with other science specials, do an admirable job on individual science topics from oceanography to astrophysics. Ratings and comments on both series testify to their success.

Altogether, these amount to several hours per week of prime-time science programs. Still, we need more of the comprehensive, iconographic series to supply the matrices into which one may fit the more specific programs, since the latter are so diverse that viewers have trouble setting them into any sort of coherent pattern. However, this is a wee cavil at worst. Public broadcasting does deal with issues (amniocentesis), it does clarify forbidding subjects (relativity), it presents nonfaddish topics (survival factors in a deer herd), it does deal with the joy and wonder of science (Cosmos), and it does so seriously.

Commercial Television. This medium gets some of the harshest criticism of all.[13,14] As one correspondent said: "Since the bulk of the public gets its news from the broadcast media, I can only conclude that the public is being poorly served."[15]

Except for newsy spectaculars and disasters such as Three-Mile Island, commercial TV treats science as if it were a carnival. Some events such as space shots, moon landings, and other planetary probes, have been heavily covered (covered to death, say some); but this effort represents primarily a news-slanted coverage of a narrow sliver of science, and one almost drowned in what the media call "human interest." ("How did you FEEL, Mrs. Carpenter?") Of course, we must grant that commercial TV's raison d'etre is to peddle other people's hardware, not to educate.

Motion Pictures. These I would pass over without mention, if it weren't that they've been major contributors to a ghastly stereotyping of the scientist as either a hopelessly impractical pipsqueak or else a terrifying paranoiac. So while the cinema has done little or nothing to solve our problem, it has certainly done more than its share to worsen it, and is still doing it. But then, the cinema's raison d'etre is to entertain, that is, peddle its own wares.

Radio. Aside from All Things Considered on National Public Radio and some few-minute science features like The University Explorer, this medium's contribution is negligible.

Book Publishing. The book publishing industry continues to turn out its share of quality material on a wide variety of scientific subjects for a readership of intelligent adults. Also, they often do something for the younger teens--for example, the Time-Life science series of a few years back. The latter venture showed outstanding use of clear, interesting writing, together with vivid, technically sound, well-designed illustrations of a type that now characterizes much science publishing in all formats.

Whenever possible, commercial and university presses use scientists as authors. The university presses in particular have a genuinely educational motive in much of their output. In deciding what to publish they do careful market research and pay close attention to such things as the reputation of the author and the quality of the writing. Because many science books are long-lasting, particularly those with a social or philosophical slant, book publishers are providing a growing volume of generally high-quality science information for the laity.

But in the end, they too follow the tastes of their public and do cater somewhat to fads. Thus we see in bookstores more Azimov and Sagan, and perhaps less earth-bound science, than we need or might be willing to read. I am, of course, aware of the argument that Sagan's celebrity status and rhapsodic style get a lot of people reading this material who otherwise might not. I think it's true and a good thing.

The News Press. On many lists for worst job of communicating science, the newspapers get top billing. However, I think their reporting is better than it used to be.[12] Since WW II, science reporting has become noticeably more sophisticated and knowledgeable. Science features with a lot of background are commoner than they were. The weekly news magazines like Time and Newsweek, with more time to do research and let things settle after a story breaks, do a still better job on science news and features and reach an enormous audience.

However, if you list all the worst sins attributed to science writing: sensationalism, exaggeration, inaccuracy, incompleteness, gullibility, sentimentality, emphasis on sex, unclearness, misquotation of sources, faddishness, irrelevance, and so on, there's no doubt that they're all committed far too often by the daily press.

Many such faults can, of course, be laid to the nature of the newspaper business. Reporters have short deadlines, their stories are cut and rewritten by others, headlines are written by others, editors have biases,[16] and so on. But while these factors may explain the existence of faults, they cannot excuse them. Science can be, and deserves to be, treated differently from hot news. But it takes a positive decision by editors, who now make no distinction between the two.[16] Newspapers make much of being responsible. I suggest that this same order of responsibility might benefit all of us greatly if it were aggressively applied to upgrading the coverage of science.

If, however, the press is unwilling to handle science any differently than it always has, then it should relinquish the field of in-depth science reporting to media less subject to daily deadlines. There is some evidence that this is happening.

The New Science Magazines. The New York Times Tuesday science section is a striking attempt by a newspaper to fill the chasm between the daily press and the monthly science magazines. It also indicates that the media now sense a great enough public appetite for science to be profitable. Couple this observation with the fact that more than half our people over 25 have had some college[17] and are thus likely to be interested in science,[16] and it's not surprising that a daily such as the NYT has decided not to surrender the field but to adapt by going weekly itself.

Is there a "science craze" in progress today, as one editor has asked in print?[18] If not, then there's at least a strong leap in interest in good science writing and reporting. It's equally interesting that Time, a major news weekly, has just launched Discover, a monthly devoted entirely to science, to compete with a number of other brand new science monthlies. These new magazines cover a range in style and level of difficulty, extending from OMNI at the "pop" end of the scale to those like Natural History at the other. Their formats are attractive, the writing generally very good, and the subject matter varied. It's especially interesting that science societies, too, are now publishing lay journals--for example, Science 81, by the American Association for the Advancement of Science, and SciQuest, by the American Chemical Society.

Couple the new entries with the specialist magazines like Star & Sky, Psychology Today, and National Wildlife; and with established journals like Nature, Scientific American, and National Geographic, or a more socially oriented journal like MIT's Science, Technology and Human Values, and we begin to see a notably catholic menu of science fare. Also, several universities have begun publishing for their own constituencies magazines dealing partly or wholly with science. For example, Minnesota's brand new Research, whose premier issue does a fine job of making difficult research subjects accessible to bright readers, and strongly emphasizes the implications of scientific research.

Altogether these monthlies exhibit extremely healthy signs. They assume a perceptive, intelligent, educated audience. They're not afraid to go into depth and detail. They avoid sensationalism and most of the other vices of cheap journalism. They feature well-known and skillful writers. They don't shy away from social issues. And they take pains to avoid jargon and simple-mindedness. There's a clear impression, strikingly illustrated by the sales letter announcing Discover, that they've sought out, considered with care, and are trying hard to respond to the major criticisms levelled at bad journalism. If there's a revolution going on in science communication, it's in the monthlies that it's most clearly reflected. Their success with readers shows in increasing circulation figures, and their reception by professionals is reflected in good marks from my correspondents.[12,19-21]

In general then, how is scientific communication to the laity going? Measurably better than previously, particularly in public television, science magazines, and book publishing. But this is not good enough. Publications for other scientists outnumber those for the laity by more than an order of magnitude. Thus, while the laity are still climbing the first hill, science is disappearing over the horizon.

III. How Can Scientific Communication Be Improved?

It would be easy to say that all we need is just to keep on doing what we're doing. Instead, I believe that what we do need is a fundamental change in the scientists' perception of their role in society, and an active acceptance of that role. They have to learn to see themselves as fully reciprocal, fully integrated participants in the community--highly specialized, highly competent, priviledged to enjoy rewarding work, yes--but not as a class set apart, exempted from any substantial obligation to the community that largely supports them.

What does such an acceptance imply? It implies embracing the duty to tell the rest of us what they're doing and what its implications are for the health and happiness of the population, including all the possible dangers. It implies being willing to enter fully into public discussions and debates about science and its products. Finally, it implies being willing to take the personal risks involved in entering the sociopolitical fray and to make the sacrifice of valuable time inherent in the effort. I know this is asking a lot and that this kind of change can't be brought about by the scientists themselves, acting alone. Nearly everybody needs to get into the act.

In particular, the attitudes of university science departments have to change. During their college years, scientists need some real exposure to the humanities.[22] Communication must be seriously taught as a part of their career responsibilities so that they come to see themselves as citizen-scientists with a major responsibility to inform the laity. Some science professors and deans do give lip service to the idea, but everybody knows that straight science and mathematics courses (with straight A's) are what get you a good job.

As it works out, this is a tragically separatist attitude, a disservice both to science and the society. How can scientists be other than babes in the wood when they try to deal with the full humane and political implications of their work, if they've learned nothing but their own narrow specialties, spend their lives buried in their laboratories, and speak only with each other? Bronowski expresses the larger issue this way: "Every thoughtful man who hopes for the creation of a contemporary culture knows that this hinges on one central problem: to find a coherent relation between science and the humanities."[6]

In the task of changing attitudes, the professional science societies, too, should assist. And they're well-suited for the effort, since they're controlled and staffed by scientists and have the ear of their membership. What can they do? They can influence their membership to accept the burden of communication and full social responsibility. They can--as a few now do--publish lay journals as outlets for their own members' writings. They can open up dialog with the media on problems in science communication. In addition to considering the questions of ethics, social involvement, and communication among themselves, special committees from various societies could sit down with one another for deliberations on such matters and develop strategies for joint efforts.

In the near future, it might be useful to initiate an annual conference on the role of science in society. Some body such as the National Science Foundation or the American Association for the Advancement of Science could sponsor it. They could invite scientists, media specialists, science writers, publishers and editors, representatives of industry and government, educators, philosophers, and other interested delegates. Such a conference could concentrate on defining the role of science and considering its implications for all aspects of human life. They could target problems and establish priorities for a program to bring science and the laity together. And, of course, a major aim would be to foster intelligible, effective communication.

CONCLUSION

Unless there's a fundamental change of the kind we've been discussing, scientific research and the integration of science into the intellectual life of the community will simply continue to be two processes advancing at radically different rates. The present gulf of nonunderstanding can only get wider and the danger of a complete breakdown of communication increase. I think it can be argued that no other field of human effort has so much genuine potential for contributing to the solution of the enormous problems facing the race--problems of overpopulation, preservation of the environment, restoration of damaged species, development of alternatives for depleted resources, exploitation of space for human advantage, conquering disease, and stamping out ignorance. Bringing science and the laity together in this task deserves our best efforts.

REFERENCES

1. Rae Goodell, The Visible Scientists, Little Brown, Boston, 1975.
2. C. P. Snow, in The Way of the Scientist, Eds. of International Science and Technology, Simon and Schuster, 1962-66.
3. Interview with Sen. Harrison Schmitt, OMNI, June 1980.
4. Ian Ridpath, science writer, letter, 1981.
5. "The Cosmic Explainer," Time, Oct. 20, 1980.
6. Jacob Bronowski and Bruce Mazlish, The Western Intellectual Tradition, Harper & Row, New York, 1960.
7. Launcelot Hogben, Science in Authority, W. W. Norton, New York, 1963.
8. John M. Coady, Exec. Dir., American Dental Association, letter, 1980.
9. Edward J. Lehman, Exec. Dir., American Anthropological Association, letter, 1980.
10. Spencer Klaw, The New Brahmins, William Morrow & Company, Inc., New York, 1968.
11. Harry Shipman, physicist, letter, 1980.
12. J. T. Bonner, biologist, letter, 1980.
13. John R. Manger, Editor, Oxford University Press, letter, 1980.
14. Louis J. Batten, Institute of Atmospheric Physics, U. of Arizona, letter, 1980.
15. J. T. Johnson, science writer, letter, 1981.
16. Hillier Krieghbaum, Science and the Mass Media, New York University Press, New York, 1967.
17. Research, editors, U. of Minnesota, statement of objectives, 1979.
18. John S. Rigdon, Editor, "Is There a Science Craze?" Am. J. Phys., 48 (11), November 1980.
19. Vickie P. Raeburn, Assoc. Exec. Editor, Columbia University Press, letter, 1980.
20. Karen Colvard (for Lionel Tiger, anthropologist), letter, 1980.
21. C. P. Idyll, Sr. Marine Analyst, National Advisory Committee on Oceans and Atmosphere, letter, 1980.
22. Alan Ternes, Editor, Natural History, letter, 1980.

RESPONSIVENESS AND RESPONSIBILITY: THE FEDERAL REGULATION WRITER'S INCREASING CHALLENGE

DONNA C. STARTZEL
MORGAN MANAGEMENT SYSTEMS, INC.

Donna C. Startzel
Morgan Management Systems
Associate
Suite 301, Clark Bldg.
Columbia, MD 21044
(301) 997-4060

Since the Federal Register Act of 1935, the Federal government has tried to make its regulations more responsive to the public tasked with compliance. From relatively simple beginnings, this regulatory reform effort has resulted in more and more complex requirements for government agencies and their writers. Through the Administrative Procedure Act, Executive Order 12044, the recent Regulatory Flexibility Act and other directives, the public has been given a much clearer role in the regulatory development process. Correspondingly, regulations writers have been given responsibilities for more sophisticated audience analysis, tighter coordination procedures, stricter language standards, and adherence to much more demanding time schedules. There is every indication that their jobs will continue to become more complex as additional regulatory reform bills now pending in Congress gain approval.

Recent government directives have profoundly impacted the process of writing Federal regulations. However, these are only the latest in a long line of major legislative steps taken toward increasing the responsiveness of the Federal government to the public, and in turn greatly affecting the role and responsibility of regulations writers.

Early Regulations

The surge of New Deal legislation during the early 1930's, with its wide-reaching social and economic programs, brought expanded responsibilities for Federal departments and agencies to issue regulations. More regulations than ever before were drafted, often very quickly. Predictably, communications problems were the result, within and among Federal agencies and between the government and the public.

The reason for this was the absence of a central publication system. No effective way had been devised for citizens to learn about the regulations directly affecting their lives.

Federal Register Act

To impose some order on the chaos, the Federal Register Act was signed into law on July 26, 1935. This first major government directive in the "regulatory reform" area had important ramifications for the individuals tasked with writing regulations. The Federal Register Act established the Office of the Federal Register and required that every new regulation be immediately filed with that Office, placed on public inspection, and published.

The Act furthermore ensured the "legal effect" of regulations thus published in the Federal Register. The regulation writer was thereby tacitly acknowledged as effectively writing "law" in the form of published regulations. Obviously the responsibility factor in the job of the regulation writer was becoming much more crystallized.

Administrative Procedure Act

The signing of the Administrative Procedure Act on June 11, 1946, brought several new additions to the system, and greatly increased the emphasis on government responsiveness to the complying citizenry. The APA established as a general requirement the right of public participation in the process of developing regulations. By law, then, the public was guaranteed a period of at least 30 days for comment on proposed rules before they became effective. It became the task of the regulations writers to reflect the revisions, based on public comments which were often numerous and complicated, in the final rules.

Recent Government Directives

With this increasing awareness of the need to make documents available and responsive to a wider public came the recognition that these documents must be more detailed and more useful to that public.

In the context of this evolution, President Carter issued Executive Order 12044, "Improving Government Regulations" in March, 1978. The Executive Order made three key policy statements. First, regulations must be written in "simple and clear" English. Second, they must achieve legislative goals effectively and efficiently. Third, they must not impose unnecessary burdens on the economy, individuals, public or private organizations, or State or local governments.

New responsibilities were outlined for all those involved in the regulatory development process. Agency heads were called upon to exercise "effective oversight," which in turn required more policy coordination among and within agencies. Meaningful alternatives to regulations had to be considered and analyzed, again emphasizing the desire for responsiveness to public needs. For the first time, agendas of significant regulations under development or review had to be published semiannually in the Federal Register, extending the requirement for advance public notice called for by the APA. Regulatory analyses were mandated for all regulations implying major economic consequences for the general economy, individual industries, regions, or levels of government. Finally, all existing regulations were required to be reviewed for continued need and clarity, or revocation. The regulations writer had an imposing new set of responsibilities, and the job of producing acceptable and effective regulations had become even more complex than before.

The most recent government directive impacting the regulations writer was the Regulatory Flexibility Act of September 19, 1980, which incorporated many of the policy and procedural issues of the Executive Order, but took even more determined steps in the direction of responsiveness to the complying public. Its objective is to ensure that agencies fit regulations to the scale and capabilities of the groups that must comply with them. To make sure this occurs, agencies are required to actively solicit the ideas and comments of small businesses, small organizations and small governmental jurisdictions, to consider flexible regulatory proposals, and to explain their final actions so that alternative proposals are given serious consideration.

New Tasks and Techniques

The consequences of sincere responsiveness to the public are often difficulties for the regulation writer, such as adherence to difficult time scheduling requirements, the review of enormous volumes of public commentary on proposed rules, substantive and often extensive and repeated revision, and the complexities of simplifying language while retaining precise meaning.

The consumer-oriented requirements are creating totally new challenges of accurate audience analysis. From both economic and stylistic perspectives, Federal writers are being forced to carefully analyze the effected public to determine economic impacts and sexist and minority concerns of language, syntax and their implications.

Clarity of thought eliminates confusion of language and presentation, but a well-thought-out regulation takes time to "gel," time that is rarely built into the schedule. To produce clear, precise, and concise regulations within ever-decreasing lead-times, agency heads and writers are developing numerous techniques to improve writing efficiency, including 1) establishing core groups of writers and reviewers, 2) applying objective management principles to the total regulation development process, 3) systematically eliminating well-established stylistic problems typified by legalese and sexist writing, and 4) providing increased training for regulation writing personnel.

What Lies Ahead?

What lies ahead for the drafters of Federal regulations? A host of even more demanding regulatory reform bills loom on the Congressional horizon. The Equal Access to Justice Act, effective last October, will encourage lawsuits against government regulations which have left themselves open to challenge. The "paperwork burden" bill, the "reg reform bill" with its sunset provision for automatically phasing out dated regulations, the "regulatory budget" bill, and others are waiting in the wings for passage or amendment. Whether successful in Congress or not, the call for regulatory reforms is a popular one. And it will continue, in one form or another, to bring increasingly complex responsibilities to the task of writing responsive regulatory documents.

SHARING TECHNICAL KNOWLEDGE – CREATOR TO CONVEYOR TO READER

Lois Kathryn Thuss
The Johns Hopkins University Applied Physics Laboratory

Lois Kathryn Thuss
Applied Physics Laboratory
Technical Writer
Johns Hopkins Road
Laurel, Maryland 20810
301-953-7100

Society for Technical Communication
Washington, D.C. Chapter

Institute of Electrical and Electronics Engineers Professional Communication Society.

Ours is a unified effort. We are – and must be – a team. Else, we fail. Those in the technological functions – science, engineering, math, and so on – are gifted and essential to progress. But, so are **we** *their transmitters – their technical writers and editors.*

That's why both parties must display mutual respect. It isn't a question of precedence. It is a mandatory factor for corporate or organizational success and the betterment of our work, our personal lives and the world at large.

One of the problems – indeed, the primary difficulty that we have in our system today – is that often whole schemes that may affect future generations are constructed on negativism, pitting one factor against another internally for petty, self-serving interests. We, as a general society, cannot succeed in that manner. We must pull together and, in the following paragraphs, few as they are, may be a lesson that could transform your lives and careers. I cannot say exactly ***how****; I only know* ***why****.*

The reason is that we all are knowledgeable, hopefully prolific, and darned dedicated. In fact, acknowledged or not, our employers are as dependent upon us as we are upon them.

The Ostrich Syndrome

Scientific, engineering and some management minds are special instruments, to be admired and, perhaps, even coveted. Yet, their very natures frequently tend to enclose themselves, to be self-contained, noncommunicative. Sometimes intentionally, sometimes not. (Sometimes just shy.)

The great thoughts that have brought about most of our ancient and modern advances in civilization have been necessarily extracted like wisdom teeth from their brilliant discoverers and capable exponents.

Why? Can these people not be geniuses and communicators, too? Creative forces and teachers, also? Many of us feel they can, and blame a great deal of the problem on a lack of inspiration or motivation to take their minds off of particular pet projects or problems long enough to share their cumulative thoughts with others.

Scientists and engineers must realize that their good and noble instincts belong to all of us as parts of our total heritage, not just to be coveted by them as their private holdings because they were blessed with a mental – biological, if not plain genetic – gift, and we are being assessed organizationally and as a total public, with the expense. They must share. And, to share, they must communicate. That's where we technical writers and editors can help most. We can "say" their ideas for them – and let them have the credit.

The Greatest Have Problems

It was Dr. Albert Einstein who fervently tried to explain his Theory of Relativity to a body of astute mathematicians, and they – at least, at first – were unable to fathom $E = mc^2$.

He didn't have in mind an Atomic Bomb. No. In fact, he was a life-long pacifist, a theoretician, basically a mathematician, and perhaps the finest physicist we'll ever know. But his initial difficulty was in transmitting his genius to his peers. In later years, he philosophized, even politicized, but died knowing technical facts he never had an opportunity to expound, but which could be to our great benefit today. Only by delving deeply are we able now to absorb or even comprehend Professor Einstein. If yet.

Let me share with you one amusing, but true, anecdote about Dr. Einstein, undoubtedly the most distinguished member of the elite Institute for Advanced Studies at Princeton University where he served from 1933 until his death. It typifies the scientific mind. Dr. Einstein was walking down the street, engrossed, as he usually was, in one of his equations. This particular day, looking for something as simplistic as an ice cream cone (Dr. Einstein loved ice cream), he walked down the middle of the street, interrupting traffic, horns blowing to his obliviousness, turned into a hardware store, and said ''Vanilla.'' The man behind the counter, recognizing Dr. Einstein immediately, directed him to the nearest drugstore, where the good doctor got his treat, and took the same motorist-exasperating route home, mid-street.

But, such are geniuses. They can be possessed of all the information in the world on a subject or theory, but it does precious little good if they don't – or won't – use their trained professional channels to share it.

The Cure

Scientists and engineers can share their information through technical writers and editors who will convey it accurately to the readers. We, as specialists can take the yokes off of the backs of technical people so they can get on with their principal work ... that of discovery, adaptation and improvement.

In fact, if it were not for good and competent technical writers and editors, such people's knowledge might never be recognized nor properly comprehended. Like Dr. Einstein, originally trying ardently to convey his theories to fellow mathematicians, most of whom also were of genius stature, vital knowledge can be lost without proper conveyance to the benefactors.

Often, those of us in the writing and editorial profession tend to place ourselves on par or even above our authors, the real hearts of the profession, the scientists and engineers who create the information we convey. That should *never* be the case.

Scientists and engineers generate information. We merely convey it, as accurately as possible. If our product ever is wrong, it should be attributable to incorrect, imprecise or ambiguous information provided to use. And whoever is the person at fault should bear the responsibility.

Both of our jobs are important, and neither of our bodies should neglect nor malign the other. We must work as a unit. Always, and together. To the happiness and success of our teams, and for the betterment of the world.

Four Cardinal Rules

To maintain mutual respect and to succeed as a team, we must remember certain fundamentals:

First, we must, as scientists, engineers, writers and editors, *understand each other* and our distinctive roles in our fields.

Second, we must *respect each other* for our different expertises, each implementing the other, always discussing, never arguing. Writers and editors must report accurately to the finest detail, what the originator has developed. By the same token, scientists and engineers must consider the professional's judgement in reaching a myriad of audiences, both media and readers, who may have varying interests. The message must be comprehensible and palatable to them all, if possible, without changing the integrity or ultimate purpose intended by the author.

Third, we must *encourage each other*. Negativism never won an argument, a war – or even a bridge game. It kills, never brings life to anything. It is defeated before it begins, perhaps before it is even conceived.

And, lastly, we must better *acquaint ourselves with each other*. Perhaps, that should have been the first point. We must know what to expect from each other. And expect nothing more. Acknowledge our limitations. Be realistic about our capabilities. None of us can be all things to all people.

Wasn't it Abraham Lincoln who said, ''No honest man can be all things to all people at all times''? A wise person. He changed the course of history with his homespun philosophies.

Why can't we?

THINK VISUALLY...COMPOSE IN SEQUENCES...WRITE FOR THE EAR...

Grace Tillinghast
Eastman Kodak Company

Grace Tillinghast
Eastman Kodak Company
Editor-writer
343 State Street
Rochester, N.Y. 14650
(716) 724-2010

Think visually...
Compose in sequences...
Write for the ear...

Exciting visuals, gripping sequences and continuity. And the images and narration complement each other perfectly. THE END.

Well, it is not that simple. A scriptwriter can envision an entire production, but it is not easy. He can also envision the development of a script sequence by sequence. People, objects and action.

Professional scriptwriters visualize their way to a script because audiovisual media are meant to be seen and heard. So they have to gear their thinking to the senses. For this, they have to forget much of what they have learned in school.

By saying "think visually", I mean that a person writing for movies and audiovisuals has to become visually fluent. That is to say, a writer must be able to visualize his way to a script.

How to become visually fluent?

First, it is important to pay attention to all visual messages around you. People react to an increasing number of emotions, orders, thoughts.

Everyday you see a great variety of visual messages that trigger a reaction.

- a stop sign
- a green light
- an elevator going up
- an arrow indicating direction

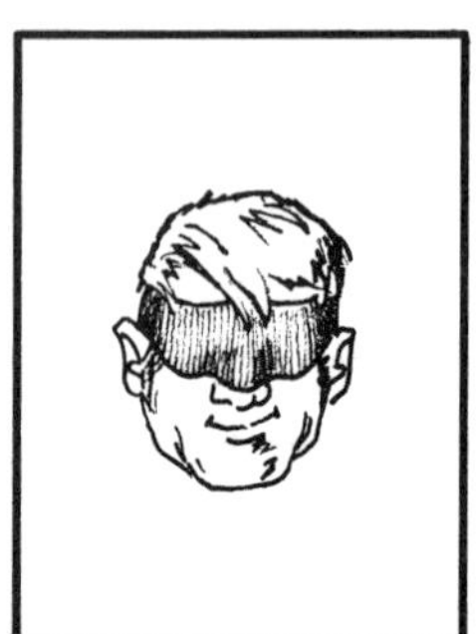

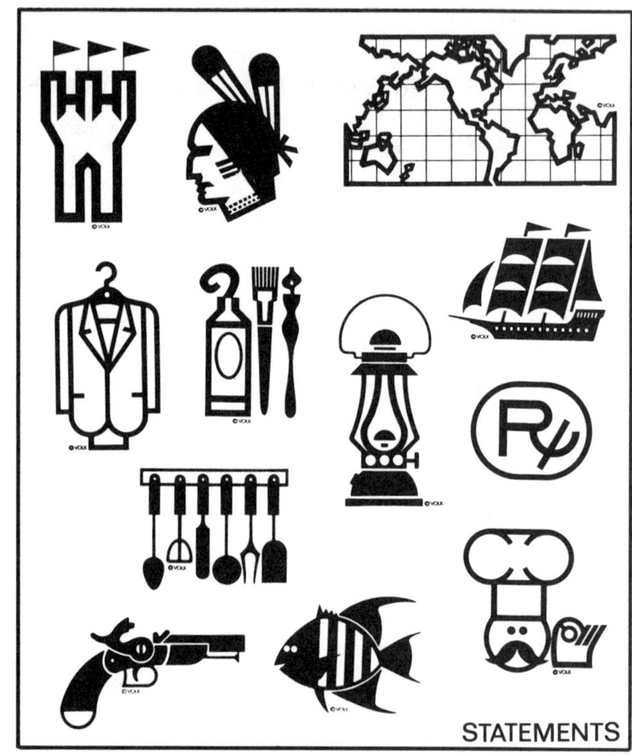

For 444 days, this meant a number of things: Iran, the shah, families waiting, our national pride, etc. Emotions and reactions differ from person to person.

Second, work with others to generate visuals. Remember that four eyes see more than two. Get together with your fellow writers and feel free to tell them that you are looking for ideas.

Third, consider maps, faces, locations, graphs, so you can identify a visual possibility.

Fourth, try different areas of interest and do not forget daydreaming. Many great ideas develop in a dream.

Keep your mind open and get into the habit of writing, clipping, recording or photographing whatever may help to increase your visual fluency.

Fifth, once you have collected a mountain of ideas, file them under two major categories:

1. Visual statements
2. Visual metaphors

Under visual statements file things, places, words, graphs. Under visual metaphors include things that are used to represent something else.

And, as it is the case with a foreign language, to maintain your visual fluency you must exercise it all the time.

Instead of thinking logically, you must do it visually. You are dealing with images that illustrate your ideas. They have to be concrete and specific images that convey a given meaning. And instead of writing in well formed sentences and paragraphs, you must write in sequences because ideas and images develop side by side. The impact comes through your ears and your eyes.

We know that audiovisual scripts are written in a two-column format. Such a format makes it easier to establish a visual relationship between words and images.

Writing for the senses demands more than thinking visually. An author like James Michener spends pages and even chapters with long descriptions and leaves it up to the reader to come up with a visual rendition of his prose. He makes you imagine. He gives you all the necessary ingredients and you mix them to your liking. But if you were to write a script for one of his books, your visuals would have to speak by themselves. And then, the challenge of the script-writer is to be careful about finding the right words and those that can be heard and understood with ease.

Because words are spoken at a rate of 125 per minute, there is no chance to go back, to pause, to reflect. People are going to see it and hear it at 60 miles per hour, so a writer wants the visual and the verbal to relate to one another.

When you like a certain passage in a book, all you have to do is to read it again. You may take as much time as you like to pause and reflect. While attending an audiovisual presentation, if you miss a word, you may miss a whole sentence.

Because of this, a professional script-writer must select words and adopt a style that can be understood: usually short, easy to understand sentences. And since style is pitched to the listener's ear, a writer should read a script aloud. If one person can read it without difficulty, a narrator will have no problem. While doing this, if a writer stumbles over a word, he should change it. Unless it is absolutely necessary, long difficult words should be avoided. While trying to comprehend a meaning, a person may lose a few seconds and many more words. If a sentence happens to be too long, he should change it. Remember, there is no chance to go back and listen to the sentence again. If a sentence does not make sense or if it does not belong in the sequence, he should delete or change it.

Being a scriptwriter is not easy because he must remember that there are other components to deal with: there is music, sound effects and narration. It means he will deal with photographers, cameramen, editors, musicians, directors and producers. How well he deals and works with them will have a definite impact on the finished product. Because of this, a writer should be involved from the very beginning of the project and recognize the visual orientation of the thinking and writing that goes into making the visual idea work. Words and music have to get along like ham and eggs.

But, what comes first, the chicken or the egg? It all depends on who you are and whom you work for. Time and economics play a major role. Sometimes, you produce a script and because you are visually fluent, you become the producer and director of the presentation. Other times, someone else will give the visual and sound effects and ask you to write accordingly. In my own experience, especially in the area of TV coverage of current events, I am often asked to write a script that will go along an already produced piece of video. To comply with such an assignment, I use the visual available to trigger a reaction. Then, I outline the facts that <u>must</u> be given and compose my sequences. Once I have written the final draft of a sequence, I read it aloud or I tape it.

I get my eyes and ears to react. Depending on their reaction, I may start all over again...

In all audiovisual presentations, narration will be 1/2 or 3/4 at most. In some instances, narration may be less depending on the value of your visuals. Then, you must decide what is the most important element in a given sequence, sound or picture. Whatever your choice, all elements should speak the same language, that is, be compatible. Of course, if you are writing about Shakespeare's production and have all necessary images, your leading music will not be a rock-and-roll tune...In many cases, if your eyes <u>have it</u>, sound effects become secondary. Some statistics reveal that people tend to remember what they have seen better (and longer) than what they have heard.

If you are trying to convey a message to technically oriented people, first you must establish how technically oriented they are. And consider that potatoes plus potatoes add easily, but other elements may not. Are all of them equally informed? Once you have got answers to a number of questions, you determine the level of your presentation: beginner, intermediate and advanced.

If a writer is visually fluent, a technical presentation presents a challenge. Certain things are rather difficult to explain and you may end up with a wordy script. But remember that images save 50% of the effort. Your battle is won if the audience can see it.

In my case, and for obvious reasons, I write in Spanish and must be visually fluent and use images that will provoke a reaction on Spanish speaking people. Luckily, many visuals are international:

Others are not:

If you are visually fluent, you can substitute, change, delete or add visuals to give a meaning. For instance, in this country, the color green often stands for money because dollars are green. In Spanish, green is sometimes applied with a peculiar connotation: a "green joke" is a dirty joke; a "green man" is a dirty old man.

A recent example: Yellow ribbons were seen everywhere until the American hostages came home. Such a visual has no meaning in Spanish because of cultural differences.

In order to be visually fluent in another culture, a person must rely on experience. Things that a person has seen, heard, lived through, or emotions that have left an impression. Communication makes it easier for a foreign writer to keep in touch with the latest developments, and, of course, periodic trips to the target country to see how people react to visuals.

Whatever your assignment may be, remember this useful ABC: think visually, compose in sequences, and write for the ear.

OPERATING PROCEDURE SYSTEM AT SAVANNAH RIVER PLANT

Charles W. Tope
E. I. du Pont de Nemours and Company

CHARLES W. TOPE

E. I. du Pont de Nemours & Co.
Savannah River Plant
Aiken, S.C. 29808
(803)-725-2872

Senior Editor: Edits operating procedures, technical reports, and technical papers.

A.B., Physics, Duke University
M.S., Physics, Georgia Institute of Technology

The Savannah River Plant (SRP) is operated by the Du Pont Company for the U.S. Department of Energy (DOE) to produce nuclear materials. A highly organized procedure system has been developed that provides positive control for all plant operations and supporting functions. Authorized written procedures convey clear, concise, up-to-date instructions to operating personnel to accomplish desired operations in a safe and efficient manner. All procedures for repetitive operations are processed and issued by the Technical Services publications group, which provides centralized authority and control of procedures for the plant. Word processors are used to expedite procedure processing.

INTRODUCTION

Virtually all operations at SRP are performed per individually authorized written procedures. These procedures are originated by supervisory personnel trained in procedure writing and employed by the various departments responsible for plant operations and supporting functions. Procedures are processed by the Technical Services publications group whose main functions are to edit, type, provide illustrations, print, and control the distribution of process descriptions and operating procedures. Details pertaining to the SRP procedure system are presented in this paper.

DISCUSSION

Procedure Description

Procedures are written instructions for operating, maintaining, and modifying plant facilities to provide systematic control of safety, quality, and yield. A procedure usually consists of a set of directions for performing a job or for managing a process or an operation. It may serve to inform, to train, to achieve adherence to job methods and product standards, or to record operational details. The following three types of procedures are the most widely used at SRP:

- DPSOL (Du Pont Savannah Operating Logsheet)
  The DPSOL is designed primarily for on-the-job use. It may be a logsheet for recording data, or it may be a detailed how-to-do-it procedure that may or may not require checkoffs or initials to indicate progress on the job. Each DPSOL is individually page-numbered and has its own sequence of revision numbers. Printed DPSOLs are supplied by Technical Services to the using department in multicopy pads.
- DPSOP (Du Pont Savannah Operating Procedure)
  A DPSOP is a book containing descriptions and/or directions for a major plant operation. Some DPSOPs are largely collections of individual DPSOLs.
- Plant Manual
  Any one of several volumes stating plant policy and providing directions for administering the policy. These manuals usually have plantwide applicability, i.e., Safety Manual, Service Manual, etc.

Originating Procedures

Procedures are written, reviewed, approved, and controlled according to the requirements set forth in DPSOP 1, SRP Operating Procedures. A change in any part of the SRP procedure system requires a revision to DPSOP 1 approved by the Plant Manager.

Basic responsibility for procedures lies with the originating department. Normally, the Department Superintendent appoints one or more supervisory employees as procedure coordinators. In some cases the department has a procedures group. The coordinator is responsible for preparing (or having others prepare) initial drafts for procedures, assigning identifying DPSOL and DPSOP numbers (in coordination with Technical Services), routing drafts for appropriate supervisory review and approval, forwarding approved procedures to Technical Services for publication, ensuring that adequate supplies of current DPSOLs are available for field use, canceling obsolete procedures, and keeping related records. Most field coordinators prepare procedures using word processing equipment.

Procedures are initiated where a need for written directions exists. In the case of a new production facility, a task group may be assigned to begin writing procedures months before the facility is to begin operation. In the case of a machine purchased from a vendor, DPSOLs may not be written until the machine is delivered to the user. In the case of a new employee benefit plan, a procedure must be added to the Service Manual to inform supervision how to administer the plan.

In all of the above cases, source material for preparing procedures will be available. For a production facility, documents will exist in the form of Technical Manuals, Technical Standards, Mechanical Standards, and drawings. For a new machine, there will be vendor-supplied information -- operating and maintenance instructions, drawings, etc. For the new benefit plan, material will be available from the Wilmington Service Manual or Management directives.

Procedure Review and Approval

Procedures are reviewed and approved at successive levels of supervision. The review/approval level required is determined by the nature of the procedure and is specified on an approval form attached to the procedure. Procedures are routed for review/approval to all departments that may be affected by the procedure.

When procedures are routed for review/approval, any questions or comments are resolved with the author. Procedures are not processed by Technical Services unless all reviewer/approver comments are resolved.

Procedure Revision

Procedure revisions are initiated as necessary for the following typical reasons: Management directives, process changes, test authorizations, test conclusions, audit results, safety considerations, equipment modifications, changes in standards, and employee suggestions. The procedure coordinator also provides for periodic review of procedures for timeliness and accuracy.

Procedure Audit

Procedures are subject to audit at any time for completeness and correctness. Routine audits involving on-the-job use of procedures are made by area supervision and by independent audit committees. DOE-SR personnel also make periodic audits. Monthly audit summaries are sent to Department Superintendents. Semiannual audit reports are submitted to Plant Management and DOE-SR. Procedures are revised as necessary based on audit results.

Procedure Cancellation

Obsolete procedures are canceled by the departmental procedure coordinator or higher authority by submitting the proper form to Technical Services. Technical Services removes the procedure from the parent DPSOP; however, the canceled DPSOL is retained in a permanent history file, which provides a cumulative record of procedure changes. This DPSOL number is not used for another DPSOL except in unusual circumstances.

Technical Services Functions

The procedures that are originated by the various departments responsible for plant operations and supporting functions are submitted to a centralized Technical Services group for processing and publication. The material is submitted as hard copy accompanied in most cases by a word processor diskette. The main functions of Technical Services are to edit, type, provide illustrations, print, and control the distribution of the submitted process descriptions and procedures.

Editing

The material received from the field coordinators is logged in using word processing equipment. The hard copy is then submitted to the appropriate Technical Services Editor. The editorial staff consists of 10 Editors. Each Editor is responsible for particular volumes (DPSOPs) that include process descriptions and/or individual procedures (DPSOLs) pertaining to main plant functions.

Each DPSOL is edited to convey clear, concise instructions to operating personnel to accomplish desired operations in a safe and efficient manner. Questionable or unclear information is resolved by consultation with the appropriate field coordinator and is revised as necessary. Editors are also responsible for verifying proper approval, references, revision numbers, and indexing. A Style Guide and a Word List have been published by Technical Services and are used by all Editors to provide plantwide uniformity to procedures.

Typing, Illustrating, and Printing

After editing, DPSOLs are retyped or revised as necessary by Manuscript Editors using word processing equipment. Manuscript Editors provide backup editing pertaining to style, format, spelling, and punctuation. All material possible is processed using the word processor and is stored on hard disks within the central processing unit. Hard copy (camera copy) is provided by the word processor for review and subsequent printing. Camera copy for diagrams, complicated data sheets, and maps is provided, as necessary, by Illustrators.

Following appropriate review by both the Editor and field coordinator, the material is printed and distributed for field use. Information pertaining to special printing requirements, number of copies, padding requirements, and color of paper is transcribed from the approval sheet to the printing order. Approval sheets are filed and retained permanently. Material recorded by the word processor is transferred from a working file to a permanent file when the DPSOL is sent to printing. This becomes the official record. Hard camera copy is not retained, except for material that cannot be recorded by the word processor, for example, diagrams, photographs, maps, etc. New hard copy is obtained from the word processor in the event that additional copies need to be printed.

A logging system using the word processor is employed that enables efficient determination of the status of a procedure, on request. The progress of a procedure is logged as the procedure moves through the various processing operations. Status reports are provided by word processor printouts on a weekly basis.

Document Control

The Technical Services Document Control group accumulates printed copies of DPSOLs and process descriptions (new or revised) for each DPSOP, and issues this material to holders of individual DPSOP books every 2 months. Obsolete procedure revisions are recalled and accounted for when later revisions are printed and distributed, thereby ensuring that individual DPSOP books contain only current information. Document Control maintains a record (official) copy for each DPSOP. This copy is used to verify user copies as necessary.

Indexing

Each Technical Services Editor is responsible for maintaining and publishing an index of the procedures in his area of responsibility. These indexes list DPSOLs in numerical order, and give the DPSOL title, the current revision number, and the DPSOP(s) in which the DPSOL is located. Word processing equipment is used to maintain up-to-date indexes. Indexes are published twice a year.

Telecommunications

Word processors are used by most field procedure groups and by the centralized Technical Services group to prepare and record procedures. Telecommunications equipment linking some of the field word processors with the central Technical Services word processor is used to expedite procedure processing.

EDITING COMPUTER-BASED EDUCATION LESSONS

Barbara J. Van Eps
Control Data Corporation, Courseware Operations Division

Barbara J. Van Eps
Control Data Corporation
Manager, Editing and Text Production
HQB01H, 8100 34th Av. S.
Bloomington, Mn. 55440
612-853-7015
Member STC; Member American Society for Training and Development.

Editing computer-based education materials is a specialized task requiring knowledge as an editor and also technical expertise in computer-based education (CBE). CBE is defined as a computerized educational delivery system; it is an educational environment marked by specific applications of educational and computer technologies to the learning process.

Control Data's educational computer delivery system is called the PLATO® system. CBE on the PLATO system consists of PLM, or PLATO learning management, and PAL, or PLATO assisted learning. The PLM manages the learner by performing such tasks as automatically keeping records, scoring tests, and routing students through the course. The PAL is designed to teach the learner. PAL on the PLATO system assumes a direct instructional role by presenting new information in an interactive mode. It is the editing of PAL, the actual instruction, that is discussed here.

I will present a four-step process editors use when editing instructional lessons delivered on the PLATO system, and then discuss different things editors look for when editing PLATO lessons. The PLATO system is unique as an instructional tool in that it is interactive. Editors must keep this in mind and attempt to ensure that student interaction will be positive and motivating. Although editors now edit on paper printouts, in the near future, they will be manually editing changes on the PLATO computer system.

Computer-based education editors have a challenge to meet. The challenge is to continue to refine editing on computer-delivered instruction so that students will enjoy and learn the materials they choose to take.

In his book *Future Shock*, Alvin Toffler states: "Today's change is so swift and relentless in the techno-societies that yesterday's truths suddenly become today's fictions, and the most highly skilled and intelligent members of society admit difficulty in keeping up with the deluge of new knowledge even in extremely narrow fields." Efficient, effective instruction is needed if people are to keep up with this explosion of information.

The topic of computer-based education is discussed in almost every instructional technology journal you happen to pick up today. And it is being hailed by many as efficient, effective instruction. However, behind the scenes, there are writers, editors, and programmers, to name a few, that are preparing the computer-based education lessons and ensuring that the lessons are of high quality. This paper discusses the role of the *editor* in the production of computer-based education materials.

I will begin by defining computer-based education and its application on the Control Data computerized system. This will be followed by identifying the four steps in the process of editing computer-based lessons at Control Data. Within these four steps there are unique considerations and criteria editors must address. Eight of these items are identified. Finally, I will discuss my perceptions concerning the future of editing computer-based lessons.

Definition

Computer-based education (CBE) is defined as a computerized educational delivery system; it is an educational environment marked by specific applications of educational and computer technologies to the learning process.

Control Data's educational courses are delivered on the PLATO computer system. CBE on the PLATO system consists of PLM, or PLATO learning management, and PAL, or PLATO assisted learning. The PLM manages the learner by performing such tasks as automatically keeping records, scoring tests, and routing students through the course. The PAL is designed to teach the learner. PAL assumes a direct instructional role by presenting new information in an interactive mode, whereby the computer asks the student for responses, and then "answers" the student, or "guides" the student through the lesson.

The editor performs the critical task of pulling the PLATO lessons together with the other parts of our multimedia courseware. Even though this paper will discuss only the editing of PLATO lessons, it is important to understand how the PLATO computer lessons fit with the other media. See the following chart.

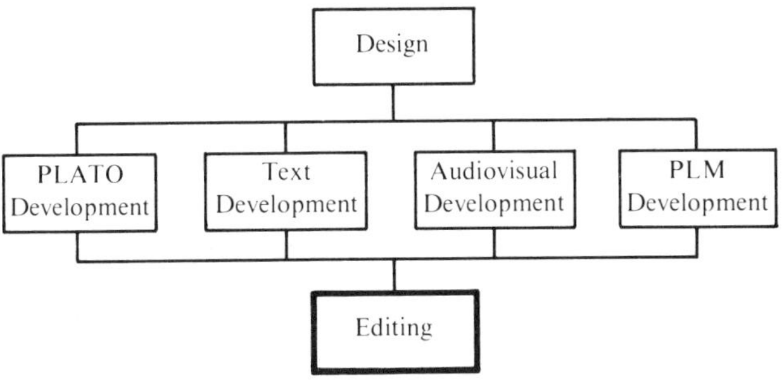

Four-Step Process

Editing the PAL lessons is accomplished in the following manner. We edit on a paper printout which is provided to us by the writer of the lesson. This paper printout is a dump of the lesson from the computer onto a line printer.

After we have edited the lesson, we return the printout to the writer. The editing changes are made to the lesson by the writer or a programmer. Then we enter the computerized system and verify that the changes were made correctly. In other words, we take the lesson on the computer as the student, with the paper printout beside us, and note on the paper printout where changes were not made or any other inconsistencies with the paper printout. We also verify that the lesson is working from a technical standpoint, such as verifying that the keys on the PLATO keyboard activate the correct answers from the system.

After the edit cycle I have just explained is complete, Control Data runs a student through the lesson as a test to evaluate its effectiveness. After this one-on-one test, a pilot evaluation is done where a number of students are run through the course. Then often there is a field test of the course following the pilot. The field test usually runs for a few months in a few selected sites. Then all the comments from these various evaluations are assessed and the course is adjusted appropriately. After this evaluation phase, we receive a paper printout with corrections that were made from the test study. We edit the paper printout and then after the changes have been programmed, we verify these changes on the computerized PLATO system. At this point, the lesson leaves our department and goes to a technical/mechanical department that does a final check and publishes the lesson.

Editing Considerations

What do editors look for when editing PAL? We look for the same things an editor looks for in traditional, printed products: clarity, readability, organization, flow, consistency, etc.

Something that is unique to the PLATO system is that it is interactive. By interactive I mean the system asks questions of the student. The student must respond with an answer in order to progress. If the wrong answer is selected, the system guides the student to learning the correct answer. In editing PAL, both on paper printouts and on the computer terminal, the following factors are considered.

- Because the PLATO system is interactive, students are often asked questions and given a chance or required to respond. Simple things such as making sure the PLATO screen displays the potential answers to the student at the right time must be verified by editors.
- The editor also checks to see that the appropriate feedback is given to the student's answer. For example, if a student answers correctly, the PLATO system congratulates the student or says something such as, "Good job." If a student answers wrong, the system asks the student to try again. It may even give the student a hint as to the correct answer at this point.
- The lessons should be visually appealing. Students sit at the computer and look at the screen. There is no instructor to keep their attention. Therefore, it is important that the screen is not crowded, that the type is easy to read, and that the format is easy to follow.
- The graphics in the lessons should have visual appeal. Editors ensure that the graphics are complete. The PLATO system can do wonderful graphics such as plotting detailed displays in front of the student. Editors check carefully that the graphics are complete, clear, and attention-grabbing as well as educational. Editors also suggest places in the lesson where a graphic might enhance or better teach a point.
- The student should not be intimidated by the computer system. Students are not usually as familiar with a computer as they are with a book. Therefore, students are easily intimidated by the technical aspects of a computer system. Primarily, editors must ensure that the directions are extra clear, helping students feel comfortable.
- More than in the normal text environment, we have to be careful that the material does not talk down to the students. Sometimes our PLATO writers have a tendency to personify the PLATO terminal. Sometimes the result is that the PLATO system will "talk" to the student in an almost childish manner. I remember once an editor was lamenting that the PLATO system had said in a lesson, "PLATO really loves to eat steak. How about you?" This may cause students to take offense, and hence give them negative feelings about computer-based education.

- As in all computer environments, the PLATO screen has space limitations. It is a small, finite space. The screen grid is made up of 512 dots. A character is 8 dots wide and 16 dots tall. Therefore, the screen is 64 characters wide and 32 characters tall. As you can see, the screen is a limited space, and screen displays can have only so much on them. For example, in order to save space, numbers are often in numerals rather than spelled out as the *Chicago Manual of Style* recommends. Editors must be aware that if they change a simple sentence it may upset the entire display and make a change impossible. This requires that the editors work closely with the PLATO writers and programmers in these cases.

Most of the above factors are unique to the PLATO system and, therefore, must be considered in addition to the standard editing criteria.

Editors must have technical expertise in certain parts of the computer language in which the lessons are programmed. The paper printouts contain computer verbiage, some of which is for the computer only. However, the computer words do appear on the printout used by the editor. The student taking the lesson on the computer will never see those computer words. Therefore, they do not need to be edited. Other computer verbiage *will* appear in the lesson to the student. Thus, editors must learn which words to edit and which to ignore. For example, the following illustration is a page from a computer printout. The left column contains "commands" to the computer. The -at- command is telling the computer at which locations on the computer screen to print the words after the -write- command. The editor only edits the material after the -write- command. The -box- command indicates that a box is drawn on the screen.

In addition to the above points that editors must consider, the lesson must also tie into the rest of the course—the text, A/V and testing—in tone and in consistency of terms. This is an important point of PAL editing. Editors must coordinate the PAL lesson with the other media. Often the text writer, PLATO writer, and A/V writer are all developing in parallel, and the editor may be the first to see all the pieces side by side.

Future

I foresee our editors actually going online and editing. By online editing I mean an editor manually making the editing changes in the program on the PLATO system. I'm sure this would have both advantages and disadvantages. It would be time-saving because the editor would make editing changes directly online, saving the programmer time. However, if the lessons are lengthy, it would have the disadvantages of not being able to page back and forth quickly and lay the pages out side by side. In any event, I can see this becoming viable in the near future. If this happens, it will require training our editors to program their editing changes on the computerized PLATO system.

Now Control Data's courses are about 60 percent PLATO assisted learning lessons, the rest, text and A/V. This percentage of PAL may go up in the future, making our editing load even more extensive. Editing in a computer-based environment is a new and challenging field which offers the potential of exciting and expanded considerations in the editing process.

```
 3  calc     datar⇐1          $$set replot all of form switch
 4  lup
 5  erase
 6  at       8Ø9
 7  write    Which background information would you like to see?
 8  at       1Ø2Ø
 9  write    1. Forecast worksheet
1Ø           2. Salary ranges
11           3. Merit increase guides
12           4. Employee background
13           5. Actual employee performance
14           6. Personnel grade verification
15  at       2611
16  write    For your convenience, the above information has
17           been included in your text.
18  box      25Ø9;276Ø;3
19  lover
2Ø  *pause    keys=numeric,funct
21  pause    keys=numeric,next,term
```

THE FINE ART OF POLITIC WRITING

DAVID K. VAUGHN
EDITORIAL CONSULTANT

David K. Vaughn
Morgan Management Systems
Editorial Consultant
Suite 301, Clark Bldg.
Columbia, MD 21044
(301) 997-4060

This presentation suggests that politic writing is common to all business and governmental organizations when success in communication depends less on getting to the point than on maneuvering around the point, on approaching it indirectly or obliquely. Although politic writing resembles two other terms, political writing and policy writing, it differs from them significantly. Politic writing is distinguished by the following characteristics: it addresses sensitive issues; it is directed to a unique audience; it is employed in a special rhetorical situation; it requires a distinctive narrative attitude; it is essentially ameliorative in nature; and it employs deductive logic. Although the style of successful politic writing may appear to violate established standards of effective writing, it is an important and necessary element in the range of writing styles available to contemporary technical writers.

This presentation will offer a brief foray into the area of politic writing in an attempt to recognize it as a unique style of writing required in special kinds of communication situations often encountered by technical writers. Although the theme of the 1981 ITCC Convention has been given as "Communication in the 80's: Getting to the Point," this discussion will occasionally offer a contrary and even slightly heretical view of the tasks of the writer, for it will suggest that there is a special kind of writing, common to all businesses and governmental organizations, in which successful communication depends less on "getting to the point" than on maneuvering around the point, on approaching it indirectly or obliquely. There are times when writers are obliged to write of their topics with deliberate vagueness, to use nouns taken from the top rung of the ladder of abstraction, to describe the subjects of their sentences in generic terms or in the passive voice, to prefer generalities to specific statements, to suggest actions and attitudes by nuance rather than with precise, definitive expressions. There are times, in short, when writers must employ a politic style.

Definition

Politic writing resembles two other kinds of writing, political writing and policy writing, but it differs from them significantly. Political writing, usually thought of as speech writing for politicians, often consists of generalized comments about vaguely stated issues; recently the expression has included connotations of misleading or even dishonest speech. Policy writing, on the other hand, is a kind of writing, sometimes academic, sometimes bureaucratic, which is concerned with statements of goals and objectives, implementation steps, and evaluative procedures. Policy writing occupies a position somewhere between these positions: it pertains to policy statements and it also pertains to statements made in a public forum. However, it also includes a certain amount of stylistic shrewdness (a term the dictionary employs in its definition of politic). Shrewdness is a valuable term to mention here for it suggests that the ability to discuss these issues tactfully is a difficult but important skill.

Purpose

The purpose of politic writing is to address sensitive issues in a discreet and indirect fashion. Typically, it conveys meanings through nuances of expression which are expected to be understood by those who read it. Instead of writing that is direct and straightforward, politic writing employs statements that are generalized or even ambivalent in nature. While we are used to hearing samples of politic expression from representatives of the Department of State, Department of Defense, or the United Nations, we can also see more familiar examples of this kind of writing in a corporate environment as well. Many sections of standard reports contain politic writing, including evaluation sections, recommendation sections, and executive summary sections. Seldom do reports and studies contain unanimous findings, conclusions, and recommendations which can be stated simply and exactly. More often than not, findings and recommendations vary widely and may even directly conflict with one another. In order to convey adequately the range of options or possibilities associated with a particular position, the writer must be able to express two opposing concepts without showing unwarranted prejudice in favor of one of them. The ability to do this is especially crucial when key individuals in the organization identify personally with a particular viewpoint.

Characteristics

The ability to express multiple viewpoints in a uniform voice is a difficult but necessary requirement. The ability to communicate in a politic style requires an awareness and effective use of these characteristics:

1. a politic style addresses sensitive issues

2. a politic style is directed to a special audience

3. a politic style is employed in a unique rhetorical situation

4. a politic style employs linguistic generalizations

5. a politic style requires a distinctive narrative attitude

6. a politic style is essentially ameliorative in nature

7. a politic style employs deductive logic

In the first place, politic writing is called for when the writer is dealing in sensitive issues, with topics that are foreign policy decisions or statements of corporate policy. Secondly, politic writing is necessary when the intended audience of the written material ranges from the specialized audience of the more technical areas of the discussion to administrators and policy-makers. The relative precision of technical detail may have to give way to larger, more encompassing language as the audience's likelihood of intimate subject familiarity decreases.

Third, politic writing usually occurs in a unique rhetorical situation. That is, the writer expects the reader to know that the subject area involves sensitive issues, and the reader is assumed to be aware that the writer is attempting to make acceptable policy statements that will summarize the sentiments of persons above the writer in the rank structure of the organization yet which will not lend the organization to unduly embarrassing critical comments from outside the organization. There is an awareness that the rhetorical environment is specifically a public environment, in which the writer's statements reflect positions or stands which represent basic organizational posture and which may be subjected to considerable scrutiny from any element in a large range of public readers.

Politic writing also calls for a distinctive narrative perspective on the part of the writer. Although the writer may professionally and emotionally identify with one particular point of view among those discussed, in passages of politic writing the writer must be especially aware of the full range of public responses that may occur. The writer must be especially sensitive to those portions of the report that could possibly create adverse reaction so that discussions of those areas can be phrased as positively and as tactfully as possible. The writer must be able to adopt an attitude of detachment from the subject under discussion.

In order to write in an impersonal and tactful manner, it may be necessary to adopt some stylistic techniques not normally associated with acceptable technical writing. Writers may have to employ vocabularies consisting of general rather than specific terms, use complex and sophisticated words instead of simple and common words, and include special-meaning words which carry symbolic or organizational connotations. In addition, politic writing may also include increased use of the passive voice, especially if it becomes necessary to conceal responsibility for unpopular or distasteful actions.

To the distinctive combination of characteristics described above should be added two more, but these are more properly characteristics of effect rather than style. The first of these is the awareness that politic writing is essentially ameliorative in nature; that is, it is usually employed in an attempt to phrase an unpleasant subject in a pleasant fashion, to put the best face possible on an awkward situation. The writer attempts to establish in print an unofficial but recognizable contract with the reader that a subject which has certain ramifications and interpretations is being set forth for examination and that both parties agree to approach the topic with this special awareness.

Closely related to the ameliorative intent of politic writing is the recognition that politic writing works on deductive logic. Most technical writing is inductive in its logic; it consists of a series of factual, concrete observations given in an understandable and sequential order designed to lead to a larger general conclusion. This is inductive reasoning, proceeding from the specific to the general. The logic of politic writing, however, reverses this process, for it employs generalizations for which the reader then has to supply the specific details which support the generalizations, details which usually have to be supplied out of personal knowledge of the subject under discussion. The kind of logic associated with politic writing relies heavily upon reader familiarity with the topic, an idea linked to the conditions of the unique rhetorical situation discussed previously.

Politic writing can present a severe challenge to the writer who prefers the straightforward style and precise expression representative of good technical writing, but politic writing is an inevitable part of much technical report writing, and should be recognized when it is called for. Although the style and characteristics of politic writing may appear to violate established standards of effective communication, it is in fact an important element in the range of writing styles required of contemporary professional technical writers.

AUTHOR INDEX

SUBJECT INDEX

The subject references found in this index were submitted by the individual authors. The page number associated with an indexed item references the first page of the paper in which the subject is covered.

For additional copies, contact:

Society for Technical Communication
815 Fifteenth St., N.W.
Washington, DC 20005